Overcoming Synthetic Challenges in Medicinal Chemistry

Mechanistic Insights and Solutions

Tongshuang Li

Author

Dr. Tongshuang Li
Wuxi AppTec
3912 Trust Way
Hayward
CA, 94545 USA

Cover Design: Wiley
Cover Images: © Unconventional/Shutterstock, Courtesy of Tongshuang Li, Adapted from Org. Process Res. Dev., 2021, 25, 1943, Elliott, L.D.; Booker-Milburn, K. I.; Lennox, A. J. J.

Typesetting by Lumina Datamatics

Library of Congress Card No.: applied for

British Library Cataloguing-in-Publication Data
A catalogue record for this book is available from the British Library.

Bibliographic information published by the Deutsche Nationalbibliothek
The Deutsche Nationalbibliothek lists this publication in the Deutsche Nationalbibliografie; detailed bibliographic data are available on the Internet at http://dnb.d-nb.de.

Print ISBN: 9783527355273
ePDF ISBN: 9783527852635
ePub ISBN: 9783527852628
oBook ISBN: 9783527852642

Printed and bound by
CPI Group (UK) Ltd, Croydon, CR0 4YY

C9783527355273_260825

Table of Contents

Preface

If you read a book or article and find yourself agreeing with every single point without a single question or doubt, it might indicate that you're not thinking critically and you have been brain washed. True growth comes from recognizing what you don't know—without acknowledging your own ignorance, progress becomes impossible.

When applying for a position as a synthetic chemist or medicinal chemist, you can expect to be questioned about reaction mechanisms during the interview process. These mechanistic problems are typically not taken directly from textbooks but are often drawn from the interviewer's own research experiences. This was certainly the case for me when I interviewed for such roles in pharmaceutical companies after completing my postdoctoral research. A solid understanding of organic reaction mechanisms is a fundamental skill for any synthetic organic chemist. It is essential not only for designing synthesis routes and experimental procedures but also for troubleshooting and resolving synthetic challenges effectively.

The fundamental reactions and named reactions in organic chemistry, along with their mechanisms, have been extensively covered in numerous textbooks and monographs. These mechanisms are also readily accessible through online platforms such as Wikipedia, Google, the Organic Chemistry Portal, and YouTube videos. However, when it comes to specific reactions in published synthetic routes, their mechanisms are not always straightforward to identify or comprehend. This often depends on the chemist's knowledge and experience. In many instances, understanding the mechanisms behind the formation of unexpected products, side products, or even impurities can significantly aid chemists in addressing challenges in their research. With these points in mind, this book compiles a selection of intriguing and challenging mechanism problems in organic reactions, particularly within the realm of medicinal chemistry, drawing from my experience and the literature I have studied.

Many of the reactions featured in this book are key steps in the synthesis of target compounds, while others involve the formation of side products, unexpected products, or impurities. I have made an effort to avoid including examples that have already been covered in other books. The primary goal of this book is to assist organic chemists, professionals, and students in deepening their understanding of reaction mechanisms. For mechanisms reported in the literature, I have redrawn them in a more accessible format to enhance clarity for readers. Additionally, I have offered my own perspectives on certain mechanisms that I find questionable in the literature. For reactions, particularly those described in patents, where mechanisms are not provided, I have proposed mechanistic explanations based on my understanding. I strongly encourage readers to attempt drawing their own mechanisms before consulting my proposed answers.

Certainly, whether derived from the literature or proposed by me, the mechanisms should not be considered the only possible explanations or the "definitive answers." You are encouraged to develop your own mechanisms, provided they offer a reasonable and logical explanation for the reaction.

Needless to say, this book is not an encyclopedia and it does not aim to cover all reaction mechanisms in organic synthesis. For example, some common, hotly cited, or frontier organic reactions, such as transition metal catalyzed coupling, catalytic asymmetric reactions, photo-promoted reactions, and electrochemical reactions, are barely included.

I would like to extend my appreciation to the original publishers for experimental procedures for granting permission to use in this book. The publishers include the American Chemical Society, Elsevier, the Royal Society of Chemistry, John Wiley & Sons, Springer Nature, the American Association for the Advancement of Science, and Oxford University Press. I would like to thank all inventors for patent literatures cited in this book.

Errors occasionally creep into any book, so it should not be surprising if you find mistakes in this one. If you do find any, feel free to discuss them with me. I would welcome any comments from readers, especially those suggestions that would make the book of greater value to the reader.

I would like to express my sincere gratitude to my mentors, students, and colleagues throughout my university studies, teaching, research, and career in the pharmaceutical industry.

I deeply appreciate Wiley editors Dr. Anne Brennführer, Priyadarshini Natarajan and Haridharini Velayoudame for helping me turning my ideas into this book.

Finally, I wish to thank my family members for their continual support and understanding.

Tongshuang Li
Hayward, CA, USA
June 2025

《付梓感怀》
菊花绽放深秋，红梅香自苦寒。蒙父母言行教诲，得各任恩师真传。集平生经历见闻，受中西思想浇灌。融古今圣贤智慧，于个人粗浅拙见。出首部学术专著，甘愿做引玉之砖。意启迪同辈后生，续科学创造新篇。限本人才疏学浅，有错误纰漏难免。望各位专家学者，不吝啬斧正纠偏！

《Reflections Upon My Manuscript's Publication》

Chrysanthemums bloom proud in autumn's snow,
Plum fragrance crystallizes winter's blow.
Nurtured by parental virtues stern and bright,
Humbled beneath mentors' scholarly light.

Ten thousand miles of wisdom fill my gaze,
East-West philosophies through my brush convey.
A hundred schools' essence within me flows,
Lone quill stirs waves where silent paper glows.

May this crude work spark jade-carved replies,
To nurture magnolia beneath new skies.
Let younger waves surge where old currents crest,
Fresh voices shake academia's hallowed nest.

Though keenly aware my borrowed insights show,
Joy finds me when errors meet your chisel's glow.

Abbreviations

3CL protease	3-chymotrypsin-like protease
5-HT	5-hydroxytryptamine or serotonin
5-HT4	5-hydroxytryptamine receptor 4
ABN	2-aminobenzonitrile
ACC inhibitor	acetyl-CoA carboxylase inhibitor
AD	Alzheimer's disease
ADC	antibody–drug conjugate
ADME	absorption, distribution, metabolism, and excretion
AI	artificial intelligence
AIDS	acquired immunodeficiency syndrome
ALK	anaplastic lymphoma kinase
AMPA	α-amino-3-hydroxy-5-methyl-4-isoxazolepropionic acid
ANRORC	addition of the nucleophile, ring opening, and ring closure
APIs	active pharmaceutical ingredients
ATC	apricitabine
ATPase	adenosine 5′-triphosphatase
ATR	ataxia telangiectasia and Rad3-related
AUC	area under the plasma drug concentration–time curve
BACE	beta-amyloid cleaving enzyme
BCC	basal-cell carcinoma
BLA	biologic license application
BMS	Bristol–Myers Squibb Company
Boc	tert-butoxycarbonyl or tert-butyloxycarbonyl
BPR	back-pressure regulator
brsm	based on recovered starting material
BSA	N,O-bis-(trimethylsilyl)acetamide
BTK	Bruton's tyrosine kinase
BTP	2-bromo-3,3,3-trifluoroprop-1-ene
BTPP	tert-butylimino-tri(pyrrolidino)phosphorane
Bus	tert-butylsulfonyl
CB1	cannabinoid receptor 1
CBC	cannanbichromene
CBTC	cannabicitran
Cbz	benzyl carbamate

Cbz-Cl	benzyl chloroformate
CCR5	cysteine-cysteine chemokine receptor 5
CD45	lymphocyte common antigen, a protein tyrosine phosphatase
CDI	carbonyldiimidazole
CFTR	cystic fibrosis transmembrane conductance regulator
cGMP	cyclic guanosine monophosphate-specific phosphodiesterase
Chk	checkpoint kinase
CMBP	cyanomethylenetributylphosphorane
CMC	chemistry, manufacturing, and controls
CNS	central nervous system
COVID-19	coronavirus disease 2019
COX	cyclooxygenase
CRF1	corticotropin-releasing factor receptor 1
CRO	contract research organization
CuAAC	Cu(I)-catalyzed azide-alkyne cycloaddition
CXCR4	C-X-C chemokine receptor type 4
CYP1A2	cytochrome P450 1A2
CYP2B6	cytochrome P450 2B6
CYP2C19	cytochrome P450 2C19
CYP2C9	cytochrome P450 2C9
CYP3A	cytochrome P450, family 3, subfamily A
DABCO	1,4-diazabicyclo[2.2.2]octane
DAST	diethylaminosulfur trifluoride
DBU	1,8-diazabicyclo[5.4.0]undec-7-ene
DCC	N,N′-dicyclohexylcarbodiimide
DCHU	N,N′-dicyclohexyl urea
DCM	dichloromethane
DDQ	2,3-dichloro-5,6-dicyano-1,4-benzoquinone
DEAD	diethyl azodicarboxylate
DCE	dichloroethane
DED	dry eye disease
Deoxo-Flour	bis(2-methoxyethyl)aminosulfur trifluoride
DEPT	distortionless enhancement by polarization transfer
DGAT-1	diacyl glycerolacyl transferase-1
DHA	dihydroartemisinin
DI	deionized
DIAD	diisopropyl azodicarboxylate
DIHD	diisopropyl hydrazine-1,2-dicarboxylate
DIPA	diisopropylamine
DIPEA	N,N-diisopropylethylamine
DMA	dimethylacetamide
DMAD	dimethyl acetylenedicarboxylate
DMAP	4-dimethylaminopyridine
DMB	2,4-dimethoxybenzyl
DMBNH2	2,4-dimethoxybenzylamine
DMF	dimethylformamide

DMF-DMA	N,N-dimethylformamide dimethyl acetal
DMMN	dimethylmalononitrile
DMPK	drug metabolism and pharmacokinetics
DMSO	dimethyl sulfoxide
DMTU	1,3-dimethylthiourea
DNA	deoxyribonucleic acid
DNBSA	2,4-dinitrobenzenesulfonamide
DOAC	direct oral anticoagulant
DOE	design of experiment
DPP4	dipeptidyl peptidase-4
DPPA	diphenylphosphoryl azide
DQF-COSY	double quantum-filtered COSY
ERR1	estrogen-related receptor 1
EUA	emergency use authorization
FAK	focal adhesion kinase
FCC	flush column chromatography
FDA	Food and Drug Administration of the United States
GAs	gibberellins
GC-MS	gas chromatography-mass spectrometry
GK-GKRP	glucokinase-glucokinase regulatory protein
GLP	Good Laboratory Practices
GMP	Good Manufacturing Practices
GPCR	G protein-coupled receptor
GPLR	G protein-linked receptor
GPR17	G protein-coupled receptor 17
H2L	hit to lead
HBTU	hexafluorophosphate benzotriazole tetramethyluronium
HCV	hepatitis C virus
HFIP	hexafluoroisopropanol
HIV	human immunodeficiency virus
HMG-CoA	3-hydroxy-3-methylglutaryl (HMG) coenzyme A
HPLC	high-performance liquid chromatography
HRMS	high-resolution mass spectrometry
HSDD	hypoactive sexual desire disorder
HTS	high-throughput screening
IND	Investigational New Drug Application
IP6K	inositol hexakisphosphate kinase
IUPAC	International Union of Pure and Applied Chemistry
JAK	Janus kinase
LED	light emitting diode
LNO	2,6-lutidine N-oxide
LO	lead optimization
LSF	late stage C—H functionalization
MCL-1	myeloid cell leukemia-1
mCPBA	meta-chloroperoxybenzoic acid
MCR	multi-component reaction

MDM2	mouse double minute 2
MEK	mitogen-activated protein kinase
MMA	methyl methacrylate
MP	melting point
MRSA	methicillin-resistant *Staphylococcus aureus*
MTBE	methyl tert-butylether, tert-butylmethyl ether
NBC	N-bromocaprolactam
NBS	N-bromosuccinimide
NCS	N-chlorosuccinimide
NDA	new drug application
NMM	N-methylmorpholine
NMP	N-methyl-2-pyrrolidone
NMPAC	National Medical Products Administration of China
NMR	nuclear magnetic resonance
NtRTI	nucleotide reverse transcriptase inhibitor
OA	orotic acid
OECD	Organization for Economic Cooperation and Development
OLM	olmesartan medoxomil
PARP	poly (ADP-ribose) polymerase
PD	pharmacodynamics
PDE IV	phosphodiesterase-4
PDE-V	phosphodiesterase V
PDGFR	platelet-derived growth factor receptors
PFA	perfluoroalkoxy, Teflon® PFA
PK	pharmacokinetics
Plk	polo-like kinase
Plk1	polo-like kinase 1
PNP	peripheral neuropathic pain
PPAR	peroxisome proliferator–activated receptor
PTFE	polytetrafluoroethylene
PTLC	preparative thin layer chromatography
QSAR	quantitative structure-activity relationship
RCM	ring-closing metathesis
RNA	ribonucleic acid
ROCK1	rho-associated kinase isoform 1
RRT	relative retention time
SAR	structure–activity relationship
SARS-CoV-2	severe acute respiratory syndrome coronavirus 2
SHP2	Src homology-2 domain-containing protein tyrosine phosphatase-2
TBAF	tetra-n-butylammonium fluoride
TBDPS	tert-butyldiphenylsilyl
TBHP	tert-butyl hydroperoxide
TBME	tert-butylmethyl ether
t-BuONSO	N-sulfinyl-O-(tert-butyl)hydroxylamine
TEA	triethylamine

TFA	trifluoroacetic acid
TFAA	trifluoroacetic anhydride
THF	tetrahydrofuran
TIPS	triisopropylsilyl
TLC	thin layer chromatography
TMP	2,2,6,6-tetramethylpiperidine
TMSCl	trimethylsilyl chloride or chlorotrimethylsilane
TMSOTf	trimethylsilyl trifluoromethanesulfonate
TPP	tetraphenylporphyrin
TPS	tert-butyldiphenylsilyl, TBDPS
TRK	tropomyosin receptor kinase
TRPV1	transient receptor potential cation channel subfamily V member 1
TV	target validation
USP	upstream processing
UV	ultraviolet
VEGFR	vascular endothelial growth factor receptor

1

Opinions and Suggestions

1.1 Be a "Doctor," Not Just a "Drug Fetcher"

Even after more than 40 years, I still vividly recall the advice my professor shared with us during my college organic chemistry experiments. He said, "When conducting a synthetic reaction, it's essential to grasp the reaction mechanism and the principles behind each step of the process. It's similar to how a physician prescribes medication based on a patient's condition. If you merely follow the prescription to prepare the medication, you'll always remain a 'drug fetcher,' never a doctor." I don't intend to diminish the role of a "drug fetcher" in any way. The key issue is that our aim in training organic chemists is to cultivate "doctors," not just "drug fetchers." Understanding the reaction mechanism allows you to master it, draw connections, design better experimental procedures, and tackle challenges in your research. For this reason, comprehending the mechanisms of organic reactions is a fundamental skill for any synthetic organic chemist.

1.2 Synthetic Organic Chemistry: Still as Much an Art as a Science

During my tenure as a university professor, one of my responsibilities was overseeing students conducting organic chemistry experiments in the laboratory. I frequently came across an intriguing scenario: despite following the same experimental procedure, students would obtain varying yields and qualities of the product. In certain extreme cases, some students failed to produce any of the intended product, resulting in a failed experiment. This highlighted the challenges of reliability and reproducibility in synthetic chemistry.

When it comes to the reliability and reproducibility of published articles, I consider the annual journal *Organic Syntheses* to be the top-ranked among organic chemistry journals. This is primarily because the procedures featured in this journal are rigorously verified by an independent group of chemists. However, even in such cases, each synthetic reaction typically yields a range of results or varying yields across different batches, rather than a fixed value.

Overcoming Synthetic Challenges in Medicinal Chemistry: Mechanistic Insights and Solutions, First Edition. Tongshuang Li.

The yield of a reaction is influenced by numerous factors, including temperature, concentration, the order and speed of reagent addition, stirring method and speed, solvent choice, catalyst, the suppliers and purities of starting materials and reagents, reaction scale, workup procedures, and purification methods. Even when all these external factors are carefully controlled, different chemists may still achieve different yields. Furthermore, for a particular chemist, yields can vary from one batch to another. This indicates that the yield of a synthesis depends not only on the reaction conditions but also on the individual performing the experiment—the chemist.

In this sense, while synthetic organic chemistry is undeniably a science, it also retains a significant degree of artistry. The skill, intuition, and experience of the chemist play a crucial role in the success of a reaction, making it as much an art as it is a scientific discipline.

1.3 Synthetic Organic Chemistry Is an Experimental Science under the Guidance of Theory

Whether an organic reaction is described detail in a textbook, in a reference book, or in a journal-published article, replicating or referencing these experimental procedures can sometimes lead to issues of varying degrees. Any seemingly insignificant error could cause a reaction failure or lead to a different result from those reported. Occasionally, some planned synthetic reactions yield unexpected products or side products. On the other hand, many groundbreaking organic reactions have been discovered serendipitously. This unpredictability makes synthetic organic chemistry both challenging and fascinating. As a result, synthetic organic chemistry remains, and will continue to be, an experimental science.

Organic chemistry is grounded in a solid theoretical framework, which has been developed and refined by skilled chemists over centuries. These theories are built upon the vast body of experimental data accumulated through the collective efforts of countless chemists over hundreds of years. Understanding and explaining organic reaction mechanisms relies heavily on these established principles. In the field of medicinal chemistry, particularly in synthetic work, a deep mastery of organic chemistry theory and a thorough familiarity with various reaction mechanisms can significantly enhance the efficiency and productivity of our synthetic designs and practices. Therefore, organic synthesis is not merely an experimental science but one that is profoundly guided by theoretical insights.

1.4 Reduce Mistakes in Your Synthetic Work

Given the above considerations, it is imperative that chemists at all levels have opportunities to enhance their knowledge, experience, and skills throughout their careers. Many of us spend more than 8 hours a day in the lab, 6 days a week. But does this alone make us productive chemists? The answer is not necessarily. While dedicating a certain number of hours to lab work each day is a necessary condition for productivity, it is not sufficient on its own. The true measure lies in the efficiency of our work.

A common question I receive from graduate students and colleagues I supervise is: How can we improve the efficiency and quality of our synthetic work? My response is straightforward: by minimizing or even eliminating mistakes in our work, thereby increasing the success rate of our synthetic reactions. If every reaction we conducted were successful, it would undoubtedly be the most efficient way to work. However, in practice, this is nearly impossible. Therefore, the best approach to enhancing productivity is to reduce errors and improve the success rate of our synthetic reactions.

To achieve this, it is crucial to completely understand the mechanism of every synthetic reaction we perform. We can drastically reduce mistakes by anticipating potential problems and devising strategies to avoid them. This proactive approach not only improves the quality of our work but also enhances overall efficiency, making us more productive chemists in the long run.

1.5 Your Knowledge, Experience, and Skills Can Never Be Too Much

For any entry-level synthetic chemist fresh out of university—whether you hold a bachelor's degree or a Ph.D. with postdoctoral research experience—working as a medicinal chemist in drug discovery, your knowledge, experience, and skills will always feel insufficient. Mistakes are inevitable in the early stages of your career. However, if you are a quick learner, you will rapidly grow by learning from your own work and the expertise of others. By striving to understand the mechanism behind every reaction you perform, your problem-solving abilities will improve significantly. Over time, as you accumulate more knowledge and experience, the frequency of errors in your synthetic work will steadily decrease.

I've heard hiring managers or HR professionals use the term "overqualified" as a reason to reject certain applicants for a position. However, when it comes to performing a synthesis, no chemist, regardless of their experience level, can ever be considered "overqualified." For even the most experienced chemist, a lack of caution during a synthetic reaction can result in failure. Experience doesn't eliminate the need for careful attention to detail—it reinforces it.

Synthetic organic chemistry is a rapidly evolving field, continuously advancing with the emergence of new synthetic methods, innovative reagents, and cutting-edge technologies. These developments are documented in a growing number of articles published across various organic chemistry journals. The volume of published research is expanding, and new journals dedicated to organic synthesis are also being established. Given the finite nature of time and energy, it is challenging to keep up with every new paper relevant to your work. Nevertheless, it is feasible to efficiently skim through titles and abstracts to identify key publications. When a paper particularly captures your interest, you can delve into it in greater detail. While the papers you read may not always directly contribute to your current projects, maintaining this habit over the long term can significantly enhance your expertise and benefit your career.

1.6 What Is a Mistake?

What is a mistake? From my perspective, it is a relative term or concept. A synthetic plan or experimental design created by a relatively junior chemist might be considered reasonable or acceptable. However, the same plan could be seen as an obvious mistake if it were designed by a more experienced, senior-level chemist. For instance, consider a medicinal chemistry project where the goal is to synthesize a series of aromatic ethers (**1**) featuring a difluoro alkyl branch, aiming for final compounds in the range of 5–10 mg. Your proposed synthesis design for compound **1** (R^2 = Me) is shown below.

1 (R^2=Me, Et, i-Pr) — Mitsunobu reaction ⟹ 2 + 3 — Selective methylation for R^2=Me ⟹ 4

The target compounds could be synthesized from phenol **2** (suppose **2** have been prepared or commercially available) and alcohol **3**. You find **3** (R^2 = Me) is commercially available, but very expensive (Sigma-Aldrich, US$1746/1 g, US$220/mmol). So, you decide to prepare **3** (R^2 = Me) from the relatively less expensive diol **4** (AK Scientific, US$162/5 g, US$3.63/mmol) by iteroselective monomethylation. You also find references (*a*, US2013/0131050, A1. *b, J. Med. Chem.*, **2021**, *64*, 14773) supporting your design. The analogs **3b** (R^2 = Et) and **3c** (R^2 = *i* − Pr) could be prepared similarly.

4 (1.121 g) — NaH, THF, MeI, rt, 15% → 3 (191 mg)
US2013/0131050, A1

5 + 6 — Ph_3P, DEAD, DCM, 20 °C, 3 h, 86% → 7
J. Med. Chem., 2021, 64, 14773

If you are a junior-level chemist and have proposed such a design, the two-step synthesis appears reasonable at first glance, making your design acceptable in principle. However, this acceptance does not necessarily mean your design is suitable for actual synthesis. While the Mitsunobu reaction step is unlikely to pose significant issues, a more thorough analysis reveals that the first step—preparing ether **3** from diol **4**—could be problematic. If you were to attempt this synthesis in the lab, you would almost certainly encounter difficulties in isolating product **3** from the reaction mixture.

The patent procedure for this step involves diluting the reaction mixture with water, extracting with ethyl acetate, evaporating the solvents under reduced pressure, and further purifying the crude product via silica gel column chromatography using an ethyl acetate-hexane eluent, followed by another round of solvent evaporation. Despite this extensive process, the yield of product **3** was only 15%. If you were to replicate this procedure, you might initially obtain no product at all. Detecting compound **3** would also be challenging, as it lacks ultraviolet (UV) absorption, making it difficult to identify using thin-layer chromatography (TLC) or high-performance liquid chromatography (HPLC).

Moreover, the situation is further complicated by the likelihood that compound **3** is volatile. Although no boiling point is reported for **3**, a structurally similar compound, 3-methoxy-1-propanol, has a boiling point of approximately 150 °C at 760 mmHg or 45 °C at 10 mmHg. This suggests that during rotary evaporation, the product could easily be lost—either carried over into the receiving flask with the solvents or even drawn into the vacuum pump and released into the air. This issue is particularly pronounced when the reaction is conducted on a small scale (e.g., <5 g), which explains the poor yield (15%) reported in the patent.

Another limitation of this design is the lack of divergent synthesis. The iteroselective monomethylation of diol **4** in the initial step prevents the synthesis of other analogs (where $R^2 = $ Et, $i-$Pr, etc.) from a shared intermediate.

In summary, while your design is conceptually sound, practical challenges in isolating and purifying compound **3** make it less viable for actual laboratory synthesis.

Therefore, if you are a senior-level chemist, creating a design as described above would be considered an "obvious mistake."

A more logical and appropriate design is presented below.

1 → Alkylation → 8 → Deprotection → 9 (OTBDPS) → Mitsunobu reaction → 10 + 2 → Selective protection → 4

The iteroselective mono protection of diol **4** with tert-butyldiphenylsilyl chloride (TBDPSCl) was reported in many patents and journal articles (e.g., *J. Med. Chem.*, **2021**, *64*, 11841).

4 → NaH, THF, TBDPSCl, rt, 72% → 10

J. Med. Chem., 2021, 64, 11841

In the synthesis of the methyl ether product **1**, although this design involves two additional reaction steps compared to the previous approach due to the introduction of the TBDPS-protecting group, the significantly higher molecular weight of compound **10** relative to ether **3** ensures that it is less prone to evaporation. This intermediate is also UV-active, as it contains two benzene rings, making it detectable by TLC and analytical HPLC. The subsequent steps, including the Mitsunobu reaction, deprotection, and methylation, are expected to proceed without issues. Additionally, this design positions alkylation as the final step, maintaining the flexibility to synthesize other homologues from the common intermediate **8**.

Nevertheless, if the goal is to synthesize hundreds of grams or even kilograms of compound **3**, the first design would be more practical. This is because, on such a scale, compound **3** could be purified by fractional distillation.

1.7 Beyond Experience: Combining Literature and Insight for Optimal Synthetic Design

In the field of medicinal chemistry, once the structures of a series of compounds to be synthesized are finalized, the subsequent step involves designing the synthesis. Even if you are a highly knowledgeable and experienced expert, it is essential to consult the literature while incorporating your own ideas. This approach ensures the development of the most rational and feasible synthetic route. Relying solely on personal experience for synthetic design often does not yield the best results.

Every stage of the reaction, from the raw materials to the final product, must be meticulously examined. It is also crucial to prepare contingency plans to address potential challenges and issues. For multi-step syntheses, the yield of each reaction step, as well as the overall yield, should be estimated. This allows for the determination of the initial reaction scale based on the required quantity of the target product.

In multi-step syntheses, it is imperative to avoid using all or most of an intermediate for the next reaction until a smaller-scale test reaction has been successfully completed. This precaution holds true even for traditional reactions that appear to be problem-free, as unexpected complications can arise.

1.8 Improve Your Decision-Making Ability

Modern online search engines like SciFinder and Reaxys are incredibly powerful tools. When searching for a specific reaction or the synthesis of a particular compound, these platforms can generate thousands of results in mere seconds. However, the challenge you face is not a lack of information but an overwhelming abundance of it. If you find yourself considering 10 out of 100 plausible options as worth trying, it essentially means you are uncertain. (Having too many options can be as paralyzing as having none!) The key is to prioritize these reaction conditions and select only two or three to test, based on the resources available in your research facility.

So, how can you enhance your decision-making process? The solution lies in developing a deep understanding of the reaction mechanism. This includes a solid understanding of the structural features, as well as the physical and chemical properties of the starting materials, reagents, catalysts, and products involved. With this knowledge, you can anticipate potential challenges, such as reaction selectivity, the need for functional group protection, compatibility between functional groups and reaction conditions, side reactions, and the intricacies of workup and purification. By addressing these factors proactively, you can refine every detail of your synthetic approach. This meticulous preparation will naturally lead to a higher success rate and improved yields in your synthetic reactions.

1.9 Know the Mechanism of the Reactions You Perform

At the start of a medicinal chemistry career, particularly for research assistants or associates, it's common to follow synthetic designs and references provided by a supervisor rather than creating your own. In such cases, you may not need to design the synthesis yourself. However, if you lack an understanding of the reaction mechanisms involved, mistakes are likely to occur during your work. While thoroughly reviewing the synthetic design and deciphering the reaction mechanisms requires extra time and effort, it is a crucial step for professional growth. Human nature often leans toward taking the path of least resistance—if one can perform tasks without deep thinking, they may avoid it. But if you are genuinely committed to enhancing your professional skills and problem-solving abilities, investing energy in understanding reaction mechanisms will significantly improve your work efficiency and accelerate your development.

If you simply follow the synthetic design and references provided by your supervisor without engaging deeply, you may blame the design or reference method when a reaction fails, claiming it "doesn't work." This passive attitude will not help you improve your problem-solving skills. Ultimately, you are responsible for your reactions. Frustration can set in after a week of unsuccessful attempts, making work feel like a burden and a source of stress.

On the other hand, if you actively engage in the synthetic design and develop a clear understanding of each reaction's mechanism, you gain control over the process. The molecules in your flask will transform into new compounds as you direct. When your reactions and syntheses proceed smoothly, yielding the desired products as planned, you'll experience a profound sense of accomplishment and success. This approach not only enhances your skills but also brings joy and satisfaction to your work, making it a rewarding and fulfilling experience.

1.10 Always Learn Something from the Reaction You Performed

Every reaction we conduct, whether it succeeds or fails, offers valuable lessons.

Understanding why a reaction worked is essential for successful reactions. Conversely, for unsuccessful ones, we need to analyze why they failed. Was there no reaction at all, or was an unexpected product formed? What caused the outcome? Identifying the root cause of a problem makes it easier to find a solution. Failure itself isn't the issue; the real problem lies in not understanding why it happened or failing to uncover the true reason behind the failure.

1.11 Knowing Reaction Mechanisms Alone Doesn't Make an Excellent Chemist—But It's Essential

Understanding the mechanisms of organic reactions is essential for becoming an outstanding synthetic medicinal chemist, but it is not enough on its own. To excel in any profession, you must first have a genuine passion for it. As a synthetic chemist, you also need to possess exceptional practical skills, a strong work ethic, and unwavering perseverance. Success belongs to those who remain steadfast in their beliefs. These qualities are the cornerstone of the scientific spirit. Understanding reaction mechanisms is like having a clear path before you, making it easier to reach your goals. Conversely, not understanding these mechanisms is like running in the dark—you are more likely to stumble or fall into obstacles along the way.

1.12 Patents Do Not Tell You the Full Story of Chemistry

As a medicinal chemist regularly engaged in synthetic work, you will frequently consult reactions documented in patents. However, it's important to recognize the significant differences between patents and journal articles. Patents typically do not include detailed reaction mechanisms. If new methods are disclosed, they often only present the results without explaining the rationale or development process behind the innovation. Additionally, patents rarely cite references for the synthetic methods they employ. The yields reported in patents are usually not optimized and may even be inaccurate, meaning they should only be used as a rough guideline rather than a reliable benchmark for calculating the yield of your own synthetic routes. In some cases, yields may not be provided at all. Furthermore, the characterization data for intermediates and final products in patents is often incomplete or insufficiently detailed.

1.13 Summarize Your Work After Completing a Synthesis

Once a synthesis is completed, it is important to promptly create a summary. Throughout the synthesis process, you may have faced challenges, encountered difficulties, or taken unnecessary detours. Now that you have gained insights into resolving these issues and avoiding similar pitfalls in the future, it is an opportune time to reflect on whether there are more efficient synthesis methods or improved approaches. Writing a summary not only enhances your writing skills but also significantly strengthens your ability to summarize, as well as your logical and analytical thinking.

1.14 Two Examples

Here I would like to present two examples to illustrate the importance of understanding reaction mechanisms in solving chemistry problems.

One example is selected from our research experience and the results were presented in a patent application [1]. In a medicinal chemistry project, we wanted to synthesize a series of platelet-derived growth factor receptor inhibitors, which structure is shown in the scheme below as compounds **11**. Those compounds consist of the core structure, named as a 2,7-naphthyridin-3(2*H*)-one. Our synthetic plan included the reductive amination of the key intermediate **16**, then cyclization, de-protection of 2,4-dimethoxybenzyl (DMB) and oxidation to form the amino pyridine **12**.

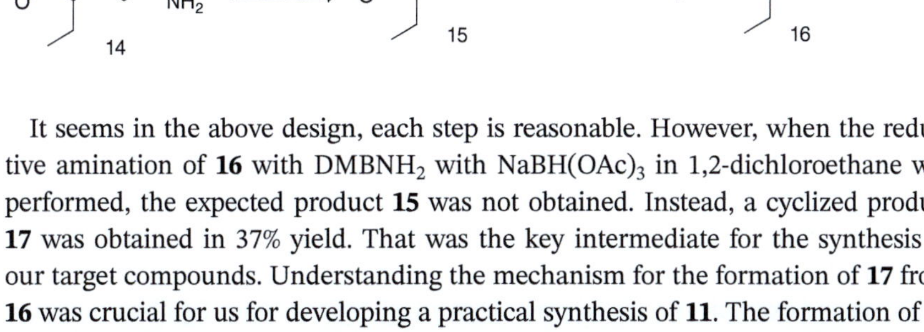

It seems in the above design, each step is reasonable. However, when the reductive amination of **16** with $DMBNH_2$ with $NaBH(OAc)_3$ in 1,2-dichloroethane was performed, the expected product **15** was not obtained. Instead, a cyclized product **17** was obtained in 37% yield. That was the key intermediate for the synthesis of our target compounds. Understanding the mechanism for the formation of **17** from **16** was crucial for us for developing a practical synthesis of **11**. The formation of **17** could be explained by the following mechanism. Since the commercially available $NaBH(OAc)_3$ always contains residual HOAc, which can be easily smelled when you open the chemical container, trace amount of HOAc catalyzed the formation of **17**.

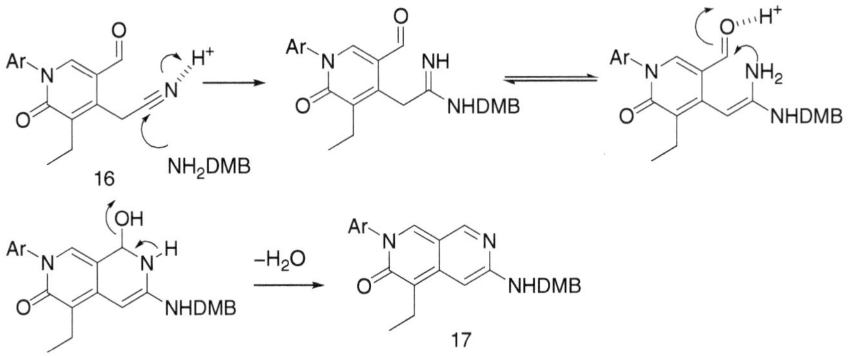

Based on this mechanism, $NaBH(OAc)_3$ should not play any role for the formation of 2,7-naphthyridin-3(2*H*)-one (**17**) and the reaction should be catalyzed by acid. Thus, a preparation of **17** from **16** and $DMBNH_2$ catalyzed by HOAc was designed and performed. To our delight, the reaction gave the desired product in quantitative yield.

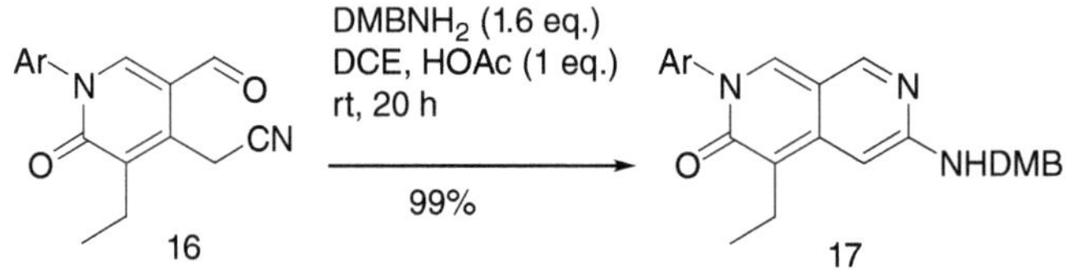

Encouraged by the above discovery, when methylamine was employed as a substrate, the desired product **11a** was produced in a single step in quantitative yield. As methylamine was used as an aqueous solution, a phase transfer catalyst, Bu_4NBr (0.25 eq), was also added.

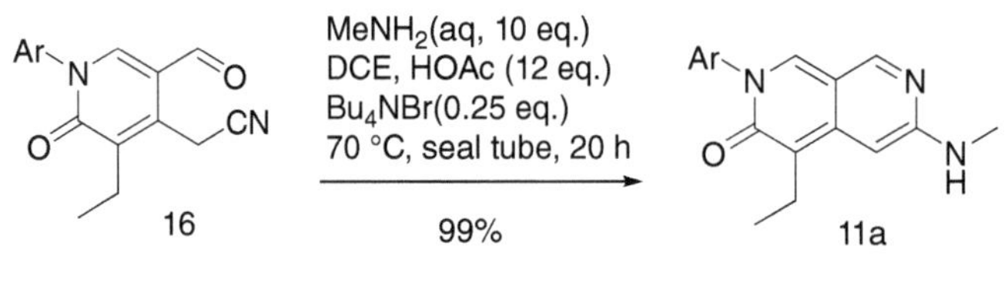

Thus, an excellent procedure for synthesis of 2,7-naphthyridin-3(2*H*)-ones was developed by understanding the mechanism of the formation of an unexpected product [1].

The second example is selected from the work of Bristol-Myers Squibb Company. Pexacerfont (**18**) is a pyrazolotriazine corticotropin-releasing factor receptor 1 (CRF1) antagonist. During a process study, chemists investigated the chlorination of **19** by $POCl_3$ catalyzed by tertiary amines [2]. In the presence of DABCO (1 mol %), the desired chloropyrazolotriazine **20** was obtained in 78% yield within 1 h at 20 °C.

While an impurity **21** (<0.5%) was also generated. Since this impurity is a potential alkylating agent and would interfere with the next reaction, its level required close monitoring in the final product. How this impurity was produced and how to avoid its formation?

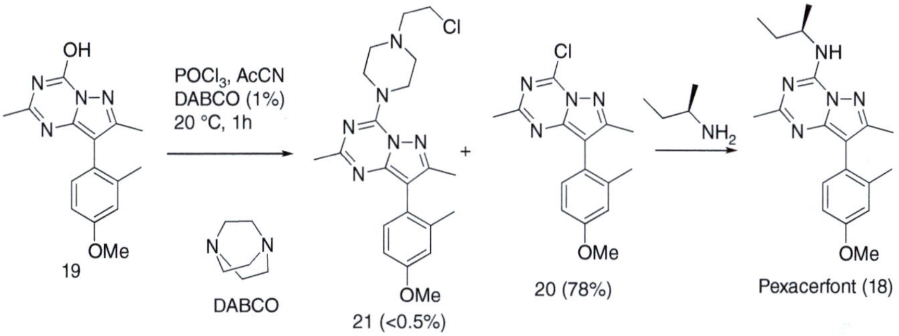

19 → (POCl$_3$, AcCN, DABCO (1%), 20 °C, 1h) → 21 (<0.5%) + 20 (78%) → Pexacerfont (18)

The formation of **21** was the result of nucleophilic attack of DABCO on the proposed *O*-phosphorylated intermediate **22**. Presumably, chloride ion displacement of DABCO (**path a**) provides the desired product **20**. Conversely, attack at either of the three α-carbons of the activated DABCO intermediate **23** (**path b**) would produce impurity **21**.

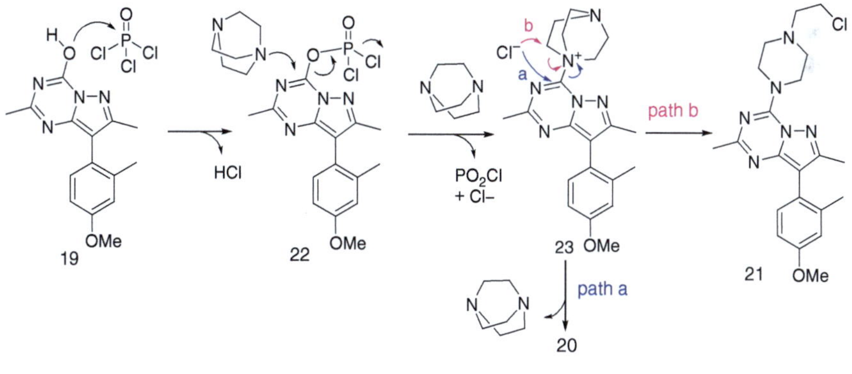

19 → (− HCl) → 22 → (DABCO; − PO$_2$Cl + Cl−) → 23 → path b → 21; 23 → path a (− DABCO) → 20

Realizing that side-reaction, the chemists from BMS used 0.5 mol % of *N*-methylmorpholine (NMM), leading to >98% conversion to **20** within 1 h at 25 °C, although a trace amount of the amination impurity **24** still formed in analogy to the DABCO catalyzed reaction. Importantly, the chlorinated impurity **B**, analogous to **21**, was not observed. Since the chloride ion would presumably attack the primary methyl group of NMM, only trace amounts of chloromethane—a volatile impurity—would form and likely evaporate during the reaction and workup. Thus, no alkylating impurity existed in the products.

NMM
HCl
PO_2Cl + Cl–
path b
MeCl
path a
19
22
A
24
20
B (not detected)

References

1 Zhang, W.; Wang, X.; Liu, H.; Zhou, Z.; Wang, Y.; Yang, L.; Li, T.; Zhao, X.; Zou, Z.; Gao, Y.; Lin, S.; Wang, W. WO 2020/063860 A1.

2 Broxer, S.; Fitzgerald, M. A.; Sfouggatakis, C.; Defreese, J. L.; Barlow, E.; Powers, G. L.; Peddicord, M.; Su, B.-N.; Tai-Yuen, Y.; Pathirana, C.; Sherbine, J. P. *Org. Process Res. Dev.*, **2011,** *15*, 343.

2

General Terms and Concepts in Synthetic Organic Chemistry

This chapter collects some terms and concepts mainly related to organic reaction mechanisms, process chemistry, and medicinal chemistry and introduces these terms and concepts as concise as possible. Almost every term or concept could be written in a book. This is indeed the case. Obviously, these introductions are not an encyclopedia of terms and concepts in synthetic organic chemistry. With the discussion of "**stereoselectivity**" in this chapter, I deliberately avoided **stereochemistry** and **asymmetric synthesis**, two important topics in organic chemistry due to two reasons. First, because these topics are too vast to explain clearly in a few short paragraphs; second, they have been discussed in detail in the literature.

2.1 Reaction Mechanism

In chemical reactions, a **mechanism** is a story that tells the detailed process of compound **A** being transformed into compound **B** under given reaction conditions. The reactions themselves may involve the interaction of atoms, molecules, ions, electrons, and free radicals. Organic reaction mechanisms form the foundation for understanding organic chemistry.

2.2 Approaches to Draw Reaction Mechanism

There may be more than one reasonable mechanism that can be drawn for a reaction. In general, a mechanism explains the detailed process from starting materials to products, which may involve some intermediates, transition states, ions, and/or free radicals. Transformations can be indicated by straight arrows between those species in a linear manner. A curve arrow represents transfer of one pair of electrons. Hook or fish hook represents single electron transfer in free radical reactions.

For noncatalytic reactions, the mechanism is usually drawn in a **linear way**. While for catalytic reactions, the mechanism can be drawn in two different ways, **linear** and **cyclic**. For example, the general catalytic cycle for Suzuki cross coupling involves three fundamental steps: oxidative addition, transmetalation, and reductive elimination as

Overcoming Synthetic Challenges in Medicinal Chemistry: Mechanistic Insights and Solutions, First Edition. Tongshuang Li.

demonstrated in the following scheme. Oxidative addition of aryl halides to Pd(0) complex is the initial step to give intermediate **I**, a Pd(II) species. With the participation of base, an organoboron compound reacts with intermediate **I** in transmetalation to afford intermediate **II**. This is followed by reductive elimination to give the desired product and regenerate the original Pd(0)-Ln, the catalyst species [1].

2.2.1 Linear Way: Catalytic and Noncatalytic Reactions

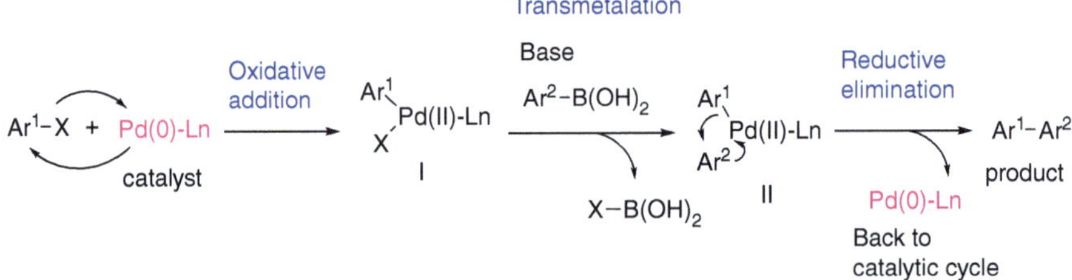

2.2.2 Cyclic Way: Catalytic Reactions

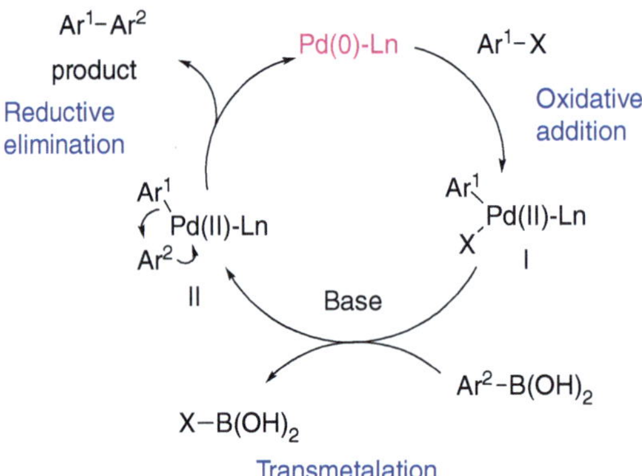

2.3 Desired Product

Desired product, also called the **expected product**, is the product expected or predicted to be obtained from a reaction.

2.4 Unexpected Product

Unexpected product is the product that is not expected to be a product from a reaction. Due to the complexity and the reactivity of starting materials and products, under the reaction conditions, unexpected or undesired products are often obtained. The formation of unexpected products is also reasonable. Interpreting the formation mechanism can sometimes lead to the discovery of a new reaction or a new synthetic method.

2.5 Side Reaction

A **side reaction** occurs simultaneously with the main reaction, to a lesser extent and results in the formation of **side products**, thereby reducing the yield of the main product. Imagine transporting oil through a pipeline: if there are leaks along the way, the amount of oil received at the destination will be less than what was initially supplied. Similarly, a side reaction, often called a competing reaction, affects the overall efficiency of the process. This concept is closely tied to reaction selectivity. If a reaction has 100% selectivity, no side products are formed, analogous to a pipeline with no leaks.

2.6 Side Product and By-product

Due to chemo-, regio-, or stereoselectivity issues, **side reactions** take place in many organic reactions. A **side product** is produced from a side reaction.

Different from a side product, a **by-product** is an inevitable component produced from the main reaction and accompanies the desired product. It is a non-negligible component of the reaction. Therefore, a by-product may also be called an **accompanying product**. There is no such a term as "by-reaction" in chemistry since the by-product is generated from the main reaction.

A **side product** can be suppressed, reduced, or eliminated under optimal reaction conditions. However, a **by-product** cannot be avoided unless the chemistry being carried out or the reagent(s) used are changed.

For example, when a mono-alkylation of aniline with phenethyl tosylate was performed, *N*-phenethylaniline was the desired product (yield 86%), *N,N*-diphenethylaniline was the side product (yield 6%), and *p*-toluenesulfonic acid was the by-product [2]. The yield of the side product, *N,N*-diphenethylaniline, can be reduced by increasing the equivalents of aniline; however, formation of the by-product, *p*-toluenesulfonic acid, cannot be avoided.

Aniline (3 eq.) + Phenethyl tosylate (1 eq.) → KI (0.1 eq.), CH_3CN, 170 °C, 10 min, microwave → *N*-phenethylaniline (desired product, yield 86%) + *N,N*-diphenethylaniline (side product, yield 6%) + TsOH (*p*-toluenesulfonic acid, by-product)

2.7 Impurity

Impurities are undesired compounds present in a desired compound. They may come from by-products, side products, starting materials, reagents, catalysts, or solvents.

2.8 Theoretical Yield, Percentage Yield (Yield), Net Yield, and Overall Yield

In a chemical reaction, reactants that are not completely used up are called **excess reagents**, and reactants that are completely consumed are called **limiting reagents**. The amount of a product formed from the complete reaction of a limiting reagent is called the **theoretical yield**, whereas the actual amount of a product produced is the actual yield. The ratio between the actual yield and the theoretical yield expressed in percentage is called the **percentage yield**, also commonly known as **yield**. Chemical reaction equations give the ideal stoichiometric relationship between reactants and products. Thus, the theoretical yield can be calculated from the reaction stoichiometry. For many chemical reactions, the actual yield is generally less than the theoretical yield, apparently due to the loss in the process, side reactions, or inefficiency of the chemical reaction.

Sometimes, a reactant is not 100% converted into a product. The reaction can be pushed toward the product by approaches such as prolonging the reaction time or raising the reaction temperature. However, these approaches often increase formation of side products. If the reactant is expensive and can be easily recovered from the reaction mixture by separation and purification process, we would rather sacrifice the conversion rate and reduce the reaction complexity. The amount of a product based on the actual consumption of the reactant is called the **net yield**.

For a multistep synthesis, the **overall yield** of the final product is calculated by multiplying the yield of each step. This overall yield is crucial for planning a multistep synthesis. For example, for an eight-step synthesis, assuming the yield of each step is 80%, then the overall yield is 80% × 80% × 80% × 80% × 80% × 80% × 80% × 80% = 17%. If the molecular weight of the starting material is 100 daltons (Da) and the final product is 400 daltons (Da), and if 1 g of the final product needs to be synthesized, then at least 1.00 g × 100/400/0.17 = 1.49 g of the starting material is needed. To account for retaining intermediates at each step and to prepare for unexpected potential issues, the first step is typically performed on a scale three times larger—requiring about 5 g of the starting material.

The **overall yield** can also be used for expressing multistep **one-pot reactions** and **telescoped reactions,** as these processes usually include more than one reaction.

2.9 Intermediate

In the synthesis or production of a compound, all the substances generated in one step and used for the succeeding step are considered **intermediates**. In descriptions of organic reaction mechanisms, certain unstable molecules—such as free radicals, carbenes, carbonium ions, and carbanions—are known or hypothesized to be intermediates, even if they have not been isolated. These intermediates are highly reactive fragments of molecules that generally remain uncombined for only very short periods.

2.10 Transition State

The **transition state** in a chemical reaction represents a specific configuration along the reaction pathway. It is characterized as the point of maximum potential energy on the reaction coordinate. According to the transition state theory, once the reactants reach this transition state and surpass it, they invariably proceed to form products.

2.11 Species

A **chemical species** is a chemical substance that can be an atom, molecule, ion, or radical. Species are often used to represent intermediates in a reaction mechanism.

2.12 Free Radical

A **free radical**, often simply referred to as a radical, is an atom, molecule, or ion that possesses at least one unpaired valence electron. In most cases, these unpaired electrons render radicals highly reactive in chemical reactions. Many radicals tend to spontaneously dimerize. Also, the majority of organic radicals exist only briefly due to their short lifetimes.

2.13 Stoichiometric Reaction

A stoichiometric reaction is one in which the quantities of reactants and products are balanced such that all of the reactants are consumed and none remain after completion of the reaction. In other words, all reactants and reagents need to be added in amounts at least matching the reaction equation.

2.14 Catalytic Reaction

Catalysis refers to speeding up a chemical reaction and lowering the temperature required for it by introducing a substance called a **catalyst**. Catalysts are not used up in the reaction, and they retain their original form throughout the reaction. In cases where the reaction is fast and the catalyst is rapidly reused, even tiny quantities of the catalyst may be sufficient. Typically, catalysts react with reactants and produce intermediate substances. These intermediates break down to form final products of the reaction, and the catalyst is restored to its initial state for continued use.

2.15 Material Balance

A material balance, often referred to as a mass balance, applies the law of mass conservation to organic reactions. This principle dictates that mass cannot be spontaneously created or destroyed. In a reaction mechanism, some small molecule by-products, such as H_2O, CO, CO_2, NH_3, HCl, etc., should not be ignored.

2.16 Electron Balance

Electron balance, also known as **charge balance**, is the conservation of electrons in an organic reaction. Since electrons are also a form of matter, in a reaction mechanism, the total number of electrons in the reactants must equal the total number of electrons in the intermediates and the products. In a reaction mechanism, some ions, salts, or radicals may be involved, but the total number of charges in the reactants must equal the total number of charges in the intermediates and the products. In a redox reaction, if one species is oxidized, another species must be reduced.

2.17 Nucleophile

A **nucleophile** is a chemical species that forms bonds by donating an electron pair. All molecules and ions with a free pair of electrons or at least one π bond can act as nucleophiles. Because nucleophiles donate electrons, they are Lewis bases. The term nucleophilic describes the affinity of a nucleophile to bond with positively charged atomic nuclei. Nucleophilicity, also called nucleophile strength, refers to a substance's nucleophilic character and is often used to compare the affinity of atoms. Nucleophiles may take part in nucleophilic substitution and nucleophilic addition.

2.18 Electrophile

Any molecule, ion, or atom that is deficient in electrons in some way is called an **electrophile**. Electrophiles can accept a pair of electrons. These are types of species that tend to attract electrons (philic). The reactions between electron donors and acceptors are described by concepts such as electrophile and nucleophile, which are the most important concepts in organic chemistry.

2.19 Acid

An **acid** is a molecule or ion capable of either donating a proton, known as a **Brønsted–Lowry acid**, or forming a covalent bond with an electron pair, known as a **Lewis acid**. Hydrogen chloride, acetic acid, and most other Brønsted–Lowry acids cannot form a covalent bond with an electron pair and are therefore not Lewis acids. Conversely, many Lewis acids are not Arrhenius or Brønsted–Lowry acids. In modern terminology, an acid is implicitly a Brønsted–Lowry acid and not a Lewis acid, since we always refer to a Lewis acid explicitly as a Lewis acid.

2.20 Base

In chemistry, there are three definitions for the word "**base**"—**Arrhenius bases**, **Brønsted–Lowry bases**, and **Lewis bases**, all of which agree that bases are substances that react with acids.

Arrhenius proposed that a base is a substance which dissociates in an aqueous solution to form hydroxide ions OH^-. According to the more general Brønsted–Lowry acid–base theory, a base is a substance that can accept hydrogen cations (H^+). The Lewis theory states that a base is an electron pair donor that can share a pair of electrons with an electron acceptor, which is a Lewis acid.

Basicity is closely related to nucleophilicity. The difference between the two is that basicity is a thermodynamic property (i.e., relates to an equilibrium state), whereas nucleophilicity is a kinetic property, which relates to rates of certain chemical reactions).

2.21 Acidity and pK_a

In chemistry, an acid dissociation constant (also known as acidity constant or acid ionization constant, denoted by K_a) is a quantitative measure of the strength of an acid in solution. It is the equilibrium constant for a chemical reaction

$$HA \rightleftharpoons A^- + H^+$$

known as dissociation in the context of acid–base reactions. The chemical species HA is an acid that dissociates into A^-, the conjugate base of the acid and a hydrogen ion, H^+.

The dissociation constant is defined by

$$K_a = \frac{[A^-][H^+]}{[HA]}$$

$$pK_a = -\log_{10} K_a = \log_{10} \frac{[HA]}{[A^-][H^+]}$$

The higher the K_a of an acid, the stronger the acidity. In other words, the higher the pK_a of an acid, the stronger the basicity of its conjugate base.

2.22 Acid–Base Reaction

An **acid–base reaction** is simply the transfer of a proton from an acid to a base. When the acid gives up a proton, the species remaining still retains the electron pair to which the proton was formerly attached. The new species can reacquire a proton and is referred to as the conjugate base of the acid. All acids have a conjugate base, and all bases have a conjugate acid. An acid–base reaction fits the equation

$$\underset{\text{Acid}_1}{A{-}H} + \underset{\text{Base}_2}{\ddot{B}} \rightleftharpoons \underset{\text{Base}_1}{\ddot{A}} + \underset{\text{Acid}_2}{H{-}B}$$

Base$_1$ is the conjugate base of **Acid$_1$**, and **Acid$_2$** is the conjugate acid of **Base$_2$**. To push the reaction to the right side of the equation, pK_a of **Acid$_2$** must be greater than pK_a of **Acid$_1$**.

Whether a compound is an acid or a base depends on the role it plays in the reaction. For example, aniline is a base in the following reaction:

$$C_6H_5NH_2 + H{-}Cl \longrightarrow C_6H_5N^+H_3Cl^-$$

Aniline base | Hydrochloric acid | Aniline hydrochloride salt

In the following reaction, aniline is an acid:

$$C_6H_5NH_2 + n\text{-BuLi} \longrightarrow C_6H_5N^-HLi^+ + n\text{-BuH}$$

Aniline acid | *n*-Butyllithium base | Lithium anilide base | *n*-Butane acid

2.23 Reaction Selectivity

Selectivity problems arise when more than one reaction is possible for a given reaction. This is one of the fundamental aspects of chemical reactions, where reactions tend to occur in specific ways, producing specific products. Selectivity in organic chemistry plays a crucial role in determining the outcome of reactions and is critical for the synthesis of complex molecules with desired properties. To achieve this goal, it is important to understand selectivity in organic chemistry and its importance in synthetic reactions to produce the desired product with the desired stereochemistry and minimize side products.

2.24 Chemoselectivity

Chemoselectivity means a chemical reagent reacts more readily with one functional group over another when multiple different functional groups are present. Another class of chemoselectivity is to control a reaction at one functional group to a desired stage, such as reducing an ester to aldehyde rather than primary alcohol.

For example, there are two carbonyl groups in 4-acetylbenzaldehyde (**1**): the aldehyde group and the ketone group. When **1** was reduced with a catalyst system consisting of $Fe(BF_4)_2 + 6H_2O$ and $P(CH_2CH_2PPh_2)_3$ [tetraphos, (PP_3)] in THF, the aldehyde group was chemoselectively reduced to the primary alcohol 1-(4-(hydroxymethyl)phenyl)ethan-1-one (**2**) [3]. However, the ketone group was preferentially reduced to 4-(1-hydroxylethyl)benzaldehyde (**3**) in 93% selectivity with diborane on silica gel by first forming an adduct of the more reactive carbonyl group with sodium bisulfite [4].

$Fe(BH_4)_2.6H_2O/PP_3$ (0.4%)
HCO_2H (1.1 eq.)
THF, 60 °C, 2 h
99%
1 → 2

1. $NaHSO_3$ (2 eq.)
dioxane-H_2O (1:2), rt, 1 h
2. Silica gel
3. Diborane
4. aq. NaOH
93%
1 → 3

There are many categories under the term "**chemoselectivity**." Examples include the Dess–Martin oxidation [5] and the Swern oxidation [6], both of which selectively convert primary alcohols to aldehydes without further oxidation to carboxylic acids.

2.25 Regioselectivity

Regioselectivity describes the propensity of a chemical reaction to favor bond formation or cleavage at a specific site over other potential locations. This principle often dictates where a reagent interacts within a molecule—such as the selective removal of a particular proton by a strong base or the positioning of a new substituent on an aromatic ring already bearing functional groups. It plays a critical role in reactions where molecules contain multiple reactive sites, guiding the process to occur predominantly at one location.

A classic illustration is the Baeyer–Villiger oxidation, in which an oxygen atom is introduced adjacent to a carbonyl group. In ketones, this insertion is directed toward the carbon atom with greater substituents, reflecting electronic or steric influences. For instance, studies on acetophenones demonstrate that oxygen incorporation is favored between the carbonyl group and the aromatic ring, resulting in phenyl acetate formation rather than methyl benzoate [7]. This outcome underscores how regioselectivity governs product distribution by prioritizing specific reaction pathways.

m-CPBA
$CHCl_3$, 32 °C

2.26 Stereoselectivity

Stereoselectivity refers to the tendency of a chemical reaction to produce stereoisomers in unequal proportions when a new stereocenter is formed non-stereospecifically or when an existing stereocenter undergoes non-stereospecific modification.

The ratio of stereoisomers generated can vary widely. While both products are theoretically possible, they typically form in distinct quantities. In some cases, depending on reaction conditions and analytical sensitivity, the minor stereoisomer may not be observed using the detection methods available.

Stereoselectivity can be classified into three basic types. The first is that an achiral or prochiral structural unit undergoes a reaction to form a chiral unit and produces unequal amounts of stereoisomers. All **asymmetric reactions** are stereoselective reactions. The asymmetric reduction of acetophenone to (*R*)- or (*S*)-(-)-1-phenylethan-1-ol is a stereoselective reduction [8]. The possible two products of this reaction are **enantiomers**. The stereoselectivity of an asymmetric reaction can be determined by **enantiomer excess** (*ee*). Enantiomeric excess (*ee*) is also a measurement of purity of chiral substances. It reflects the degree to which a sample contains one enantiomer in greater amounts than the other. A **racemic mixture** has an *ee* of 0%, while a single completely pure enantiomer has an *ee* of 100%.

$)_2BCl$

(–)-Ipc$_2$BCl
THF, –25 °C
ee 98%
yield 72%

Acetophenone

(S)-1-phenylethan-1-ol
(99%)

+

(R)-1-phenylethan-1-ol
(1%)

The Diels–Alder reaction is both regioselective and stereoselective, although it typically produces a racemic mixture of two enantiomers. For example, in a stereocontrolled total synthesis of (+/−)–tabersonine, the Diels–Alder cycloaddition reaction between diene **1** and ethacrolein **2** proceeded with complete regiocontrol and excellent *endo* selectivity to produce adduct **3** in 97% yield [9].

MeO_2C N
TBSO
1
+
OHC
2
Toluene
85 °C
97%
MeO_2C N
CHO
TBSO
3 (racemic)
Steps
N
N H
CO_2Me
(+/−)-tabersonine

The second type of stereoselectivity is the reaction of symmetrical cyclic compounds to produce unequal amounts of **cis-trans isomers**. A typical example is the reduction of 4-(*tert*-butyl)cyclohexan-1-one to *cis*- and *trans*-4-(*tert*-butyl) cyclohexan-1-ol. With the use of lithium methyl borohydride as a reducing agent, the proportion of more stable trans product reached 98.3%, whereas with $NaBH_4$, the trans/cis ratio was only 86:14 [10].

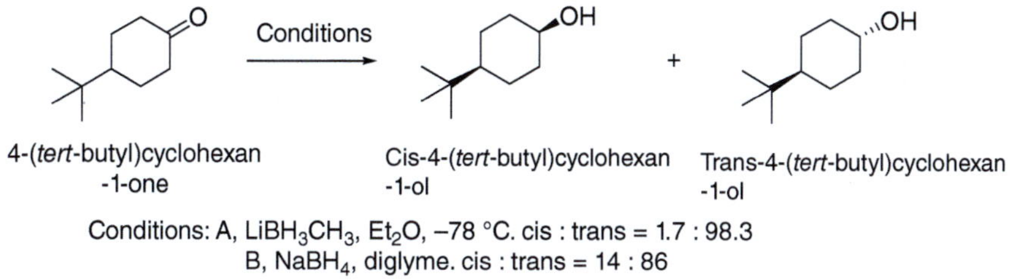

Conditions

4-(*tert*-butyl)cyclohexan-1-one → Cis-4-(*tert*-butyl)cyclohexan-1-ol + Trans-4-(*tert*-butyl)cyclohexan-1-ol

Conditions: A, $LiBH_3CH_3$, Et_2O, −78 °C. cis : trans = 1.7 : 98.3
B, $NaBH_4$, diglyme. cis : trans = 14 : 86

The third type of stereoselectivity is a reaction that produces a C=C or C=N double bond and produces unequal amounts of *Z*/*E*-isomers, also called **geometric isomers**. The dehydrohalogenation of 2-iodobutane demonstrates moderate stereoselectivity, producing *trans*-2-butene as the major product (60%) and *cis*-2-butene as the minor product (20%) [11].

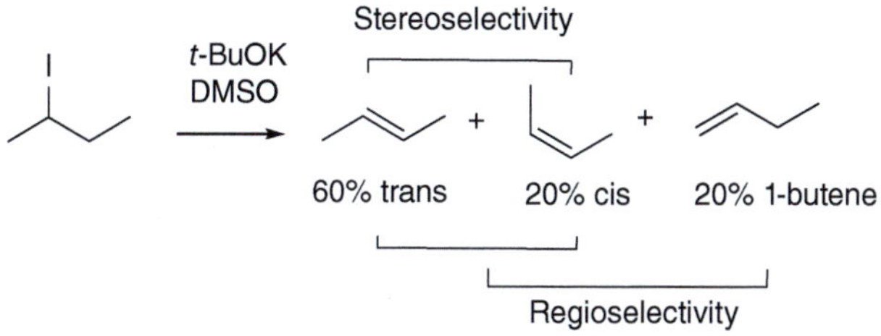

Stereoselectivity

t-BuOK
DMSO

60% trans + 20% cis + 20% 1-butene

Regioselectivity

The Wittig reaction is generally stereoselective, with the geometry of the resulting alkene dependent on the type of ylide employed. The reaction of aldehydes with unstabilized ylides (where R^2 = alkyl groups) favors the formation of the (*Z*)-alkene with moderate-to-high selectivity. This preference can be amplified to near-exclusive *Z*-selectivity when conducted in dimethylformamide (DMF) with lithium iodide or sodium iodide as additives. Conversely, stabilized ylides (where R^2 = electron-withdrawing groups such as esters or ketones) predominantly yield the (*E*)-alkene with high selectivity [12].

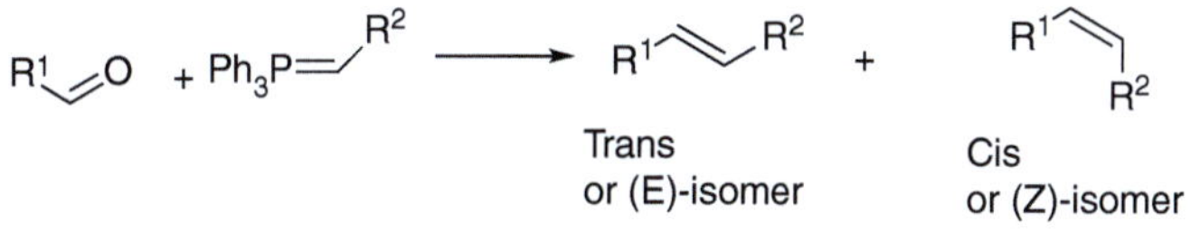

R^1CHO + Ph_3P=CHR^2 → R^1CH=CHR^2 (Trans or (E)-isomer) + R^1CH=CHR^2 (Cis or (Z)-isomer)

2.27 Iteroselectivity

Iteroselectivity describes the selectivity when a given reaction can occur at least twice on a substrate (e.g., alkylation of a symmetric diol and protection of a symmetric diamine) but selectively stops after a given number of iterations. This is a long-standing and challenging practical problem. Roy Lavendomme and Ivan Jabin recently named this type of selectivity "**iteroselectivity**" and their work was published in *Cell Reports Physical Science* [13].

This new selectivity concept is defined and compared with the classical chemo-, regio-, and stereoselectivities encountered in chemical synthesis. Examples of iteroselective reactions range from very common reactions such as electrophilic aromatic substitution to advanced methods involving large supramolecular complexes.

Unlike traditional isomers, **iteromers** are not considered isomers because they have different molecular formulas.

The comparison between regioselectivity and iteroselectivity is shown in the following example of bromination of anisole.

Bromination of anisole with photodecomposed brominated methanes could iteroselectively provide 1-bromo-4-methoxybenzene [14]. In contrast, using 1.5 equivalents of *N*,*N*-dibromo-*p*-toluenesulfonamide ($TsNBr_2$), 1,3,5-tribromo-2-methoxybenzene was prepared in 77% yield [15].

Methylation of benzene-1,3,5-triol (phloroglucinol) is another example of iteroselectivity [16].

Monoalkylation of primary amines is a challenging problem in traditional chemistry due to the formation of significant amounts of overalkylated side products. The canonic case of monomethylation is the most difficult since nucleophilicity of the amines increases according to the number of methyl groups without steric counterbalance. For example, reaction of a primary amine **1** with MeI yields an inextricable mixture of methylated products **2** and **3**, along with the quaternary ammonium salt **4** as a major product [17]. In 2014, Julien Legros' group developed an iteroselective monomethylation of primary amines by using MeOTf as methylation agent and HFIP as solvent [18].

Iteromers

$$\underset{1}{R{-}NH_2} \xrightarrow{MeI} \underset{2}{R{-}NHMe} + \underset{3}{R{-}NMe_2} + \underset{4}{R{-}N^+Me_3I^-}$$

$$\xrightarrow{MeOTf,\ HFIP} 2\ (52\text{–}96\%\ selectivity)$$

Monoprotection of diamine was a potential issue and often encountered in medicinal chemistry. A Korean group of chemists developed an iteroselective mono-Boc protection of diamine by using 1 equivalent of HCl and 1 equivalent of Boc_2O in MeOH-H_2O [19]. For example, monoprotected ethylene diamine was prepared in 87% yield by this method.

H_2N–CH_2CH_2–NH_2 → H_2N–CH_2CH_2–NHBoc (87%)

1. HCl (1 eq.), MeOH, H_2O, 0 °C - rt
2. Diamine
3. Boc_2O (1 eq.)

2.28 One-pot Synthesis/Telescoping Process

Chemists frequently employ **one-pot synthesis**, a method in which multiple sequential reactions are conducted in a single reactor without isolating intermediates. This strategy enhances reaction efficiency by eliminating time-consuming isolation and purification steps for intermediate compounds, thereby conserving resources, reducing labor, and often enhancing overall reaction yields. One-pot synthesis is particularly advantageous when intermediates are unstable under standard workup conditions or prone to volatility during solution concentration, as it circumvents these challenges by maintaining reaction continuity within the controlled environment of the reactor.

The **telescoped process** is to perform multiple transformations (including quenching and other workups) without directly isolating intermediates. Telescoped solutions of intermediates can be extracted, filtered (as long as the product remains in the filtrate), and solvent exchanged, but the intermediate ultimately remains in the solution and undergoes subsequent transformations. Therefore, the telescoped process can be carried out in multiple pots.

Whether it's the **one-pot synthesis** or the **telescoped process**, they are all within the scope of process chemistry. These processes are based on successful step-by-step synthesis or have defined reaction mechanisms.

An example of a one-pot synthesis is the preparation of molnupiravir starting from cytidine [20]. Molnupiravir, which exerts its anti-SARS-CoV-2 effect through lethal mutagenesis, was approved for emergency use authorization in the United States. In 2022, Aisa and Shen's group reported a one-pot process for preparing molnupiravir from cytidine. In that process, three reactions were carried out in one flask. It was not necessary to separate and purify intermediates **1** and **2**.

Cytidine
DMF-DMA, THF
65 °C, 18 h
1
Evaporation of THF
solvent exchange to DCM
Isobutyric anhydride
Et_3N, DMAP, 0–25 °C, 2 h
2
Evaporation of DCM
Hydroxylamine sulfate
70% aq. *i*-PrOH
78 °C, 18 h
Molnupiravir
63% crystallization yield

The large-scale synthesis of an ampakine-type active pharmaceutical ingredient (API) **5** is based on a telescoped regioselective double amidation [21]. Compound **5**

3
DMSO/MeCN
CDI (1.1 eq.)
60 °C, 4 h
80 °C, 18 h
CDI (1.1 eq.)
25 °C, 9 h
Cyclopropanamine
(1.1 equiv)
25 °C, 1 h
4
Overall yield from 3, 65%
3 steps
5

has been selected for clinical trials, which can increase AMPA receptor activation and is therefore considered a "high-impact" AMPA receptor potentiator. In a process of **5**, the key intermediate **4** was prepared from compound **3** in the telescoped process [21].

2.29 Cascade Reaction

A **cascade reaction** (also termed as **domino reaction** or **tandem reaction**) is a process involving two or more sequential reactions, where each step is triggered exclusively by functional groups generated in the preceding transformation. Intermediates are not isolated, as each transformation occurs spontaneously under consistent reaction conditions, with no additional reagents introduced after the initial step. By strict definition, cascade reactions maintain unchanged parameters (e.g., temperature, solvent, catalysts) throughout all stages.

In comparison, one-pot synthesis broadly encompasses conducting multiple reactions consecutively in a single vessel without isolating intermediates, but it allows modifications such as adding reagents or altering conditions (e.g., temperature, pH) between steps. Thus, while all cascade reactions qualify as one-pot processes, not all one-pot syntheses meet the stringent criteria of cascade reactions. Although cascade sequences are predominantly intramolecular, they can proceed intermolecularly, overlapping with **multicomponent reactions** when multiple starting materials react in a single orchestrated sequence.

For example, an important class of complex multifunctional bicyclic lactams (such as **3**) were produced by a continuous **flow process** from aryl enamides. The second stage is a thermo cascade process, which includes the ring-opening and Diels–Alder reaction [22].

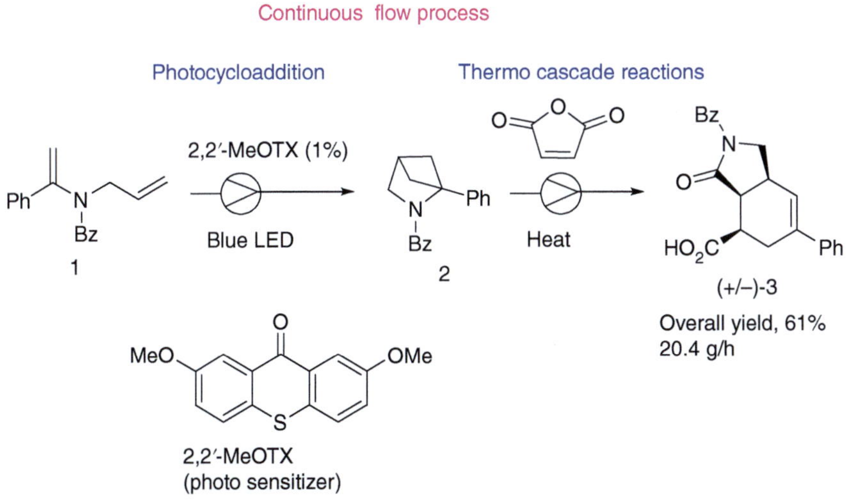

Continuous flow process

Photocycloaddition

Thermo cascade reactions

2,2′-MeOTX (1%)

Blue LED

Heat

1

2

(+/–)-3

Overall yield, 61%
20.4 g/h

2,2′-MeOTX
(photo sensitizer)

The mechanism of the process is shown here [22]:

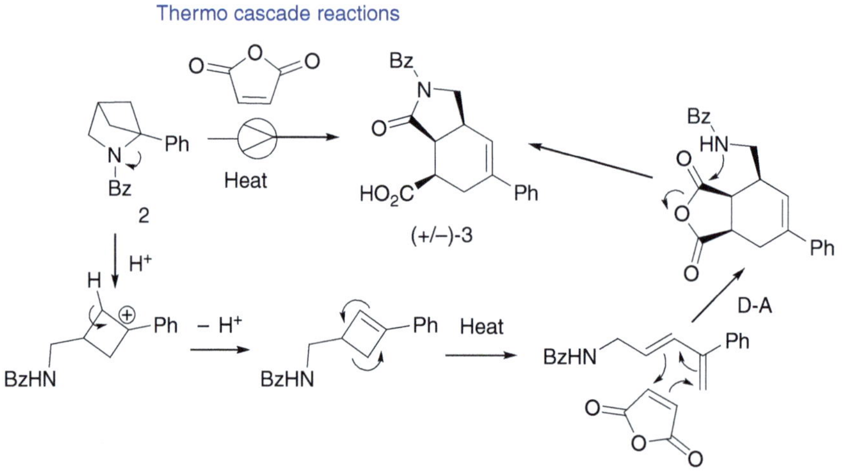

In steroidal chemistry, the backbone rearrangement is another example of the cascade reaction. Under the catalysis of the acidic clay mineral, montmorillonite K-10, cholest-5-enes undergoes a series of Wagner–Meerwein rearrangement to give a mixture of an equal amount of 20R- and 20S-5β,14β-dimethyl-18,19-dinor-8α,9β,10α-cholest-13(17)-enes [23].

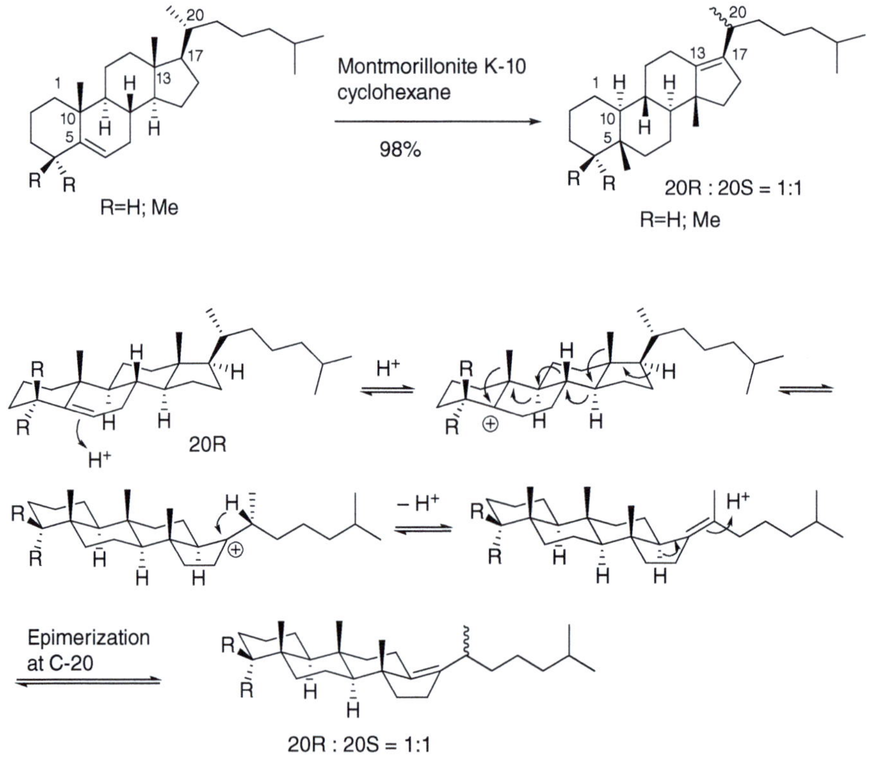

2.30 Multicomponent Reaction

A **multicomponent reaction (MCR)** is a chemical reaction in which three or more compounds react to form a single product. By definition, an MCR is one in which more than two reactants are combined in a sequential manner to produce a highly selective product that retains most of the atoms of the starting materials. Multicomponent reactions have been known for over 170 years. The first recorded MCR was the synthesis of α-aminocyanide by Strecker in 1850, from which α-amino acids could be derived [24].

Numerous MCRs have been developed, with isocyanide-based variants being the most extensively documented. Other notable types are free-radical-mediated, organoboron compound-based, and metal-catalyzed MCRs, among others [25]. Some typical examples include Biginelli reaction [26], Bucherer–Bergs reaction [27], and Ugi reaction [28]. The Ugi reaction is a four-component reaction involving a ketone or aldehyde, an amine, an isocyanide, and a carboxylic acid to form a bis-amide.

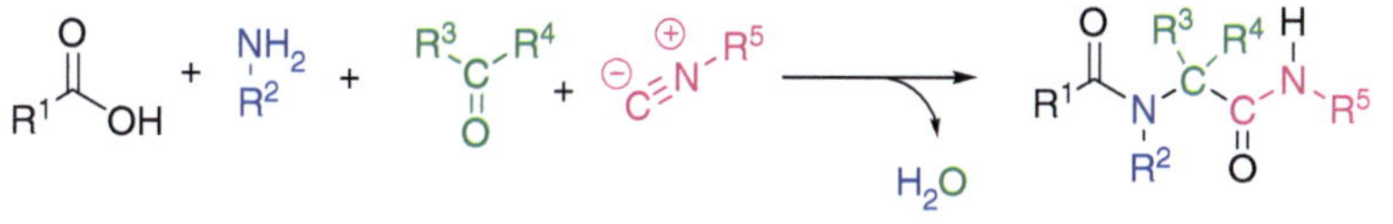

The mechanism of Ugi reaction is depicted as follows:

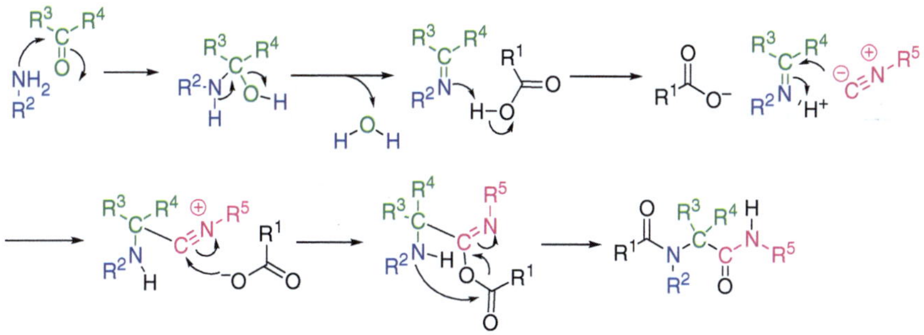

2.31 Flow Chemistry

Flow chemistry, also known as **continuous flow processing**, is a method in which chemical reactions are conducted within a continuously moving stream of fluid, as opposed to traditional batch production. In this system, reactants are pumped through a reactor—typically a network of interconnected tubes or channels—where they mix and interact. When reactive substances come into contact under controlled conditions, a chemical reaction is initiated and sustained as the mixture flows through the system.

Flow chemistry has been used in petrochemical and bulk chemical production for more than 100 years. The original Haber–Bosch process was continuous in 1913. But it is only in the past 10 years that continuous processes have garnered significant interest in the pharmaceutical and fine chemical industries. The production of advanced intermediates for fine chemicals and APIs relies primarily on traditional batch processing. Simple batch reactors were used because the synthesis of these intermediates often involves multiple reactions, and simple stirred tank reactors offer great versatility and flexibility.

However, batch reactors also have some significant limitations. One driving force behind the adoption of continuous manufacturing as a technological innovation is the ability to perform reactions that are more difficult to perform in batch mode. Some of these new reactions could only be performed in flows [29, 30]. Controlled oxidation of primary alcohols to aldehydes using Jones reagent is not possible in traditional stirred reactors. In a continuous flow reactor, Wiles et al. were able to selectively oxidize a series of primary alcohols to aldehydes or carboxylic acids depending on the flow rate employed [31].

OH
0.01 M in DCM
0.150 g silica-supported Jones' reagent
in borosilicate glass flow reactor
(0.3 cm (i.d.) · 3.0 cm (length))
O
CO_2H
flow rate: 0.650 mL/min 100% 0%
flow rate: 0.050 mL/min 0% 100%

2.32 Atom Economy

Atom economy (atom efficiency/percentage) is the conversion efficiency of a chemical process in terms of all atoms involved and the desired products produced. The simplest definition was introduced by Barry Trost in the 1990s [32] and is equal to the ratio between the mass of desired product and the total mass of all products, expressed as a percentage. Atom economy stands as a cornerstone principle within green chemistry [33], serving as one of the most prevalent metrics for evaluating the sustainability and environmental efficiency of chemical processes or syntheses.

A high atom economy indicates that the majority of reactant atoms are effectively used in forming the desired product, with minimal production of undesired by-products or side products. This efficiency lowers not only waste generation but also the economic and environmental burdens associated with waste management.

Atom economy is calculated as follows:

$$\text{Atom economy} = \frac{\text{molecular weight of desired product}}{\text{molecular weight of all products}} \times 100\%$$

An atom economy of 100% represents the ideal scenario, where all reactant atoms are incorporated into the desired product.

A Diels–Alder reaction serves as an excellent example of a highly atom-efficient process. For instance, the atom economy of the following Diels–Alder reaction is 100% [34].

O + O O O
Neat grinding
100%
H O O O H O

In contrast, the Mitsunobu reaction is not a good atom economic reaction because it produces a lot of wastes including triphenylphosphine oxide and dialkyl hydrazine-1,2-dicarboxylate. In the following example, the atom economy is only 36% based on 87% yield [35].

MW: 167.1 + MW: 186.2 + Ph_3P MW: 262.3 + DIAD, MW: 202.2 — THF, 0 °C - rt, 2 h, 87% →

Desired product MW: 335.3 + $Ph_3P{=}O$ MW: 278.3 + MW: 204.2 (Wastes)

$$\text{Atom economy} = 335.3 \div (335.3 + 278.3 + 204.2) \times 87\% \times 100\% = 35.7\%$$

2.33 Green Chemistry

Green chemistry, similar to **sustainable chemistry** or **circular chemistry**, is an area of chemistry and chemical engineering focused on the design of products and processes that minimize or eliminate the use and generation of hazardous substances.

In 2000, Paul Anastas and John C. Warner published a set of principles to guide the practice of green chemistry [36]. Since then, more than a dozen textbooks and reference books have been published in the field [37]. There are also journals that publish research and review articles in the field of green chemistry [38]. The 12 principles propose a series of ways to reduce the environmental and health impacts of chemical production and identify research priorities for the development of green chemistry technologies.

The 12 principles of green chemistry are [36]:

1. **Prevention:** Preventing waste is better than treating or cleaning up waste after it is generated.
2. **Atom Economy:** Synthetic methods should try to maximize the incorporation of all materials used in the process into the final product.
3. **Less Hazardous Chemical Syntheses:** Synthetic methods should avoid using or generating substances toxic to humans and/or the environment.
4. **Designing Safer Chemicals:** Chemical products should be designed to achieve their desired function while being as nontoxic as possible.
5. **Safer Solvents and Auxiliaries:** Auxiliary substances should be avoided wherever possible, and as nonhazardous as possible when they must be used.
6. **Design for Energy Efficiency:** Energy requirements should be minimized, and processes should be conducted at ambient temperature and pressure whenever possible.

7. **Use of Renewable Feedstocks:** Whenever it is practical to do so, renewable feedstocks or raw materials are preferable to nonrenewable ones.
8. **Reduce Derivatives:** Unnecessary generation of derivatives—such as the use of protecting groups—should be minimized or avoided if possible.
9. **Catalysis:** Catalytic reagents that can be used in small quantities to repeat a reaction are superior to stoichiometric reagents (ones that are consumed in a reaction).
10. **Design for Degradation:** Chemical products should be designed so that they do not pollute the environment; when their function is complete, they should break down into nonharmful products.
11. **Real-time Analysis for Pollution Prevention:** Analytical methodologies need to be further developed to allow real-time, in-process monitoring and control before hazardous substances form.
12. **Inherently Safer Chemistry for Accident Prevention:** Whenever possible, the substances in a process, and the forms of those substances, should be chosen to minimize risks such as explosions, fires, and accidental releases.

In 2024, Kai Rossen, Krishna Ganesh, and Kai Oliver Donsbach suggested that **greenhouse gas emissions** generated throughout the entire drug production process should be incorporated into the framework of green chemistry [39].

2.34 Partial Synthesis, Total Synthesis, and Formal Synthesis

Partial synthesis, also known as **semi-synthesis**, involves starting with a naturally occurring compound or a simpler molecule and then modifying it through chemical reactions to obtain the desired product. This approach is often used when the target compound is complex and difficult to create from scratch. By using partial synthesis, chemists can take advantage of existing chemical structures and modify them to achieve the desired compound.

For example, artesunate is a more hydrophilic artemisinin derivative developed by Amivas [40]. It was granted Food Drug Administration (FDA) approval for the initial treatment of severe malaria. The active dihydroartemisinin is the metabolite of artesunate. Artesunate was semi-synthesized from natural product artemisinin, which could be isolated from *Artemisia annua*. Researchers from Jiangsu Swithin Biological Medicine Engineering Research Center reported a process-scale (6.4 kg) synthesis of artesunate [41].

Artemisinin → (KBH_4, dioxane, 0 °C) → Dihydroartemisinin → (succinic anhydride, DMAP, 35 °C, 94%, one-pot) → Artesunate

Total synthesis, on the other hand, involves creating the target compound from simple starting materials. This approach requires a deep understanding of the reactions employed and often involves multiple steps to build the desired molecule from the ground up. Total synthesis is a powerful tool for chemists to demonstrate their abilities to create complex molecules that may not be readily available in large quantities from natural resources. The term "total synthesis" is often used to describe the total synthesis of structurally and stereochemically complex natural products.

A contrasting example of total synthesis is the total synthesis of artemisinin developed by Zhu and Cook at Indiana University as shown here [42].

1. Me_2Zn, L* $Cu(OTf)_2$ toluene 2. Br 61%; 1. $TsNHNH_2$ MeOH 2. BuLi 3. DMF 72%; TIPSO OMe Et_2AlCl DCM 98%; TIPSO OMe; $PdCl_2$ H_2O_2 61%; TIPSO OMe; 1. $(NH_4)_2MoO_4$ H_2O_2, t-BuOH 2. p-TSA, DCM 42% $^1O_2$ generated in situ; Artemisinin; L*= Ph O P–N O Ph

Partial synthesis and total synthesis can be understood like this. A partial synthesis is similar to convert an office building into a library, while a total synthesis is to build a new library from bricks, wood, steel bars, cement, and other materials on the ground.

A **formal synthesis** is the synthesis of a complex intermediate en route to the natural product or pharmaceutical compound. This means that you can intersect someone else's route and say that the molecule can be made from there or that the molecule is known through isolation studies to degrade or be transformed into the desired product via some pathway. The key here is that you don't make the actual product and thus you're relying on someone else's report for the final steps.

For example, atropurpuran is a non-alkaloidal diterpene natural product isolated from the roots of *Aconitum hemsleyanum* [43]. Suzuki et al. reported the formal synthesis of atropurpuran [44]. They completed the synthesis of the key intermediate (**Xu's intermediate**) and claimed the formal synthesis of atropurpuran, since **Xu's intermediate** was further converted to the target compound atropurpuran by Xu's group [45].

OMe; ref. 5 21 steps; OH OMe OMe Me; Xu's intermediate; ref. 6 3 steps; OH Me; Atropurpuran

2.35 Linear Synthesis and Convergent Synthesis

When a multistep synthesis is conducted, two strategies need to be considered: a **linear synthesis** or a **convergent synthesis**. A linear synthesis is one in which all the reagents and building blocks are added in a sequential fashion. In a convergent synthesis, two or more separate pathways may be followed with the pieces joined at a later time to form the final product.

The difference in efficiency between linear synthesis and convergent synthesis can be demonstrated through the following example. Consider a target compound composed of eight structural units, commonly referred to as **building blocks** (A-B-C-D-E-F-G-H). In a linear synthesis approach, the process would proceed as follows.

A + B → A–B (80%); + C → A-B-C (80%); + D → A-B-C-D (80%); + E → A-B-C-D-E (80%)

+ F → A-B-C-D-E-F (80%); + G → A-B-C-D-E-F-G (80%); + H → A-B-C-D-E-F-G-H (80%)

There are seven synthetic steps. If the yield for each step is 80%, the overall yield will be 80% × 80% × 80% × 80% × 80% × 80% × 80% = 21%.

However, if you perform the synthesis in a convergent pathway, there will be only as less as three steps. If the yield for each step is 80%, the overall yield will be 80% × 80% × 80% = 51%.

A + B → A–B (80%)
C + D → C–D (80%)
A–B + C–D → A-B-C-D (80%)
E + F → E–F (80%)
G + H → G–H (80%)
E–F + G–H → E-F-G-H (80%)
A-B-C-D + E-F-G-H → A-B-C-D-E-F-G-H (80%)

You can regard a multistep linear synthesis as completing a marathon (26 miles) run. While a convergent synthesis of the same molecule is four runs to complete the total 26 miles. For an amateur runner who has basic running training and usually only runs about 3 miles each time, completing a marathon within 4 hours is not an easy work. He or she may need more than half-year of systematic training to achieve this goal. However, if we divide these 26 miles into four sections, each section being about 6.5 miles, it will be a much easier job for an amateur runner to complete a total distance of 26 miles in four runs and each run to be finished within an hour.

Ineleganolide (**1**) is a diterpenoid natural product isolated from the Formosan soft coral Sinularia inelegans by Duh and coworkers in 1999 [46]. In 2022, Wood and Tuccinardi reported the first total synthesis of ineleganolide [47]. That synthesis can be considered a linear synthesis. They started from cyclopentenone **2** and furan **3**, and the final product was synthesized in 23 steps with 0.3% overall yield.

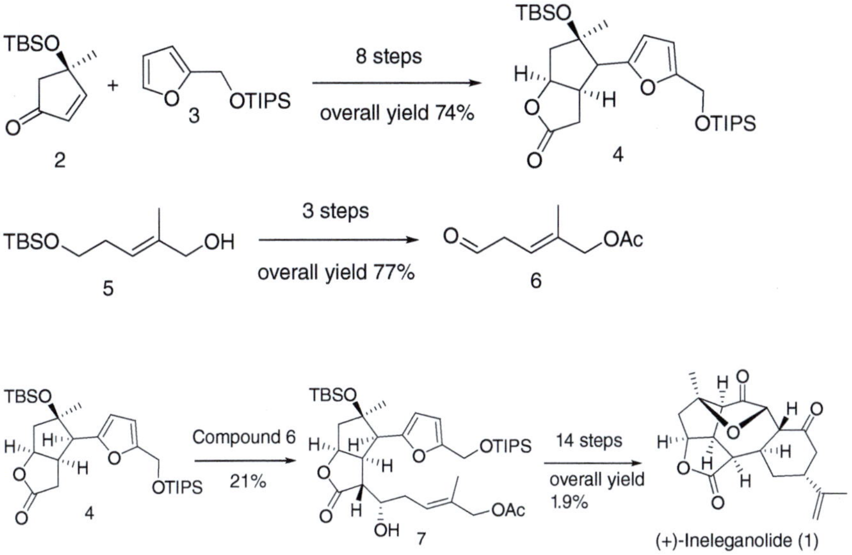

Shortly thereafter, Gross et al. reported a convergent synthesis of ineleganolide (**1**) [48]. The synthesis was started from compound **8** and **10** to prepare the two key building blocks **9** and **11**. The ester **12** was prepared by coupling of **9** and **11** by esterification. After further six steps were performed, the target compound (+)-ineleganolide (**1**) was synthesized in 16 steps, counted from the longest route from **8**, and the overall yield was 1.5%, five times higher than the linear synthesis [47].

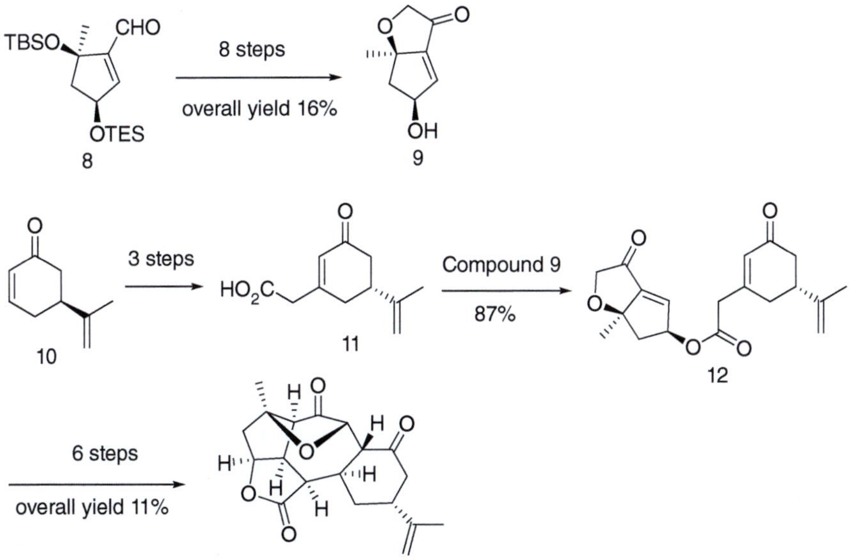

(+)-Ineleganolide (1)

2.36 Divergent Synthesis

Divergent synthesis is a strategy designed to create a diverse library of chemical compounds by initially reacting a core molecule with a set of reactants. The first generation of compounds is then subjected to further reactions, producing the second generation of compounds. This process rapidly expands, leading to a large number of new compounds. For example, a starting molecule (**A**) generates compounds **A1**, **A2**, and **A3** in the first generation. Subsequently, **A1** generates **A1-1**, **A1-2**, and **A1-3** in the second generation, and so on. In the context of medicinal chemistry, divergent synthesis typically involves the creation of a central scaffold, from which a series of new compounds are systematically derived.

One example is selected from our previous work. In a medicinal chemistry project, chemists from Alectos Therapeutics performed a divergent synthesis of a series of cyclohexane derivatives [49]. Those compounds were screened for OGA inhibition. The scaffold **3** was synthesized from starting materials, furan (**1**) and methyl acrylate (**2**) [50, 51], in 13 steps. From scaffold **3**, dozens of compounds with structure diversified were further prepared.

1 + 2 (CO$_2$Me) → 13 Steps → 3

Aminocarbooxazoline

Aminocarbothiazoline

6.6-Diflourooxazoline

Carbooxazoline

Trans-Oxazine

Cis-Oxazine

Carbooxazoline

2.37 Click Chemistry

Click chemistry refers to a category of straightforward, atom-efficient reactions widely employed to connect two selected molecular entities. Rather than denoting a single specific reaction, click chemistry represents an approach to creating products inspired by natural processes, where substances are formed by linking small, modular units. In numerous applications, click reactions are used to attach a biomolecule to a reporter molecule.

These reactions typically take place in a single vessel, are unaffected by the presence of water, and produce few, harmless by-products. They are characterized as "spring-loaded," meaning they possess a strong thermodynamic driving force that propels the reaction rapidly and irreversibly toward a high yield of a single product. Additionally, click reactions often exhibit high specificity, sometimes demonstrating both regio- and stereo-specificity.

The term "click chemistry" was first introduced in 1998 by Jan Dueser [52], the wife of K. Barry Sharpless, and was later comprehensively described by Sharpless, Kolb, and Finn in 2001 [53]. In 2022, the Nobel Prize in Chemistry was awarded jointly to Carolyn R. Bertozzi, Morten P. Meldal, and K. Barry Sharpless for their contributions to the development of click chemistry and bioorthogonal chemistry.

A classic example of click chemistry is the copper-catalyzed reaction between an azide and an alkyne, which forms a 1,2,3-triazole through a process known as Cu(I)-catalyzed azide-alkyne cycloaddition (CuAAC) [54]. The first synthesis of a triazole, achieved by reacting diethyl acetylenedicarboxylate with phenyl azide, was reported by Arthur Michael in 1893 [55]. This family of 1,3-dipolar cycloaddition reactions gained prominence in the mid-twentieth century, largely due to the extensive studies on their reaction kinetics and conditions by Rolf Huisgen, after whom the reactions were later named [56].

The copper(I)-catalyzed version of the Huisgen 1,3-dipolar cycloaddition was discovered simultaneously and independently by two research groups: one led by Valery V. Fokin and K. Barry Sharpless at the Scripps Research Institute in California [57], and the other by Morten Meldal at the Carlsberg Laboratory in Denmark [58]. Unlike the non-catalyzed reaction originally described by Huisgen, which produces a mixture of 1,4- and 1,5-isomers, proceeds slowly, and requires elevated temperatures (around 100 °C) [59], the copper-catalyzed variant selectively yields only the 1,4-isomer.

2.38 Parallel Synthesis

Parallel synthesis significantly reduces time and enhances compound differentiation by conducting multiple synthetic experiments concurrently. This approach is widely employed to expedite the discovery of novel compounds and to identify optimal process conditions. In medicinal chemistry, parallel synthesis plays a crucial role in the discovery and development of potential drug candidates. For instance, it facilitates the creation of libraries containing diverse chemical structures, which can then be screened for potential biological activity.

Parallel synthesis is applied across various stages, including lead generation, lead optimization, and the screening of optimal reaction conditions. In process scale-up and development, parallel synthesis accelerates process optimization by offering deeper insights into the impact of key reaction variables. These variables include solvent systems, optimal temperatures and concentrations, appropriate reagents, reaction times, and the selection of catalysts. By enabling the simultaneous exploration of these factors, parallel synthesis streamlines the optimization process and enhances efficiency in chemical development.

2.39 Solid-phase Synthesis

Solid-phase synthesis is a technique where molecules are covalently attached to a solid support material and synthesized sequentially in a single reaction vessel using selective protecting group chemistry. This approach is widely employed for the synthesis of peptides [60, 61], deoxyribonucleic acid (DNA), ribonucleic acid (RNA), and other molecules that require precise sequential assembly [62]. In recent years, it has also found applications in combinatorial chemistry and various synthetic processes. The method was pioneered in the 1950s and 1960s by Robert Bruce Merrifield, who developed it specifically for synthesizing peptide chains [63]. His groundbreaking work earned him the 1984 Nobel Prize in Chemistry.

In peptide synthesis, an amino-protected amino acid is first attached to a solid-phase material or resin, typically low cross-linked polystyrene beads, forming a covalent bond—often an amide or ester linkage—between the carbonyl group of the amino acid and the resin [64]. The amino group is then deprotected and reacted with the carbonyl group of the next *N*-protected amino acid, resulting in a dipeptide bound to the solid support. This cycle is repeated iteratively to build the desired peptide chain. Once the synthesis is complete, the peptide is cleaved from the resin.

The process can be illustrated through the synthesis of a dipeptide, as outlined in the following scheme [60].

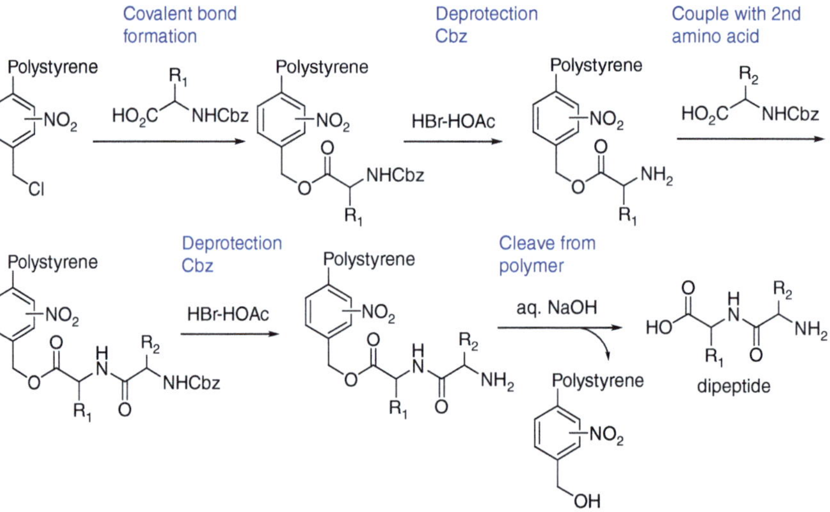

Covalent bond formation; Deprotection Cbz; Couple with 2nd amino acid; Deprotection Cbz; Cleave from polymer

2.40 Combinatorial Chemistry

Combinatorial chemistry involves chemical synthesis techniques designed to produce a vast number of compounds—ranging from tens to thousands, or even millions—in a single procedure. This approach was pioneered by Árpád Furka from Eötvös Loránd University in Budapest, Hungary. The core principle of combinatorial chemistry is to synthesize a multicomponent compound mixture, known as a combinatorial library, through a single stepwise process. This library is then screened to identify potential drug candidates or other useful compounds. A key innovation of this method is the use of mixtures during both synthesis and screening, which significantly enhances the efficiency and productivity of the process. The motivations behind the development of combinatorial chemistry were detailed in a publication in 2002 [65].

Combinatorial chemistry began to gain significant traction in the industrial sector starting in the 1990s [66] and has since become a crucial component of early-stage drug discovery for over 20 years. To date, only one drug developed through de novo combinatorial chemistry synthesis has received FDA approval for clinical use: sorafenib, a multikinase inhibitor used to treat advanced renal cell cancer [67].

2.41 Process Chemist versus Medicinal Chemist

Process chemistry plays a crucial role in medicinal chemistry, yet the responsibilities of **process chemists** and **medicinal chemists** (also known as **discovery chemists**) differ significantly. These differences can be summarized in the following key aspects:

- **Number of Synthesized Compounds:** Medicinal chemists focus on synthesizing a large number of compounds within a given project to screen for bioactivity and conduct structure-activity relationship (SAR) studies. In contrast, process chemists typically synthesize a single compound, which is intended for use in clinical studies or as an API in a marketed drug.
- **Quantity of Synthesized Compounds:** Medicinal chemists usually produce small quantities of each target compound, typically in the range of 5–10 mg, which is sufficient for preliminary bioactivity screening. Process chemists, however, work on a much larger scale, synthesizing compounds in kilogram or even ton quantities to meet the demands of clinical trials and commercial production.
- **Synthetic Methods and Reaction Conditions:** Medicinal chemists prioritize speed and diversity in synthesis, often employing divergent strategies to generate numerous compounds from common intermediates (scaffolds). The synthetic routes and reaction conditions are generally not optimized, and factors such as yield, purification convenience, cost, and green chemistry are not their primary concerns. If a compound lacks bioactivity, the elegance of its synthesis becomes irrelevant. In contrast, process chemists focus on optimizing synthetic routes to ensure high yield, safety, cost-effectiveness, labor efficiency, environmental sustainability ("greenness"), and robustness. As a result, the final synthetic route developed by process chemists often differs significantly from that used by medicinal chemists.
- **Involvement in Drug Discovery Stages:** Medicinal chemists are primarily involved in the early stages of drug discovery, synthesizing a large number of compounds for activity screening. Once a promising drug candidate is identified, the synthesis must be scaled up to gram or hundred-gram quantities for

preclinical studies. At this stage, process chemists step in to ensure the drug candidate is synthesized in the required quantities and quality, adhering to project timelines. Additionally, process chemists play a critical role in developing robust synthetic processes for producing the drug candidate or API on a kilogram scale during clinical studies and commercial production.

- **Quality Requirements for Final Products:** The purity requirements for compounds synthesized by medicinal chemists are relatively modest, typically around 98% (and sometimes as low as 95%) for preliminary screening purposes. In contrast, process chemistry demands much higher purity standards for its products [68]. Certain impurities must be controlled to levels below specified thresholds. For APIs, if an impurity exceeds 0.1%, its structure must be identified and characterized.

In summary, while medicinal chemists and process chemists both contribute to drug development, their roles, priorities, and methodologies differ significantly, reflecting the distinct requirements of early-stage discovery versus late-stage development and production.

References

1 Miyaura, N.; Suzuki, A. *Chem. Rev.*, **1995,** *95*, 2457.
2 Romera, J. L.; Cid, J. M.; Trabanco, A. A. *Tetrahedron Lett.*, **2004,** *45*, 8797.
3 Wienhöfer, G.; Westerhaus, F. A.; Junge, K.; Beller, M. *J. Organomet. Chem.*, **2013,** *744*, 156.
4 Chihara, T.; Wakabayasi, T.; Taya, K.; Ogawa, H. *Can. J. Chem.*, **1990,** *68*, 720.
5 Dess, D. B.; Martin, J. C. *J. Org. Chem.*, **1983,** *48*, 4155.
6 Omura, K.; Swern, D. *Tetrahedron*, **1978,** *34*, 1651.
7 Palmer, B. W.; Fry, A. *J. Amer. Chem. Soc.*, **1970,** *92*, 2580.
8 Brown, H. C.; Chandrasekharan, J.; Ramachandran, P. V. *J. Am. Chem. Soc.*, **1988,** *110*, 1539.
9 Kozmin, S. A.; Rawal, V. H. *J. Am. Chem. Soc.*, **1998,** *120*, 13523.
10 Kim, S.; Lee, S. J.; Rang, H. J. *Synth. Comm.*, **1982,** *12*, 723.
11 Bartsch, R. A.; Pruss, G. M.; Bushaw, B. A.; Wiegers, K. E. *J. Am. Chem. Soc.*, **1973,** *95*, 3405.
12 Bergelson, L. D.; Shemyakin, M. M. *Angew. Chem. Int. Ed.*, **1964,** *3*, 250.
13 Lavendomme, R.; Jabin, I. *Cell Reports Physical Science*, **2022,** *3*, 101121.
14 Kawakami, K.; Tsuda, A.; *Chem. Asian J.*, **2012,** 7, 2240.
15 Saikia, I.; Chakraborty, P.; Sarma, M. J.; Goswami, M.; Phukan, P. *Synth. Comm.*, **2015,** *45*, 211.
16 Crauste, C.; Vigor, C.; Brabet, P.; Picq, M.; Lagarde, M.; Hamel, C.; Durand, T.; Vercauteren, J. *Eur. J. Org. Chem.*, **2014,** 4548.
17 Salvatore, R. N.; Yoon, C. H.; Jung, K. W. *Tetrahedron*, **2001,** *57*, 7785.
18 Lebleu, T.; Ma, X.; Maddaluno, J.; and Legros, J. *Chem. Commun.*, **2014,** *50*, 1836.
19 Lee, D. W.; Ha, H.-J.; Lee, W. K. *Synth. Comm.*, **2007,** *37*, 737.
20 Hu, T.; Xie, Y.; Zhu, F.; Gong, X.; Liu, Y.; Xue, H.; Aisa, H. A.; Shen, J. *Org. Process Res. Dev.*, **2022,** *26*, 358.

21 Hardouin, C.; Pin, F.; Giffard, J.-F.; Hervouet, Y.; Hublet, J.; Janvier, S.; Penloup, C.; Picard, J.; Pinault, N.; Schiavi, B.; Zhang, P.; Zhao, W.; Zhu, X. *Org. Process Res. Dev.*, **2019,** *23*, 1932.

22 Elliott, L.D.; Booker-Milburn, K. I.; Lennox, A. J. J. *Org. Process Res. Dev.*, **2021,** *25*, 1943.

23 a) Blunt, J. W.; Hartshorn, M. P.; Kirk, D. N. *Tetrahedron*, **1969,** *25*, 149. b) Sieskind, O.; Albrecht, P. *Tetrahedron Lett.*, **1985**, *26*, 2135. c) Li, T.; Yang, Y.; Li, Y. *J. Chem. Research (S)*, **1993,** 28.

24 Strecker, A. *Ann. Chem. Pharm.* **1850,** *75* 27.

25 "*Multicomponent Reactions*" Edited by Zhu, J.; Bienayme, H., **2005**, WILEY-VCH Verlag GmbH & Co. KGaA, Weinheim.

26 Biginelli, P. *Chem. Ber.*, **1891,** *24*, 1317.

27 *a)* Bucherer, H. T.; Fischbeck, H. T. *J. Prakt. Chem.*, **1934,** *140*, 69. *b*) Bergs, H. Ger. pat. 566,094, **1929.**

28 Ugi, I.; Meyr, R.; Fetzer, U.; Steinbrückner, C. *Angew. Chem.*, **1959,** *71*, 386.

29 Naber, J. R.; Kappe, C. O.; Pesti, J. A. *Org. Process Res. Dev.*, **2020,** *24*, 1779.

30 Baumann, M.; Moody, T. S.; Smyth, M.; Wharry, S. *Org. Process Res. Dev.*, **2020,** *24*, 1802.

31 Wiles, C.; Watts, P.; Haswell, S. J. *Tetrahedron Lett.*, **2006,** *47*, 5261.

32 Trost B. M. *Angew. Chem. Int. Ed.*, **1995,** *90*, 259.

33 Dicks, A. P.; Hent, A. "*Green chemistry metrics: a guide to determining and evaluating process greenness*" Springer Cham. **2014**. ISBN 2191-5407, eBook.

34 Yelgaonkar, S. P.; Swenson, D. C.; MacGillivray, L. R. *Chem. Sci.*, **2020,** *11*, 3569.

35 Zhou, H.; Sun, H.; Yang, W.; Gan, Z. *Org. Prep. Proc. Int.*, **2024,** *59*, 264.

36 Anastas, P. T.; Warner, J. C. "*Green chemistry: theory and practice*" Oxford [England]; New York: Oxford University Press, **2000.**

37 Examples are shortlisted; *a)* Lancaster, M. "*Green Chemistry: An Introductory Text*", RSC Publishing, Cambridge, UK, **2010**. *b*) Clark, J. H. (Editor), Macquarrie, D. J. (Editor) "*Handbook of Green Chemistry and Technology*", Wiley, **2007**. *c*) Ahluwalia, V. K. "*Green Chemistry, Environmentally Benign Reactions*" 3rd edition, Springer, **2021**. *d*) Sheldon, R. A.; Arends, I.; Hanefeld, U. "*Green Chemistry and Catalysis*" Wiley, **2007.**

38 *a)* "*Green Chemistry*", RSC Publishing, **1999**. *b*) "*Green Chemistry Letters and Reviews*", Taylor & Francis online, **2007**. *c*) "*Current Research in Green and Sustainable Chemistry*", Elsevier, **2020**. *d*) "*ACS Sustainable Chemistry & Engineering*". ACS publications, **2013.**

39 Rossen, K.; Ganesh, K.; Donsbach, K. O. *Org. Process Res. Dev.*, **2024,** *28*, 3753.

40 Adebayo, J. O.; Tijjani, H.; Adegunloye, A. P.; Ishola, A. A.; Balogun, E.A.; Malomo, S.O. *Parasitol. Res.*, **2020**, *119*, 2749.

41 Dong, P. X.; Mei, Z.; Zhao, Z. J.; Yi, Y. Y. CN102887908, **2011.**

42 Zhu, C.; Cook, S. P. *J. Am. Chem. Soc.*, **2012,** *134*, 13577.

43 Tang, P.; Chen, Q.-H.; Wang, F.-P.; *Tetrahedron Lett.*, **2009,** *50*, 460.

44 Suzuki, T.; Koyama, T.; Nakanishi, K.; Kobayashi, S.; Tanino, K. *J. Org. Chem.*, **2020,** *85*, 10125.

45 Xie, S.; Chen, G.; Yan, H.; Hou, J.; He, Y.; Zhao, T.; Xu, J. *J. Am. Chem. Soc.*, **2019,** *141*, 3435.

46 Duh, C. Y.; Wang, S. K.; Chia, M. C.; Chiang, M. Y. *Tetrahedron Lett.*, **1999,** *40*, 6033.
47 Tuccinardi, J. P.; Wood, J. L. *J. Am. Chem. Soc.*, **2022,** *144*, 20539.
48 Gross, B. M.; Han, S.-J.; Virgil, S. C.; Stoltz, B. M. *J. Am. Chem. Soc.*, **2023,** *145*, 7763.
49 Li, T.-S.; McEachern, E. J.; Vocadlo, D. J.; Zhou, Y.; Zhu, Y.; Liu, K.; Selnick, H. G. US 9,120,781 B2, **2015**.
50 Ogawa, S.; Tonegawa, T. *Carbohydrate Res.*, **1990,** *204*, 51.
51 Ogawa, S.; Nishi, K.; Shibata, Y. *Carbohydrate Res.*, **1990,** *206*, 352.
52 Nobel Prize lecture: Barry Sharpless, Nobel Prize in Chemistry 2022.
53 Kolb, H. C.; Finn, M. G.; Sharpless, K. B. *Angew. Chem. Int. Ed.*, **2001,** *40*, 2004.
54 Kolb, H. C.; Sharpless, B. K. *Drug Discov. Today*, **2003,** *8*, 1128.
55 Michael, A. *J. Prakt. Chem.*, **1893,** *48*, 94.
56 Huisgen, R. "Centenary Lecture– 1,3-Dipolar Cycloadditions". *Proceedings of the Chemical Society of London*, **1961,** 357.
57 Rostovtsev, V. V.; Green, L. G; Fokin, V. V.; Sharpless, K. B. *Angew. Chem. Int. Ed.*, **2002,** *41*, 2596.
58 Tornoe, C. W.; Christensen, C.; Meldal, M. *J. Org. Chem.*, **2002,** *67*, 3057.
59 Huisgen, R. *Angew. Chem., Int. Ed.*, **1963,** *2*, 565.
60 Merrifield, B. A. *J. Am. Chem. Soc.*, **1963,** *85*, 2149.
61 Palomo, J. M. *RSC Adv.*, **2014,** *4*, 32658.
62 Krchňák, V.; Holladay, M. W. *Chem. Rev.*, **2002,** *102*, 61.
63 Merrifield, B. *Science*, **1986,** *232*, 341.
64 Guillier, F.; Orain, D.; Bradley, M. *Chem. Rev.*, **2000,** *100*, 2091.
65 Furka Á. *Drug Discov. Today*, **2002,** *7*, 1.
66 Rasheed, A.; Farhat, R. *Int. J. Pharm. Sci. Res.*, **2013,** *4*, 2502.
67 Newman, D.; Cragg, G. *J. Nat. Prod.*, **2007,** *70*, 461.
68 Finotti, C. C.; Lopardi, F. L.; Teixeira, C. D.; Bonfilio, R. *Crit. Rev. Anal. Chem.*, **2024**, 1.

3

General Terms and Concepts in Medicinal Chemistry

Some basic concepts, principles, and approaches in medicinal chemistry and drug discovery are collected in this chapter.

3.1 What Is a Drug?

A drug is any chemical substance that, when taken, causes changes in the physiology of an organism, including its psychology (if applicable) [1, 2]. Typically drugs are molecules used as drugs or pharmaceutical ingredients to diagnose, cure, mitigate, treat, or prevent disease. Medications are generally distinguished from foods and other substances that provide nutritional support. The drug may be consumed by inhalation, injection, smoking, ingestion, eye or nose drops, via skin patches, suppositories, or by dissolving under the tongue. Traditionally, drugs were obtained through extraction from medicinal plants, but in modern times an increasing number of drugs are obtained through organic synthesis [3]. Medications can be used for a limited time, or on a regular basis to treat chronic conditions.

3.2 Drug Discovery

Drug discovery is the process of discovering new drugs.

Historically, drugs were discovered by identifying active ingredients from traditional remedies or by serendipitous discovery. More recently, chemical libraries of synthetic small molecules, natural products, or extracts are screened in intact cells or whole organisms to identify substances with desirable therapeutic effects in a process known as classical pharmacology. After the sequencing of the human genome allowed rapid cloning and synthesis of large quantities of purified proteins, high-throughput screening (HTS) using large libraries of compounds against isolated biological targets that were hypothesized to ameliorate disease has become common practice in a process known as reverse pharmacology. The hits from these screens are then tested in cells and then in animals for efficacy.

Overcoming Synthetic Challenges in Medicinal Chemistry: Mechanistic Insights and Solutions,
First Edition. Tongshuang Li.

In modern drug discovery, target molecules, such as proteins, related to disease processes are identified. Once a target molecule is validated, scientists initiate a discovery process to find other molecules or substances that act on the target. The process of finding molecules that act on a target (i.e., lead molecules) can be arduous; thousands of compounds may need to be screened to find one or more promising leads. Once a lead molecule is identified, it can be chemically modified to increase its efficacy against the target molecule.

Drug discovery is a multifaceted process that begins with the identification of potential screening hits [4], followed by medicinal chemistry [5] efforts to refine these hits. The goal is to optimize their properties, including affinity, selectivity (to minimize side effects), efficacy/potency, metabolic stability (to extend half-life), and oral bioavailability. Once a compound meets these criteria, it advances to the drug development phase. If it proves viable, clinical trials are initiated.

Modern drug discovery is a highly capital-intensive endeavor, requiring substantial investments from pharmaceutical companies and governments alike. Despite technological advancements and a deeper understanding of biological systems, the process remains lengthy, costly, and challenging, with a low success rate for discovering new treatments [6]. In 2018, the research and development costs for a new molecular entity ranged from approximately 318 million to 2.8 billion [7]. In the twenty-first century, early-stage discovery research is primarily funded by governments and philanthropic organizations, while later-stage development is typically financed by pharmaceutical companies or venture capitalists. To gain approval for market release, a drug must successfully navigate multiple phases of clinical trials and undergo a rigorous regulatory review process, such as the New Drug Application (NDA) in the United States.

3.3 Medicinal Chemistry

Medicinal chemistry is an interdisciplinary field that bridges chemistry and pharmacy, focusing on the design, discovery, and development of therapeutic drugs. This discipline encompasses the identification, synthesis, and optimization of new chemical compounds for medical use, as well as the investigation of the biological effects and structure–activity relationships (SAR) of existing drugs. These studies often involve quantitative structure–activity relationship (QSAR) analyses to understand how chemical structures influence biological activity [8, 9].

As a highly integrative science, medicinal chemistry draws on principles and techniques from organic chemistry, biochemistry, computational chemistry, pharmacology, molecular biology, statistics, and physical chemistry.

The compounds utilized as drugs can be classified into several broad categories, each with distinct characteristics and applications.

1. Organic compounds
  - **1a.** Small organic molecules (e.g., simvastatin, dexamethasone, imatinib)
  - **1b.** "Biologics" (erythropoietin, insulin glargine, talquetamab), which are usually protein pharmaceuticals (natural and recombinant antibodies, hormones, etc.)

2. Inorganic and organometallic compounds [10, 11], commonly known as metallodrugs, include agents such as cisplatin (platinum-based), lithium carbonate (lithium-based), and gallium nitrate (gallium-based).

Medicinal chemistry, particularly in its most prevalent form—centered on small organic molecules—encompasses synthetic organic chemistry, natural products, and computational chemistry. It is deeply intertwined with chemical biology, enzymology, and structural biology, collectively striving to discover and develop new drugs. This field involves investigating the synthetic and computational dimensions of both existing drugs and those in development, with a focus on their bioactivity and properties. A key aspect is understanding SARs to optimize drug efficacy. Additionally, medicinal chemistry plays a critical role in pharmaceutical quality control, ensuring the safety and effectiveness of drug products [12].

3.4 Drug Target

A drug target is a molecule in the body, usually a protein, that is intrinsically associated with a particular disease process and that could be addressed by a drug via interaction with the target to produce a desired therapeutic effect. Drug targets can be classified as **enzymes**, **receptors**, and **nucleic acids**. In modern drug discovery, most of the new approved drugs are targeted drugs. For example, in 2023, the Food and Drug Administration (FDA) approved 55 novel drugs, 47, or 85.5%, were targeted therapies, reflecting a clear trend toward more personalized and effective treatment options [13].

3.5 Enzyme

Enzymes are specialized proteins that function as biological catalysts—substances that accelerate chemical reactions without being consumed in the process. Without enzymes, the chemical reactions within cells would either proceed too slowly or fail to occur at all. The specific molecule upon which an enzyme acts is referred to as a substrate, and the enzyme transforms the substrate into different molecules called products. Nearly all metabolic processes in cells depend on enzyme catalysis to proceed at rates sufficient to support life. Metabolic pathways are driven by enzymes, which facilitate multiple steps in these processes. The scientific study of enzymes is known as enzymology.

Enzyme activity may be affected by other molecules: **inhibitors** are molecules that reduce enzyme activity, and **activators** are molecules that increase enzyme activity.

Drugs that function as enzyme inhibitors constitute a significant portion of the orally bioavailable therapeutic agents that are in clinical use today. Likewise, much of drug discovery and development efforts at present are focused on identifying and optimizing drug candidates that act through inhibition of specific enzyme targets [14, 15]. Lovastatin is an inhibitor of 3-hydroxy-3-methylglutaryl (HMG) coenzyme A (HMG-CoA) reductase used in the treatment of hypercholesteremia [16].

Protein kinases are enzymes that phosphorylate proteins (add a phosphate or PO_4 group) and regulate their function and are important drug targets for the treatment of certain diseases, especially cancer. As of September 2021, the FDA has approved 73 small-molecule kinase inhibitor drugs [17].

Paxlovid is a medication specifically designed to treat COVID-19 [18]. It consists of two antiviral drugs, nirmatrelvir and ritonavir, co-packaged together. Nirmatrelvir works by inhibiting the main protease of SARS-CoV-2, the virus responsible for COVID-19. Ritonavir, on the other hand, acts as a potent CYP3A inhibitor, which slows down the metabolism of nirmatrelvir, thereby enhancing its effectiveness.

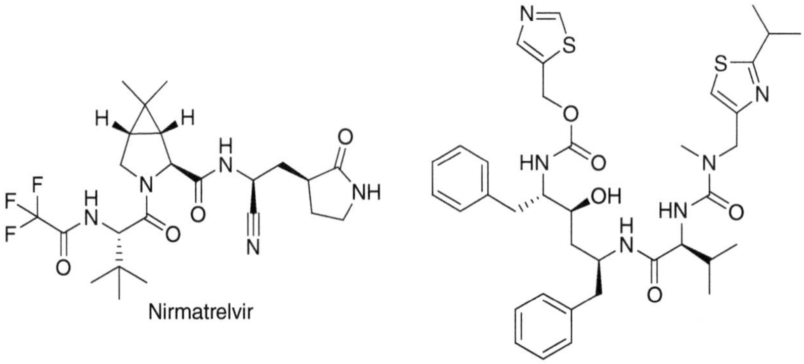

Nirmatrelvir

Ritonavir

3.6 Receptor

In the fields of biochemistry and pharmacology, **receptors** are protein-based chemical structures that receive and transmit signals, which can then be integrated into biological systems. These signals are usually chemical messengers that bind to receptors, triggering physiological responses, such as alterations in a cell's electrical activity. The actions of receptors can be categorized into three primary functions: signal relay, amplification, and integration. Signal relay involves transmitting the signal forward, amplification enhances the impact of a single ligand, and integration enables the signal to be incorporated into other biochemical pathways. Receptors are, by a significant margin, the most critical drug targets in medical science [19].

Receptor proteins are categorized based on their location within the cell. **Cell surface receptors**, also referred to as transmembrane receptors, encompass ligand-gated ion channels, **G protein-coupled receptors** (GPCRs), and **enzyme-linked hormone receptors**. Intracellular receptors, on the other hand, are located inside the cell and include **cytoplasmic receptors** and **nuclear receptors**. A **ligand** is a molecule that binds to a receptor, which can be a protein, peptide, or a small molecule such as a neurotransmitter, hormone, pharmaceutical drug, toxin, calcium ion, or components of a virus or microbe's outer structure.

A naturally occurring substance that binds to a specific receptor is known as its endogenous ligand. For example, acetylcholine is the endogenous ligand for the nicotinic acetylcholine receptor, though this receptor can also be activated by nicotine and inhibited by curare. Each type of receptor is connected to distinct cellular biochemical pathways that align with the signal it receives. Although most cells contain a variety of receptors, each receptor is selective and will only bind to ligands with a specific structure. This specificity is often likened to a lock that only opens with a uniquely shaped key. When a ligand binds to its corresponding receptor, it either activates or inhibits the receptor's associated biochemical pathway, which is often highly specialized in its function.

GPCRs and G protein-linked receptors are a class of cell surface receptors that sense extracellular molecules and initiate intracellular signaling pathways. These receptors are associated with G proteins, which facilitate the transmission of signals into the cell. GPCRs represent a critical target for pharmaceutical development, and as of 2018, around 34% of all drugs approved by the U.S. FDA act on 108 members of this receptor family [20].

Drugs that mimic the natural messengers and activate receptors are known as **agonists**. Drugs that block receptors are known as **antagonists**. Antagonist still binds to the receptor, but it does not activate it. However, because it is bound to the receptor, it prevents the natural messenger from attaching and activating it.

Cortisol is used as anti-inflammatory agent and acts as an agonist at the glucocorticoid receptor. Raloxifene is an antagonist of the estrogen receptor and is used for the treatment of hormone-dependent breast cancer.

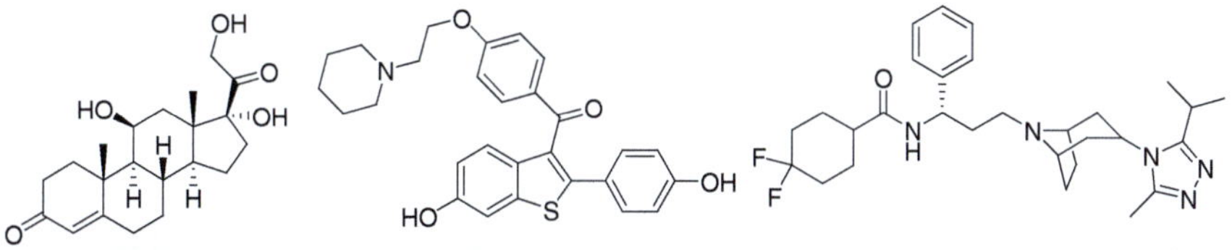

Cortisol   Raloxifene   Maraviroc

Chemokine receptor 5 (CCR5) and CXCR4 serve as the primary receptors facilitating the entry of human immunodeficiency viruses (HIV) into cells. These receptors are part of the GPCR family. CCR5 receptor antagonists are a group of small molecules designed to block the CCR5 receptor. As a result, these antagonists act as entry inhibitors and hold promise for therapeutic use in managing HIV infections. Maraviroc, which was discovered and developed by Pfizer, became the first CCR5 antagonist to receive FDA approval in August 2007 [21].

3.7 Inhibitor

An **enzyme inhibitor** is a molecule that attaches to an enzyme and disrupts its function. Enzymes are proteins responsible for speeding up and enabling the chemical reactions essential for life, where substrates are transformed into products. They achieve this by binding substrates to their active site, a specialized area on the enzyme that accelerates the most challenging steps of the reaction. Enzyme inhibitors hinder this process by either binding to the enzyme's active site (preventing the substrate from attaching) or attaching to another part of the enzyme, thereby blocking its catalytic activity.

Many pharmaceutical drugs function as enzyme inhibitors, targeting either malfunctioning human enzymes or those vital for the survival of pathogens like viruses, bacteria, or parasites. For instance, methotrexate is used in chemotherapy and to manage rheumatoid arthritis, while protease inhibitors, such as darunavir, are employed in the treatment of HIV/AIDS.

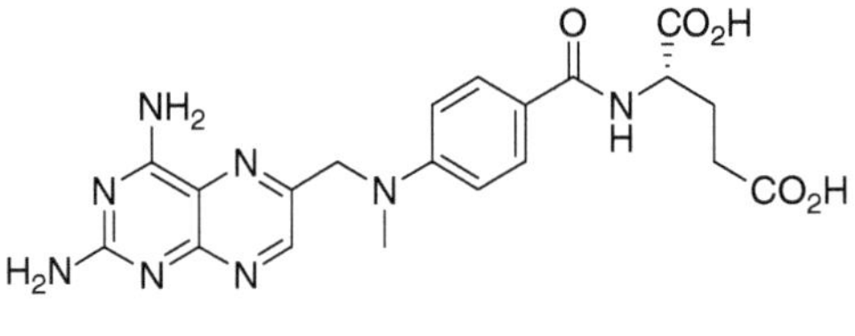

Methotrexate

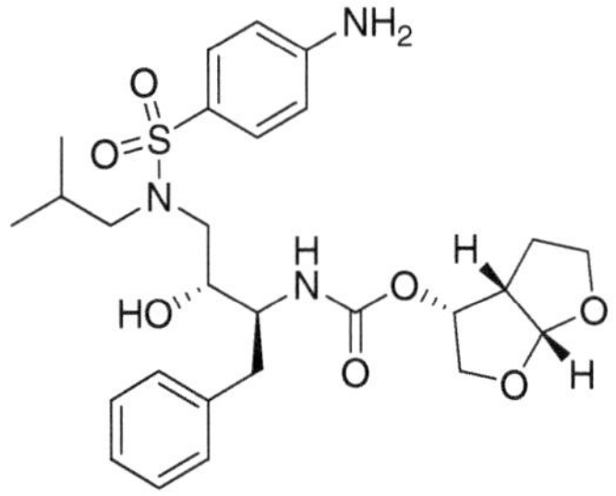

Darunavir, a protease inhibitor

3.8 Activator

Activators are molecules that increase the activity of enzyme. While the majority of drug discovery efforts are mainly focused on enzyme inhibition, small-molecule activators of enzymes remain a relatively less explored area. From therapeutic point of view, it is often beneficial to enhance the expression of certain enzymes whose low expression is responsible for the observed ailment. Small molecules as activators of several enzymes have great biological potential as antimicrobial and anticancer agents, for the treatment of diabetes, obesity, metabolic disorders, and for the treatment of neurological disorders including Alzheimer's disease [22].

3.9 Drug Design

Drug design, also referred to as rational drug design or simply rational design, is the innovative process of discovering new drugs by leveraging knowledge about biological targets. Typically, the drug is a small organic molecule that either activates or inhibits the function of biomolecules, such as proteins, to deliver therapeutic benefits to patients. At its core, drug design involves creating molecules whose shape and charge complement the biomolecular targets they interact with, enabling them to bind effectively. While not always required, drug design frequently utilizes computer modeling techniques, a practice known as **computer-aided drug design**. When the process is guided by the three-dimensional structure of biomolecular targets, it is termed as **structure-based drug design**. Beyond small molecules, biopharmaceuticals—including peptides and, notably, therapeutic antibodies—have emerged as a significant class of drugs. Computational methods have also been developed to enhance the affinity, selectivity, and stability of these protein-based therapeutics.

The process of identifying small molecules that interact with a specific target typically starts with screening a library of potential drug compounds. This screening can be conducted using various assays. Furthermore, if the target's structure is known, virtual screening of drug candidates can also be employed. Ideally, these candidate compounds should exhibit "drug-like" properties, meaning they should have characteristics that promote oral bioavailability, sufficient chemical and metabolic stability, and low toxicity. Various methods are available to assess drug likeness, including **Lipinski's rule of five** and other scoring systems, such as lipophilic efficiency [23]. Additionally, numerous approaches for predicting drug metabolism have been proposed in scientific research [24].

Given that numerous drug properties need to be optimized simultaneously during the design process, multi-objective optimization techniques are often utilized. However, due to the constraints of current activity prediction methods, drug design remains significantly reliant on serendipity and bounded rationality [25].

In an ideal scenario, computational methods would accurately predict the binding affinity of a compound before it is synthesized, theoretically eliminating the need for multiple iterations and significantly reducing both time and cost. However, the reality is that current computational approaches are not flawless and, at best, can only offer qualitatively reliable estimates of affinity. As a result, the drug discovery process typically involves numerous cycles of design, synthesis, and testing before the optimal compound is identified. Despite these limitations, computational methods play a crucial role in accelerating discovery by minimizing the number of required iterations and often suggesting novel molecular structures [26, 27]. Recent advancements in this field have been comprehensively reviewed in two articles, which highlight progress in drug design and its practical applications [28, 29].

Artificial intelligence (AI) is revolutionizing every stage of the drug discovery process [30, 31, 32]. In drug design, AI is employed to predict protein 3D structures, drug–protein interactions, and drug activity, enabling the creation of novel molecules from scratch. In pharmacology, AI facilitates the design of targeted molecules and multi-target drugs. Within chemical synthesis, AI aids in planning synthesis pathways, forecasting reaction yields, and deciphering reaction mechanisms. Additionally, AI excels at repurposing existing drugs for new therapeutic applications. Undoubtedly, AI plays an indispensable role in drug screening by predicting toxicity, biological activity, absorption, distribution, metabolism, and excretion (ADME) properties, and physicochemical characteristics, among other critical factors.

3.10 Ligand-based Drug Design

Ligand-based drug design, also known as indirect drug design, utilizes information about molecules that interact with a specific biological target. These molecules serve as the foundation for developing pharmacophore models, which outline the essential structural characteristics required for a molecule to bind effectively to the target [33]. By analyzing the binding properties of these molecules, a model of the biological target can be constructed, enabling the design of new compounds that can interact with the target. Another approach involves the use of QSARs, which establish correlations between the molecular properties of compounds and their experimentally measured biological activities. These QSAR models can then be applied to predict the biological activity of new, structurally similar compounds [34].

3.11 Structure-based Drug Design

Structure-based drug design, also known as direct drug design, depends on understanding the three-dimensional structure of a biological target, typically determined through techniques like X-ray crystallography or nuclear magnetic resonance spectroscopy [35]. When the experimental structure of the target is unavailable, a homology model of the target can be constructed using the known structure of a related protein. With the target's structure in hand, medicinal chemists can employ interactive visualization tools and their expertise to design potential drug candidates that are expected to bind to the target with high affinity and selectivity. Additionally, automated computational methods can be utilized to propose novel drug candidates [36].

Current structure-based drug design methods can be broadly divided into three main categories [37]. The first approach is to identify new ligands for a given receptor by searching large databases of small-molecule 3D structures to find ligands that fit into the receptor binding pocket using fast approximate docking programs. This method is called virtual screening. The second category is the de novo design of new ligands. In this method, ligand molecules are constructed within the constraints of the binding pocket by stepwise assembly of small fragments. These fragments can be single atoms or fragments of molecules. The main advantage of this approach is that novel structures not included in any database can be proposed [38, 39, 40]. A third approach is to optimize known ligands by evaluating proposed analogs within the binding cavity [37].

3.12 Lipinski's Rule of Five

Lipinski's Rule of Five, also known as **Pfizer's Rule of Five** or simply the **Rule of Five** (RO5), is a rule of thumb for evaluating druglikeness or determining whether a compound with a certain pharmacological or biological activity has the following chemical and physical properties: likely making it an orally active drug for humans. This rule was developed by Christopher A. Lipinski in 1997 and is based on the observation that most oral drugs are relatively small and moderately lipophilic molecules [41, 42].

The rules describe the molecular properties important for the pharmacokinetics (PK) of drugs in humans, including their ADME. However, this rule does not predict whether a compound is pharmacologically active.

Lipinski's rule states that, in general, an orally active drug has no more than one violation of the following criteria [41]:

- No more than five hydrogen bond donors (the total number of nitrogen–hydrogen and oxygen–hydrogen bonds)
- No more than 10 hydrogen bond acceptors (all nitrogen or oxygen atoms)
- A molecular mass less than 500 Daltons
- A calculated octanol–water partition coefficient (Clog P) that does not exceed 5

In the drug discovery process, it is crucial to adhere to this rule when optimizing pharmacologically active lead compounds in a stepwise manner. The goal is to enhance both the activity and selectivity of these compounds while ensuring that their physicochemical properties remain consistent with drug-like characteristics, as outlined by Lipinski's rule. Drug candidates that comply with the RO5 generally exhibit lower attrition rates in clinical trials, increasing their likelihood of successfully reaching the market.

It is crucial to recognize that not all small-molecule drugs conform to this principle. The RO5 operates under the assumption that passive diffusion is the primary mechanism for cellular drug uptake, overlooking the significant role of transporters. In reality, only approximately 50% of orally administered new chemical entities adhere to this rule [43].

Furthermore, research has shown that certain natural products, such as macrolides and peptides [44, 45, 46], defy the chemical criteria established by Lipinski's filters.

3.13 Lead Compound

In drug discovery, a **lead compound** is a chemical substance that exhibits pharmacological or biological activity with potential therapeutic value. However, its structure may not be optimal, necessitating modifications to better align with the target. A lead compound serves as a foundation for further development and provides a basis for creating backup candidates for future drug launches. The chemical structure of a lead compound acts as a starting point for structural optimization to enhance properties such as potency, selectivity, or pharmacokinetic performance.

Moreover, newly discovered pharmacologically active compounds may lack sufficient drug-like characteristics, requiring chemical adjustments to make them suitable for biological or clinical evaluation. Prior to identifying lead compounds, it is essential to select appropriate targets for rational drug design, based on biological relevance. Alternatively, targets can be identified by screening potential lead compounds against a range of biological targets to determine their efficacy and specificity [47].

Drug libraries are frequently evaluated using **HTS** methods, where compounds are tested for their capacity to either inhibit (as antagonists) or activate (as agonists) a specific receptor. Compounds that demonstrate activity are identified as "**hits**," and their selectivity is further assessed during this process.

A lead compound can originate from multiple sources. These compounds are typically discovered through the characterization of natural products, the use of combinatorial chemistry, or molecular modeling techniques such as rational drug design. Additionally, chemicals identified as hits during HTS may also serve as potential lead compounds. Once a lead compound is identified, it undergoes lead optimization (LO) to enhance its "drug-like" properties. During this process, Lipinski's RO5 is often applied to evaluate its suitability. Other considerations, such as the feasibility of scaling up the chemical's production, are also critical factors in the optimization process.

3.14 Hit-to-Lead

A compound that exhibits the desired magnitude of effects in a HTS is referred to as a **hit**. The procedure of identifying these hits is known as hit selection. Following this, the hit-to-lead (H2L) stage, also termed lead generation, is a critical phase in early drug discovery. During this stage, small-molecule hits identified from HTS are assessed and undergo preliminary optimization to pinpoint promising lead compounds. These lead compounds are then subjected to more comprehensive refinement in the subsequent phase, known as LO. The drug discovery pipeline typically follows a sequential pathway that includes the H2L stage as a key step.

The process typically follows these stages: Target Validation (TV) → Assay Development → HTS → H2L → LO → Preclinical Development → Clinical Development.

The H2L stage begins with the confirmation and evaluation of initial screening hits, followed by the synthesis of analog compounds (hit expansion). Typically, the initial hits exhibit binding affinities for their biological target in the micromolar range (10^{-6} M). Through focused H2L optimization, the binding affinities of these hits are often enhanced by several orders of magnitude, reaching the nanomolar range (10^{-9} M). Additionally, the hits undergo targeted optimization to improve metabolic stability, enabling testing in animal disease models, and to enhance selectivity against other biological targets. This selectivity optimization helps minimize potential off-target interactions that could lead to undesirable side effects.

On average, just one out of every 5000 compounds that progress from drug discovery to preclinical development ultimately gains approval as a marketed drug.

3.15 High-Throughput Screening

HTS is a method for scientific discovery especially used in drug discovery and relevant to the fields of biology and chemistry [48, 49]. Using robotics, data processing/control software, liquid handling devices, and sensitive detectors, HTS allows a researcher to quickly conduct millions of chemical, genetic, or pharmacological tests. Through this process, one can quickly recognize active compounds, antibodies, or genes that modulate a particular biomolecular pathway. The results of these experiments provide starting points for drug design and for understanding the noninteraction or role of a particular location.

3.16 Pharmacophore

In the fields of medicinal chemistry and molecular biology, a **pharmacophore** represents an abstract framework of the essential molecular characteristics required for the recognition of a ligand by a biological macromolecule. According to IUPAC, a pharmacophore is defined as "an ensemble of spatial and electronic features that are crucial for achieving optimal supramolecular interactions with a specific biological target, thereby eliciting (or inhibiting) its biological response [50]." Pharmacophore models provide insights into how structurally varied ligands can interact with a shared receptor site. Furthermore, these models are valuable tools for discovering new ligands that can bind to the same receptor, through either de novo design or virtual screening approaches.

Historically, the modern notion of a pharmacophore gained prominence through the work of Lemont Kier, who introduced the concept in 1967 [51] and later formally used the term in a book published in 1971 [52].

3.17 Affinity, Efficacy, and Potency

Affinity is the tendency of one molecule to bind to another molecule. The affinity of a drug refers to its ability to bind to its biological target (receptor, enzyme, transport system, etc.). For pharmacological receptors, it can be thought of as the frequency at which the drug will reside at the minimum free energy position within that receptor's force field when the drug enters the vicinity of the receptor by diffusion.

Efficacy is a measure of the maximum biological effect that a drug can produce as a result of receptor binding. A compound with high affinity does not necessarily mean high efficacy.

The **potency** of a drug refers to the amount of drug required to achieve a specific biological effect—the smaller the dose needed, the more potent the drug. Potency is a comparative rather than an absolute expression of drug activity. Drug potency depends on affinity and efficacy. Thus, two agonists could be equipotent but have different intrinsic efficacies with compensating differences in affinity.

An enzyme (**E**)-catalyzed biological reaction is the conversion of a substrate (**S**) to a product (**P**).

$$E + S \rightleftharpoons E{\cdot}S \rightleftharpoons E{\cdot}P \rightleftharpoons E + P$$

When enzyme inhibitor drug (**I**) is added, it will compete with the substrate for the enzyme's active binding site. For an agonist (or for an antagonist), the numerical representation of affinity is the reciprocal of the equilibrium dissociation constant (k_d) of the ligand–receptor complex denoted k_i, calculated as the rate constant for offset (k_{off}) divided by the rate constant for onset (k_{on}). k_i is also referred as **inhibition constant**.

$$E + I \underset{k_{off}}{\overset{k_{on}}{\rightleftharpoons}} E{\cdot}I$$

$$k_i = k_{off}/k_{on}$$

The smaller the k_i value for **I**, the more potent the inhibitor. Another common measure of inhibition is **IC$_{50}$**, which is the concentration of inhibitor that produces 50% enzyme inhibition in the presence of substrate.

The half-maximal effective concentration (**EC$_{50}$**) is a measure of the concentration of a drug, antibody, or toxicant that induces a biological response between baseline and maximum after a specified exposure time. More simply, EC$_{50}$ can be defined as the concentration required to obtain 50% effect. EC_{50} is often used as a measure of drug potency.

3.18 Structure–Activity Relationship

The **SAR** is the relationship between the chemical structure of a molecule and its biological activity. The idea was first proposed at least as early as 1868 by A. C. Brown and T. R. Fraser [53, 54].

SAR analysis enables the identification of chemical groups responsible for causing a target biological effect in an organism. This allows the effect or potency of a bioactive compound (usually a drug) to be altered by changing its chemical structure. Medicinal chemists use chemical synthesis techniques to insert new chemical groups into biomedical compounds and test the biological effects of the modifications.

This method has been modified to establish a mathematical relationship between chemical structure and biological activity, called a **QSAR** [55].

In medicinal chemistry practice, once a pharmacophore is identified, the SAR is usually performed by synthesizing several series of analogs of the lead compound.

3.19 Partition Coefficient and Log *P*/Clog *P*

The **partition coefficient**, abbreviated P, is defined as a particular ratio of the concentrations of a solute between the two solvents (a biphase of liquid phases), specifically for un-ionized solutes, and the logarithm of the ratio is thus **log *P***. When one of the solvents is water and the other is a nonpolar solvent, then the log *P* value is a measure of **lipophilicity** or **hydrophobicity**. The defined precedent is for the lipophilic and hydrophilic phase types to always be in the numerator and denominator, respectively; for example, in a biphasic system of *n*-octanol (hereafter simply "octanol") and water:

$$\log P_{\text{oct/water}} = \log_{10} ([\text{solute}]_{\text{oct}}/[\text{solute}]_{\text{water}})$$

The distribution coefficient of a drug plays a critical role in determining its ability to reach the intended target within the body, the potency of its effect upon reaching the target, and the duration it remains in its active form. As a result, the log *P* value of a molecule is a key parameter considered by medicinal chemists during preclinical drug discovery, particularly when evaluating the druglikeness of potential drug candidates.

Measuring the log *P* of a compound requires its physical synthesis, which makes it impossible to determine this property beforehand. To address this, computational tools such as ClogP® v4.0 (BioByte Corp., 1999) and ACD/logPdb® v7.0 (Advanced Chemistry Development Inc., Toronto, Ontario, Canada, 2003) have been developed to predict the log *P* of compounds. The log *P* value estimated by these programs is referred to as **Clog *P***. According to Lipinski's RO5, one of the criteria for an orally active drug is that its Clog *P* should not exceed 5.

3.20 Drug Candidate

A **drug candidate** is a compound that has shown potential to become a therapeutic agent for the treatment of a specific disease or medical condition, whether because it has sufficient target selectivity, drug-like properties, or sufficient potency. Once a potential drug candidate is identified, rigorous preclinical studies are conducted to evaluate its safety (toxicity), efficacy, PK, and pharmacodynamics (PD). This preclinical stage involves testing drug candidates in labs and animal models to collect data on their effects and potential risks. If a drug candidate successfully passes preclinical testing, clinical trials may be conducted, which involve testing the drug in human subjects to further evaluate its safety and effectiveness. Ultimately, if it shows sufficient safety and efficacy in clinical trials, it may gain regulatory approval to be marketed and used as a therapeutic drug.

Drug candidates are critical to healthcare because they represent entirely new treatments or the potential to improve existing treatments. They can address unmet needs and bring hope to diseases or conditions for which there are currently no effective treatment options. New drugs can also address limitations of existing drugs, such as reducing side effects, improving efficacy, or targeting specific patient groups. However, only a small percentage of drug candidates make it through the rigorous testing and approval process to become commercialized drugs. While the journey from drug candidate to approved drug is challenging, their potential to improve patient outcomes and advance medical knowledge makes them an important part of healthcare progress.

3.21 Preclinical Studies

In the process of drug development, **preclinical** or **nonclinical studies** represent a crucial phase of research conducted prior to clinical trials (which involve testing in humans). During this stage, essential data on feasibility, iterative testing, and drug safety are gathered, typically through experiments conducted on laboratory animals.

Drugs may undergo **PD** (what the drug does to the body), **PK** (what the body does to the drug), **ADME**, and **toxicology** testing. These data allow researchers to allometrically estimate safe starting doses of drugs for use in human clinical trials. Most preclinical studies must adhere to **Good Laboratory Practices** (**GLP**) in the **International Council for Harmonization of Technical Requirements for Pharmaceuticals for Human Use** guidelines before they can be submitted to regulatory agencies such as the U.S. FDA.

Typically, both **in vitro** and **in vivo** testing will be performed. Drug toxicity studies include which organs the drug targets and whether there are any long-term carcinogenic or disease-causing toxic effects.

The information collected from these studies is vital so that safe human testing can begin. Typically, in drug development studies, animal testing involves two species. The most commonly used models are murine and canine, although primate and porcine are also used.

3.22 Toxicity

One of the initial steps in evaluating a new drug candidate is assessing its potential **toxicity**. This process usually starts with in vitro testing using genetically engineered cell cultures and/or in vivo testing on genetically modified mice to study its impact on cell reproduction and identify possible carcinogenic effects. Additionally, the drug is tested for acute toxicity by administering a dose large enough to induce toxic effects or death within a short timeframe. These studies often involve multiple animal species, and the animals are later dissected to determine if specific organs were affected.

Following this, further acute toxicity studies are conducted over several months, during which the drug is given to experimental animals at doses designed to cause toxicity without being lethal. Throughout this period, blood and urine samples are collected and analyzed. At the end of the study, the animals are euthanized, and pathologists examine their tissues for signs of cellular damage or cancer.

Finally, long-term toxicology trials are carried out over several years at lower dose levels to evaluate the drug for chronic toxic effects, carcinogenicity, specific toxicological impacts, mutagenicity, and reproductive abnormalities. These comprehensive studies are critical to ensuring the safety and efficacy of the drug before it progresses to human trials.

There are two types of data used to measure drug toxicity. One is the **LD_{50}** value, which is the lethal dose required to kill 50% of a group of animals. The other is the **ED_{50}**, which is the dose required to produce the desired effect in 50% of the animals tested. The ratio of LD_{50} to ED_{50} is called the **therapeutic ratio** or **therapeutic index**.

3.23 Pharmacokinetics and Pharmacodynamics

PK is a field within pharmacology focused on understanding how the body processes a drug following its administration. It involves the study of chemical metabolism and traces the journey of a drug from the moment it enters the body until it is fully eliminated. PK relies heavily on mathematical models, with a particular focus on the relationship between the concentration of a drug in the bloodstream and the time that has passed since the drug was administered.

PD examines the effects of a drug on the body and how these effects are produced. Together with PK, PD plays a crucial role in determining drug dosing, therapeutic benefits, and potential adverse effects, as illustrated in PK/PD models.

3.24 Absorption, Distribution, Metabolism, and Excretion

ADME is primarily used in the fields of PK and pharmacology. These four terms represent key processes that describe how a drug interacts with the body over time.

Absorption/administration describes how a drug enters the body. Absorption relates to the movement of a drug from the administration site to the bloodstream. There are four main routes of administration:

- Ingestion through the digestive tract;
- Inhalation via the respiratory system;
- Dermal application to the skin or eye;
- Injection through direct administration into the bloodstream.

Only injected compounds enter the systemic circulation directly. For drugs to be administered by ingestion, inhalation, or skin contact, the chemical must cross a membrane before entering the bloodstream.

Distribution: Once a drug is absorbed, it moves from the site of absorption to surrounding tissues in the body. This distribution from one part of the body to another is usually accomplished through the bloodstream, but can also occur from cell to cell. Researchers examine where chemicals travel, how quickly they reach certain locations, and how widely they are distributed to help determine efficacy. Some compounds move easily, while others do not. Factors such as blood flow, lipophilicity, tissue binding, and molecular size influence distribution.

Drug **metabolism** refers to the process in which drugs are biotransformed through organs or tissues (mainly liver, kidney, skin, or digestive tract) so that drugs are excreted from the body. To facilitate elimination through feces or urine, drug compounds are altered to be more soluble in water. Chemical metabolism can lead to toxicity, such as the production of damaging by-products or toxic metabolites. Scientists map out specific metabolic pathways of candidate drugs, called adverse outcome pathways (AOPs). AOPs provide the data needed to determine the potential safety or toxicity of a drug.

Excretion is the process of elimination of the metabolized drug compounds from the body. The researchers wanted to know how quickly the drug was excreted and how it was excreted from the body. Most drugs are excreted in feces or urine. Other excretion methods include excretion of sweat molecules through the lungs or through the skin and affects the cargo excretion pathway. Not all drug compounds are completely excreted. Adverse effects may occur when chemical or metabolic by-products bioaccumulate. Lipid-soluble compounds are more likely to bioaccumulate than soluble compounds.

DMPK is the abbreviation of **drug metabolism** and **pharmacokinetics** studies.

3.25 Bioavailability

Bioavailability is a specific aspect of absorption, representing the percentage (%) of an administered drug that successfully enters the systemic circulation. By definition, when a drug is administered intravenously, its bioavailability is 100%. However, when drugs are delivered through other routes (e.g., oral, intramuscular, etc.), their bioavailability is reduced due to factors such as absorption through the intestinal epithelium and first-pass metabolism. Mathematically, bioavailability is calculated as the ratio of the **area under** the plasma drug concentration–time **curve** (**AUC**) for an extravascular formulation compared to the AUC for an intravascular (intravenous) formulation. AUC is used as a measure because it is directly proportional to the amount of drug that reaches the systemic circulation.

For dietary supplements, herbs, and other nutrients whose route of administration is almost always oral, bioavailability usually simply represents the amount or fraction of the ingested dose that is absorbed.

3.26 Pharmacology

Pharmacology is the scientific study of drugs and medications, encompassing their origins, chemical composition, PK, PD, therapeutic applications, and toxicological effects. It focuses on understanding the interactions between living organisms and chemicals that influence normal or abnormal biochemical processes. A substance is classified as a drug if it possesses medicinal properties. This field explores various aspects, including drug composition and properties, functions, sources, synthesis, drug design, molecular and cellular mechanisms, organ and system-level effects, signal transduction and cellular communication, molecular diagnostics, chemical interactions, chemical biology, therapeutic and medical uses, and antipathogenic capabilities. Pharmacology is broadly divided into two primary areas: PD, which examines the effects of drugs on the body, and PK, which studies how the body processes drugs.

3.27 Formulation

In pharmacy, **formulation** is the process of combining different chemical substances, including **active pharmaceutical ingredients** (**API**), to produce the final medicinal product. The term formulation is often used in a way that includes dosage forms.

Formulation research involves the development of drug formulations that are stable and acceptable to patients. For oral medications, this usually involves incorporating the drug into tablets or capsules. It is important to distinguish that tablets contain a variety of other potentially inert substances in addition to the drug itself, and studies must be conducted to ensure that the encapsulated drug is compatible with these other substances and will not cause harm, either direct or indirect.

Preformulation involves the characterization of the physical, chemical, and mechanical properties of the drug in order to select which other ingredients (excipients) should be used in the formulation. When working with protein preformulation, an important aspect is to understand the solution behavior of a given protein under various stress conditions (e.g., freeze/thaw, temperature, shear stress, etc.) to identify degradation mechanisms and thus mitigate their effects.

Formulation studies then take into account factors such as particle size, polymorphism, pH, and solubility, as all of these can affect bioavailability and therefore the activity of the drug. The drug must be mixed with the inactive ingredients in a way that ensures a consistent amount of drug is present in each dosage unit, for example, per tablet. The dose should have uniform appearance, acceptable taste, tablet hardness, and capsule disintegration.

Formulation studies are often still ongoing when clinical trials commence. As a result, basic formulations are typically developed for phase I clinical trials. These initial formulations usually involve manually filled capsules containing a small quantity of the drug mixed with a diluent. Since these formulations are intended for short-term use and will be tested within a few days, there is no requirement to demonstrate their long-term stability. However, it is crucial to consider the "drug load," which refers to the ratio of the active drug to the total content of the dose. Low drug loading can lead to issues with homogeneity, while high drug loading, particularly in compounds with low bulk density, may result in flow problems or necessitate the use of larger capsules.

By the time a drug advances to phase III clinical trials, its formulation should closely resemble the final version intended for market use. At this critical stage, understanding the drug's stability is essential, and appropriate conditions must be established to ensure the formulation remains stable. If the drug demonstrates instability, the clinical trial results could be compromised, as it would be impossible to determine the actual dose administered to participants.

3.28 Active Pharmaceutical Ingredient

The **API** is the primary component in a medication responsible for producing its intended therapeutic effect. Some drugs may contain multiple APIs, each contributing to the desired outcome through different mechanisms of action in the body.

On the other hand, inactive ingredients, commonly referred to as **excipients** in the pharmaceutical industry, play a supportive role. The main excipient used as a carrier or medium to deliver the active ingredient is often termed a vehicle. Examples of common vehicles include petrolatum and mineral oil. It's important to note that while these ingredients are labeled as "inactive," this does not imply that they are inert or without any function; they often serve critical roles in the formulation and effectiveness of the drug.

3.29 Drug Stability

Drug stability refers to the extent to which a drug substance or product retains the same properties and characteristics as when it was manufactured, within specified limits and throughout its storage and use. Stability studies are performed to test if there are any effects of temperature, humidity, oxidation, or photolysis (ultraviolet or visible light), and the formulation is analyzed to see if any degradation products are formed.

Stability types are generally divided into chemical, physical, microbiological, therapeutic, and toxicological stabilities. Drug stability can be divided into pre-market stability and commercial (marketed product) stability. Pre-market stability supports clinical trials, in which a drug product is stored under different conditions for assessment of safety and effectiveness, typically throughout clinical trials and during application. Commercial stability is an ongoing assurance for post-approval batches and is used to monitor the long-term stability of a drug product. Drug stability assessment typically involves testing the drug substance or drug product using stability-indicating methods to determine the retest period (for pre-market stability) and shelf life (for commercial stability).

3.30 Prodrugs

A **prodrug** is defined as a chemical that undergoes transformation before exerting a pharmacological effect. In other words, when you take a prodrug, it changes in your body before it starts to take effect. Prodrugs are often designed to increase bioavailability when the drug itself is poorly absorbed from the gastrointestinal tract. Prodrugs can be used to improve the selectivity of a drug for interacting with cells or processes other than its intended target. This can reduce the unwanted or unintended effects of the drug, which is especially important in treatments such as chemotherapy, which can have serious unintended and unwanted side effects.

There are two main types of prodrugs. These categories are based on how the prodrug is converted in the body. Type I prodrugs are converted into the active form within the cell. Type II prodrugs are turned into their active forms extracellularly, such as in blood or other fluids.

Different types of prodrugs can also impact drug interactions. Certain interactions may prevent a prodrug from becoming active.

Aspirin, acetylsalicylic acid, is a synthetic prodrug of salicylic acid [56].

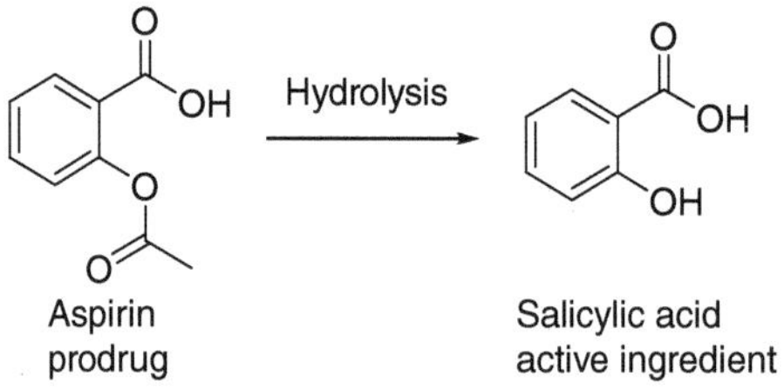

Hydrolysis

Aspirin
prodrug

Salicylic acid
active ingredient

Clopidogrel (Plavix) is a medication that can prevent heart attacks and strokes. It's a prodrug that's absorbed in the intestine and activated in the liver. Acid-reducing medications like omeprazole (Prilosec) can make clopidogrel less effective by preventing it from becoming active. Clopidogrel is activated in two steps: first is the oxidation catalyzed by the enzymes CYP2C19, CYP1A2, and CYP2B6, then hydrolyzed by CYP2C19, CYP2C9, CYP2B6, and CYP3A [57].

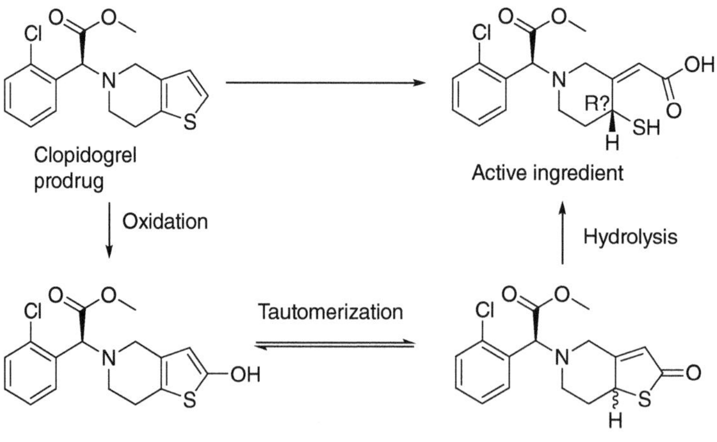

Clopidogrel
prodrug

Active ingredient

Oxidation

Hydrolysis

Tautomerization

Around 10% of all drugs marketed globally are classified as prodrugs. Some recently approved examples of prodrugs include benzhydrocodone (a prodrug of hydrocodone, approved in 2018) [58], serdexmethylphenidate (a prodrug of dexmethylphenidate, approved in 2021) [59], and omidenepag isopropyl (a prodrug of omidenepag, approved in 2022) [60].

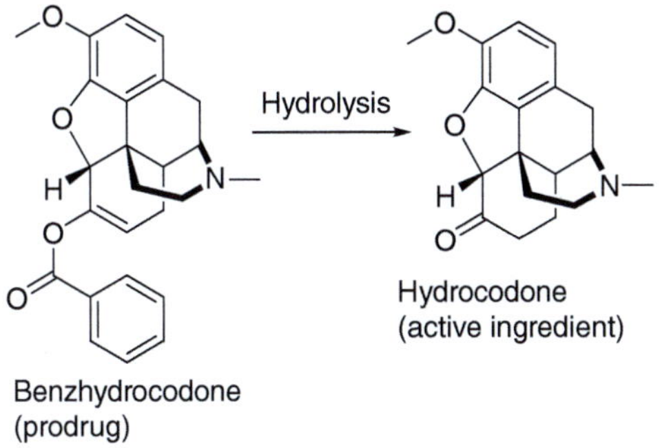

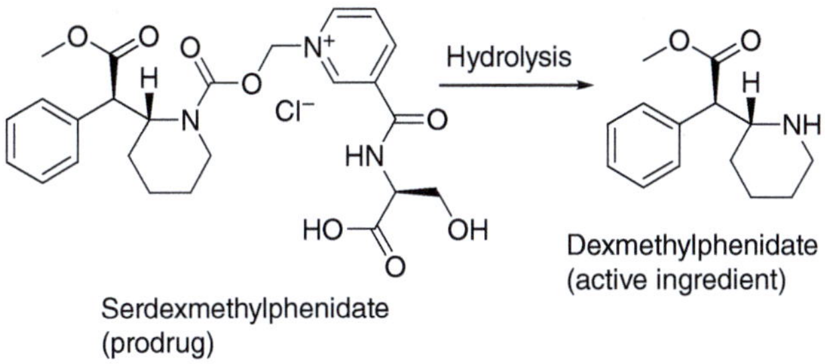

Benzhydrocodone (prodrug) → Hydrolysis → Hydrocodone (active ingredient)

Serdexmethylphenidate (prodrug) → Hydrolysis → Dexmethylphenidate (active ingredient)

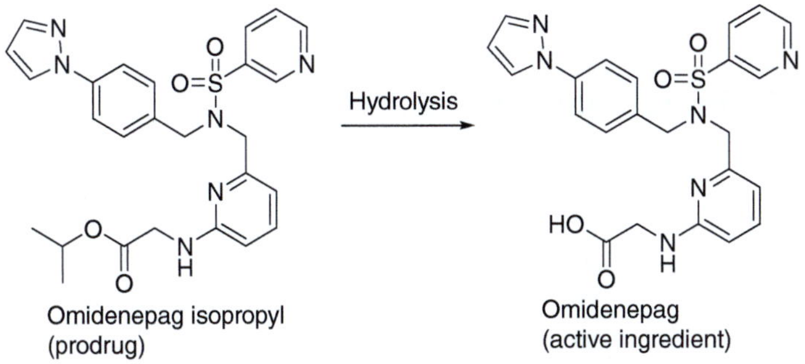

Omidenepag isopropyl (prodrug) → Hydrolysis → Omidenepag (active ingredient)

3.31 Deuterium-containing Drug

Deuterated drugs are small-molecule pharmaceuticals where one or more hydrogen atoms in the drug molecule are replaced by deuterium, a heavier stable isotope of hydrogen. This substitution can lead to kinetic isotope effects, which may significantly slow the drug's metabolic rate, resulting in an extended half-life.

The concept of deuterated drug candidates originated in the 1970s, building on early research involving deuterated metabolites. Despite this early start, it wasn't until 2017 that the first deuterated drug, Austedo® (deuterated tetrabenazine), gained FDA approval [61]. Chemically, deuterated tetrabenazine is an isotopic isomer of tetrabenazine, with six hydrogen atoms replaced by deuterium atoms. This deuterium incorporation reduces the drug's metabolic breakdown, allowing for less frequent dosing [62].

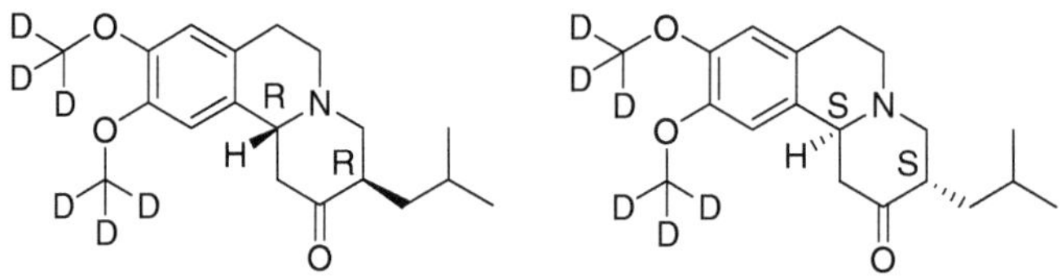

Deutetrabenazine

Deucravacitinib is a deuterated drug launched by Bristol-Myers Squibb for the treatment of moderate to severe plaque psoriasis. It received FDA approval in September 2022 [63].

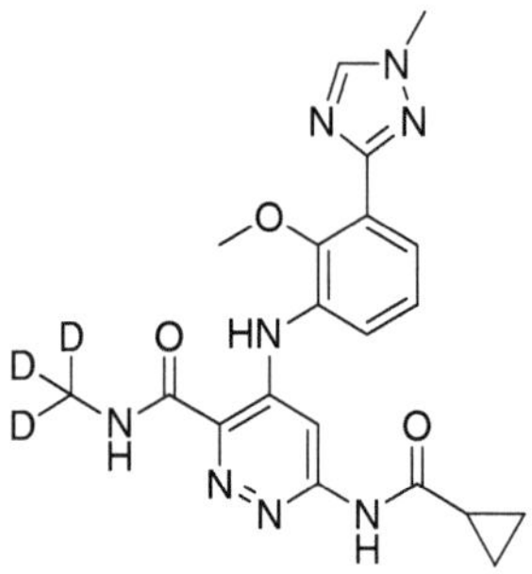

Deucravacitinib

3.32 Antibody Drug Conjugates

Antibody drug conjugates (**ADCs**) are an innovative family of drugs assembled by covalently linking a cytotoxic drug (payload) to a monoclonal antibody (mAb) and then delivering it to tumor tissue expressing its specific antigen, which theoretically has the advantage of improving the therapeutic ratio [64, 65, 66, 67, 68, 69]. As of March 2024, 13 ADCs have been approved by the FDA and are on the market. Unlike chemotherapy, ADCs are designed to target and kill tumor cells while sparing healthy cells. ADCs are examples of bioconjugates and immunoconjugates.

ADCs combine the targeting properties of monoclonal antibodies with the anticancer capabilities of cytotoxic drugs and are designed to differentiate between healthy and diseased tissue.

3.33 Good Laboratory Practice

The principles of **GLP** define a framework of rules and standards for a quality management system that governs organizational processes and conditions related to the planning, execution, monitoring, documentation, reporting, and archiving of nonclinical research on health and environmental safety. These principles are applicable to the nonclinical safety testing of substances used in a wide range of products, ensuring the reliability, consistency, and integrity of safety data that is submitted to regulatory authorities globally.

The GLP principles established by the Organization for Economic Cooperation and Development (OECD) apply to the testing of chemicals or chemical products in nonclinical environments, including laboratory settings or controlled environmental conditions like greenhouses and field experiments. These principles do not apply to research involving human subjects. Additionally, depending on the regulations of individual OECD member countries, the OECD GLP principles may also be applied to nonclinical safety testing of other regulated products, such as medical devices.

Studies conducted under GLP in OECD member countries encompass a wide range of research areas, including:

- Physical-chemical testing;
- Toxicity studies;
- Mutagenicity studies;
- Environmental toxicity studies on both aquatic and terrestrial organisms;
- Investigations into behavior in water, soil, and air, as well as bioaccumulation;
- Residue studies to determine pesticide levels in food or animal feedstuffs;
- Research on the effects on mesocosms and natural ecosystems;
- Analytical and clinical chemistry testing.

3.34 Good Manufacturing Practice

Good Manufacturing Practices (**GMP**, also known as "**cGMP**" or "**Current Good Manufacturing Practices**") is an aspect of quality assurance that ensures that pharmaceutical products are consistently produced and controlled to quality standards suitable for their intended use and in accordance with product specifications.

GMP defines quality measures for production and quality control and defines general measures to ensure that the processes required for production and testing are clearly defined, verified, reviewed, and documented, and that personnel, sites, and materials are suitable for the production of medicinal and biologicals including vaccine products. GMP also has a legal component that covers distribution responsibilities, contract manufacturing and testing, and response to product defects and complaints. Specific GMP requirements relevant to product categories such as sterile pharmaceutical products or biomedical products are provided in a series of annexes to the general GMP requirements.

All guidelines are based on a set of fundamental principles:

- **Clean and Hygienic Area:** Manufacturing facilities must maintain a clean and hygienic manufacturing area.
- **Controlled Environmental Conditions:** Manufacturing facilities must maintain controlled environmental conditions in order to prevent cross-contamination from adulterants and allergens that may render the product unsafe for human consumption or use.

- **Clearly Defined and Controlled Processes:** Manufacturing processes must be clearly defined and controlled. All critical processes are validated to ensure consistency and compliance with specifications.
- **Evaluate and Validate Changes:** Manufacturing processes must be controlled, and any changes to the process must be evaluated. Changes that affect the quality of the drug are validated as necessary.
- **Clarity and Precision in Documentation:** Instructions and procedures should be written in clear, unambiguous language, adhering to good documentation practices.
- **Training and Compliance:** Operators must receive proper training to execute and accurately document procedures.
- **Accurate Record-Keeping:** Records, whether manual or electronic, must be maintained during production to confirm that all defined procedures and instructions were followed. These records should verify that the quantity and quality of the food or drug met the expected standards. Any deviations must be thoroughly investigated and documented.
- **Traceability and Retention:** Manufacturing and distribution records must be retained in a comprehensible and accessible format to ensure the complete history of each batch can be traced.
- **Risks Minimize:** The distribution of products must ensure that risks to their quality are minimized at all times.
- **Robust Recalling System:** A robust system must be established to facilitate the recall of any batch from sale or supply if necessary.
- **Procedures for Complaints:** Complaints regarding marketed products must be thoroughly reviewed, the root causes of quality issues must be investigated, and appropriate corrective actions must be implemented to address defective products and prevent future occurrences.

Regulatory agencies have recently shifted their focus toward more essential quality metrics for manufacturers, moving beyond mere adherence to basic GMP regulations. The U.S. FDA has observed that manufacturers who adopt quality metrics programs gain deeper insights into employee behaviors that influence product quality.

3.35 Chemistry, Manufacturing, and Controls

Chemistry, manufacturing, and controls (**CMC**) is one of the most important activities in pharmaceutical development. It occurs at all stages of the drug development cycle and ensures quality and consistency during the production of pharmaceutical products.

CMC is a critical component that cannot be ignored. It ensures that medicines have consistent formulations so that the medicines used in clinical trials are the same as those on the market. Because CMC applies to all stages of drug development and focuses on establishing consistency in drug product stability, release, and manufacturing, it must comply with FDA guidance. Without a CMC, the product would be considered unsafe and would not be approved.

3.36 Contract Research Organization

In the life sciences sector, a **contract research organization** (**CRO**) is a company that offers specialized research services to the pharmaceutical, biotechnology, and medical device industries on a contractual basis. CROs deliver a wide range of services, including biopharmaceutical development, bioassay development, commercialization, clinical development, clinical trial management, pharmacovigilance, outcomes research, and real-world evidence generation.

CROs are structured to help companies reduce costs, particularly those developing new drugs and therapies in niche markets. Their primary objective is to facilitate smoother drug market entry and optimize the development process, eliminating the necessity for large pharmaceutical companies to handle all aspects of drug development internally. Beyond serving corporate clients, CROs also collaborate with government agencies, foundations, research institutions, and universities to support their research initiatives.

3.37 Investigational New Drug Application

An **Investigational New Drug** (**IND**) **Application** is an application to the U.S. FDA for authorization to use an investigational drug or biological product in human clinical trials. Any new drug or biologic that does not have an approved NDA or Biologics/Product License Application must obtain IND authorization prior to interstate transportation and administration.

The U.S. FDA's IND program is the means by which a pharmaceutical company obtains permission to start human clinical trials and to ship an experimental drug across state lines (usually to clinical investigators) before a marketing application for the drug has been approved. Regulations are primarily found at 21 CFR 312. Similar procedures are followed in the European Union, China, Japan, and Canada.

3.38 Clinical Trials

A **clinical trial** is a type of research that tests new medical products and evaluates their impact on human health outcomes. People volunteer to participate in clinical trials to test medical interventions, including drugs, cell and other biological products, surgery, radiation therapy, devices, behavioral treatments, and preventive care.

Clinical trials are carefully designed, reviewed, and completed, and require approval before they can begin. People of all ages can participate in clinical trials, including children.

Biomedical clinical trials are divided into four phases:

- **Phase I** studies typically test a new drug for the first time in a small group of people (usually 20–100 healthy volunteers or people with the disease/condition) to assess safe dosage ranges and determine side effects. Phase I studies usually take several months. About 70% of drugs advance to the next phase.

- **Phase II-a** trials are primarily focused on evaluating dosing requirements, specifically determining the appropriate amount of the drug to be administered. In contrast, **Phase II-b** trials aim to assess the drug's efficacy, typically involving 100–300 participants with the targeted disease or condition, to evaluate how effectively the drug works when used as prescribed. These trials also help establish therapeutic dose ranges and monitor for potential side effects. Phase II studies generally span from a few months up to 2 years. Approximately 33% of the drugs that enter this phase advance to the next stage of clinical development.
- **Phase III** trials involve large groups of participants, typically ranging from 1000 to 3000 individuals, who have the targeted disease or condition. These trials aim to confirm the drug's efficacy, assess its effectiveness, monitor potential side effects, compare it to existing standard treatments, and gather comprehensive safety data to ensure its safe use. Participants in Phase III trials may be recruited from various regions or countries, and this phase is often the final step before a new treatment receives regulatory approval. The duration of Phase III studies generally spans 1–4 years. Approximately 25–30% of drugs successfully advance to the next phase. The probability of FDA approval after submitting a NDA or Biologics License Application is 83.2% (based on a sample size of 659). Success rates for lead indication development pathways are consistently higher than those for all indication development pathways across every phase of clinical trials.
- **Phase IV** studies are conducted after national approval and require further testing in a broad population over a longer period of time. Therefore, they persist throughout the life cycle of active medical use of the drug.

3.39 New Drug Application

For decades, regulation and control of new drugs in the United States has been based on **NDA**. Since 1938, every new drug must be approved by an NDA before it can be commercialized in the United States. An NDA is a vehicle through which a drug sponsor formally proposes FDA approval of a new drug for sale and marketing in the United States. Data collected during animal studies and human clinical trials of an IND become part of the NDA.

The goal of an NDA is to provide sufficient information to allow FDA reviewers to make the following key decisions:

- Whether patent information for the drug is provided.
- Whether the drug is safe and effective for its intended use, and whether the drug's benefits outweigh its risks.
- Whether the recommended labeling (package insert) for the drug is appropriate and what should be included in it.
- Whether the methods used to produce the drug and the controls used to maintain the quality of the drug are adequate to maintain the identity, strength, quality, and purity of the drug.
- Whether the drug is susceptible to abuse.

The documents required in an NDA should tell the entire story of the drug, including what happened during clinical trials, what the drug's ingredients are, the results of animal studies, how the drug behaves in the body and what it does, how it is manufactured, processed, and packaged.

References

1 "Drug". Drug Definition & Meaning. The American Heritage Science Dictionary. Houghton Mifflin Company.

2 "Drug Definition". Stedman's Medical Dictionary.

3 Atanasov, A. G.; Waltenberger, B.; Pferschy-Wenzig, E. M.; Linder, T.; Wawrosch, C.; Uhrin, P.; Temml, V.; Wang, L.; Schwaiger, S.; Heiss, E. H.; Rollinger, J. M.; Schuster, D.; Breuss, J. M.; Bochkov, V.; Mihovilovic, M. D.; Kopp, B.; Bauer, R.; Dirsch, V. M.; Stuppner, H. *Biotechnol. Adv.*, **2015,** *33*, 1582.

4 Helleboid, S.; Haug, C.; Lamottke, K.; Zhou, Y.; Wei, J.; Daix, S.; Cambula, L.; Rigou, G.; Hum, D. W.; Walczak, R. *J. Biomol. Screening*, **2014,** *19*, 399.

5 Herrmann, A.; Roesner, M.; Werner, T.; Hauck, S. M.; Koch, A.; Bauer, A.; Schneider, M.; Brack-Werneret, R. *Sci. Rep.*, **2020,** *10*, 1326.

6 Anson, D.; Ma, J.; He, J. Q., *Genetic Engineering & Biotechnology News. Tech Note*, **2009,** *29* (9), 34.

7 Rennane, S.; Baker, L.; Mulcahy, A. *INQUIRY: The J. Health Care Organ., Prov., and Fin.*, **2022,** *58*, 1.

8 Davis, A.; Ward, S. E. eds. "*The Handbook of Medicinal Chemistry*". **2015,** Royal Society of Chemistry.

9 Barret, R. "*Medicinal Chemistry: Fundamentals*". **2018,** London: Elsevier.

10 Hanif, M.; Yang, X.; Tinoco, A. D.; Plażuk, D. *Front. Chem.*, **2020,** *8*, 453.

11 Anthony, E. J.; Bolitho, E. M.; Bridgewater, H. E.; Carter, O. W.; Donnelly, J. M.; Imberti, C.; Lant, E. C.; Lermyte, F.; Needham, R. J.; Palau, M.; Sadler, P. J.; Shi, H.; Wang, F.-X.; Zhang, W.-Y.; Zhang, Z. *Chem. Sci.*, **2020,** *11*, 12888.

12 Roughley, S. D.; Jordan, A. M. *J. Med. Chem.*, **2011,** *54*, 3451.

13 Zheng, L.; Wang, W.; Sun, Q. *Springer Nature, Signal Transduction and Targeted Therapy*, **2024,** *9*, 46.

14 Copeland, R. A.; Harpel, M. R.; Tummino, P. J., *Expert Opin. Ther. Targets*, **2007,** *11*, 967.

15 Copeland, R. A., "*Evolution of enzyme inhibitors in drug discovery, A guide for medicinal chemists and pharmacologists*", Wiley, **2013.**

16 Williams, O.; Jacks, A. M.; Davis, J.; Martinez, S., "*Case 10: Merck(A): Mevacor*". In Allan Afuah (ed.). "*Innovation Management - Strategies, Implementation, and Profits*", **1998,** Oxford University Press.

17 Ayala-Aguilera, C. C.; Valero, T.; Lorente-Macías, A.; Baillache, D. J.; Croke, S.; Unciti-Broceta, A. *J. Med. Chem.*, **2022,** *65*, 1047.

18 "*FDA Authorizes First Oral Antiviral for Treatment of COVID-19*" (Press release). U.S. Food and Drug Administration. 22 December 2021.

19 Williams, M.; Rita, R. "*Receptors as drug targets*". *Current Protocols in Pharmacology*, **2006,** 1.1.1-1.1.18, Wiley & Sons, Inc.

20 Hauser, A. S.; Chavali, S.; Masuho, I.; Jahn, L. J.; Martemyanov, K. A.; Gloriam, D. E.; Babu, M. M. *Cell*, **2018,** *172*, 41.

21 Flexner, C., *Nature Rev. Drug Discov.*, **2007,** *6*, 959.

22 Hameed, A.; al-Rashida, M.; Alharthy, R. D.; Uroos, M.; Mughal, E. U.; Ali, S. A.; Khan, K. M. *Expert Opin Ther Pat.*, **2017,** *27*, 1089.

23 Hopkins, A. L. "*Chapter 25: Pharmacological space*". In Wermuth C. G. (editor), "*The Practice of Medicinal Chemistry*" (3rd edition). **2011,** Academic Press. pp. 521–527.

24 Kirchmair, J. "*Drug Metabolism Prediction. Wiley's Methods and Principles in Medicinal Chemistry*" **2014,** Vol. 63, Wiley-VCH.

25 Ban, T. A. *Dialogues in Clinical Neuroscience*, **2006,** *8*, 335.

26 Singh, J.; Chuaqui, C. E.; Boriack-Sjodin, P. A.; Lee, W. C.; Pontz, T.; Corbley, M. J.; Cheung, H.-K.; Arduini, R. M.; Mead, J. N.; Newman, M. N.; Papadatos, J. L.; Bowes, S.; Josiah, S.; Ling, L. E. *Bioorg. Med. Chem. Lett.*, **2003,** *13*, 4355.

27 Becker, O. M.; Dhanoa, D. S.; Marantz, Y.; Chen, D.; Shacham, S.; Cheruku, S.; Alexander, H.; Pradyumna, M.; Fichman, M.; Sharadendu, A.; Nudelman, R.; Kauffman, M.; Noiman, S. *J. Med. Chem.*, **2006,** *49*, 3116.

28 Zhou, S.-F.; Zhong, W.-Z. *Molecules*, **2017,** *22*, 279.

29 Doytchinova, I. *Molecules*, **2022,** *27*, 1496.

30 Hessler, G.; Baringhaus, K.-H. *Molecules*, **2018,** *23*, 2520.

31 Schneider, P.; Walters, W. P.; Plowright, A. T.; Sieroka, N.; Listgarten, J.; Goodnow, R .A., Jr.; Fisher, J.; Jansen, J. M.; Duca, J. S.; Rush, T. S.; Zentgraf, M.; Hill, J. E.; Krutoholow, E.; Kohler, M.; Blaney, J.; Funatsu, K.; Luebkemann, C.; Schneideet, G.; *Nat. Rev. Drug Discov.*, **2020,** *19*, 353.

32 Paul, D.; Sanap, G.; Shenoy, S.; Kalyane, D.; Kalia, K.; Tekade, R. K. *Drug Discov. Today*, **2021,** *26*, 80.

33 Guner, O. F. "*Pharmacophore Perception, Development, and use in Drug Design*". **2000**, La Jolla, California, International University Line.

34 Tropsha, A. "*QSAR in Drug Discovery. In: Structure and Ligand-Based Drug Design*". Merz, K.; Ringe, D.; Reynolds, C. L., Eds. Cambridge University Press, New York, **2010,** Chapter 10, pp. 151-164.

35 Leach, A. R.; Harren, J. "*Structure-based Drug Discovery*", Berlin: Springer, **2007**.

36 Mauser, H.; Guba. W. *Curr. Opin. Drug Discov. Devel.*, **2008,** *11*, 365.

37 Klebe, G. *J. Mol. Med.* **2000,** *78*, 269.

38 Wang, R.; Gao, Y.; Lai, L. *J. Mol. Modeling*, **2000,** *6*, 498.

39 Schneider, G.; Fechner, U. *Nature Rev. Drug Discov.*, **2005,** *4*, 649.

40 Jorgensen, W. L. *Science*, **2004,** *303*, 1813.

41 Lipinski, C.A.; Lombardo, F.; Dominy, B. W.; Feeney, P. J. *Adv. Drug Delivery Rev.*, **1997,** *23*, 3.

42 Lipinski, CA. *Drug Discovery Today: Technologies*, **2004,** *1*, 337.

43 O Hagan, S.; Swainston, N.; Handl, J.; Kell, D. B. *Metabolomics*, **2015,** *11*, 323.

44 Doak, B. C.; Over, B.; Giordanetto, F.; Kihlberg, J. *Chem. Biol.*, **2014,** *21*, 1115.

45 de Oliveira, E. C. L.; Santana, K.; Josino, L.; eLima, A. H. L.; de Souza de Sales Júnior, C. *Scientific Reports*, **2021,** *11*, 7628.

46 Doak, B. C.; Kihlberg, J. *Expert Opin. Drug Discov.*, **2017,** *12*, 115.

47 Hughes, J. P.; Rees, S.; Kalindjian, S. B.; Philpott, K. L. *Br. J. of Pharmacol.*, **2011,** *162*, 1239.
48 Inglese, J.; Auld, D. S. *"Application of High Throughput Screening (HTS) Techniques: Applications in Chemical Biology"* in *"Wiley Encyclopedia of Chemical Biology"*, Wiley & Sons, Inc., Hoboken, NJ, **2009,** *Vol 2*, pp 260–274.
49 Macarron, R.; Banks, M. N.; Bojanic, D.; Burns, D. J.; Cirovic, D. A.; Garyantes, T.; Green, D. V.; Hertzberg, R. P.; Janzen, W. P.; Paslay, J. W.; Schopfer, U.; Sittampalam, G. S. *Nat. Rev. Drug Discov.*, **2011,** *10*, 188.
50 Wermuth, C. G.; Ganellin, C. R.; Lindberg, P.; Mitscher, L. A. *Pure Appl. Chem.*, **1998,** *70*, 1129.
51 Kier, L. B. *Mol. Pharmacol.*, **1967,** *3*, 487.
52 Kier, L. B. *"Molecular orbital theory in drug research"*, **1971,** Boston: Academic Press, pp. 164–169.
53 Brown, A. C.; Fraser, T. R. *Transactions of the Royal Society of Edinburgh*, **1868,** *25*, 151.
54 Brown, A. C.; Fraser, T. R. *Transactions of the Royal Society of Edinburgh*, **1869,** *25*, 693.
55 Tong. W.; Welsh, W. J.; Shi, L.; Fang, H.; Perkins, R. *Environ. Toxicol. Chem.*, **2003,** *22*, 1680.
56 Moriarty, L. M.; Lally, M. N.; Carolan, C. G.; Jones, M.; Clancy, J. M.; Gilmer, J. F. *J. Med. Chem.*, **2008,** *51*, 7991.
57 Cattaneo, M. *J. Thrombosis and Haemostasis*, **2012,** *10*, 32.
58 Mustafa, A. A.; R.; Suarez, J. D.; Alzghari, S. K. *Cureus*, **2018,** *10*, e2844.
59 Mickle, T. *Drug Development & Delivery*, **2019,** *March*, 24.
60 Iwamura, R.; Tanaka, M.; Okanari, E.; Kirihara, T.; Odani-Kawabata, N.; Shams, N.; Yoneda, K. *J. Med. Chem.*, **2018,** *61*, 6869.
61 Schmidt, C. *Nature Biotechnology*, **2017,** *35*, 493.
62 Coppen, E. M.; Roos, R. A. *Drugs*, **2017,** *77*, 29.
63 Mullard, A. *Nature Rev. Drug Discov.*, **2022,** *21*, 623.
64 Hofland, P.; Portillo, S. *ADC Review, J. Antibody-Drug conjugates*, **2019,** March 22.
65 Beck, A.; Goetsch, L.; Dumontet, C.; Corvaïa, N. *Nature Rev. Drug Discov.*, **2017,** *16*, 315.
66 Chau, C. H.; Steeg, P. S.; Figg, W. D. *The Lancent*, **2019,** *394*, 793.
67 Joubert, N.; Beck, A.; Dumontet, C.; Denevault-Sabourin, C.; *Pharmaceuticals*, **2020,** *13*, 245.
68 Gogia, P.; Ashraf, H.; Bhasin, S.; Xu, Y. *Cancers (Basel)*, **2023,** *15*, 3886.
69 Goundry, W. R. F.; Parker, J. S. *Org. Process Res. Dev.*, **2022,** *26*, 2121.

4

Mechanism Problems from Reactions Give Expected Products

Drawing from my research and professional experience in medicinal chemistry, as well as the literature I have reviewed, this chapter compiles 121 cases of reaction mechanisms pertinent to synthesis in medicinal chemistry. These mechanisms are both intriguing and challenging. Given that most mechanism problems encompass multiple reactions, categorizing them by reaction type proves difficult. Consequently, the mechanisms are organized based on the structure of the reaction products. In the experimental and compound structural characterization sections of each problem, unless otherwise stated, the compound numbers have been renumbered to align with the layout of this book. This same approach has been applied to Chapter 5 as well.

4.1 Formation of Noncyclic Compounds

Problem 1: *t*-BuOK Promoted Isomerization of Terminal Alkyne to Internal Alkyne

AMG-3969 was discovered as a compound capable of disrupting the interaction between glucokinase and glucokinase regulatory protein (GK-GKRP), showing potential for use in diabetes treatment in humans [1]. During the scale-up synthesis of AMG-3969, a crucial step involved the isomerization of terminal alkyne **2** to alkyne **3** using *t*-BuOK [2].

Please suggest a mechanism for this reaction.

Experimental Procedure (from *J. Org. Chem.*, **2014**, *79*, 3684)

(3*S*)-1-Benzyl-3-(1-propyn-1-yl)piperazine (3). To a solution of (3*S*)-1-benzyl-3-(propan-2-yl)piperazine (**2**, 38.8 g, 181 mmol) in tetrahydrofuran (THF) (200 mL) was added potassium *tert*-butoxide (40.6 g, 362 mmol) portionwise. After the addition was complete, the reaction mixture was stirred at room temperature for 30 min and then quenched with water (500 mL) and diluted with EtOAc (1 L). The organic phase was dried (Na_2SO_4), filtered, and concentrated to give a solid. Purification via silica gel chromatography (0–7% MeOH in CH_2Cl_2) followed by

Overcoming Synthetic Challenges in Medicinal Chemistry: Mechanistic Insights and Solutions, First Edition. Tongshuang Li.

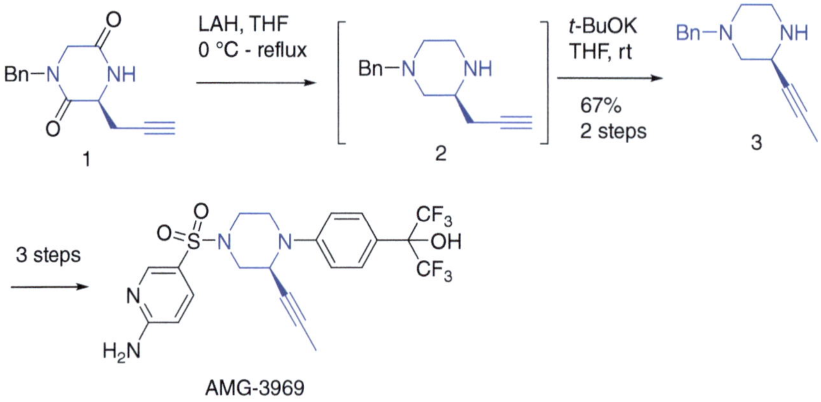

recrystallization from 100% hexanes provided (3*S*)-1-benzyl-3-(propan-2-yl)piperazine **3** (26.0 g, 67%) as tan crystalline solid: ^{1}H NMR (400 MHz, $CDCl_3$) δ 7.36–7.18 (m, 5 H), 3.59 (br d, J = 8.8 Hz, 1 H), 3.51 (s, 2 H), 2.98 (d, J = 11.9 Hz, 1 H), 2.88–2.76 (m, 2 H), 2.65 (d, J = 11.0 Hz, 1 H), 2.22–2.05 (m, 2 H), 1.84–1.76 (s, 3 H), 1.73 (br s, 1 H); ^{13}C NMR (101 MHz, CD_3OD) δ 138.5, 130.7, 129.5, 128.5, 80.3, 78.9, 64.0, 59.8, 53.8, 48.0, 45.5, 3.2; HRMS (ESI-TOF) m/z $[M + H]^+$ calcd for $C_{14}H_{19}N_2$ 215.1548, found 215.1551; mp 65–68 °C, $[\alpha]_D^{20} = -55.4$ (c = 2.9, MeOH).

The Mechanism

It is well-known that a nonterminal alkyne can isomerize into a terminal alkyne under strong basic conditions—a process known as the alkyne zipper reaction [3]. The current case isomerization can be understood as a retro-alkyne zipper reaction. Similar rearrangement was also observed by Grubbs' group when they investigated the silylation of terminal alkynes at basic conditions [4]. Tsurugi and Mashima's group discovered that terminal alkynes could be converted into allenes and further to internal alkynes catalyzed by organomagnesium complexes under mild conditions [5].

Under strong basic condition in THF, the terminal alkyne-H can be deprotonated to form propynylpotassium **A**. Then a 1,3-H shift occurs to give allenylpotassium **B**. At this stage, species **B** could be protonated by *t*-BuOH to afford allene **C**. While, a second 1,3-H shift would lead to the prop-2-yn-1-ylpotassium **D**, which is a much stronger base than *t*-BuOH, and should be readily protonated by *t*-BuOH to provide the product **3**. I believe all the transformations are reversible and the distribution of **2**, **3**, and **C** depends greatly on the reaction conditions.

Problem 2: Dimerization of Terminal Aziridines to (*E*)-but-2-ene-1,4-diamines

In 2006, D. M. Hodgson's group reported a dimerization of terminal aziridines to give *N*-protected but-2-ene-1,4-diamines with complete *E*-olefin selectivity when starting with enantiopure substrates [6]. The stereochemistry of the double bond and the two chiral centers was proven unambiguously by X-ray crystallographic analysis of diamine **(*S*,*S*)-2**. The reaction was applied to the synthesis of Cobicistat, a drug for the treatment of human immunodeficiency virus type 1 (HIV-1) infection [7]. The mechanism was discussed in Hodgson's paper.

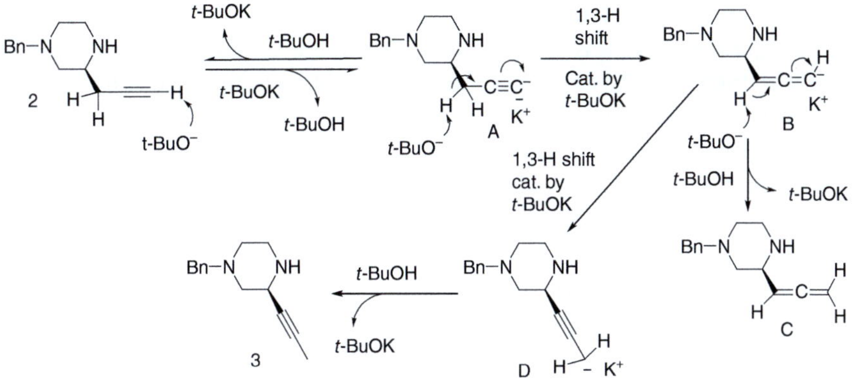

Please provide your own mechanism for the reaction.

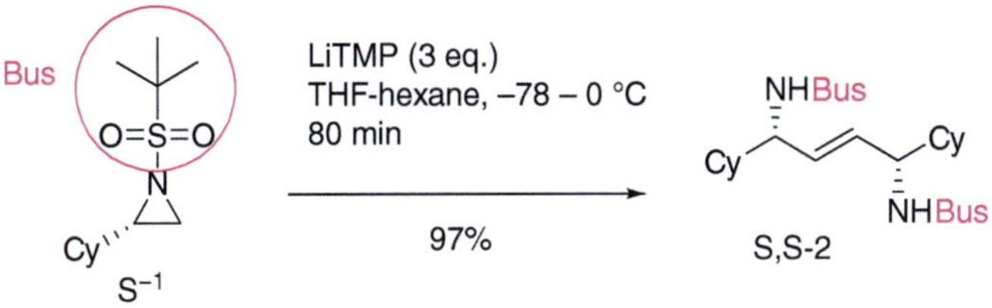

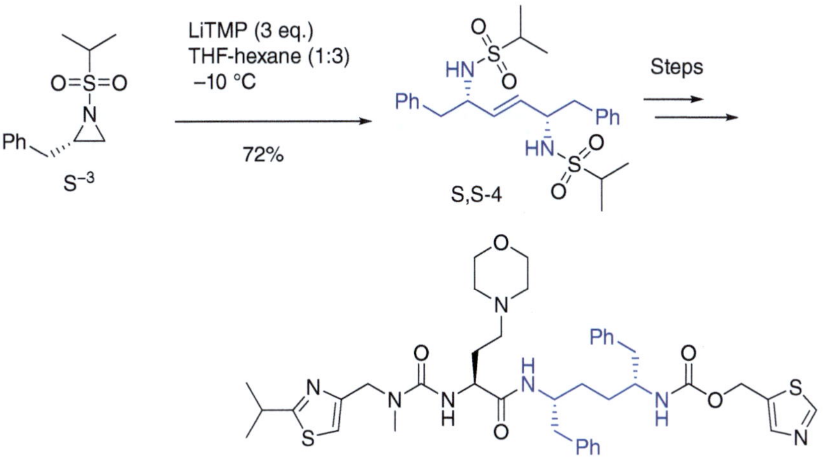

Cobicistat

Experimental Procedure (from *J. Org. Chem.*, **2007**, *72*, 10009)

General Procedure D: Synthesis of 2-ene-1,4-diamines from *N*-Bus aziridines. *n*-BuLi (1.6 M in hexanes, 1.88 mL, 3.0 mmol) was added dropwise to a stirred solution of 2,2,6,6-tetramethylpiperidine (0.51 mL, 3.0 mmol) in THF (0.4 mL) at −78 °C. The mixture was warmed to 0 °C over 15 min, then re-cooled to −78 °C before dropwise addition of the aziridine (1.0 mmol) in THF (0.8 mL). The mixture was stirred at −78 °C for 20 min, then at 0 °C for 1 h, before the addition of

MeOH (0.8 mL), sat. aq. NH_4Cl (8 mL) and Et_2O (16 mL). The layers were separated, and the aqueous phase was extracted with Et_2O (16 mL). The combined organic phase was dried ($MgSO_4$) and concentrated. Purification of the residue by column chromatography (petroleum ether/Et_2O, SiO_2) gave the 2-ene-1,4-diamine.

Following **General Procedure D**, aziridine **(*S*)-1** (31 mg, 0.13 mmol) gave 2-ene1,4-diamine **(S,S)-2** as a white solid (30 mg, 97%). $[\alpha]^{D}_{25} = -3.5$ (c 1.0, $CHCl_3$); (mp = 208 °C); R_f 0.18 (petroleum ether/Et_2O 3:7); IR (film) 3270br.s, 2986 s, 2926 s, 2850 s, 1479 m, 1447 s, 1323 s, 1312 s, 1304 s, 1284 s and 1130 s cm^{-1}; ^1H NMR (500 MHz, $CDCl_3$) δ 5.57–5.54 (2H, m), 4.01 (2H, d, J 10), 3.81–3.73 (2H, m), 1.81–1.62 (10H, m), 1.51–1.37 (20H, m), 1.26–1.02 (10H, m); ^{13}C NMR (125 MHz, $CDCl_3$) δ 131.4 (2 × CH), 61.4 (2 × CH), 59.8 (2 × C), 43.9 (2 × CH), 29.3 (2 × CH_2), 28.9 (2 × CH_2), 26.2 (2 × CH_2), 26.2 (2 × CH_2), 26.0 (2 × CH_2), 24.3 (6 × CH_3); MS CI m/z (rel. int.) 508 ($M + NH_4^+$, 100), 491 ($M + H^+$, 5), 371 (10), 354 (35), 287 (20), 232 (40), 150 (25); HRMS m/z calcd for $C_{24}H_{50}N_3O_4S_2$ requires 508.3243, found 508.3242; Anal found C 58.66, H 9.55, N 5.80, $C_{24}H_{46}N_2O_4S_2$ requires C 58.74, H 9.45, N 5.71.

The Mechanism

The preferential formation of the *E*-alkene isomer from enantiopure substrates can be explained by a mechanism involving an initial diastereoselective trans-lithiation, followed by the nucleophilic attack of one *R*-lithiated aziridine on another, which acts as an electrophile. This step may proceed through a 1,2-metallate shift. The subsequent *syn* elimination, driven by the steric constraints imposed by the aziridine substituents and the *N*-Bus groups, results in the observed *E*-alkene products.

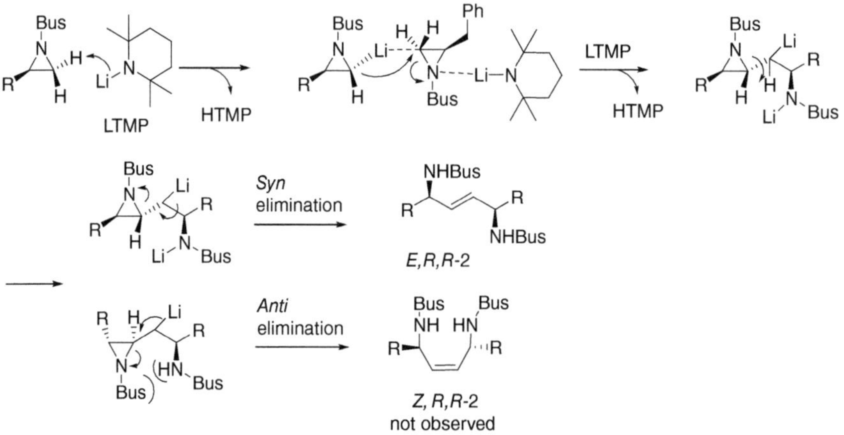

Problem 3: Thiourea-Mediated Conversion of Alcohols to Alkylhalides

Benzyl halides are an important class of organic compounds and have numerous industrial and synthetic uses. Some halides are key building blocks for preparation of several pharmaceuticals [8–11]. For example, 2-(bromomethyl) benzonitriles (**1a, 1b**) are starting materials for synthesizing alogliptin (**2**) [8] and fotagliptin (**3**) [9]. Therefore, conversion of relatively easier obtained benzyl alcohols to benzyl halides is an important reaction. Recently, a group of Israel chemists reported a straightforward methodology for the direct transformation of a wide scope of alcohols to alkyl bromides and chlorides using thioureas and *N*-halosuccinimides (NXS), in a single step under mild conditions [12].

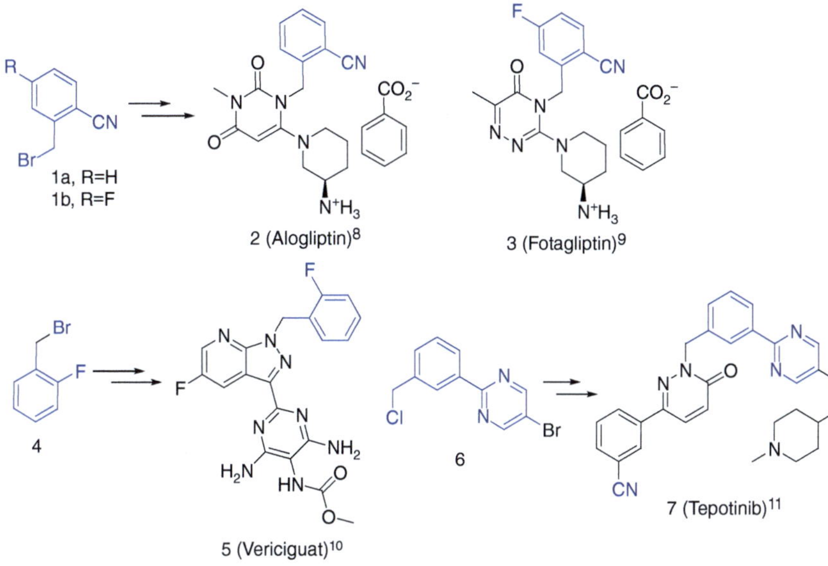

A representative reaction is shown below. Please propose a mechanism for this reaction.

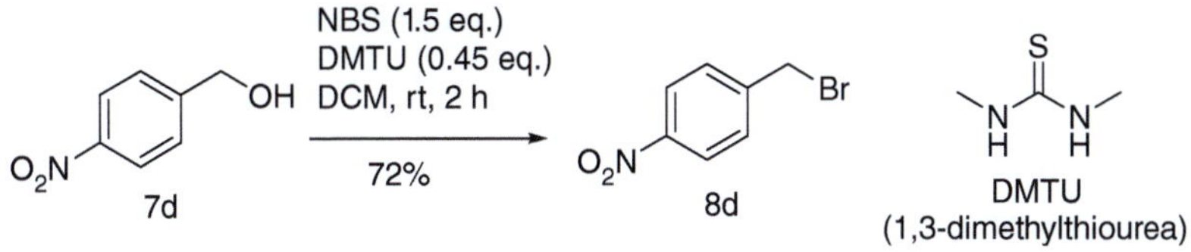

Experimental Procedure (from *J. Org. Chem.*, **2007**, *85*, 12901)

General Procedure for Bromination of Alcohols. Alcohols **7** (0.9–1.1 mmol, 1 eq.) and *N,N′*-dimethylthiourea (DMTU) (0.45 eq.) in dry dichloromethane (DCM) (4 mL) were stirred at room temperature until the starting materials were completely dissolved. The reaction mixture was vigorously stirred and *N*-bromosuccinimide (NBS) (1.5 eq.) was added in a single portion. After completion of reaction, the mixture was diluted with DCM (5 mL) and an aliquot (1 μL) injected to the gas chromatography-mass spectrometry (GC-MS). Then, the organic layer was concentrated over rotary evaporator to produce crude product, which was purified by silica gel flash chromatography over n-hexane (100%) as mobile phase to afford the corresponding bromides, **8** (*vide infra*). The yields of known bromides **8a,c,f-h,k,n,t,w,x**, **8zc** and **8zd** were determined by GC-MS (see Supporting Information Figures S43–S82).

1-(bromomethyl)-4-nitrobenzene (**8d**). Reaction time: 2 h; obtained as pale yellow solid (162 mg, 72%). ^{1}H NMR (400 MHz, CD_2Cl_2) δ 4.53 (s, 2H), 7.56 (d, $J = 8$ Hz, 2H), 8.17 (d, $J = 8$ Hz, 2H); ^{13}C NMR (100 MHz, CD_2Cl_2) δ 147.7, 145.0, 130.1, 129.8, 124.0, 123.8, 31.2; GC-MS (ESI) m/z calcd for $C_7H_6BrNO_2$ [M] 214.9, found: 214.9.

The Mechanism

The mechanism was studied by the Israel group and it was concluded that the reaction was a free radical process [12]. As you can see below, one equivalent of DMTU could generate a maximum of three equivalents of alkyl radical (R$^•$) and three equivalents of bromine radical (Br$^•$); thus, substoichiometric amounts (0.45 eq) of DMTU is enough.

NBS, DMTU → A + B (Br, R–OH); B + ROH → HBr; → C + R$^•$

C + NBS → A; + ROH → HBr; → D + R$^•$; D + NBS → A; + ROH → HBr; → E + R$^•$

A + H–Br → NH (succinimide) + Br$^•$ + R$^•$ → R–Br (product)

Problem 4: Oxidative Decarboxylation of *N*-Aroylglycines with Lead(IV) Acetate

In 1986, Threadgill's group reported an oxidative decarboxylation of *N*-aroylglycines to *N*-(acetoxymethy1)benzamides with lead (IV) acetate [13]. In the case of (4-nitrobenzoyl)glycine **1d**, the yield of product **2d** was 72%. The yield from other substrates was relatively low (2.5–39%). When catalytic amount of $Cu(OAc)_2$ was added, the reaction yield was improved to 40–90% [14]. The scope of the reaction was expanded to aliphatic analogs [15] and applied to the synthesis of the marketed drug deruxtecan (**5**) [16]. Deruxtecan is a derivative of exatecan that acts as topoisomerase I inhibitor. It is available as an antibody drug conjugate, linked to a specific monoclonal antibody—such as trastuzumab deruxtecan, which is used to treat breast cancer, as well as gastric or gastroesophageal adenocarcinoma.

Please propose a mechanism for this reaction.

1d → 2d: $Pb(OAc)_4$, Ac_2O, AcOH, 60 °C, 10 min, 72%

$Pb(OAc)_4$
AcOH, THF
91%
Steps
3
4
Deruxtecan (5)
Antibody–drug conjugate
Trastuzumab deruxtecan
sold under the brand name Enhertu

Experimental Procedure (from *J. Org. Chem.*, **1986**, *51*, 3196)

***N*-(acetoxymethyl)-4-nitrobenzamide(2d).** *N*-(4-nitrobenzyl) glycine **(la)** (5.60 g, 25 mmol) was treated with $Pb(OAc)_4$ (22.15 g, 50 mmol) as for the preparation of **2b** above except that chromatography was replaced by recrystallization from ethyl acetate/light petroleum (bp 60–80 °C). The acetoxymethylbenzamide **2d** (4.28 g, 72%) was obtained as a very pale greenish yellow solid: mp 94 °C (lit. mp 120 °C); IR 3370, 1760, 1670 cm^{-1}; NMR ($CDCl_3$) δ 2.10 (3H, s, $COCH_3$), 5.45 (2H, d, $J = 7$ Hz, NCH_2O), 7.90 (1H, ca. t, $J = 7$ Hz, NH), 8.05 (2H, d, $J = 8$ Hz) and 8.35 (2H, d, $J = 8$ Hz) [Ar H].

(From WO 2019044947 A1)

({*N*-[(9H-fluoren-9-ylmethoxy) carbonyl] glycyl} amino) methyl acetate (**4**). THF (9.75 L) and acetic acid (1.95 L) were added to *N*-[(9H-fluoren-ylmethoxy) carbonyl] glycylglycine (**3**, 650.0 g, 1.834 mol), and dissolved by heating to 40 °C. Lead tetraacetate (1301.3 g, 2.935 mol) was added and refluxed for about 1.5 h. The mixture was cooled to room temperature, the insolubles were filtered off, and the filtered off insolubles were washed with ethyl acetate (3.25 L) and matched with the filtrate. To the resulting solution was added a 20 (w/v)% aqueous solution of trisodium citrate dihydrate (3.25 L), and the mixture was stirred, separated, and the aqueous layer was removed. The obtained organic layer was washed twice with 20 (w/v)% aqueous trisodium citrate dihydrate (3.25 L), and then the organic layer was concentrated to 6.5 L under reduced pressure. Water (1.95 L) was added followed by

({*N*-[(9H-fluoren-9-ylmethoxy) carbonyl] glycyl} amino) methyl acetate (0.65 g) and stirred at room temperature for about 1 hour. Water (6.5 L) was added dropwise, cooled to 0–5 °C. and stirred for about 3 hours. The precipitate was filtered off and the filtered powder was washed with cold 30 (v / v)% aqueous THF (2.6 L). The obtained powder was dried at 40 °C under reduced pressure, and ({*N*-[(9H-fluoren-9-ylmethoxy) carbonyl] glycyl} amino) methyl acetate (**4**, 617.1 g, 1.675 mol, yield 91.3% Got).

^{1}H NMR (400 MHz, $CDCl_3$) δ 2.06 (3H, s), δ 3. 90 (2H, d, 4.9 Hz), 4.23 (1H, t, 6.7 Hz), 4.45 (2H, d, 6.7 Hz), 5.25 (2H, d, 7.3 Hz), 5.39 (1H, brs), 7.05 (1H, brs), 7.30–7.34 (2H, m)), 7.41 (2H, t, 7.3 Hz), 7.59 (2H, d, 7.3 Hz), 7.77 (2H, d, 7.3 Hz).

^{13}C NMR (100 MHz, $CDCl_3$) δ 20.8, 44.4, 47.0, 63.9, 67.2, 120.0, 125.0, 127.1, 127.7, 141.3, 141.3 143.6, 156.6, 169.8, 171.7.

MS (ESI) (m/z): 369 ([M + H$^+$]).

The Mechanism

The reaction mechanism was studied by Threadgill's group [13]. The results indicated that the process involved initial ligand exchange at lead, followed by *N*-acetoxylation, decarboxylation, elimination, and a final re-addition of acetic acid to *N*-aroylimines to yield the products.

Ligand exchange: $Pb(OAc)_4$, −HOAc; *N*-acetoxylation: −$Pb(OAc)_2$; Decarboxylation elimination: −CO_2, −HOAc; Addition HOAc → 2

Problem 5: Preparation of Aromatic Amines Promoted by DDQ and Ph_3P

Amide and amine formation are the current most frequently used synthetic reactions in medicinal chemistry [17]. In 2009, Iranpoor et al. reported selective mono- and di-*N*-alkylation of aromatic amines with alcohols and acylation of aromatic amines using Ph_3P/DDQ [18]. A representative reaction is shown below.

1 (1 eq.) + 2 (1.2 eq.) → [Ph_3P (1.2 eq.), DDQ (1.2 eq.), CH_2Cl_2, rt, 1 min; 86%] → 3

Please propose a mechanism for this reaction.

Experimental Procedure (from *Tetrahedron*, **2009**, *65*, 3893)

Typical Procedure for *N*-benzylation of 4-amino-2,3-dimethyl-1-phenyl-3-pyrazolin-5-one. To a flask containing a stirred mixture of Ph_3P (1.2 mmol, 0.314 g) and DDQ (1.2 mmol, 0.272 g) in DCM (5 mL) was added 4-amino-2,3-dimethyl-1-phenyl-3-pyrazolin-5-one (1.2 mmol, 0.244 g) at room temperature. Benzyl alcohol (1.0 mmol, 0.1 mL) was then added to the reaction mixture. Thin-layer chromatography (TLC) monitoring showed the completion of the reaction after 1 min. The solvent was evaporated and the residue was chromatographed on a silica gel column using *n*-hexane/ethyl acetate (4:1) as eluent. 4-(*N*-Benzylamino)-2,3-dimethyl-1-phenyl-3-pyrazolin-5-one was obtained in 86% yield (0.30 g). IR (neat) 3403, 3054, 2986, 2928, 1662, 1594, 1455, 1359 cm^{-1}; ^{1}H NMR (250 MHz, $CDCl_3$): δ (ppm) = 1.86 (3H, s), 2.06 (1H, s), 2.60 (3H, s), 5.18 (2H, s), 7.07–7.42 (10H, m); ^{13}C NMR (62.9 MHz, $CDCl_3$): δ (ppm) = 9.75, 36.32, 57.81, 113.45, 123.34, 126.04, 126.82, 127.65, 127.93, 128.10, 128.82, 129.00, 139.83, 154.59. Anal. Calcd for $C_{18}H_{19}N_3O$: C, 73.69; H, 6.53; N, 14.32%. Found: C, 73.46; H, 6.57; N, 14.22%.

The Mechanism

The authors [18] proposed a mechanism to account for this transformation. Initially, the known quaternary phosphonium salts (i) are formed via the addition of Ph_3P to DDQ. The negatively charged oxygen in the hydroquinone moiety of this adduct can function as a base, deprotonating the amine without the need for an additional base, thereby facilitating the generation of an amine anion. The resulting complex (ii) then reacts with the alcohol to yield the alkoxyphosphonium salt (iii), which undergoes an SN-type displacement to produce the desired alkylated amine.

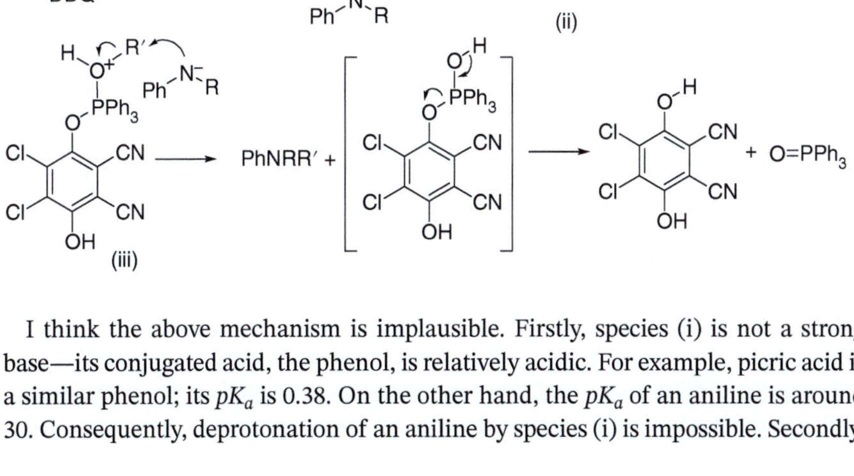

I think the above mechanism is implausible. Firstly, species (i) is not a strong base—its conjugated acid, the phenol, is relatively acidic. For example, picric acid is a similar phenol; its pK_a is 0.38. On the other hand, the pK_a of an aniline is around 30. Consequently, deprotonation of an aniline by species (i) is impossible. Secondly, even if species (ii) were formed, the protonated alkyl-phosphine ether would be very acidic. I predict its pK_a to be in the range of $-6 \sim 0$. It would likely be immediately deprotonated by the strong base PhN^-R, resulting in the release of free aniline. The aniline anion (PhN^-R) had no chance to attack the alcoholic carbon in (iii). Furthermore, that mechanism could not explain that amides were less active than amines, since amides are more acidic than amines and should be easier to be deprotonated than amines.

Based on above considerations, I propose a different mechanism as shown below. The first step is still the addition of Ph_3P to DDQ to form onium salt (**A**). Then the oxygen of the hydroxyl in the alcohol as a nucleophile attacks the quaternary phosphonium and gives the protonated phosphonyl ether (**B**). The proton on the ethereal oxygen would be quickly transferred to the phenolic oxygen to generate the neutral molecule (**C**). Thus, the alcohol is activated by forming a phosphonate, which is a good leaving group. Finally, aniline as a nucleophile attacks the activated carbon to provide the product along with triphenylphosphine oxide and 4,5-dichloro-3,6-dihydroxyphthalonitrile. This mechanism is similar to the mechanism of Mitsunobu reaction, in which the DDQ plays the role of diethyl azodicarboxylate (DEAD).

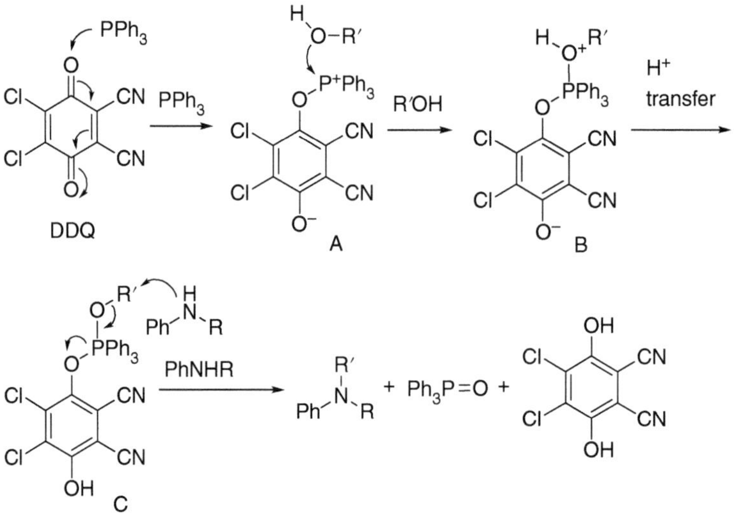

However, this mechanism still cannot explain the results from the paper [18], which indicated that alkylamines (no reaction) were less reactive than arylamines. Obviously, the results and the mechanism are contradictory.

Problem 6: Conversion of Alkyl Aryl Ketones to α-Arylalkanoic Acids Using DPPA

In 1978, Shioiri and Kawai reported a preparation of α-arylalkanoic acids (**4**) from alkyl aryl ketones (**1**) by using pyrrolidine and diphenylphosphoryl azide (DPPA) [19]. The reactions are illustrated as follows in three steps. The method was successfully applied to the synthesis of ibuprofen, the anti-inflammatory drug.

Please provide mechanisms for those transformations.

Pyrrolidin benzene, reflux 79% DPPA, THF rt- reflux 80% KOH, glycol reflux 91%

1. Pyrrolidine, benzene, reflux
2. DPPA, THF, rt- reflux
3. KOH, glycol, reflux
overall yield 62%

Ibuprofen

Experimental Procedure (from *J. Org. Chem.*, **1978**, *43*, 2936)
A typical procedure is as follows. DPPA (4.95 g) was added with stirring to pyrrolidine enamine **2** (3.05 g) in THF (45 mL). The mixture was stirred at room temperature for 1 h, at 40 °C for 1 h, and then refluxed for 2 h. After dilution with ethyl acetate and benzene (1:1, 150 mL), the mixture was successively washed with 5% aqueous citric acid, water, saturated aqueous sodium chloride, saturated aqueous sodium bicarbonate, water, and saturated aqueous sodium chloride. The dried solution was evaporated and the residue was purified by column chromatography on silica gel with ethyl acetate and benzene (1:5) to give the *N*-phosphorylated amidine **3** (5.68 g, 80%).

The Mechanism
The first reaction is the conversion of alkyl aryl ketones **1** to pyrrolidine enamine **2**. Then 1,3-dipolar/[3+2] cycloaddition of DPPA to enamine **2** followed by aryl migration with concomitant evolution of nitrogen from labile triazoline provides intermediate **3**. Finally, hydrolysis of the resulting *N*-phosphorylated amidine **3** yields the product **4**.

Pyrrolidine H_2O DPPA [3+2] cycloaddition N_2 KOH hydrolysis KOH hydrolysis

Problem 7: Synthesis of β-Aryl-α-Keto Acid from Benzaldehyde and Hydantoin

Rupintrivir (**1**, AG-7088) is a peptidomimetic antiviral drug that acts as a 3C and 3CL protease inhibitor [20]. (R)-3-(4-fluorophenyl)-2-hydroxy propionic acid (**2**) is one of the building blocks for synthesizing **1**. The compound **3**, precursor of **2**, is prepared from 4-fluorobenzaldehyde (**4**) and hydantoin (**5**) by the reaction shown below [21].

Please provide a mechanism for this reaction.

Rupintrivir (1, AG-7088, Rupinavir)   2   3

4 + 5 → 6 (Na+ salt)

1. H_2O, 1-amino-2-propanol, reflux
2. NaOH, reflux, 23 kg, 77–82%

Experimental Procedure (from *Org. Process Res. Dev.*, **2002**, *6*, 520)

Sodium 3-(4-fluorophenyl)-2-oxo-propionate (3). 4-Fluorobenzaldehyde **(4)** (3.72 kg, 30 mol), hydantoin **(5)** (3.00 kg, 30 mol), 1-amino-2-propanol (225 g, 3.0 mol), and water (7.5 L) were added to a 50-L reactor equipped with a temperature probe, reflux condenser, agitator, and cooling coils. The resulting mixture was heated and refluxed for approximately 10 h. The reaction was monitored by ^{1}H NMR and was deemed complete upon disappearance of the hydantoin (**5**) proton signal at δ 3.9 (s, 2H, CH_2) and the appearance of condensed intermediate olefin proton δ 6.4 (s, 2H, CH). Aqueous sodium hydroxide (6.00 kg in 30.0 L) was then added to the bright yellow slurry, and reflux continued until completion as shown by high-performance liquid chromatography (HPLC), leading to a transparent orange solution. The mixture was cooled to $20 \pm 5\,°C$, and sodium chloride (3.51 kg, 60.0 mol) was added with agitation. The pH of the mixture was adjusted to 8.0 using concentrated HCl and the suspension was stirred for 4 h until a pale yellow slurry was obtained. The solids were filtered off and purified via slurrying in methanol (30.0 L), followed by filtration. Upon drying under house vacuum at ambient temperature for 4 days, 5.47 kg of solids (**6**) was obtained with a yield of 82% and HPLC purity above 80%. ^{1}H NMR (D2O): 4.72 (s, 2H), 7.02–7.19 (m, 4H). Anal. Calcd for $C_9H_6O_3FNa{\bullet}H_2O$: C, 48.66; H, 3.63; Found: C, 48.64, H 3.74.

The Mechanism

A similar reaction was reported previously [22]. The first step is a condensation of the 4-flourobenzaldehyde **4** and hydantoin **5** catalyzed by the aminoalcohol. Then hydrolysis of the formed 4-flourobenzalhydantoin **A** at basic condition yields the sodium salt of **6**.

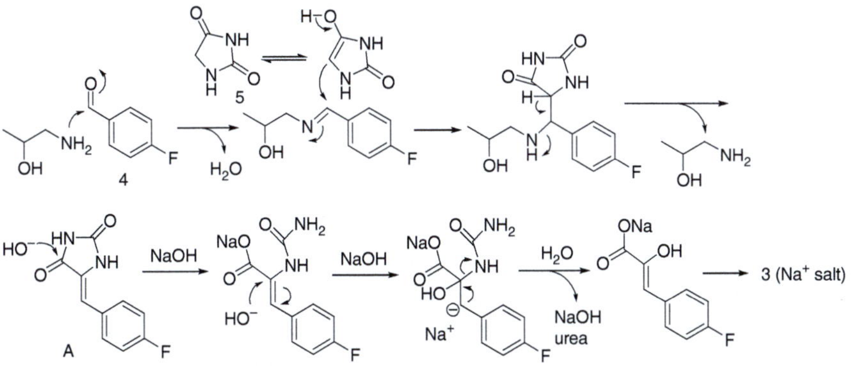

Another possible mechanism for the hydrolysis of **A** could be as follows. The hydroxyl anion adds to the C=C bond via a Michael addition followed by hydrolysis of the amide in hydantoin **B** to form mono substituted urea **C**. Subsequently, elimination of the urea yields the epoxide **D** which would rearrange to the product **6**.

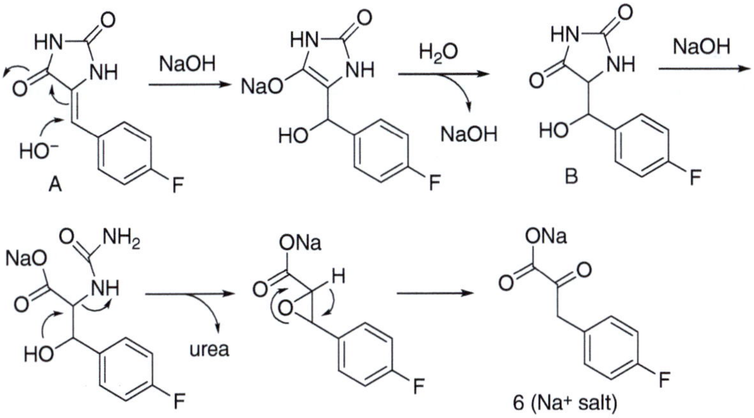

6 (Na+ salt)

Problem 8: α-Amination of Amides

α-Amino acids and peptides are becoming an increasing subset of commercialized drugs [23, 24]. Therefore, developing synthetic methods for non-natural α-amino acids is an attractive area for medicinal chemistry. Maulide's group reported a stereoselective α-amination of amides with azides under mild conditions [25]. One example is shown below.

Please provide a mechanism for the reaction.

1 + Ph⁄⁄N$_3$ (2, 2 eq.) → 1. Tf_2O (1 eq.), 2-F-Py (2 eq.) DCM, 0 °C - rt, 1 h; 2. aq. $NaHCO_3$, 85% → 3

Experimental Procedure (from *J. Am. Chem. Soc.*, **2016**, *138*, 8348, Supporting Information)

General Procedure for α-Amination of Amides. To a mixture of amide (0.3 mmol) and 2-fluoropyridine (0.6 mmol, 2 equiv, 58.3 mg, 51.6 μL) in DCM (1 mL), triflic anhydride was added dropwise (0.6 mmol, 2 equiv, 84 mg, 51 μL) at 0 °C under Ar. The mixture was stirred for 15 minutes at this temperature. Then a solution of azide (0.6 mmol, 2 equiv) in 0.5 mL of DCM was added and the mixture was brought to room temperature. N_2 release was observed after addition of the azide. After 30 min, 2 mL of a saturated solution of $NaHCO_3$ was added and the mixture was further stirred for 1 h. The biphasic mixture was then diluted with DCM and washed with a 20 mL of $NaHCO_3$. The combined organic layers were dried over $MgSO_4$ and the solvent removed under reduced pressure. Purification through column chromatography DCM/dimethylamine (DMA) (0–60% DMA) afforded the products.

2-(Phenethylamino)-1-(pyrrolidin-1-yl)butan-1-one (3ba). ^{1}H NMR (400 MHz, $CDCl_3$) δ = 7.30–7.26 (m, 2H), 7.22–7.19 (m, 3H), 3.57–3.39 (m, 4H), 3.31 (t, J = 6.5 Hz, 1H), 2.89–2.80 (m, 2H), 2.78–2.73 (m, 1H), 2.29–2.65 (m, 1H), 1.93 (dd, J = 12.9 and 6.5 Hz, 2H), 1.85 (dd, J = 12.9 and 6.4 Hz, 2H), 1.62 (dt, J = 14.8, 7.3 Hz, 2H), 0.94 (t, J = 7.5 Hz, 3H) ppm. ^{13}C NMR (100 MHz, $CDCl_3$) δ = 172.2, 140.1, 128.9, 128.5, 126.3, 61.5, 50.1, 46.4, 45.9, 37.0, 26.6, 26.3, 24.3, 10.5 ppm. HRMS (ESI) m/z calculated for $[M + H^+]$ = 261.1961, found 261.1967. Attenuated total reflectance (ATR)-FTIR (cm^{-1}): 3502, 3297, 3026, 2875, 1633, 1428, 1340, 1226, 1150, 1031, 751, 700.

The Mechanism

The amide **1**, in its enol form, is first transformed into the triflate **A**. Following this, the triflate group (TfO) is substituted with 2-fluoropyridine, yielding the reactive intermediate **B**. Notably, density functional theory calculations indicate the participation of a distinct keteniminium species **C** in this process. This intermediate **C** undergoes nucleophilic attack by the azide **2**, leading to the formation of intermediate **D**. Subsequently, a molecule of N_2 is eliminated, resulting in the aziridinium intermediate **E**. Finally, an aqueous workup step liberates the product **3**.

In Maulide's paper, one special case was reported. Treatment of mixed adipic acid ester/amide **4** with azide **2** led to the bicyclic product **5**.

The mechanism is shown as follows.

Could you predict the products of the reaction below?

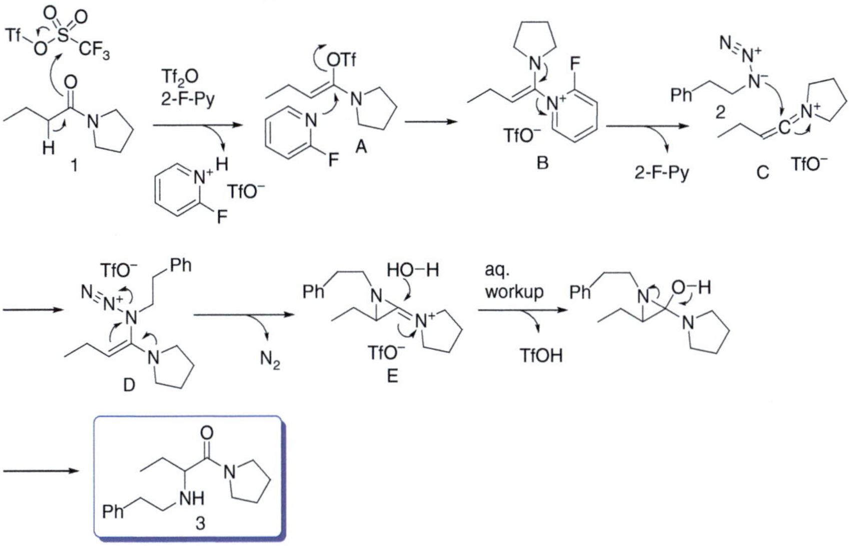
Tf₂O
2-F-Py
A
B
C
D
E
TfO⁻
2-F-Py
N₂
aq.
workup
TfOH
1
2
3

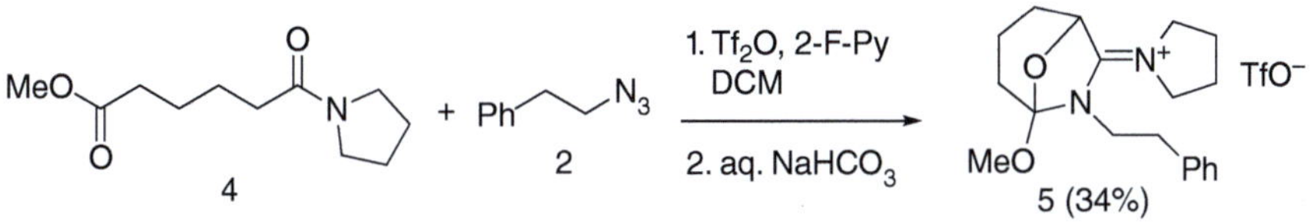
1. Tf₂O, 2-F-Py
DCM
2. aq. NaHCO₃
4
2
5 (34%)

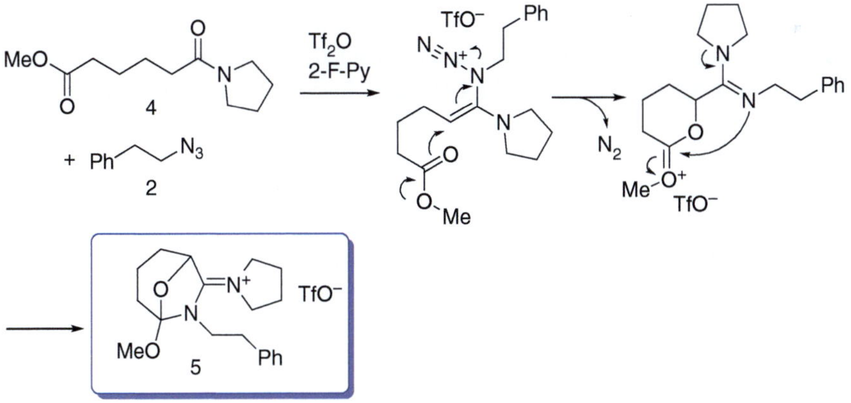
Tf₂O
2-F-Py
N₂
4
2
5

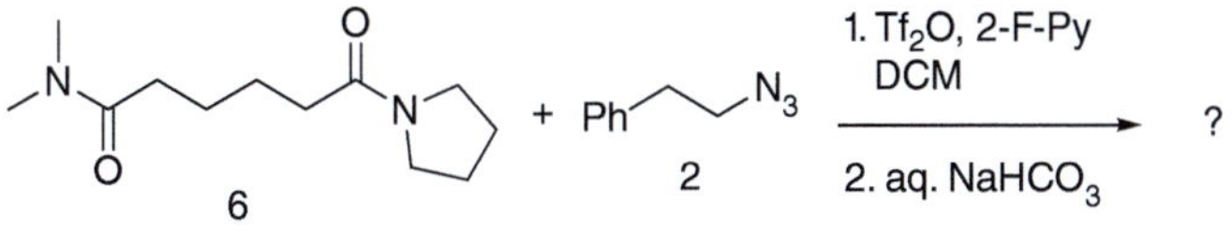
1. Tf₂O, 2-F-Py
DCM
2. aq. NaHCO₃
6
2
?

Problem 9: α-Oxidation of Amides

In 2017, Maulide's group reported an α-oxidation of amide by using 2,6-lutidine *N*-oxide (LNO) in the presence of Tf_2O [26]. The method was applied to the synthesis of the antitumor active compound **5** [26, 27].

1. Tf_2O (1.1 eq.), LNO (2.2 eq.) DCM, MS 3A, 0 - 23 °C, 3 h
2. NaOH (1 M, 3 eq.), 23 °C, 30 min

1 → 2, 37–78%

3 → 4, 62%

Steps → 5, antitumor active

Please provide a mechanism for this oxidation.

Experimental Procedure (from *J. Am. Chem. Soc.*, **2017**, *139*, 6758, Supporting Information)

***α*-Ketoamides from LNO-adduct.** To a solution of the amide (1 equiv) and LNO (2.2 equiv) in DCM (0.1 M) with activated molecular sieves (3 Å) at 0 °C was added Tf_2O (1.1 equiv) and the resulting mixture was allowed to warm to room temperature over the course of 3 h. After this, the molecular sieves were removed by filtration and an aqueous solution of sodium hydroxide (1 M, 3.0 equiv) was added. The biphasic mixture was vigorously stirred at room temperature for 30 min, after which saturated aqueous ammonium chloride was added and the phases were separated. The aqueous phase was extracted with DCM and the combined organic phases were dried over anhydrous sodium sulfate. The dried solution was filtered and the filtrate was concentrated under reduced pressure on a rotary evaporator. The crude residue was purified by flash column chromatography on silica gel (heptane/ethyl acetate) to afford the title compounds.

Methyl 8,9-dioxo-9-(pyrrolidin-1-yl)nonanoate (2g). (63%). ^{1}H-NMR (400 MHz, $CDCl_3$): δ 3.66 (s, 3H), 3.59 (t, J = 6.8 Hz, 2H), 3.51 (t, J = 6.8 Hz, 2H), 2.83 (t, J = 7.4 Hz, 2H), 2.30 (t, J = 7.5 Hz, 2H), 1.98–1.83 (m, 4H), 1.67–1.57 (m, 4H), 1.39–1.30 (m, 4H); ^{13}C-NMR (100 MHz, $CDCl_3$): δ 201.1, 174.3, 163.2, 51.6, 47.4, 46.4, 39.3, 34.1, 29.0, 28.9, 26.5, 24.9, 23.8, 23.0; IR (neat) ν_{max}: 2923, 2855, 1736, 1716, 1635, 1438, 1366, 1200, 1171; HRMS (ESI+): exact mass calculated for $[M + Na]^+$ ($C_{14}H_{23}NO_4Na$) requires m/z 292.1519, found m/z 292.1515.

The Mechanism

The dipolar form of the amide is triflated to give the *O*-triflate as the salt **A**. Due to the excellent leaving ability of the TfO group, it is substituted by LNO to generate the intermediate **B**, which could be isomerized to the enol form. The second LNO adds to the α-position, with elimination of 2,6-lutidine to yield **C**. Deprotonation of the α-H at basic condition provides **D**, and finally elimination of another 2,6-lutidine affords the α-ketoamide **2**.

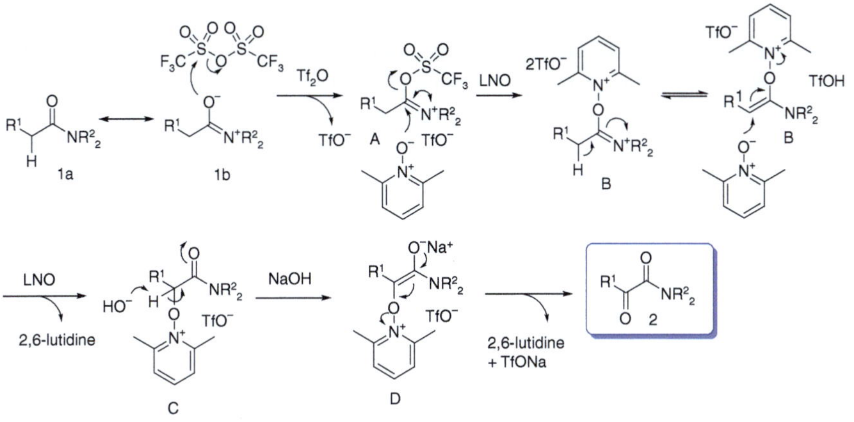

Problem 10: Tsunoda Reagent for Mitsunobu Reaction

In 1994, Tsunoda's group reported 2-(tributylphosphoranylidene)acetonitrile or cyanomethylenetributylphosphorane (CMBP, **1**) as a novel reagent for Mitsunobu reaction. This reagent works better than normal azodicarboxylates such as DEAD and diisopropyl azodicarboxylate when nucleophiles have a weak acidity. For example, benzylation of *N*-benzyl-2,2,2-trifluoroacetamide (**3**, pK_a 13.6) by benzyl alcohol (**2**) under the normal Mitsunobu condition provided the desired product **3** in only 3% yield. Whereas, a good yield of **4** (77%) was reached by using CMBP [28].

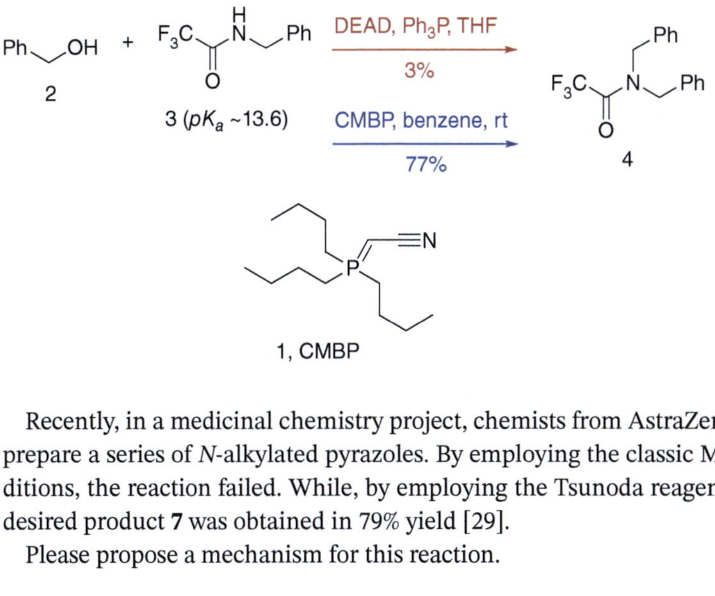

Recently, in a medicinal chemistry project, chemists from AstraZeneca needed to prepare a series of *N*-alkylated pyrazoles. By employing the classic Mitsunobu conditions, the reaction failed. While, by employing the Tsunoda reagent (CMBP), the desired product **7** was obtained in 79% yield [29].

Please propose a mechanism for this reaction.

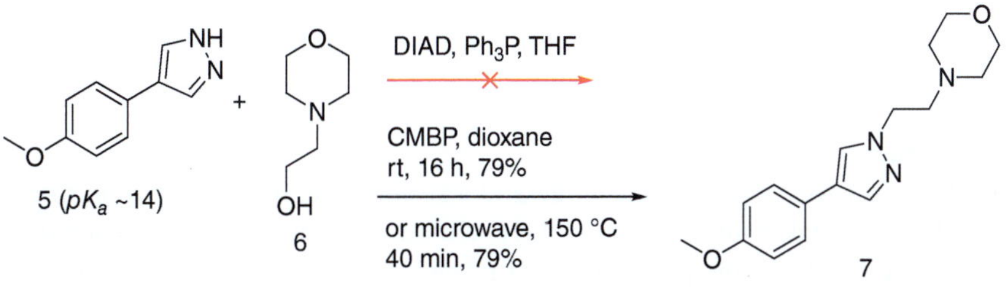

Experimental Procedure (from *Tetrahedron Lett.*, **1994**, *35*, 5081)
In a typical experiment, an alcohol (1 mmol) and HA (the nucleophile, 1.5 mmol) were successively dissolved in dry benzene (5 mL) with stirring under an argon atmosphere, and CMBP (1.5 mmol) was added all at once, using a syringe. The reaction mixture was heated at 100 °C for 24 h with stirring in a sealed tube. The product was purified by silica gel column chromatography after evaporation of the solvent in vacuum.

(from *Tetrahedron Lett.*, **2018**, *59*, 1708)

CMBP (2 eq.)
dioxane, microwave
150 °C, 30 min

Pd(dppf)Cl_2
K_2CO_3, dioxane-H_2O
120 °C, 20 min

69%, 2 steps

Typical experimental procedure: (1) (Cyanomethylene)tributylphosphorane (262 μL, 1.00 mmol) was added to a solution of (tetrahydrofuran-3-yl)methanol (51 mg, 0.50 mmol) and 4-(4,4,5,5-tetramethyl-1,3,2-dioxaborolan-2-yl)-1H-pyrazole (97 mg, 0.5 mmol) in degassed 1,4-dioxane (2 mL) sealed in a microwave tube at rt under nitrogen. The solution was heated to 150 °C for 30 min in the microwave reactor and cooled to rt. 2) 1-Bromo-4-methoxybenzene (94 mg, 0.50 mmol), potassium carbonate (207 mg, 1.50 mmol), and [1,1-Bis(diphenylphosphino) ferrocene]dichloropalladium(II) complex with CH_2Cl_2 (40.8 mg, 0.05 mmol) were added to the solution. The tube was sealed, evacuated, and backfilled with nitrogen. Degassed water (1 mL) was added under nitrogen. The resulting mixture was stirred at 120 °C for 20 min. The reaction mixture was diluted with EtOAc (25 mL) and water (15 mL), the layers were separated, and the aqueous layer was extracted with EtOAc (15 mL). The combined organic layers were washed with saturated brine (15 mL). The organic layer was dried with $MgSO_4$, filtered, and evaporated to afford the crude product. The crude product was purified by preparative HPLC (Waters XSelect CSH C18 ODB column, 5 m silica, 30 mm diameter, 100 mm length), using decreasingly polar mixtures of water (containing 1% by volume NH_4OH (28–30% in H_2O)) and MeCN as eluents to afford 4-(4-methoxyphenyl)-1-((tetrahydrofuran-3-yl)methyl)-1H-pyrazole (89 mg, 69%) as a beige solid.

The Mechanism
The CH in CMBP is a good nucleophile and a strong base. It gains a proton from even relatively weaker nucleophilic acid (H—Nu) to form the ions pair **A**. Then the alcohol oxygen attacks the phosphine and releases acetonitrile. The deprotonated

nucleophile (Nu^-) attacks the alcoholic carbon to provide the product and tributylphosphine oxide.

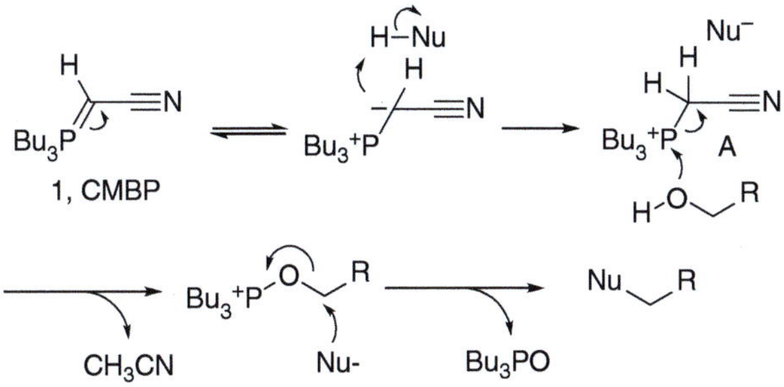

Problem 11: Preparation of Arylacetonitriles from Arylbromides

In 2011, Velcicky et al. reported a preparation of arylacetonitriles (**3**) from arylbromides (**1**) through Suzuki coupling and base-induced fragmentation as shown below [30]. The arylacetonitrile unit is found as a structural motif in several important drugs, including verapamil, isoaminile, anastrozole, and cilomilast [31].

Could you please provide a mechanism for the reaction without checking the reference?

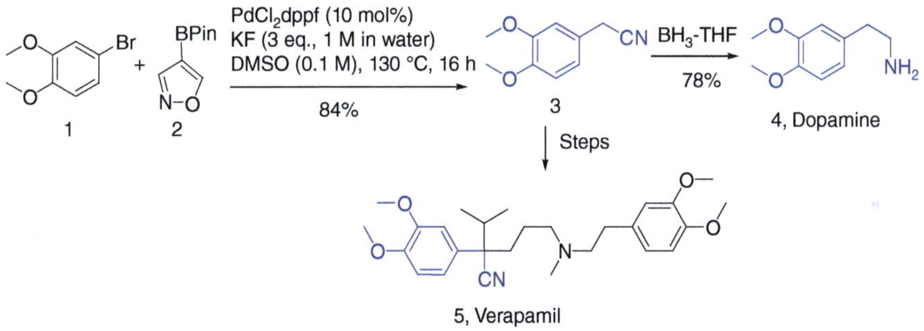

Experimental Procedure (from *J. Am.Chem. Soc.*, **2011**, *133*, 6948, Supporting Information)

Representative Procedure for the Cyanomethylation of Aryl Halides: (2-Naphthalen-2-yl)acetonitrile (**3**) according to Table 1, entry 25). A silicon/PTFE screw-capped vial containing boronate **1** (40.0 mg, 0.205 mmol, 1.2 equiv), 2-bromonaphthalene **2** (35.0 mg, 0.171 mmol, 1eq) in dimethyl sulfoxide (DMSO) (1.21 mL, 0.017 mmol, 0.1 M) was charged with potassium fluoride (30.0 mg, 0.513 mmol, 3 equiv) and pure water (0.51 mL, 0.513 mmol, 3 equiv). After 5 min degassing in an ultrasonic bath under argon atmosphere, $PdCl_2dppf$ (14.0 mg, 0.017 mmol, 10 mol%) was added. The vial was sealed and flushed with argon once again. After stirring in an aluminum block at 130 °C for 16 h, the crude reaction mixture was cooled to room temperature and filtered over Celite. NaCl (saturated solution, 25 mL) was added to the dark solution. After extraction with EtOAc (5 × 25 mL), the organic phase was dried (Na_2SO_4) and concentrated under reduced pressure. The residue was purified by flash chromatography on silica gel (EtOAc/CycHex = 1:9) to provide compound **3** (24.4 mg, 88%). The characteristic data of **3** were identical with those described below.

Beige solid, $R_f = 0.49$ (EtOAc/CycHex = 1:2); ^{1}HNMR: δ_H (300 MHz, $CDCl_3$) = 3.92 (s, 2H), 7.38 (dd, $J = 8.4$ Hz, $J = 1.5$ Hz, 1H), 7.49 – 7.56 (m, 2H), 7.82 – 7.87 (m); ^{13}CNMR: δ_C (75 MHz, $CDCl_3$) = 23.9, 118.0, 125.5, 126.5, 126.8, 126.9, 127.2, 127.7, 127.8, 129.1, 132.7, 133.4; IR-ATR: v_{max} (film) = 2250 (m, -CN), 1597 (m), 1507 (m), 1408 (s), 1365 (m), 960 (m), 927 (m), 861 (s), 825 (s), 755 (s); MS (EI, 70 eV): *m/z* (%) = 167 (100) [M^+], 139 (49), 126 (8), 115 (17), 99 (3), 83 (10), 74 (14), 63 (16), 51 (9), 39 (8); HRMS (EI, 70 eV): determined mass: 167.073 u, exact mass: 167.0735 u.

The Mechanism

The reaction begins from a Suzuki coupling between **1** and **2** to give **A**. Then the isoxazole ring is opened by F^- to release the aldehyde **B**. Retro-aldol reaction of **B** provides the product **3**.

Problem 12: Transformation of Aromatic Bromides into Aromatic Nitriles

Aromatic nitrile subunit exists in many drugs, such as citalopram hydrobromide, periciazine, fadrozole, letrozole, bicalutamide and etravirine [32]. In 2011, Togo's group reported a one-pot preparation of aromatic nitriles (**2**) from aromatic bromides (**1**) by the following reaction [33].

Could you please provide a mechanism for the reaction without checking the reference?

Experimental Procedure (from *Tetrahedron Lett.*, **2011**, *52*, 2404)

Typical Experimental Procedure for the Transformation of Aromatic Bromides into Aromatic Nitriles: To a flask containing Mg turnings (0.28 g, 14 mmol) was added *p*-bromotoluene (1.38 g, 8.0 mmol) in THF (8 mL) at room temperature. After stirring for 2 h, DMF (1.3 mL, 12 mmol) was added to the reaction mixture. The obtained mixture was stirred for 2 h at room temperature. Then, aq NH_3 (7 mL, 28–30%) and I_2 (4.06 g, 1.6 mmol) were added to the reaction mixture. After stirring overnight, the reaction mixture was poured into aq sat. Na_2SO_3 solution and extracted with $CHCl_3$ (3 × 30 mL). The organic layer was dried over

Na_2SO_4 and filtered. After the solvent was removed, the residue was purified by short column chromatography on silica gel (eluent: hexane /ethyl acetate = 9:1, v/v) to provide pure *p*-tolunitrile (0.77 g) in 67% yield. Most aromatic nitriles mentioned in this work are commercially available and were identified by comparison with the authentic samples.

4-Methylbenzonitrile: Mp 26–28 °C (commercial, Mp 26–28 °C); IR: 2227 cm^{-1}; ^1H NMR ($CDCl_3$: 500 MHz) $\delta = 2.41$ (s, 3H), 7.26 (d, $J = 8.1$ Hz, 2H), 7.52 (d, $J = 8.1$ Hz, 2H); ^{13}C NMR (125 MHz, $CDCl_3$) $\delta = 21.7$, 109.2, 119.1, 129.7, 131.9, 143.6.

The Mechanism

The first stage is the formylation of the Grignard reagent with DMF to give the aldehyde **A**. Then addition of ammonia to the aldehyde, after elimination of H_2O, provides imine **B**. Finally, the imine **B** is oxidized by I_2 to yield the nitrile product **2**.

Br Mg THF MgBr N OMgBr NMe2 H2O HOMgBr O H NMe2 1
HNMe2 O NH3 A O H NH H H2O H N H I–I B
HI H N I C HI N 2

Problem 13: Preparation of Nitriles from Carboxylic Acid by Smiles Rearrangement

In the pursuit of developing novel ionophores capable of effectively and selectively complexing and separating metal cations, Huber and Bartsch synthesized a series of lariat ethers **2**, where X represents an alkyl, perfluoroalkyl, or aryl group [34]. The synthesis began with the conversion of carboxylic acid **1** into its corresponding acyl chloride using oxalyl chloride. This acyl chloride was then reacted with a sulfonamide, which had been pre-treated with potassium hydride (KH) in THF, to yield the desired *N*-sulfonyl carboxamide [35].

Interestingly, when 2,4-dinitrobenzenesulfonamide (DNBSA) was used as the sulfonamide, the expected lariat ether **3** was not obtained. However, when trimethylamine was employed as the base instead, lariat ether **3** was produced, albeit in low yield (<25%), while the corresponding nitrile **4** was isolated in 52% yield. This unexpected outcome led to the development of a new method for synthesizing nitriles directly from carboxylic acids using DNBSA.

4-Methoxybenzonitrile (**5**) was prepared in 76% yield by using this method, which is an important starting material for synthesis of a series of ROCK1 inhibitors, such as **6** [35].

Please provide a mechanism for the preparation of nitriles from carboxylic acids.

1. $(COCl)_2$, benzene rt, 6 h
2. H_2NSO_2X, KH, THF rt, 3–12 h
2. or DNBSA, TEA THF, reflux

1 → 2, X = CF_3; 3, X = 2,4-$(NO_2)_2C_6H_3$ → 4

DNBSA; THF, TEA reflux, 24 h; 76% → 5 → Steps → 6

Experimental Procedure (from *Tetrahedron*, **1998**, *54*, 9281)

General Procedure for Conversion of Carboxylic Acids into the Corresponding Nitriles. Oxalyl chloride (0.6 mL, 6.20 mmol) was added to a mixture of carboxylic acid (1.0 mmol) in benzene (25 mL). The mixture was stirred at room temperature under nitrogen for 6 h and evaporated *in vacuo*. The acid chloride was dissolved in THF (15 mL) and a solution of DNBSA (0.30 g, 1.20 mmol) and TEA (1.4 mL, 16 mmol) was added in dry THF (10 mL). The resulting orange mixture was refluxed under nitrogen for 24 h. EtOAc (50 mL) was added and the resulting solution was washed with 1 N HCl (2 × 25 mL) and aq. 10% K_2CO_3 (2 × 25 mL). The dark organic solution was dried over $MgSO_4$ and evaporated *in vacuo* to give a red residue which was purified by column chromatography.

The Mechanism

This is a special Smiles rearrangement. Acylation of sulfonamide **DNBSA** with the acid chloride provides the sulfonylbenzamide **A** in which the hydrogen on the NH is deprotonated to form anion **B**. Due to the strong electron-withdrawing property of the dinitro group, anion **B** is readily cyclized to the spirobicyclic intermediate **C**. The five-membered ring in **C** is opened at basic conditions to produce the sulfenate **D**. Final α-elimination of the sulfenate **D** affords products 4-methoxybenzonitrile (**5**), SO_2, and 2,4-dinitrophenol (next page).

Problem 14: Synthesis of Disubstituted Malononitriles

Arylmalononitriles serve as crucial intermediates in the synthesis of JAK2 inhibitors [36]. Additionally, disubstituted malononitriles have been utilized as antiparasitic agents in mammals [37]. Mills and Rousseaux developed an efficient one-pot synthesis method for both arylmalononitriles and disubstituted malononitriles [38]. In this process, primary nitriles are first treated with a Grignard reagent, followed by the addition of dimethylmalononitrile (DMMN). After an aqueous workup, monosubstituted malononitriles are obtained in high yields. If an electrophile is introduced before the aqueous workup, the reaction yields disubstituted malononitriles as the final product.

Please provide mechanism for those transformations.

MeMgBr (1.1 eq.)
LiCl (1.1 eq.)
THF, rt, 30 min
then DMMN (1.1 eq.)
80 °C, 6 h

2 (99%)

DMMN

MeMgBr (1.1 eq.)
LiCl (1.1 eq.)
THF, rt, 30 min
then DMMN (1.1 eq.)
80 °C, 6 h

$PhCH_2Br$ (1.2 eq.)
THF/DMF (1:1)
80 °C, 16 h

3 (99%)

Experimental Procedure (from *Tetrahedron*, **2019**, *75*, 4298)

General Procedure A: Synthesis of malononitriles from primary nitriles using MeMgBr Step 1: To an 8-mL culture tube with stir bar was added lithium chloride (1.1 equiv) and the flask was flame-dried under vacuum and cooled under an atmosphere of N_2. THF (1.0 M) was added, followed by the appropriate nitrile (1.0 equiv). While stirring, methylmagnesium bromide (1.1 equiv of a solution in Et_2O) was added dropwise at room temperature, and the solution was stirred at room temperature for 30 min under an atmosphere of N_2. A 1.1 M stock solution of DMMN or dibenzylmalononitrile in THF was prepared and added at room temperature (1.1 equiv of a 1.1 M solution in THF, reaction volume ¼ 0.50 M with respect to nitrile starting material). The reaction was stirred at 80 °C for 6 h under an atmosphere of N_2. Step 2: The reaction was cooled to room temperature and DMF was added to bring the reaction solvent to a 1:1 THF/DMF ratio (0.25M of a 1:1 THF/DMF mixture with respect to nitrile starting material). The desired electrophile (1.2 equiv) was added in a single portion. If the electrophile was a solid, it was added with the DMF as a 0.60M stock solution. The reaction was stirred at 80 °C for 16 h, or until complete conversion was achieved as judged by TLC. The reaction was cooled to room temperature, opened to air, quenched with 1M aq. HCl, and extracted with EtOAc (×3). The organic fractions were combined, dried over $MgSO_4$, and concentrated. The crude residue was purified by flash column chromatography to yield the desired malononitrile.

2-Benzyl-2-phenylmalononitrile (2d). According to **General Procedure A**, **2d** was prepared using the following amounts of reagent: Step 1: lithium chloride (19 mg, 0.44 mmol, 1.1 equiv), THF (0.40 mL), benzyl cyanide (46 mL, 0.40 mmol, 1.0 equiv), methylmagnesium bromide (0.21 mL of a 2.1 M solution in Et_2O, 0.44 mmol, 1.1 equiv), and DMMN (0.40 mL of a 1.1 M solution in THF, 0.44 mmol, 1.1 equiv). Step 2: DMF (0.80 mL; total reaction solvent ¼ 1.6 mL of a 1:1 DMF/THF mixture, 0.25 M) and benzyl bromide (57 mL, 0.48 mmol, 1.2 equiv). The crude residue was purified by flash column chromatography (gradient of 0–15% EtOAc/hexanes) to yield **2d** as a white solid (93 mg, 0.40 mmol, 99%). Analytical data: ^{1}H NMR (400 MHz, $CDCl_3$, 298 K): δ_H 7.52–7.43 (m, 5H), 7.39–7.26 (m, 3H), 7.17–7.10 (m, 2H), 3.46 (s, 2H) ppm; ^{13}C NMR (100 MHz, $CDCl_3$, 298 K): δ_C 131.7, 131.6, 130.6, 130.1, 129.7, 129.0, 128.8, 126.3, 114.8, 48.7, 44.2 ppm; R_f (9:1 hexanes/EtOAc): 0.41.

The Mechanism

Due to the acidity of the benzylic proton in 2-phenzyl acetonitrile **1**, it is deprotonated by methyl magnesium bromide and then the carbanion adds to the nitrile group in DMMN to form intermediate **A**. After eliminative transnitrilation, **A** would be cleaved to 2-phenylmalonitrile (**2**) and Grignard species **B**. As **2** is even more acidic than **1**, anion exchange would quickly take place between **2** and **B**, to give isobutyronitrile (**C**) and tertiary Grignard species **D**. Protonation of **D** at acidic aqueous condition will provide the product **2**. Adding an electrophile such as benzyl bromide before the aqueous workup leads to the coupling product **3**.

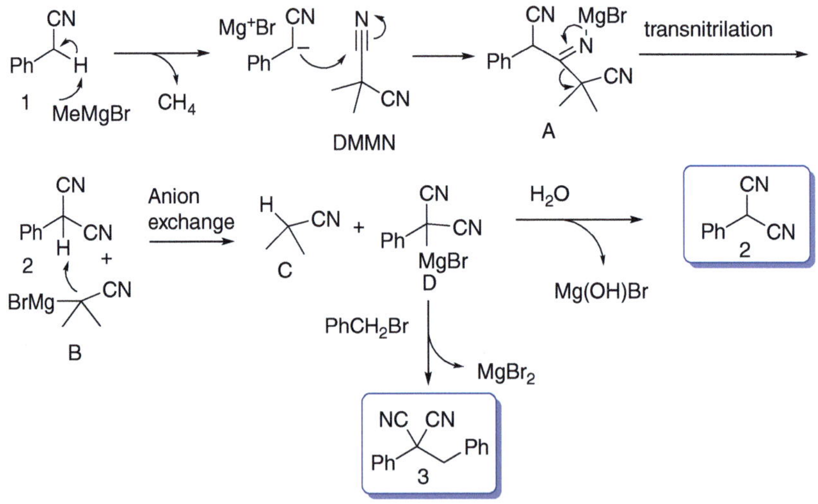

Problem 15: Preparation of 2-Arylacetonitriles from Aromatic Ketones

2-(3,4-Dimethoxyphenyl)-3-methylbutanenitrile (**2**) is a key intermediate for synthesizing verapamil, a calcium channel blocker medication. The preparation of **2** from aromatic ketone **1** was described in a patent by a telescoped process as shown below [39].

Could you please propose a stepwise mechanism for the reactions?

1. *t*-BuOK, toluene, −10 °C
$ClCH_2CO_2Et$, rt, 2 h
2. KOH, MeOH, rt, 6 h
3. HOAc
4. $NH_2OH.HCl$, 60 °C, 1 h
5. HOAc, Ac_2O, NaOAc
75 °C, 6 h

1 → 2, 81% → Steps → Verapamil

The Experimental Procedure (from US 5,097,058, **1992**)

EXAMPLE 6. α-(1-Methylethyl)-3,4-dimethoxybenzeneacetonitrile. 26 g (0.125 moles) of isobutyryl-3,4-dimethoxybenzene are put into 100 mL of toluene and then 35 g (0.31 moles) of potassium *tert*-butoxide are added. After cooling to −10 °C, the reaction mixture is added with 33.3 mL (0.31 moles) of ethyl chloroacetate in about 2 h. Subsequently, the temperature is raised to the value of the room temperature and after another 2 h, the reaction mixture is added with a solution containing 22 g (0.35 moles) of 90% potassium hydroxide (KOH) in 100 mL of methanol and is kept under stirring for 6 h. The reaction mixture is then added with 10 mL of acetic acid, and subsequently with 8.68 g (0.125 moles) of hydroxylamine hydrochloride dissolved in 20 mL of water. The reaction mixture is heated to 60 °C for 1 h, then it is acidified to pH 4 by means of concentrated aqueous hydrochloric acid, kept another hour at 60 °C and finally added with water until complete dissolution of the undissolved salts. The layers are separated, the aqueous phase is extracted three times with 50 mL of toluene and then is discarded. The organic phase, together with the toluene extracts, is washed with water, dried over anhydrous sodium sulfate and evaporated to dryness. The oily residue is dissolved in 80 mL of glacial acetic acid and added, under nitrogen atmosphere, with 19.5 mL (0.20 moles) of acetic anhydride. After about 30 min, the reaction mixture is added with 10 g (0.122 moles) of anhydrous sodium acetate and is heated to 75 °C for 6 hours. The acetic acid is evaporated under vacuum and the residue is treated with a mixture of 100 mL of water and 100 mL of toluene. The mixture is brought to pH 9 by means of a 30% (w/w) aqueous solution of sodium hydroxide, then the layers are separated and the aqueous phase is twice extracted with 75 mL of toluene and then is discarded. The organic phase is added with the toluene extracts and then is twice washed with 100 mL of water, dried over anhydrous sodium sulfate, filtered and evaporated to dryness under vacuum. The obtained oily residue is distilled under vacuum obtaining 22.3 g of pure nitrile with a yield of 81.4% of the theoretical.

The Mechanism

Step 1. The first step should be a Darzens reaction to form an epoxide **A**.

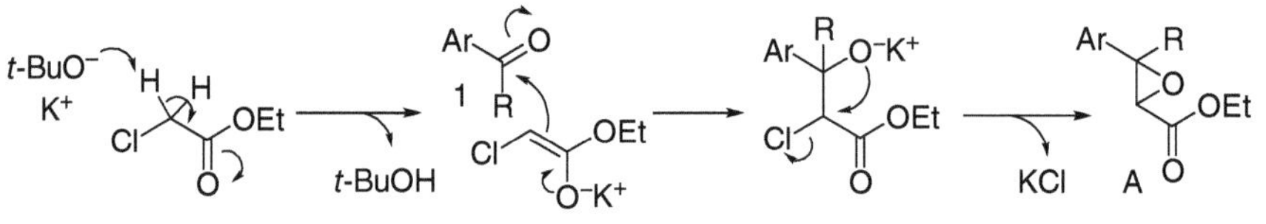

Step 2. The second step should be the hydrolysis of the ester, then decarboxylation followed by rearrangement to give the aldehyde **B**.

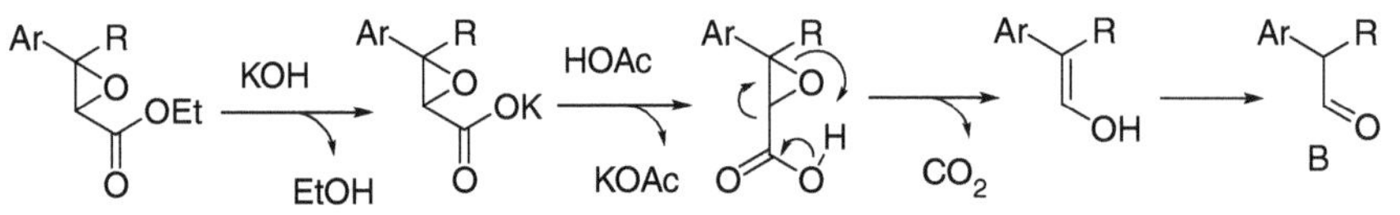

Step 3. Reaction of hydroxylamine with aldehyde **B** provides oxime **C**.

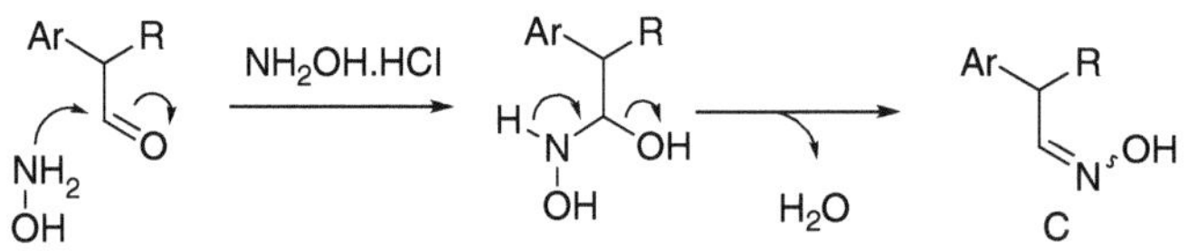

Step 4. The final step is a dehydration of the oxime, assisted by Ac_2O, to produce nitrile **2**.

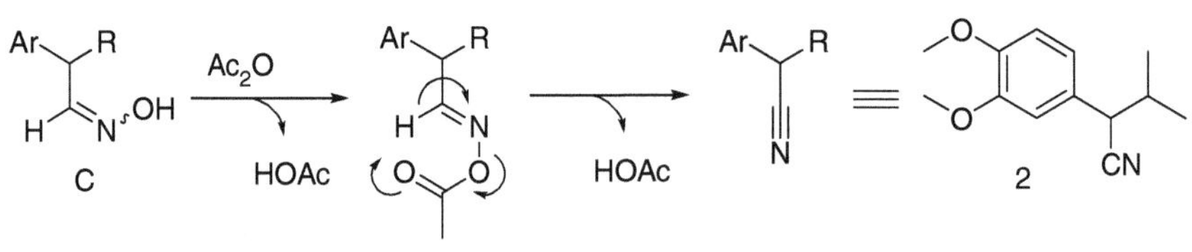

By using the Van Leusen reaction, the arylacetonitrile **4** could be prepared in a single step [40].

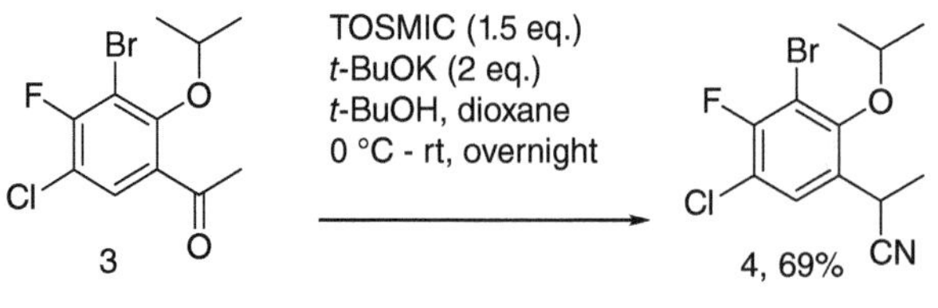

Problem 16: Synthesis of Aminophosphonate by Birum–Oleksyszyn Reaction

UAMC-00050 (**1**) was recently developed at the University of Antwerp as a new treatment of dry eye disease [41]. The core aminophosphonate structure in the key intermediate **2** was assembled by the one-pot three-component Birum–Oleksyszyn reaction [42] from aldehyde **3**, benzyl carbamate **4**, and phosphite **5**, using yttrium triflate as the catalyst. [41] Please suggest a mechanism for this reaction.

O NHBoc 3 + Cbz H_2N 4 + NHAc O)₃P 5 — $Y(OTf)_3$ MeCN -THF TFAA, rt, 44% → AcHN, NHAc, O–P(=O)–O, NHCbz, BocHN 2 — 2 steps → AcHN, NHAc, O–P(=O)–O, NHCbz, NH, H_2N, N H, .HCl 1 (UAMC-00050)

The Experimental Procedure (from *Org. Process Res. Dev.*, **2022**, *26*, 2937)

***tert*-Butyl(4-(2-(((benzyloxy)carbonyl)amino)2(bis-(4acetamidophenoxy) phosphoryl)ethyl)phenyl) carbamate (2).** To a 500 mL flask equipped with a magnetic stirrer compound **3** (10.0 g, 0.043 mol, 1.0 equiv), yttrium triflate (2.30 g, 4.3 mmol, 0.1 equiv) dissolved in dry-MeCN (130.0 mL) (water content <0.001%), benzyl carbamate (**4**) (6.50 g, 0.043 mol, 1.0 equiv), tris(4- acetamidophenyl) phosphite (**5**) (23.00 g, 0.043 mol, 1.0 equiv), dry THF (130.0 mL) (water content <0.005%), and trifluoroacetic anhydride (5.98 mL, 0.043 mol, 1.0 equiv) were added under argon. The mixture was stirred for 4 h at 20–25 °C. The solvent was removed, and the residue was dissolved in a solution of ethyl acetate/ethanol (4:1 v/v) (500 mL). The organic phase was washed with NaOH 0.5 M (4 × 500 mL) and brine (500 mL). The organic layers were collected together and dried on Na_2SO_4 (300 g), and the solvent was removed with vacuum. A silica pad with silica gel (200 g) was packed in a 500-mL glass filter. The residue was dissolved in ethyl acetate/ethanol (3:1 v/v), celite (20 g) was added, and the solvent was removed. The solid mixture was placed on top of the filter and washed with ethyl acetate/heptane (2:1 v/v) (2000 mL) to collect fraction 1, the collection flask was changed, the silica pad was washed with ethyl acetate/ethanol (3:1 v/v) (1000 mL) to collect fraction 2, and the solvent was removed from fraction 2 to give a yellow foamy solid. The crude product was dissolved in acetone (100 mL), and a solution of $NaHCO_3$ 0.5% (200 mL) was added dropwise. The solid was filtered, washed with methyl *tert*-butyl ether (MTBE) (100 mL), and dried in vacuum overnight. The dry solid was suspended in ethyl acetate/acetone (19:1 v/v) (300 mL), stirred for 24 h, filtered, washed with ethyl acetate (100 mL), and dried in vacuum (5 mbar) overnight to get a white solid with a yield of 44%. 98.2% AN by HPLC. HRMS (ESI+): *m/z* calculated for $C_{37}H_{41}N_4O_9PNa$ $[M + Na]^+$: 739.2509, found 739.2519. ^{1}HNMR (400 MHz, DMSO-d_6) δ: 10.00 (s, 2H), 9.32 (s, 1H), 8.11 (d, J = 8 Hz, 1H), 7.57 (m, 4H), 7.39 (d, J = 8 Hz, 2H,), 7.30 (m, 3H), 7.19 (d, J = 8 Hz, 2H), 7.12 (m, 6H), 4.97 (dd, J = 32, 12 Hz, 2H), 4.42 (q, J = 12 Hz, 1H), 3.18 (d, J = 12 Hz, 1H),

2.90 (m, 1H), 2.04 (s, 6H), 1.49 (s, 9H). ^{13}CNMR: (101 MHz, DMSO-d_6) δ 168.70, 156.39, 153.27, 145.74, 145.46, 138.60, 137.45, 137.01, 130.92, 129.83, 128.71, 128.02, 127.59, 121.29, 121.03, 120.62, 118.36, 79.40, 65.87, 50.51, 34.04, 28.62, 24.37. ^{31}PNMR: (162 MHz, DMSO-d_6) δ: 18.35.

The Mechanism

The reaction mechanism that was first published by Birum in 1974 [42] is illustrated below. The aldehyde is activated by acid and attacked by urea to form adduct **A**. Then a nucleophilic substitution occurs between the trialkyl/triaryl phosphite and **A** to give the intermediate **B**. Final elimination of alcohol/phenol leads to the aminophosphonate product **C**.

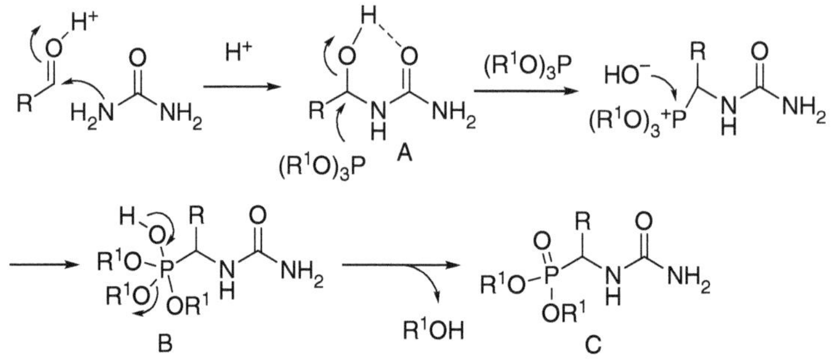

While, in the current case, trifluoroacetic anhydride (TFAA) participates in the catalysis of the reaction. The aldehyde **3** is activated by the Lewis acid $Y(OTf)_3$ and thus attacked by H_2NCbz (**4**) to form an intermediate **D**. The triflic acid is released and trapped by TFAA to give the mixed anhydride **E** and trifluoroacetic acid (TFA). The TFA activates the intermediate **D** and is then attacked by another molecule H_2NCbz (**4**) to provide the biscarbamates **F**. **F** is activated by TFA and is then attacked by phosphite **5** to give an intermediate **G**. Finally, TFA attacks the trifluoroacetyl carbon in **G** to provide the product **2**, simultaneously one molecule of ArOH is eliminated and one molecule of TFAA is released.

Problem 17: Synthesis of Primary Sulfonamide by using *N*-Sulfinyl-*O*-(*tert*-butyl)hydroxylamine

Primary sulfonamide is present in several marketed drugs, such as celecoxib **4**. Willis' group reported a preparation of primary sulfonamide (**2**) by using *N*-sulfinyl-*O*-(*tert*-butyl)hydroxylamine (*t*-BuONSO) (**1**) with Grignard or organolithium reagent. [43] Celecoxib **4** was prepared by this method in 55% yield.

Please provide a mechanism for the reaction.

Experimental Procedure (from *Org. Lett.*, **2020**, *22*, 9495, Supporting Information)

General Procedure A for Primary Sulfonamide Synthesis

O-(*tert*-Butyl)-*N*-sulfinylhydroxylamine (*t*-BuONSO **1**, 40.5 mg, 0.30 mmol, 1.0 equiv.) was dissolved in THF (1.2 mL) and cooled to −78 °C. The corresponding organometallic reagent was added. The reaction was then taken out of the dry-ice-acetone bath and allowed to warm to room temperature and stirred for 18 h. The reaction mixture was purified directly by flash chromatography to afford the desired primary sulfonamide **2**.

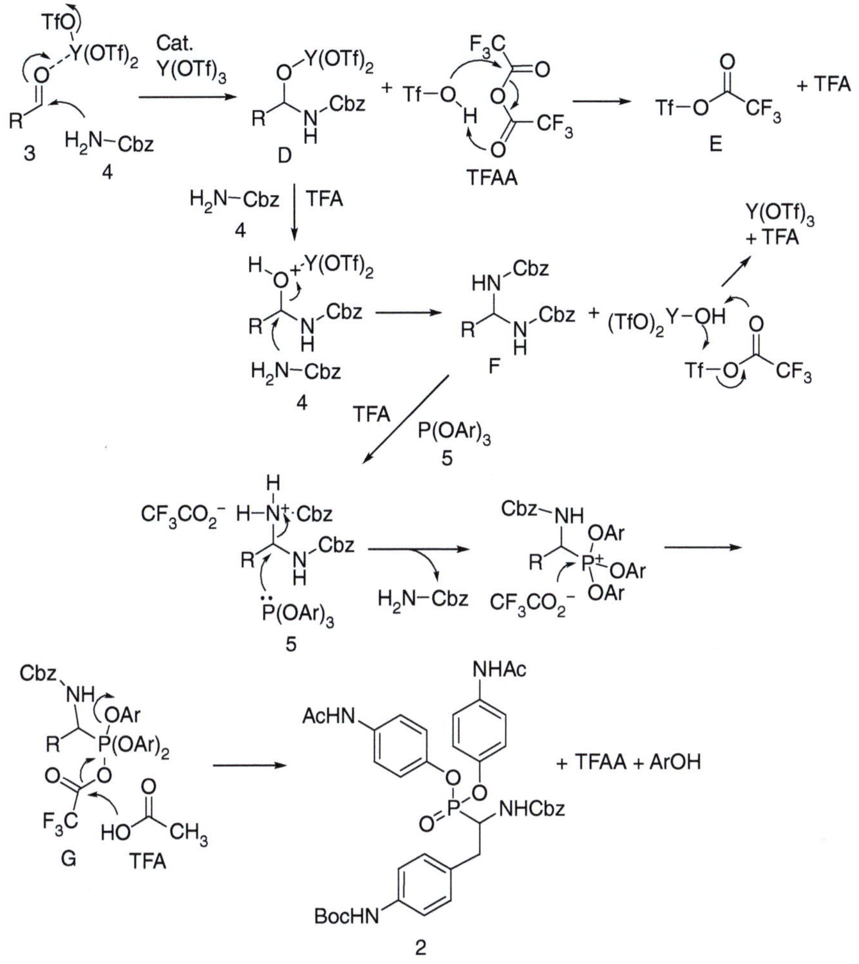

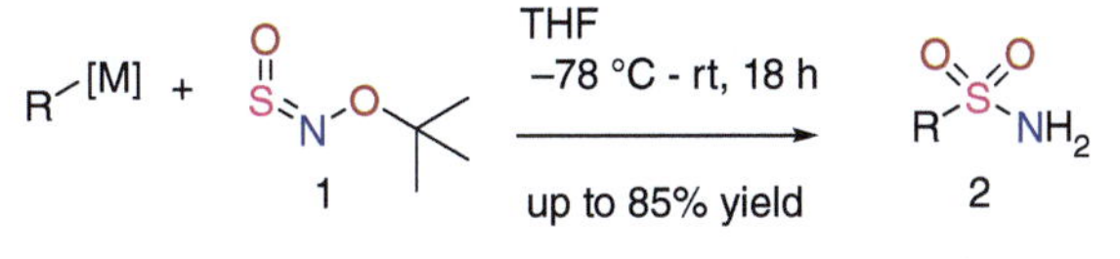

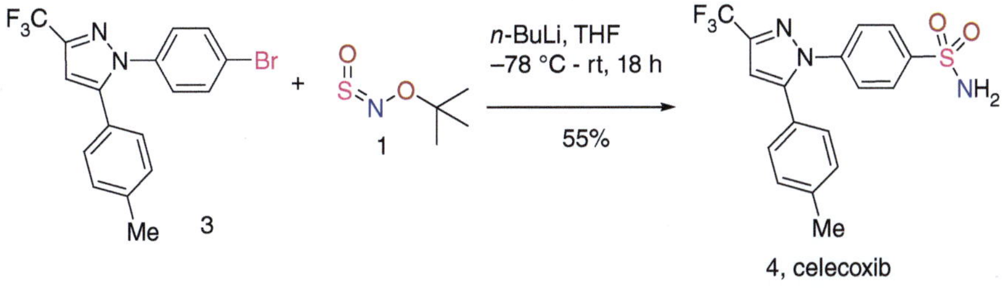

Note: the reaction can be quenched by the addition of isopropanol, or SiO_2 for base-sensitive products. When carried out on a small scale, the reaction could be added directly to the top of a silica gel column. For reactions with larger quantities of solvent, quenching with SiO_2 and evaporation on a rotary evaporator is recommended before addition to the column.

Note: when the reaction was run on gram scale, a pressure build-up was observed. Use of an outlet is therefore recommended when running the reaction on large scale.

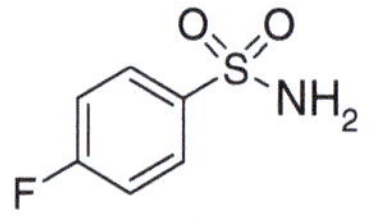

Prepared according to **general procedure A**, using 4-fluorophenylmagnesium bromide (340 μL, 0.30 mmol, 0.88 M in THF, 1.0 equiv). Purification by flash chromatography (SiO_2, heptane/ethyl acetate, gradient 9:1 to 0:1) afforded sulfonamide **2a** as a white solid (42 mg, 80% yield). The data was consistent with the literature.

Note: the same procedure was carried out on 1 mmol scale to give 124 mg of **2a** (71% yield) and on gram scale (1.08 g (8 mmol) of *t*-BuONSO **1**) to give 862 mg **2a** (62% yield). ^{1}H NMR (Acetone-d_6, 400 MHz) 8.01 (dd, 2H, J = 8.5, 5.0 Hz, Ar-*H*), 7.36 (app. t, 2H, J = 8.5 Hz, Ar-*H*), 6.69 (br. s, 2H, NH_2); ^{13}C NMR (Acetone-d_6, 100 MHz) 165.3 (d, $^1J_{CF}$ = 251.0 Hz), 141.3 (d, $^4J_{CF}$ = 3.0 Hz), 129.8 (d, $^3J_{CF}$ = 9.5 Hz), 116.6 (d, $^2J_{CF}$ = 23.0 Hz); ^{19}F NMR (Acetone-d_6, 376 MHz,) −109.1 (app. ddd, J = 14.0, 9.0, and 5.3 Hz, Ar-F); LRMS *m/z* (ESI$^-$) [M − H]$^-$ 174.0.

The Mechanism

The mechanism outlined in the literature [43] suggests that the addition of a metal reagent to *t*-BuONSO (**1**) generates sulfinamide **A**, which subsequently transforms into sulfonimidate ester **B**. An intramolecular proton transfer to the nitrogen atom then occurs, leading to the elimination of isobutene and the formation of sulfonamide anion **C**. Following an aqueous workup, the final product, sulfonamide **2**, is obtained. This proposed mechanism is supported by two key observations: the absence of ^{18}O incorporation when the reaction was quenched with ^{18}O-labeled water, and the detection of ^{1}H NMR signals corresponding to isobutene in an aliquot of the crude reaction mixture.

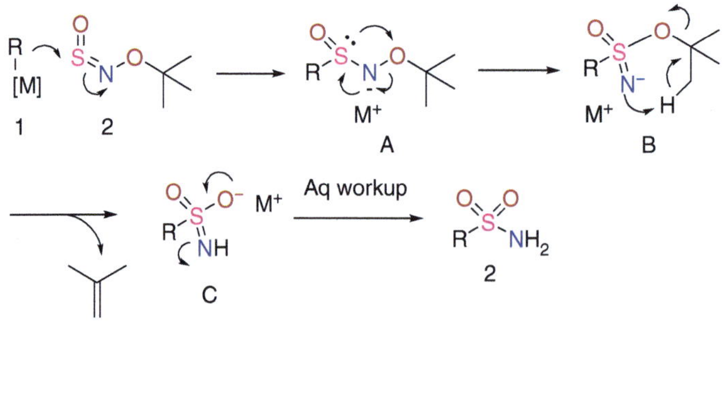

4.2 Formation of Hydrazine Derivatives

Problem 18: Preparation of Hydrazines from Amines by Using Oxaziridines
Hydrazine subunit is an important building block in construction of pharmaceutical compounds, such as galectins inhibitors (e.g. compound **7**) [44]. In 1991, Collect's group reported that primary and secondary amines, such as cyclohexylamine **1**, could be converted to the corresponding carbazates, such as **3**, by reaction with *N*-methoxycarbonyl-3-phenyloxaziridine **2** [45]. Later on, related oxaziridines were investigated in detail regarding their utilities in synthesis by the same group [46].

Please provide a mechanism for the formation of **3**.

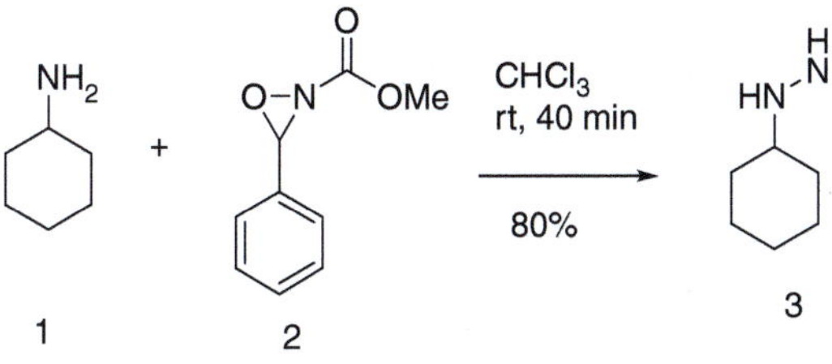

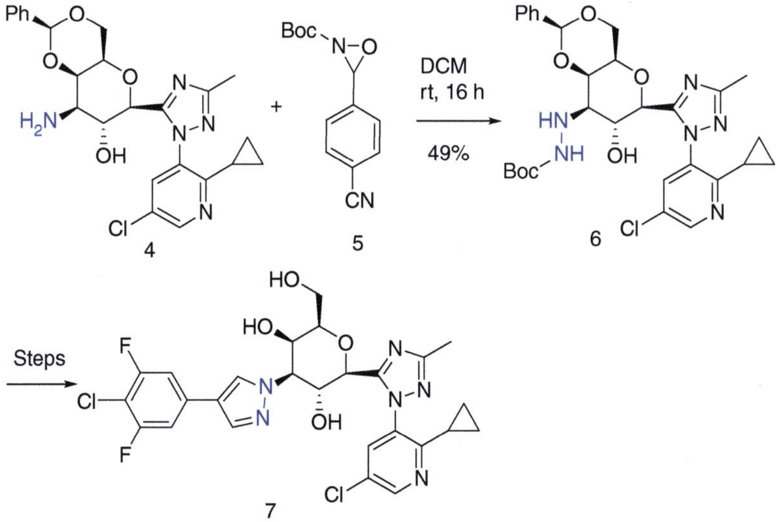

$CHCl_3$
rt, 40 min
80%

1 2 3

DCM
rt, 16 h
49%

4 5 6

Steps

7

Experimental Procedure (from *Chem. Eur. J.*, **1997**, *3*, 1691)
General Procedure for the Amination of Amines by *N*-Moc oxaziridine (2a) or ***N*-Boc oxaziridine (5a):** A solution of the required amine (2 mmol) in Et_2O or $CHCl_3$ (2 mL) was treated at 0 °C by a solution of **2a** (376 mg, 2.1 mmol) or **5a** (516 mg, 2.1 mmol) in the same solvent (2 mL). At the end of the addition, the cooling bath was removed. The reaction was monitored by TLC (secondary amines) or by NMR (primary amines). The reaction product was either recrystallized or chromatographed over silica gel to give the corresponding N_β-alkyloxycarbonylhydrazine.

***N*-(Methoxycarbonylamino)cyclohexylamine (3):** The reaction of cyclohexylamine (0.230 mL) with **2a** in $CHCl_3$ for 40 min according to the procedure above gave a 85:15 mixture of **3** and ***N*-(benzylidene)cyclohexylamine** (δ(CHN) = 3.14). Chromatography (16 g silica gel; Et_2O/hexane 20:80) afforded **3** (275 mg, 80%). Colorless crystals, m.p. 63 °C (DSC; ref. [59]: 63.5–64.5 °C); ^{1}HNMR ($CDCl_3$): δ = 1.03–1.33 (m, 5H), 1.58–1.83 (m, 5H), 2.79 (m, 1H), 3.50 (brs, 1H), 3.70 (s, 3H), 6.16 (brs, 1H).

The Mechanism

Amine **1** attacks the oxaziridine in **2** followed by elimination of benzaldehyde to give the carbazates **3**.

Armstrong's group developed a novel oxaziridine **8** performing the amination of amines to provide *N*-Boc hydrazines [47].

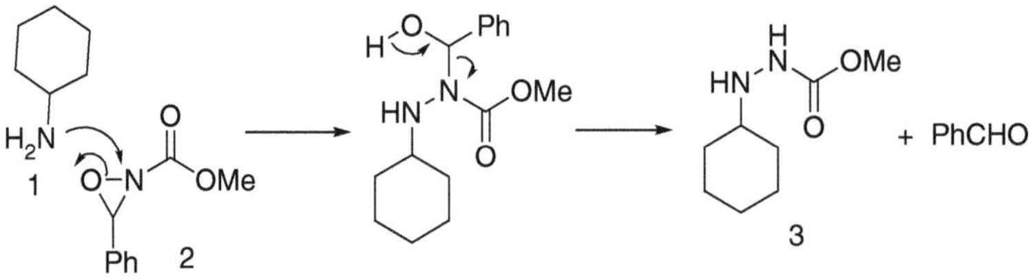

Oxaziridine **8** was initially used for amination of sulfides to sulfimides [48].

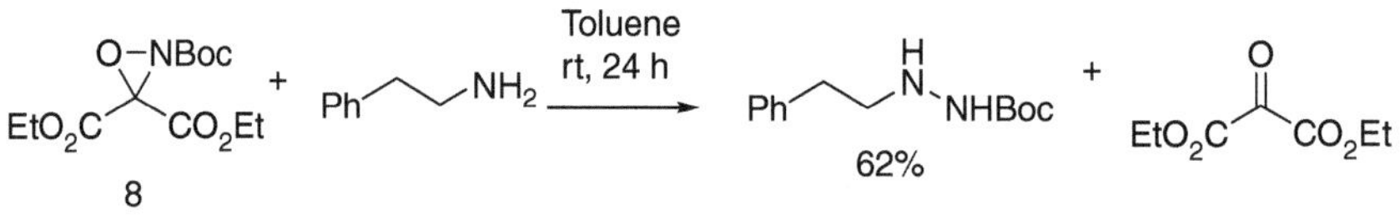

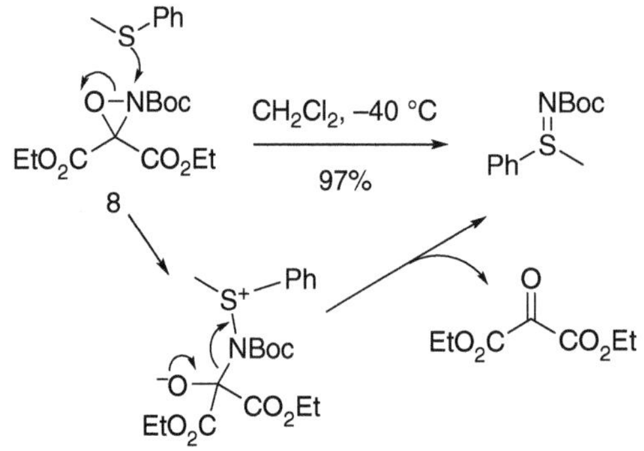

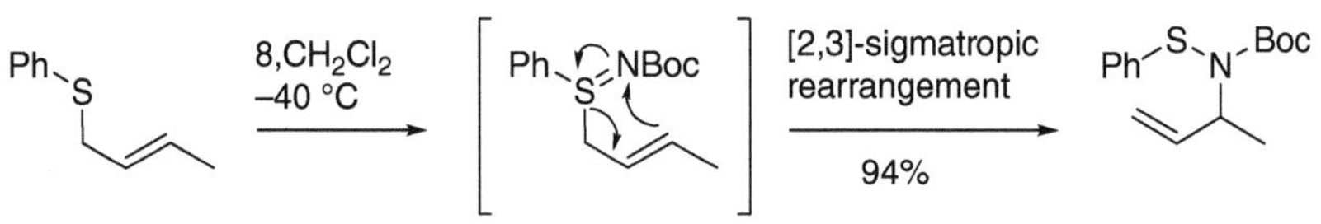

Problem 19: Preparation of Hydrazines from Alcohols

Hydrazine **4** is a key building block for construction of PDE IV inhibitor **5** [49]. In 1981, Wierenga et al. reported the direct conversion of alcohol to hydrazine under

Mitsunobu reaction condition [50]. The reaction works for primary and secondary alcohols. That reaction was applied for the preparation of cyclopentylhydrazine dihydrochloride **4** in > 7 kg scale in 69% yield [49].

Please propose a mechanism for the reaction.

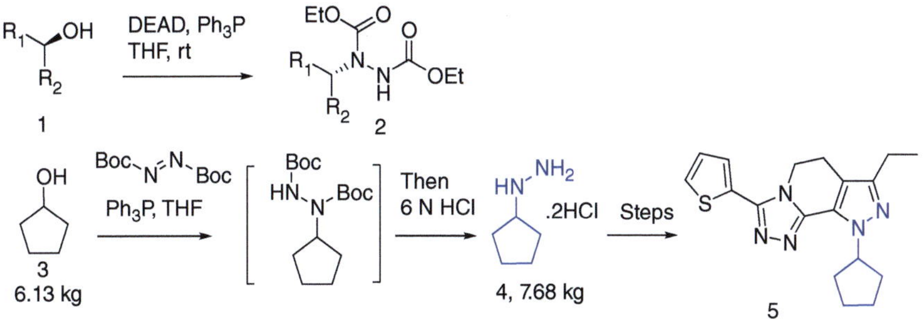

Experimental Procedure (from *Org. Process Res. Dev.*, **2001**, *5*, 575)

Cyclopentylhydrazine Dihydrochloride (9). Cyclopentanol (6.13 kg, 71.1 mol) and triphenylphosphine (18.67 kg, 71.25 mol) were dissolved in THF (151 L) in a nitrogen-purged 100-gal glass-lined tank and was cooled to 5 °C. A solution of di-*tert*-butyl azodicarboxylate **21** (14.9 kg, 64.7 mol) in THF (36 L) was added over about 2 h, keeping the temperature <6 °C. The reaction was stirred for 5 h as the temperature was allowed to slowly increase to 20–25 °C. A solution of 6 N HCl (26.5 L) was added to the reaction at 20 °C. The reaction was stirred for 24 h at 20–25 °C at which point the starting material had been consumed. Water (37.85 L) was added, and the THF was removed by vacuum distillation. During the concentration, triphenylphosphine oxide precipitated, and an additional 75.7 L of water was added. The reaction was cooled, and methylene chloride (113.6 L) was added. The layers were separated, and the aqueous layer was extracted twice more with methylene chloride (37.85 L). The aqueous was concentrated by distillation. As the volume was reduced, 2-propanol (3 × 75.7 L) was added to azeotrope the residual water. The resulting slurry was filtered and the solids were dried under vacuum at room temperature to give 7.68 kg (68.6% theory) over multiple crops. This material was characterized to be the dihydrochloride salt: mp 189–194 °C. ^{1}HMR (DMSO-d_6, 300 MHz) δ 3.48 (m, 1), 1.79 (m, 2), 1.64 (m, 4), 1.49 (m, 2).

The Mechanism

This is a Mitsunobu reaction without additional nucleophile. The resulting di-Boc hydrazine **B** is the nucleophile attacking the active species **A** to yield the di-Boc alkylhydrazine (see next page).

Problem 20: Synthesis of Arylacetohydrazonoyl Cyanide

Azo coupling is an old reaction used for the production of dyes. The addition of active methylene compounds as nucleophiles to diazonium compounds to form azo derivatives has also been reported [51, 52]. Recently, this reaction was applied for the preparation of *N*-arylhydrazonoyl cyanides. These compounds are important intermediates for the synthesis of some protein kinase inhibitors [53, 54]. For example, (*E*)-*N*-(4-phenoxyphenyl)cyclopentanecarbohydrazonoyl cyanide (**3**) was prepared by the following reactions [53]. The chemistry was applied to synthesis of a series of Bruton's tyrosine kinase inhibitors such as compound **4**.

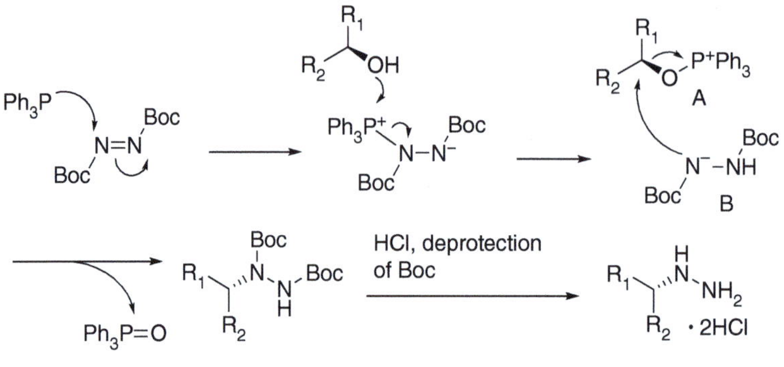

Please provide mechanisms for the formation of **3** from **1** and **2**.

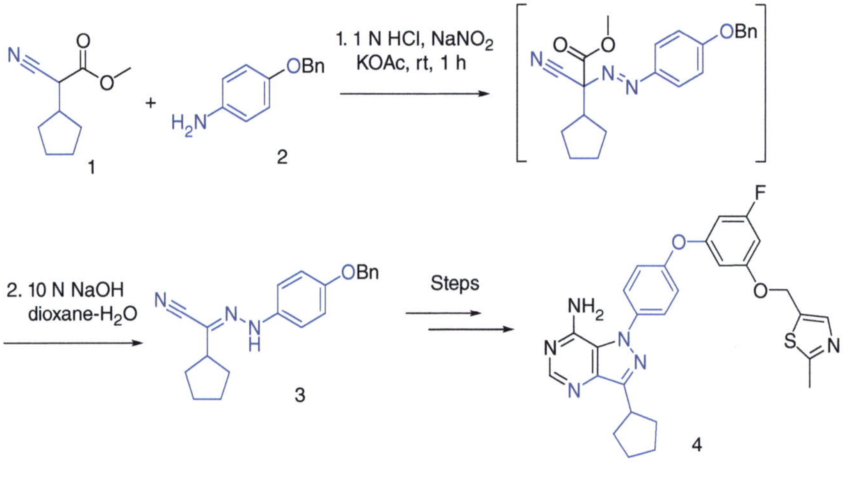

Experimental Procedure (from WO**2014**/005217, A1)

To a solution of 4-(benzyloxy)aniline hydrochloride (10.0 g, 42.4 mmol) in 1 N HCl (60.6 mL) was added dropwise a 1.0 M solution of sodium nitrite (41.9 mL, 41.9 mmol) in water at room temperature. The mixture was stirred for 1 h, filtered and then added dropwise to an ice cooled solution of methyl 2-cyano-2-cyclopentylacetate (5.0 g, 29.9 mmol) in ethanol (16.2 mL) and water (222.0 mL). The pH was maintained at 7 by adding potassium acetate portionwise. The mixture was stirred at 0 °C for 3 hours and at room temperature for 1 hour. A saturated aqueous solution of ammonium chloride and ethyl acetate were added, the organic layer was separated, washed with brine, and dried over $MgSO_4$, filtered and concentrated under reduced pressure to provide intermediate, methyl (*E*)-2-((4-(benzyloxy)phenyl) diazenyl)-2-cyano-2-cyclopentylacetate, as a beige oil.

To a solution of methyl (*E*)-2-((4-(benzyloxy)phenyl)diazenyl)-2-cyano-2-cyclopentylacetate (10.0 g, 26.5 mmol) in a 1:1 mixture of 1,4-dioxane/water (265.0 mL) cooled to 0 °C was added NaOH 10 N (53.0 mL, 530.0 mmol) and the reaction was stirred at room temperature for 1 hour. A saturated aqueous solution of ammonium chloride and ethyl acetate was added, the organic layer was separated, washed with brine, and dried over $MgSO_4$, filtered and concentrated under reduced pressure. Purification by silica gel chromatography provided intermediate *N*-(4-(benzyloxy)phenyl)cyclopentanecarbohydrazonoyl cyanide as a yellow solid.

The Mechanism

The first stage is a Sandmeyer diazotization of the aniline **2** to form the diazonium salt **A**.

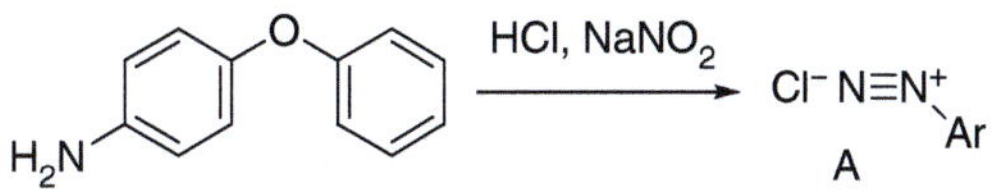

HCl, $NaNO_2$ — A

Then the cyanoester **1** as a nucleophile adds to the diazonium compound **A** to form the diazene product **B**, which can be isolated.

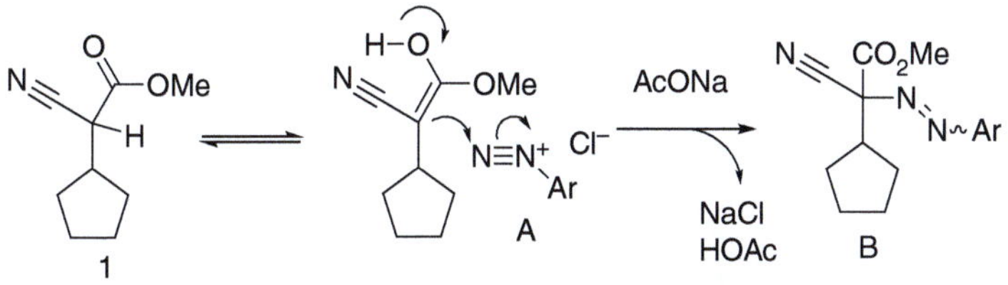

1 — A — AcONa — NaCl, HOAc — B

The final step is the hydrolysis of the ester **B** followed by decarboxylation to release the product **3**.

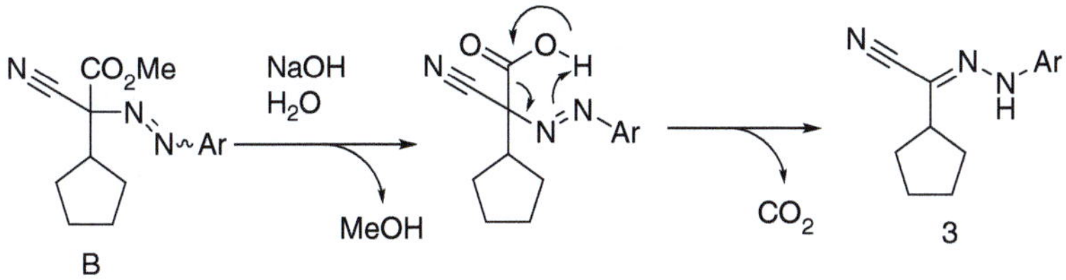

B — NaOH, H_2O — MeOH — CO_2 — 3

4.3 Formation of Cyclic Alkane and Derivatives

4.3.1 Cyclopropane

Problem 21: Preparation of Methyl (1S,5R)-2-oxo-3-oxabicyclo[3.1.0]hexane-1-carboxylate

In 1989, Pirrung's group reported a synthesis of methyl (1*S*,5*R*)-2-oxo-3-oxabicyclo[3.1.0]hexane-1-carboxylate **3** from (*R*)-2-(chloromethyl)oxirane **1a** and dimethyl malonate **2** in the presence of MeONa [55]. Compound **3** is a key starting material for the synthesis of dipeptidyl peptidase-4 inhibitor, carmegliptin **4** [56].

Please explain the mechanism for the formation of **3**.

For 1a MeONa, MeOH reflux, 16 h, 36%

For 1b *t*-BuOK, THF 60 °C, 16 h, 60%

1a, X=Cl
1b, X=ONs (ee=99.6%)

3 (ee=99.8%, from 1b)

Steps

4, Carmegliptin

Experimental Procedure (from *Helv. Chem. Acta*, **1989**, *72*, 1301)

Methyl (*3aS,4aR*)-3,3a,4,4a-tetrahydro-3-oxo-1*H*-cyclopropa[c]furun-3a-carboxylate ((-)-3). To abs. MeOH (30 mL) under dry N_2, cooled to 0° in an ice bath, was added Na (900 mg, 39.2 mmol) in three portions over 15 min. The ice bath was removed after complete dissolution, and dimethyl malonate (**2**, 5.1 mL, 5.95 g, 45.0 mmol) was added at once. (*R*)-Epichlorohydrin ((*R*)-**1a**; 3.90 g, 40.0 mmol) was added dropwise with vigorous stirring over 10 min at room temperature. After an additional 5 min at room temperature, the solution became cloudy and was heated at reflux for 16 h. The resulting suspension was cooled to room temperature, the precipitated salts were filtered, and the filtrate was evaporated: yellow oil/white solid. This residue was taken up in Et_2O (50 mL) and stirred for 5 min. The Et_2O layer was decanted and the remaining polymer dissolved in H_2O (40 mL) and extracted with Et_2O (3 × 40 mL) and CH_2Cl_2 (3 × 40 mL). The combined org. phase was dried (Na_2SO_4) and evaporated. The resulting semisolid was chromatographed with Et_2O. Recrystallization from Et_2O at −78° gave white plates (1.798 g, 29.4%). Evaporation of the mother liquor and two recrystallizations gave another 386 mg, together 2.18 g (35.7%) of **(-)-3.** GC: t_R 3.20 min. TLC: R_f 0.5 (Et_2O). M.p. 46–47°. $[\alpha]_D^{25} = -163.8 \pm 0.2$ (c = 1.32, CH_2Cl_2). IR ($CHCl_3$): 3020w (br.), 2960w, 2910w, 2410w, 1780s, 1770s, 1720m, 1440m, 1380m, 1370w, 1315m, 1270m, 1200w, 1115m, 1090m, 1045m, 1000m, 970w, 930w. ^{1}H-NMR (400 MHz, $CDCl_3$): 4.37 (dd, $J = 9.5$ and 4.8 Hz, 1H-C(1)); 4.20 (dd, $J = 9.5$ and 1.5 Hz, 1H-C(1)); 3.82 (s, CH_3O); 2.76 (dddd, $J = 8.1$, 5.5, 4.8, and 1.5 Hz, H-C(4a)); 2.10 (dd, $J = 8.1$ and 4.8 Hz, 1H-C(4)); 1.42 (dd, $J = 5.5$ and 4.8 Hz, 1H-C(4)). ^{13}C NMR (100 MHz, $CDCl_3$): 170.5, 167.3 (2 s, C(3), $COOCH_3$); 67.0 (t, C(1)); 52.9 (q, CH_3O); 29.3 (s, C(3a)); 28.0 (d, C(4a)); 20.9 (t, C(4)). HRMS: 156.0423 (8.8, M^+, calc. 156.0423, error <0.1 p (0.6 ppm)), 126.0319 (92.5, M^+-CH_2O), 125.0234 (63.6, M^+-CH_2OH). GC-MS (4.96 min): 156 (13.1, M^+), 127 (9.0), 126 (100, M^+-CH_2O), 125 (54.9, 124 (11.5), 108 (18.0), 100 (23.7), 98 (25.3), 97 (14.1), 95 (14.1), 83 (39.7), 69 (35.1), 68 (38.2), 59 (60.6), 55 (16.6), 54 (12.5), 53 (97.1), 52 (10.3), 51 (11.4), 45 (12.7), 42 (14.5), 41 (44.9). Anal. calc. for $C_7H_8O_4$: C 53.85, H 5.16; found: C 53.68, H 5.11.

The Mechanism

The deprotonated dimethyl malonate as a nucleophile displaces the chloride in **1** and the alkylated intermediate is further deprotonated to give anion **A**. The anion cyclizes along with opening the epoxide to produce the cyclopropane **B**. After acidic aqueous work, the lactonization/transesterification occurs to release the product **3**. Only cis-methyl ester cyclize.

(R)-1 + 2 (CO_2Me / CO_2Me) —MeONa, MeOH, NaCl→ (S)-A —→ (S)-B —Transesterification, MeOH→ 3

When glycidyl 3-nitrobenzenesulfonate was employed in place of 2-(chloromethyl)oxirane and using CsF as a base, the yield of the product was improved to 63% [57], while the reaction time was 110 h. By employing *t*-BuOK as a base and in refluxing THF, the product was obtained in 60% as crystalline solid [56].

Problem 22: Synthesis of Cyclopropylamines

Cyclopropanols and cyclopropylamines serve as crucial building blocks in both organic synthesis and medicinal chemistry. Cyclopropylamine, for instance, is a key component in several synthetic drugs, including anlotinib and danoprevir. The Kulinkovich reaction, first reported by Oleg Kulinkovich and colleagues [58, 59], provides a method for synthesizing substituted cyclopropanols. This reaction involves the interaction of esters with dialkyl-dialkoxy-titanium reagents, which are generated in situ from Grignard reagents containing a hydrogen in the β-position and titanium(IV) alkoxides, such as titanium isopropoxide.

Variations of this reaction have been developed to expand its utility. In the de Meijere variation, amides are used instead of esters, yielding aminocyclopropanes as the reaction product [60]. Similarly, the Szymoniak variation employs nitriles as substrates, resulting in cyclopropanes bearing a primary amine group [61–64]. These adaptations highlight the versatility and significance of the Kulinkovich reaction in synthetic organic chemistry.

A representative example is the synthesis of *tert*-butyl (1-(hydroxymethyl)cyclopropyl)carbamate (**2**) from *tert*-butyl (cyanomethyl) carbonate (**1**),6 which is a key building block for constructing anlotinib [65].

Szymoniak's group also developed the synthesis of bicylic cyclopropylamines from unsaturated nitriles, such as from **3** to **4** [64].

Please provide a mechanism for the reaction from **3** to **4**.

Experimental Procedure (from *Tetrahedron Lett.*, **2003**, *44*, 2485)

Typical procedure for the synthesis of cyclopropylamines: To a solution of nitrile (1 mmol) in Et_2O (5 mL), $Ti(Oi\text{-}Pr)_4$ (0.33 mL, 1.1 mmol) and cyclohexylmagnesium chloride (1.1 mL, 2.2 mmol, 2 M in ether) were added successively at room temperature. After stirring for 0.5 h, $BF_3{\cdot}OEt_2$ (0.25 mL, 2 mmol) was added at once.

Kulinkovich reaction

1. $Ti(OiPr)_4$ (5–10%)
2. H_3O^+

de Meijere variation

1. $Ti(OiPr)_4$ (>1 eq.)
2. H_3O^+

Szymoniak variation

1. $Ti(OiPr)_4$
2. BF_3-OEt_2
3. H_3O^+

NaCN + HCHO Boc_2O → 1 (72%) → EtMgBr (2.2 eq.), $Ti(OiPr)_4$ (0.2 eq.), Et_2O, 0 °C, 1 h → 2 (58%)

Steps → Anlotinib

1. $Ti(OiPr)_4$, Et_2O cyclohexylmagnesium chloride
2. Et_2O-BF_3

3 → 4 (67%)

Stirring was continued over a period of 30 min. A solution of 10% NaOH (ca. 1 mL) was added and the mixture was extracted with ether. The combined ether layers were dried over anhydrous sodium sulfate, filtered and concentrated under reduced pressure. The resulting crude material was purified by flash chromatography on silica gel.

3-Benzyl-5-methyl-3-azabicyclo[3.1.0]hept-1-ylamine (**4**): ^1H NMR ($CDCl_3$, 250 MHz) δ: 0.25 (m, 1H), 1.10 (m, 4H), 1.50 (s, 2H), 2.20 (d, $J = 8.4$ Hz, 1H), 2.25 (d, $J = 8.4$ Hz, 1H), 2.90 (d, $J = 8.3$ Hz, 1H), 3.05 (d, $J = 8.3$ Hz, 1H), 3.55 (s, 2H), 7.25–7.40 (m, 5H); ^{13}C NMR ($CDCl_3$, 250 MHz) δ: 14.2, 20.6, 26.1, 43.0, 59.2, 60.6, 68.8, 126.7, 128.0, 128.5, 139.1; MS (EI) m/z (%): 202 ($M^{+\bullet}$, 13), 147 (2), 120 (4), 111 (14), 91(100).

**Comments on the ^1H NMR data: The date presented in the literature seems wrong.*

The Mechanism

The Grignard reagent as a nucleophile attacks the titanium isopropoxide to give the dicyclohexyltitanium diisopropoxide (**A**). After one molecule of cyclohexane is eliminated, the cyclohexene-titanium complex (**B**) is formed. Then alkene exchange with the C=C bond of the nitrile yields the complex (**C**), which in turn reacts with the nitrile in the intramolecular reaction to afford the cyclopropylamine product.

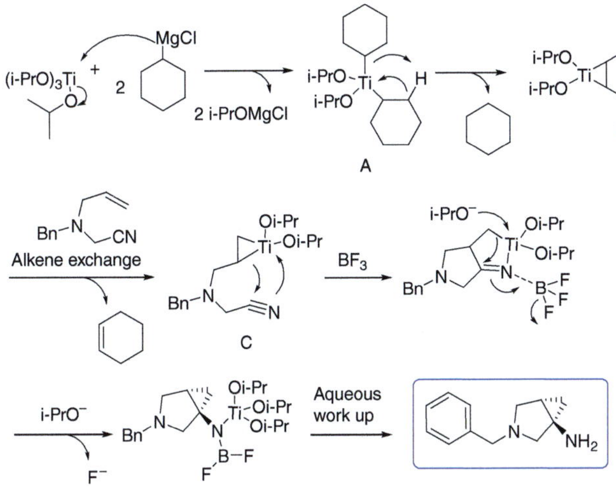

Problem 23: Formation of Cyclopropane by Dialkylation of Active Methylene

(1*R*,2*S*)-1-Amino-2-vinylcyclopropanecarboxylic acid (vinyl-ACCA) is a key building block in the synthesis of potent inhibitors of the hepatitis C virus NS3 protease such as BILN 2061.

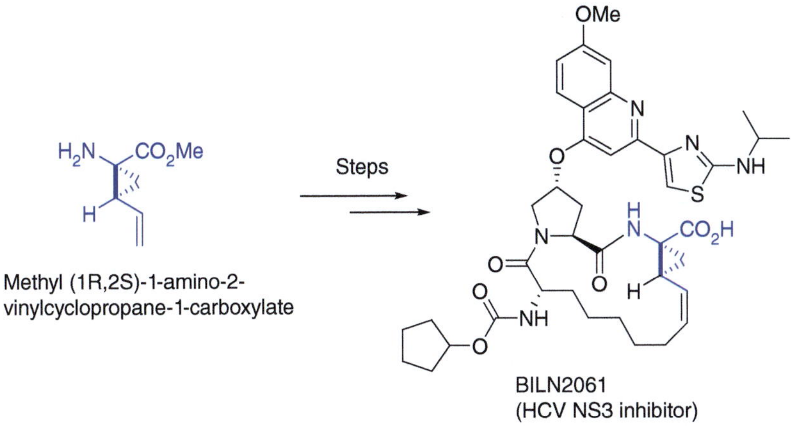

Methyl (1R,2S)-1-amino-2-vinylcyclopropane-1-carboxylate

BILN2061
(HCV NS3 inhibitor)

Chemists from Boehringer Ingelheim developed a scalable process that delivered derivatives of this unusual amino acid in >99% *ee* [66]. In an initial study, they carried out a reaction between ethyl (*E*)-2-(benzylideneamino) acetate (**1**) and (*E*)-1,4-dibromobut-2-ene (**2**) under the conditions shown below. A mixture of **3** and

4 (3:1) was detected in the reaction mixture. After the acid-base aqueous workup, the amino ester **5** was obtained in 45% yield.

Could you please provide a mechanism for the formation of **3** and **4**? Why compound **5** was the only product obtained after the acid-base aqueous workup?

Ph⁀N⁀CO$_2$Et (1) + Br⁀⁀Br (2) → KOH, toluene, $PhCH_2Et_3N^+Cl^-$, rt, 18 h → (+/–)-3 + (+/–)-4
Ratio 3 : 4 = 3 : 1
→ 1. H_3^+O 2. HO^- → (+/–)-5 (H_2N, CO_2Et)

Experimental Procedure (from *J. Org. Chem.*, **2005**, *70*, 5869)

General Procedure for the Dialkylation of Imine 1: Optimization Studies Using Hydroxide Bases and Phase Transfer Catalysis (Table 1). The following procedure using KOH is representative. Powdered KOH (0.336 g, 6.00 mmol, 3 equiv) was added to a stirred mixture of imine **1** (0.574 g, 3.00 mmol, 1.5 equiv), *trans*-1,4-dibromo-2-butene (0.428 g, 2.0 mmol, 1 equiv), toluene (10 mL), and triethylbenzylammonium chloride (0.068 g, 0.3 mmol, 0.15 equiv). The mixture was stirred for 18 h at room temperature. An aliquot (2 mL) was then diluted with *tert*-butylmethyl ether (TBME, 5 mL) and washed with saturated aqueous $NaHCO_3$ (2 mL). The organic portion was dried ($MgSO_4$), concentrated under reduced pressure, and analyzed by ^{1}H NMR in $CDCl_3$ (see below for complete assignments) to determine the ratio of the desired vinyl-ACCA imine intermediate **3** (δ 2.27, q = 1H or δ 2.00, dd = 1H or 1.70, dd = 1H) and side product **4** (δ 6.04–5.95, m = 2 olefinic Hs). In this case, a 3:1 ratio of **3** to **4** was obtained. Similar procedures were used for other hydroxide bases, replacing KOH with NaOH, $CsOH{\bullet}H_2O$, $LiOH{\bullet}H_2O$, or tetramethylammonium hydroxide.

Isolation and Characterization of 7-Phenyl-6,7-dihydro-1*H*-azepine-2-carboxylic Acid Ethyl Ester 4. Imine **1** (5.737 g, 30.0 mmol, 1.5 equiv) was alkylated with *trans*-1,4-dibromo-2-butene (4.278 g, 20.0 mmol, 1 equiv) and KO*t*Bu (6.172 g, 55 mmol, 2.75 equiv) in THF (100 mL) using the general procedure described above. Following acidic cleavage of the imine, the organic phase containing benzaldehyde and **4** was separated from the aqueous phase containing vinyl-ACCA-OEt **3**. The solution was washed with 1 N HCl (50 mL) and brine (50 mL) and dried over Na_2SO_4. After removal of volatiles under vacuum at 60 °C to remove benzaldehyde, the residual oil was purified by flash chromatography using 0–5% EtOAc in hexane as an eluent. Dihydroazepine **4** was obtained as a colorless oil (0.300 g), which slowly turned orange on prolonged exposure to air: TLC R_f)

0.55 (10% EtOAchexane); ^{1}H NMR (400 MHz, $CDCl_3$) δ 7.39–7.33 (m, 2H), 7.32–7.24 (m, 3H), 6.04–5.96 (m, 2H), 5.93–5.84 (m, 1H), 5.22 (broad s, 1H), 4.23 (q, $J = 7.0$ Hz, 2H), 4.15 (broad d, $J = 7.4$ Hz, 1H), 2.88–2.78 (m, 1H, part of AB), 2.68 (broad dd, $J = 17.4$ and 7.4 Hz, 1H, part of AB), 1.31 (t, $J = 7.1$ Hz, 3H); ^{13}C NMR (100 MHz, $CDCl_3$) δ 165.7, 144.0, 135.4, 131.4, 128.8, 127.5, 126.6, 124.7, 103.2, 61.6, 58.2, 42.7, 14.1; HRMS (FAB) m/z $C_{15}H_{18}NO_2$ (MH^+) calcd 244.1338, found 244.1338.

Crude imine **3** had the following spectral characteristics: ^{1}H NMR (400 MHz, $CDCl_3$) δ 8.36 (s, 1H), 7.60 (m, 2H), 7.42 (m, 3H), 5.83–5.72 (m, 1H), 5.27 (broad d, $J = 17.0$ Hz, 1H), 5.13 (dd, $J = 10.2$ and 1 Hz, 1H), 4.25 (q, $J = 7.2$ Hz, 2H), 2.27 (dd, $J = 8.8$ and 8.6 Hz, 1H), 2.00 (dd, $J = 7.8$ and 5.5 Hz, 1H), 1.70 (dd, $J = 9.4$, 5.5 Hz, 1H), 1.30 (t, $J = 7.1$ Hz, 3H).

Intermediate Vinyl-ACCA-OEt (Racemic-5): ^{1}H NMR (400 MHz, $CDCl_3$) δ 5.71 (ddd, $J = 19.4$, 10.2, 9.2 Hz, 1H), 5.21 (ddd, $J = 17.2$ and 1.9, 0.6 Hz, 1H), 5.03 (dd, $J = 10.5$ and 1.9 Hz, 1H), 4.21–4.14 (m, 2H), 2.5 (broad s, 2H), 2.05–1.98 (m, 1H), 1.55 (dd, $J = 7.6$, 4.8 Hz, 1H), 1.32 (dd, $J = 9.2$, 4.8 Hz, 1H), 1.28 (t, $J = 7.2$ Hz, 3H).

The Mechanism

The first step is a normal alkylation of the imine ester **1** with dibromide **2** under basic condition. The second alkylation should give a mixture of *cis*-(**4a**) and *trans*-(**3**). The *cis*-**4a** could cyclize to the seven-membered product **4** via aza-Cope rearrangement [67]. However, the *trans*-(**3**) would not cyclize by this rearrangement. Under the acidic aqueous workup, compound **3** will hydrolyze to the amine **5a** as the salt form, which is soluble in aqueous phase. Whereas the product **4** and the by-product, benzaldehyde, will remain in organic phase.

KOH

KBr

Aza-Cope rearrangement

[3,3]-sigma tropic rearrangement

[1,3]-H shift

4, in organic phase

Acidic aq. workup

PhCHO In organic phase

5a, in aq. phase

Basified

Problem 24: A Tandem Cyclopropanation Ring-Expansion Reaction

Saridegib, also known as IPI-926 (**2**), is a novel, semisynthetic analog of cyclopamine (**1**, a natural product), with potent activity on the Hedgehog pathway, significant antitumor activity, and attractive pharmaceutical properties [68, 69]. The key step for this semisynthesis is the cyclopropanation ring-expansion from **3** to **4**. Please suggest a mechanism for the reactions.

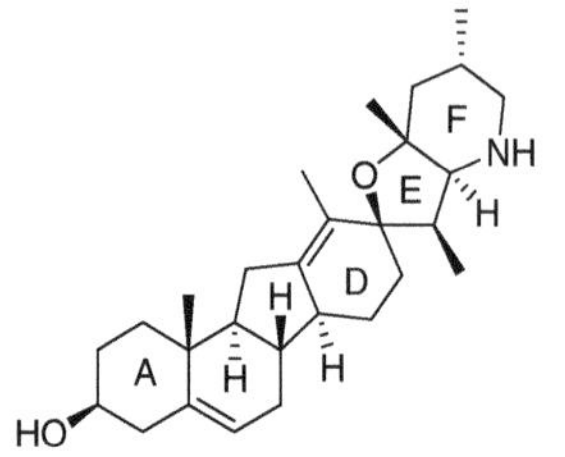

1 (cyclopamine)

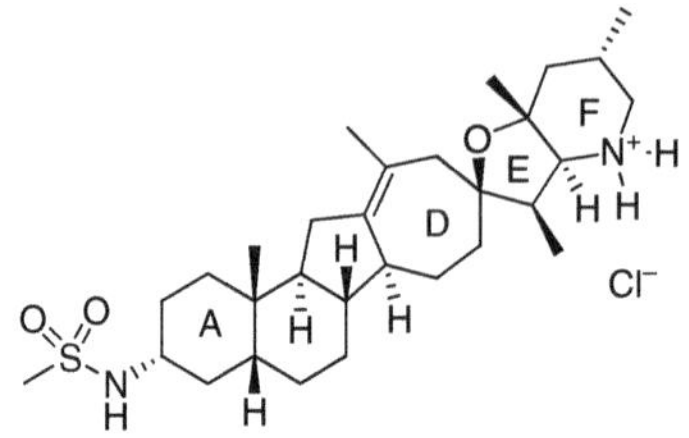

2 (IPI-926)

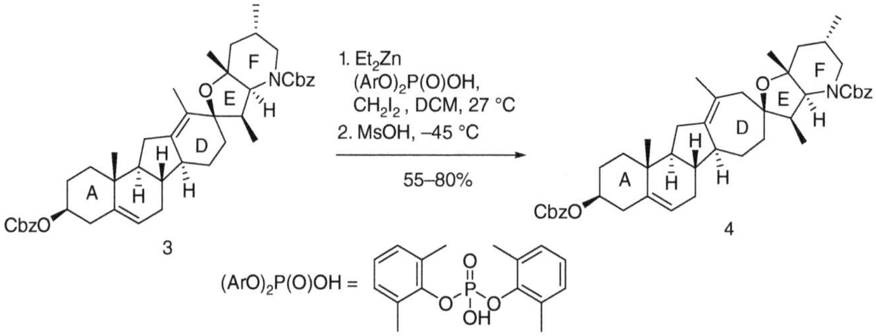

1. Et_2Zn, $(ArO)_2P(O)OH$, CH_2I_2, DCM, 27 °C
2. MsOH, −45 °C

55–80%

3 → 4

$(ArO)_2P(O)OH$ =

Experimental Procedure (from *Org. Process Res. Dev.*, **2016**, *20*, 786)

Compound 4. Equipment: Reaction Vessel A 2500 L (for diethylzinc solution and reaction) is an appropriately sized, cryogenic capable jacketed reactor equipped with a temperature control unit, a nitrogen/vacuum port, a charge port, an overhead stirrer, and condenser. Reagent Vessel B (for bis(2,6-dimethylphenyl)phosphoric acid [BDMPP] solution) is an appropriately sized jacketed reactor equipped with a temperature control unit, a nitrogen/vacuum port, a charge port, an overhead stirrer, and condenser. Reagent Vessel C (for starting material compound **3** solution) is an appropriately sized jacketed reactor equipped with a temperature control unit, a nitrogen/vacuum port, a charge port, an overhead stirrer, and condenser. Reagent Vessel D (for methanesulfonic acid solution) is an appropriately sized jacketed reactor equipped with a temperature control unit, a nitrogen/vacuum port, a charge port, an overhead stirrer, and condenser. Reaction setup: Vessel B is charged with BDMPP (45.46 kg, 1.39 wt, 3.10 equiv), then sealed and pressurized with nitrogen to 1.7 bar. The pressure is released, and a vacuum to 0.3 bar is pulled. This step is repeated twice; then the reactor is backfilled with nitrogen to 1 bar. Using residual vacuum and utilizing anhydrous techniques, charge anhydrous DCM (400 L, 13 volumes) to Vessel B. The batch is agitated at 20 ± 5 °C until all the solid has dissolved. Vessel B is sealed and pressurized with nitrogen to 1.7 bar. The pressure is released to a minimum of 0.3 bar and held for ≥5 min; then nitrogen is backfilled to 1 bar. This step is repeated three more times. The diethylzinc cylinder is connected to the transfer manifold and purged accordingly (see Supporting Information). The charge manifold is connected to Vessel A via a transfer line employing Swagelok fittings, and also to a cylinder of anhydrous DCM, and a nitrogen source. A nitrogen sweep (2–5 psi) is applied through the charge manifold, transfer line, and reaction

Vessel A for a minimum of 12 h. Using residual vacuum and utilizing anhydrous techniques, anhydrous DCM (500 L, 15 volumes) is charged to Vessel A, which is then sealed and pressurized with nitrogen to 1.7 bar. The pressure is released, and a vacuum is pulled to 0.3 bar. This step is repeated twice, then finished with nitrogen backfill to 1 bar. The batch is agitated under nitrogen, then cooled to -10 ± 5 °C. Using low pressure nitrogen, diethylzinc (18.03 kg, 0.55 wt, 3.05 equiv) is charged into Vessel A. The manifold and transfer line are rinsed with anhydrous DCM three times; the rinses (30 L) are directed into the reaction (Vessel A). After each rinse the lines are flushed with nitrogen for ≥2 min. The phosphate solution is slowly transferred from Vessel B into the reaction (Vessel A), while maintaining the temperature ≤5 °C under nitrogen. This transfer typically takes 4 h and must not exceed 7 h. Diiodomethane is charged (39.75 kg, 1.22 wt, 3.10 equiv) to the reaction (Vessel A), while maintaining the temperature $\leq -10 \pm 5$ °C. Vessel C is charged with compound 3b (32.55 kg, 1 wt, 1.0 equiv), then pressurized with nitrogen to 1.7 bar. The pressure is released and a vacuum is pulled to 0.3 bar. This step is repeated twice, then finished with a nitrogen backfill to 1 bar. Using residual vacuum and utilizing anhydrous techniques, anhydrous DCM is charged (65 L, 2 volumes) to Vessel C, and the material is agitated at 20 ± 5 °C until it dissolves. The DC solution of compound 3b is transferred from Vessel C into the reaction (Vessel A), while maintaining the temperature $\leq -10 \pm 5$ °C with stirring. Using residual vacuum and utilizing anhydrous techniques, Vessel C is rinsed with anhydrous DCM (32 L, 1.33 wt, 1 volume). Once the transfer is complete, the batch is warmed to 7.5 ± 2.5 °C and held with a nitrogen purge to the scrubber for 1 h in order to allow dissolved ethane to vent off. The batch is then warmed to 30 ± 5 °C while monitoring the internal pressure monitored every 5 °C. If the reactor pressure increases by 1 bar above atmospheric pressure, suspend the warm up, and vent the reactor with a nitrogen purge to the scrubber over 1 h. The batch is agitated at 30 ± 5 °C for ≥4 h. After the reaction is found to be complete, the batch is cooled to -60 ± 5 °C. Ring-expansion: Using residual vacuum, methanesulfonic acid (13.9 kg, 0.424 wt, 3.0 equiv) is charged to Vessel D. Using residual vacuum and utilizing anhydrous techniques, anhydrous DCM (55.6 kg, 1.72 wt, 1.3 volumes) is charged to Vessel D. When the batch in Vessel A reaches ≤−55 °C, the methanesulfonic acid solution in Vessel D is transferred to the batch (Vessel A), while maintaining the temperature at ≤−50 °C. Using residual vacuum and utilizing anhydrous techniques, anhydrous DCM is transferred (11.7 kg, 0.36 wt, 0.27 volumes) to Vessel D, then the rinse is transferred to the batch (Vessel A). Quench: Once the ring-expansion reaction is complete, morpholine (42.1 kg, 1.28 wt, 1.29 volumes, 10 equiv) is charged to the batch (Vessel A), while maintaining the temperature at ≤−35 °C. The batch (Vessel A) is then warmed to 20 °C and held for ≥1 h. The batch is washed with an aqueous solution of 2N hydrochloric acid (336 kg, 10.3 wt, 10 volumes, 13.6 equiv) for ≥60 min at 20–25 °C. The agitation is stopped, and the organic layer (bottom layer) is separated from the aqueous layer (top layer). The organic layer is transferred back into the reactor (Vessel A), and the washing step with 2N hydrochloric acid (336 kg, 10.3 wt, 10 volumes, 13.6 equiv) is repeated for ≥15 min at 20–25 °C. The organic layer is washed with an aqueous solution of

4.8 wt % sodium carbonate (338 kg, 10.5 wt, 10 volumes, 3.2 equiv) for ≥15 min at 20–25 °C. The agitation is stopped, and the organic layer (bottom layer) is separated from the aqueous layer (top layer). The organic layer is transferred back into the reactor (Vessel A). The batch is washed with an aqueous solution of 4.8 wt % sodium sulfite (339.4 kg, 10.5 wt, 10 volumes, 2.7 equiv) for ≥15 min at 20–25 °C. The agitation is stopped, and the organic layer (bottom layer) is separated from the aqueous layer (top layer). The organic layer is transferred back into the reactor (Vessel A). The batch is washed with purified water (326 kg, 10.5 wt, 10 volumes, 5.9 equiv) for ≥15 min at 20–25 °C. The agitation is stopped, and the organic layer (bottom layer) is separated from the aqueous layer (top layer). The organic layer is polished, filtered, and transferred into a 630-L reactor. The filtrate is distilled under vacuum to 2 ± 0.25 volumes, while maintaining the batch temperature at ≤35 °C during the distillation. To the concentrated filtrate is charged 2-propanol (97.7 kg, 3.0 wt, 3.82 volumes), and the batch is agitated at 20 ± 5 °C until the solution is homogeneous. Methanol (49.0 kg, 1.5 wt, 1.90 volumes) is then charged into the batch over ≥10 min at 20 ±5 °C. The mixture is agitated for 30 ± 5 min at 20 ± 5 °C to initiate nucleation. Then, more methanol (179.2 kg, 5.5 wt, 6.95 volumes) is charged over 45 ± 15 min at 20 ± 5 °C. The batch is agitated at 20 ±5 °C for ≥4 h, then at 0 ± 5 °C for ≥4 h. The solid is filtered off over a 30 μm nylon filter cloth and deliquored with nitrogen. The solid is washed with methanol (130.4 kg, 4 wt, 5.1 volumes), soaked for ≥10 min, then deliquored with nitrogen. The solid is washed with heptane (130 kg, 4 wt, 5.8 volumes), soaked for ≥10 min, then deliquored with nitrogen. The solid is dried with nitrogen for ≥4 h; then warm nitrogen (40 °C) for ≥12 h while periodically stirring the filter cake. The solid is dried under vacuum at 40 °C for ≥8 h while periodically stirring the filter cake. This procedure delivered 18.3–26.7 kg of **4** (N = 7 on 33 kg scale, range 55–80% yield). Mp = 148 °C; ^{1}H NMR (400 MHz, pyridine-d_5) δ 7.63–7.50 (m, 4H), 7.47–7.30 (m, 6H), 5.48–5.32 (m, 5H), 4.93 (s, 3H), 4.69 (tt, J = 11.3, 4.7 Hz, 1H), 3.95 (dd, J = 12.8, 4.3 Hz, 1H), 3.64 (td, J = 10.8, 4.1 Hz, 1H), 3.24 (dd, J = 10.1 and 6.1 Hz, 1H), 2.88 (m, 1H), 2.70–2.56 (m, 2H), 2.52 (d, J = 14.1 Hz, 1H), 2.42 (td, J = 12.4, 2.9 Hz, 1H), 2.23 (dtd, J = 14.5, 6.7, 6.1, and 3.2 Hz, 2H), 2.16–1.92 (m, 5H), 1.90–1.53 (m, 7H), 1.47–1.28 (m, 2H), 1.26–1.00 (m, 6H), 0.90–0.77 (m, 6H); ^{13}C NMR (101 MHz, pyridine-d_5) δ 158.20, 155.46, 141.67, 140.71, 138.10, 136.94, 129.43, 129.41, 129.20, 129.10, 129.02, 128.91, 124.58, 82.42, 78.53, 75.70, 69.87, 67.72, 63.06, 53.29, 52.75, 50.67, 50.46, 46.86, 44.37, 38.42, 38.38, 38.34, 38.14, 37.37, 32.42, 31.16, 29.37, 29.28, 28.24, 24.55, 19.81, 18.95, 11.18; LRMS: m/z = 694.7 $[M + H]^+$; HPLC purity (Method D) range of 95.6–97.2 a/a %.

The Mechanism

The first reaction is a Simmons-Smith reaction [70] to give the cyclopropane **A**. At acidic conditions, under the participation of the neighbor THF, the bicyclo[4.1.0]heptane rearranges to cycloheptane **4**.

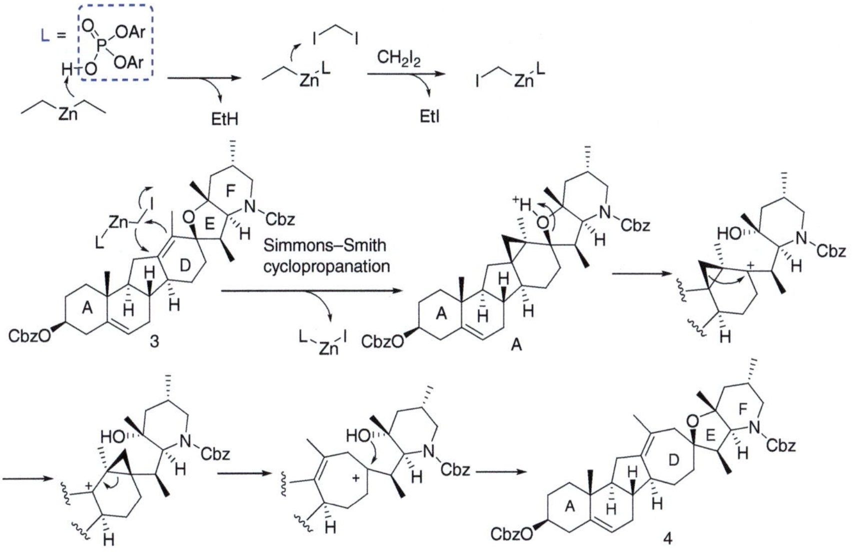

4.3.2 Cyclobutane

Problem 25: Synthesis of 3-Ethylbicyclo[3.2.0]hept-3-en-6-one

Mirogabalin besylate was developed by Daiichi Sankyo and was first approved in Japan for the treatment of peripheral neuropathic pain (PNP), including diabetic PNP and postherpetic neuralgia. In a patent application [71], as shown below, chemists from Daiichi Sankyo described a synthesis of the key intermediate, 3-ethylbicyclo[3.2.0]hept-3-en-6-one **5**.

Please provide mechanisms for the reactions of **2** to **3** and **4** to **5**.

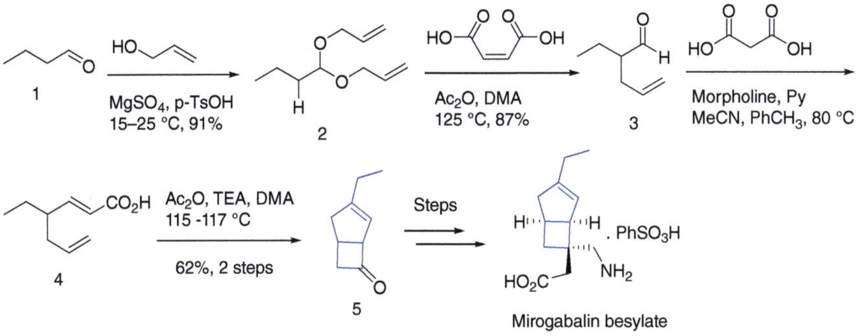

Experimental Procedure (from patent US2014/0094623, **2014**, A1)

2-Ethylpent-4-enal (3). 1,1-Bis(allyloxy)butane (102.15 g, 0.60 mol) was dissolved in *N,N*-dimethylacetamide (306 mL) under a nitrogen atmosphere. To the solution, acetic anhydride (170 mL, 1.80 mol) and maleic acid (3.48 g, 0.03 mol) were added, and the mixture was stirred. The reaction was warmed to 120 to 125 °C, stirred at this temperature for 24 h, and then cooled to 10 °C or lower. To this reaction mixture, toluene (410 mL) and water (410 mL) were added, and a 25% aqueous sodium hydroxide solution (586 mL) was slowly added with stirring to separate an organic layer. The aqueous layer was subjected to extraction with toluene (210 mL), and the extract was combined with the organic layer and then washed with water (102 mL) and 20% saline (102 mL) in this order. The organic layer was filtered to remove insoluble matter. Then, the obtained product was used in the next step without being concentrated or purified. The solution of the crude product thus obtained was analyzed by gas chromatography and consequently contained 2-ethylpent-4-enal (58.44 g, yield: 87%). ^{1}H NMR (400 MHz, $CDCl_3$) d 0.93 (t, 3H, $J = 7.4$ Hz), 1.53–1.61 (m, 1H), 1.64–1.73 (m, 1H), 2.21–2.34 (m, 2H), 2.37–2.44 (m, 1H), 5.04–5.11 (m, 2H), 5.75 (ddt, 1H, $J = 10.0$, 17.2, and 7.0 Hz), 9.62 (d, 1H, $J = 2.0$ Hz).

3-Ethylbicyclo[3.2.0]hept-3-en-6-one (5). (2E)-4-Ethylhepta-2,6-dienoic acid (100.57 g) obtained by the method described above was dissolved in *N,N*-dimethylacetamide (255 mL) under a nitrogen atmosphere. To the solution, acetic anhydride (108 mL, 1.14 mol) and triethylamine (79 mL, 0.57 mol) were added. The reaction mixture was warmed and stirred at 115 to 117 °C for 5 hours. The reaction mixture was cooled to room temperature, and n-hexane (510 mL) and water (714 mL) were added thereto to separate an organic layer. The aqueous layer was subjected to two extractions with hexane (each with 255 mL), and all organic layers were combined and then washed with a 5% aqueous sodium bicarbonate solution (102 mL) and water (102 mL) in this order. The obtained organic layer was concentrated under reduced pressure, and the residue was distilled (90–100 °C, approximately 25 mmHg) to obtain the title compound (50.92 g, colorless oil substance) (overall yield from 1,1-bis(allyloxy)butane: 62%). ^{1}H NMR (400 MHz, $CDCl_3$) δ 1.07 (t, 3H, $J = 7.4$ Hz), 2.14 (q, 2H, $J = 7.4$ Hz), 2.28–2.34 (m, 1H), 2.75–2.86 (m, 1H), 2.21–2.34 (m, 1H), 3.16–3.25 (m, 1H), 4.16–4.22 (m, 1H), 5.20–5.24 (m, 1H).

The Mechanism

The acid-catalyzed elimination of allylic alcohol from **2** gives vinyl allyl ether **A** and then a Claisen rearrangement leads to the aldehyde **3**.

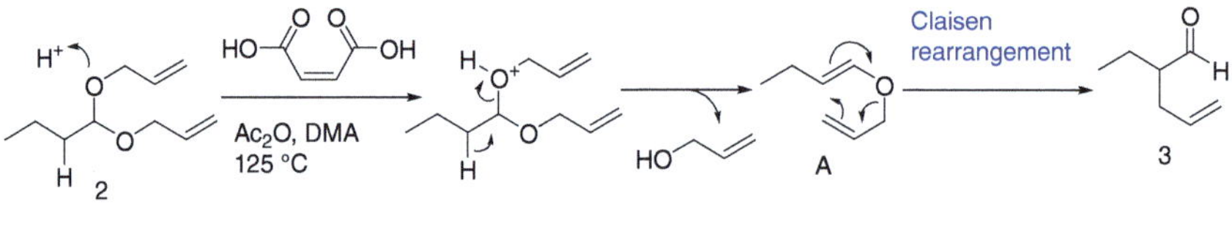

Reaction of the acid **4** with Ac_2O provides the mixed anhydride **B**. In the presence of TEA, one molecule of acetic acid is eliminated to form a mixture of cis and trans α,β-unsaturated ketenes **C**. The cis isomer undergoes an intramolecular [2 + 2] cycloaddition to produce the product (**5**). In the conventional method, the ketene intermediate **C** was produced from corresponding acid chloride and the yield of the cycloaddition was lower (≤50%) [72].

Another possible mechanism is that the mixed anhydride **B** undergoes an intramolecular synergistic cyclization to yield the product **5**. This mechanism explains the yield of the product is higher than the conventional method [72].

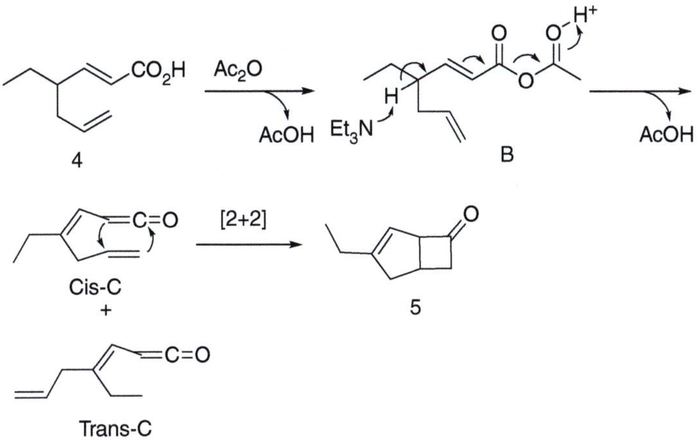

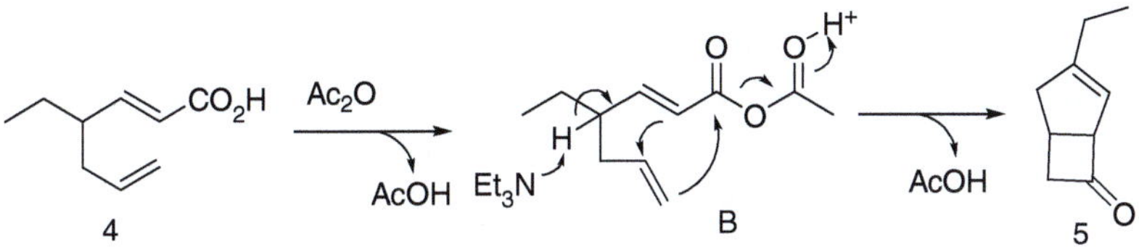

4.3.3 Cyclopentane

Problem 26: Preparation of *cis*-2-(Methoxycarbonyl)cyclopentane-1-carboxylic Acid

Compound **3** is a diacyl glycerolacyl transferase-1 inhibitor with potential application for treatment of obesity and dyslipidemia [73]. In a synthesis of **3**, chemists from Abbott Laboratories employed (±)-2-(methoxycarbonyl)cyclopentane-1-carboxylic acid (**2**) as a starting material [73]. The synthesis of **2** was described in a patent application starting from ethyl 2-oxocyclohexane-1-carboxylate (**1**) by the reaction sequence as shown below [74].

Please explain the mechanism for the formation of **2** from **1**.

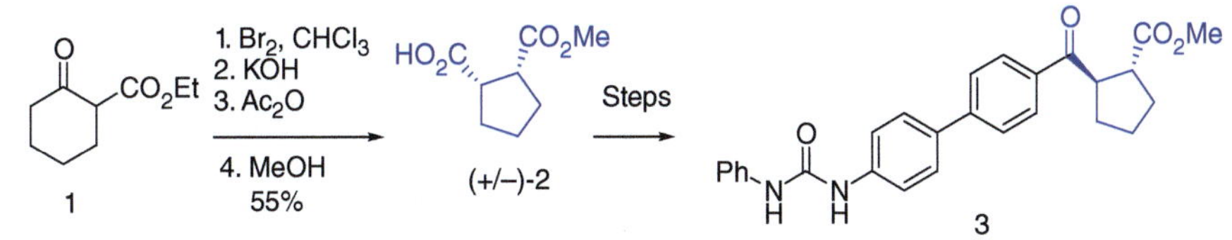

Experimental Procedure (from WO2013/095275 A1)

(1,2-Cis)-cyclopentane-1,2-dicarboxylic acid (BB2-a). Br_2 (112.8 g, 705.6 mmol) was added dropwise at 0 °C to a solution of ethyl 2-oxocyclohexane carboxylate (120 g, 705.6 mmol) in chloroform (360 mL). The resulting reaction mixture was stirred at room temperature overnight, then washed with $NaHCO_3$ (2 × 200 mL) and brine (2 × 100 mL). The organic layer was dried over Na_2SO_4 and concentrated and the residue was added dropwise to an ice-cold solution of KOH (168 g, 3 mol) in water (960 mL). The reaction mixture was stirred for 6 h at 0 °C, then extracted with diethyl ether. The aqueous phase was acidified with 4.0 M HCI (pH 5) and extracted with diethyl ether. The combined organic layers were washed with brine (200 mL), dried, filtered and concentrated. The afforded yellow oil was crystallized (EtOAc/diethyl ether) which gave the title compound as colorless crystals (70 g, 63%). MS (ESI): 159 $[M + H]^+$.

(3a,6a-Cis)-tetrahydro-1H-cyclopenta[c]furan-1,3(3aH)-dione (BB2-b). A solution of the dicarboxylic acid **BB2-a** (90 g, 580 mmol) in acetic anhydride (1080 mL) was heated at reflux. After 20 h, the acetic anhydride was removed by distillation under reduced pressure. The oily residue was distilled which gave the title compound as a solid (55 g, 89%). MS (ESI): 141 $[M + H]^+$ and 158 $[M + NH_4]^+$.

(1S,2R)-2-(Methoxycarbonyl)cyclopentanecarboxylic acid (BB35-a). Methanol (2 g, 64 mmol) was added dropwise at −65 °C to a stirred suspension of **BB2-b** (3 g, 21 mmol) and quinine (7.6 g, 24 mmol) in a 1:1 mixture of toluene and tetrachloromethane. The mixture was stirred at this temperature for 60 h, then concentrated *in vacuo* to dryness. The afforded residue was dissolved in EtOAc and the solution was washed with 2 N HCl. The organic layer was dried over Na_2SO_4, filtered, and concentrated which gave the title compound (3 g, 81%). MS (ESI): 173 $[M + H]^+$.

The Mechanism

Step 1: Bromination of **1**.

The bromination should initially occur at the tertiary carbon in **1** in a regioselective manner to give the tertiary bromide **4a**, which is easily isomerized to secondary bromide **4b** under the acidic condition (HBr).

It is worth noting that the product is the regiomer **4a** at the NBS condition [75].

Step 2: Favorskii Rearrangement [76].

Step 3: Formation of Cyclic Anhydride, Only cis-diacid Cyclizes.

Step 4: Methanolysis of the Cyclic Anhydride.

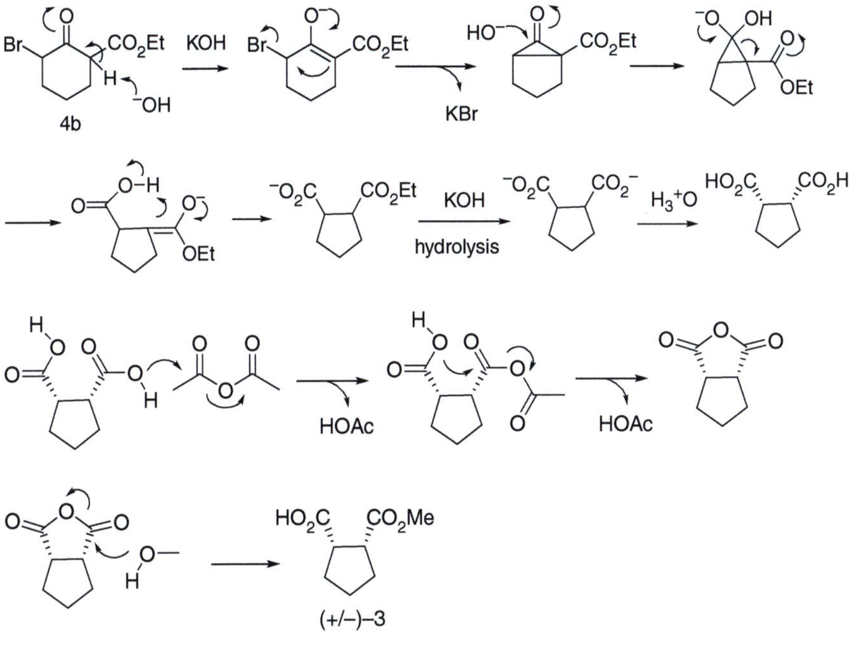

Problem 27: Rearrangement of Furfuryl Alcohol to 4-Hydroxy-2-cyclopentenone

Furfuryl alcohol (**1**) is a starting material for the large-scale synthesis of the carbocyclic nucleoside MDL 201449A (**3**) which has the potential for the treatment of multiple inflammatory diseases [77]. The first step is a rearrangement of **1** to 4-hydroxy-2-cyclopentenone (**2**) under acidic conditions [77, 78].

Please suggest a mechanism for this reaction.

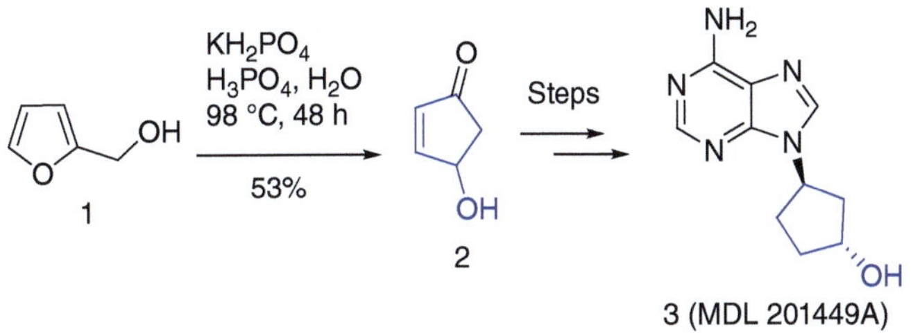

Experimental Procedure (from *Tetrahedron*, **1997**, *53*, 1983)

A solution of furfuryl alcohol (**1**, 125 g, 1.27 mol) in H_2O (3.7 L) was treated with KH_2PO_4 (6.3 g, 46.3 mmol). The solution was adjusted to pH = 4.1 (pH meter) with H_3PO_4, then heated to 99 °C for 40 h. The cooled solution was washed with CH_2Cl_2 (2 × 500 mL). The combined organic layers were extracted with H_2O (2 × 500 mL), and the H_2O layers combined and evaporated (70 °C, 20 mmHg) to give a red oil. The red oil was dissolved in CH_2Cl_2 (1 L), dried ($MgSO_4$), filtered, and the filtrate evaporated *in vacuo* (40 °C, 20 mmHg) to give **2** as a dark red oil, 66.5 g, 53%. t_R (method A) = 4.55 min; ^{1}H NMR ($CDCl_3$) 7.61 (dd, 1H, J = 5.6, 4.8 Hz), 6.20 (d, 1H, J = 5.6 Hz), 5.0 (m, 1H), 3.6 [s(broad), 1H], 2.75 (dd, 1H, J = 18.5, 3.2 Hz).

2.26 (dd, 1H, J = 18.5, 6.0 Hz); ^{13}C NMR ($CDCl_3$) 207.4, 164.0, 134.7, 70.1, 44.1; IR (neat) ν_{max} 3387, 2974, 1711 cm^{-1}; EIMS m/e (% relative intensity) 98 (M^+, 100). Anal calcd, for $C_5H_6O_2$•0.16 H_2O (116.79): C, 59.47; H, 6.30. Found: C, 59.56; H, 6.52.

The Mechanism

Acid catalyzes the full process of this rearrangement. Firstly, furfuryl alcohol (**1**) is rearranged to 5-methylene-2,5-dihydrofuran-2-ol (**A**) catalyzed by acid. Subsequently, (**A**) is aromatized to 2-hydroxy-5-methylfuran (**B**). Addition of H_2O to (**B**) gives the 2,5-dihydroxy-2,5-dihydrofuran (**C**). Elimination of water generates the ring-opened product, (Z)-4-oxopent-2-enal (**D**), which is equilibrium with enol form (**D′**). Finally, intramolecular aldol reaction leads to the product **2**.

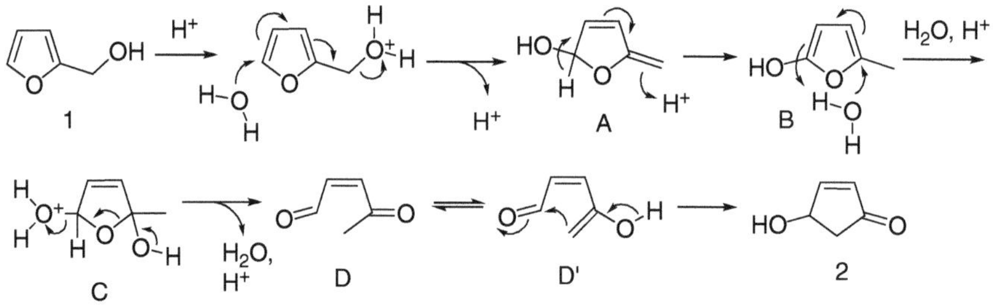

4.3.4 Cycloheptane

Problem 28: Preparation of Homobenzylic Ketones by 1,2-shift in 1,1′-disubstituted Olefins

Chemists from Teva pharmaceuticals developed a new method for the preparation of homobenzylic ketones by catalytic hypervalent iodine-mediated oxidation of 1,1′-disubstituted olefins [79]. A representative reaction is shown below. The method was applied for the synthesis of **TEV-37440**, a dual inhibitor of FAK and ALK [80]. Conventionally, this transformation was performed by employing stoichiometric $Tl(NO_3)_3$ [81]. However, the cost and the highly toxic nature of $Tl(NO_3)_3$ limit its use in the pharmaceutical industry.

Please provide a mechanism for the reaction from **1** to **3**

Experimental Procedure (from *Org. Lett.*, **2013**, *15*, 1650, Supporting Information) **1-Methoxy-5,7,8,9-tetrahydrobenzo[7]annulen-6-one (3c′).** The general procedure was performed by using 500 mg (2.87 mmol) of 5-methoxy-1-methylene-1,2,3,4-tetrahydronaphthalene (**1**) followed by 51 mg (~10 wt%) of Triton X-405, 160 mg (0.72 mmol) of catalyst (2-iodo-5-methylbenzenesulfonic acid), 4.0 mL of MTBE, 4.0 mL DI Water and 2.3 mL (7.5 equiv, 21.52 mmol) of isopropanol. To the vigorously stirred mixture was charged 1.32 g (0.75 equiv, 2.15 mmol) of Oxone and the heterogeneous mixture stirred for 24 h. After workup, the 0.85 g of crude was dissolved in 5.1 mL of isopropanol and charged with 1.4 mL of 7 M solution of aqueous sodium bisulfite followed by 0.6 mL of DI water. After stirring for 18 hours, the slurry was cooled to 10 °C, filtered and dried for 2.5 h to result in 742 mg of bisulfite adduct **3c′**.

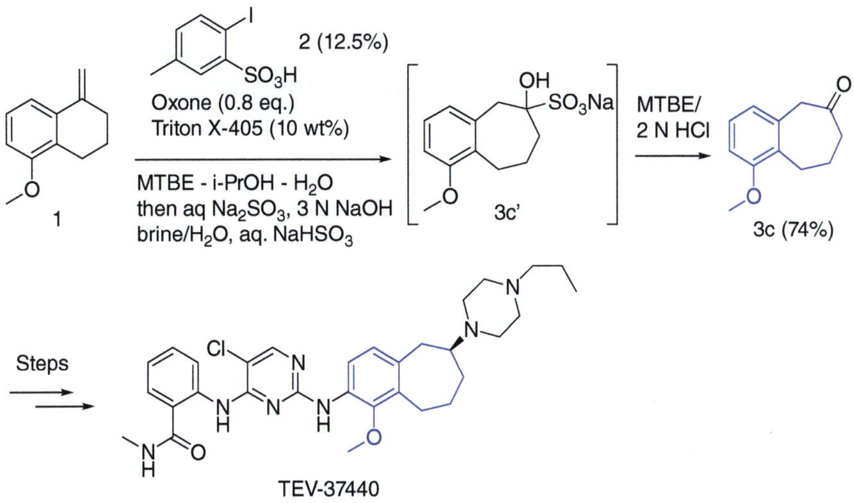

1-Methoxy-6-methyl-5,7,8,9-tetrahydrobenzo[7]annulen-6-ol•Sodium bisulfite salt (3c′). ^{1}H NMR (DMSO-d_6, 400 MHz) δ 1.31–1.40(m, 1H), 1.71–1.75 (m, 1H), 1.97–2.09(m, 2H), 2.14(t, 1H), 3.12(d, 1H), 3.22(d, 1H), 3.26–3.29(m, 1H), 3.72(s, 3H), 4.40(s, 1H), 6.68(d, 1H), 6.77(d, 1H), 6.98(t, 1H); ^{13}C NMR (DMSO-d_6, 100 MHz) δ 21.55, 24.04, 36.53, 41.06, 55.64, 84.05, 109.07, 123.89, 125.60, 130.18, 139.34, 155.59.

3c′ Was hydrolyzed according to the typical procedure to give 403 mg of **3c** as an oil in 74% yields. ^{1}H NMR ($CDCl_3$, 400 MHz) 1.93–1.96(m, 2H), 2.49–2.53(m, 2H), 2.98–3.02(m, 2H), 3.69(s, 2H), 3.81(s, 3H), 6.75(d, 1H), 6.81(d, 1H), 7.12(t, 1H); ^{13}C NMR (100 MHz) 23.26(1C), 25.10(1C), 43.14(1C), 50.0(1C), 55.75(1C), 109.88(1C), 121.98(1C), 127.27(1C), 128.48(1C), 135.43(1C), 156.89(1C), 209.16(1C); LRMS (GC-MS+) Calcd for $C_{12}H_{15}O_2$ $[M + H]^+$:191.1. Found: 191.1.

The Mechanism

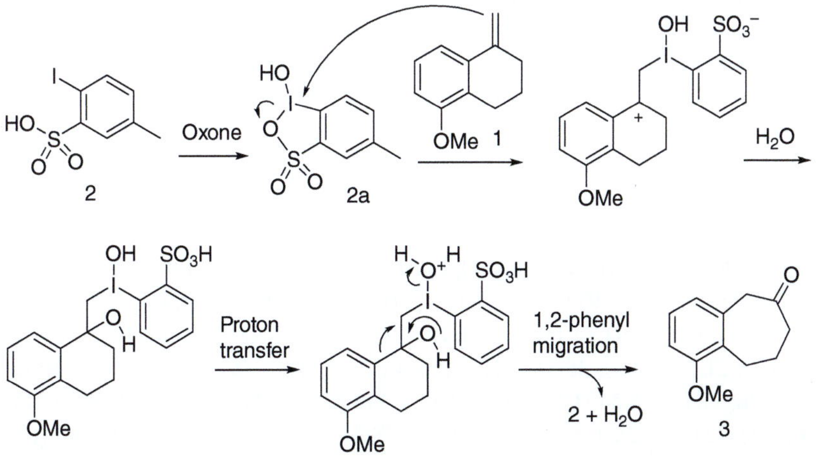

The reaction mechanism has been proposed in the literature [79]. The reaction is catalyzed by hypervalent iodine species **2a** which is generated by Oxone oxidation of **2** in situ. Subsequently, the alkene attacks the hypervalent iodine and the benzylic carbocation is captured by water. Finally, an oxocarbenium assisted 1,2-phenyl migration under acidic conditions regenerates the precatalyst **2** and releases the product **3**.

4.4 Formation of Aromatic Compounds

Problem 29: Formation of Hexasubstituted Benzene Ring by Cyclization
Mycophenolic acid **1** is a natural product discovered in 1893. The sodium salt of **1** (mycophenolate sodium) was approved in many countries as an oral immunosuppressive agent for the prevention of kidney rejection during transplantation. A reported total synthesis of **1** involved the key step for building the benzene ring core structure as shown below [82]. In this reaction, only 33% yield of the desired product **4** was obtained. The side products, identified as the diastereomeric ketoalcohols **5a** and **5b**, were also obtained in 62% yield in around 1:1 ratio.

Please suggest mechanisms for the formation of **4**, and **5a/5b**.

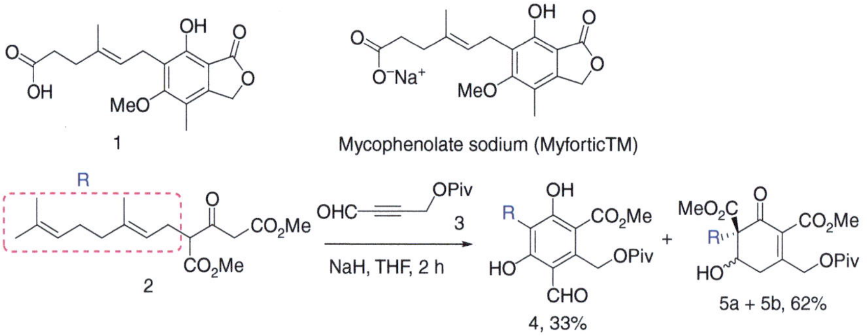

1

Mycophenolate sodium (MyforticTM)

R OHC OPiv 3 NaH, THF, 2 h CO$_2$Me 2 4, 33% 5a + 5b, 62%

Experimental Procedure (from *Tetrahedron*, **2003**, *59*, 1989)
Geranyl-6-formyl-4-methoxycarbonyl-5-pivaloyloxymethyl resorcinol (4) (3.1 g, 10 mmol) of (**2**) was added dropwise to a suspension of NaH (60%, 13.5 mmol, 0.54 g) in dry THF (31 mL) with magnetic stirring. To the resulting solution, 4-(pivaloyloxy)-2-butynal **3** (15 mmol, 2.55 g) was added and after 2 h, the mixture was poured into dilute HCl (30 mL). The organic phase was separated and the aqueous layer was extracted with ethyl acetate. The combined organic phase was washed with brine, dried over anhyd. Na_2SO_4, and concentrated. The residue was subjected to column chromatography, the resorcinol **4** (1.47 g, 33% yield) being eluted with 5% ethyl acetate in hexane (100 g of silica gel), and the diasteromeric ketols **5a,b** in 62% yield. **4** Mp 62–63 °C, IR (KBr) 3347, 2976, 1739, 1676, 1135 cm^{-1}; ^{1}H NMR δ ppm 1.16 (s, 9H), 1.54 (s, 3H), 1.60 (s, 3H), 1.75 (s, 3H), 1.90–2.10 (m, 4H), 3.37 (d, 2H, J = 6.3 Hz), 3.92 (s, 3H), 5.01 (t, 1H, J = 6.3 Hz), 5.18 (t, 1H, J = 7.2 Hz), 5.57 (s, 2H), 10.11 (s, 1H), 11.66 (s, 1H), 12.84 (s, 1H); ^{13}C NMR δ ppm 16.12 (CH_3), 17.56 (CH_3), 21.51 (CH_2), 25.53 (CH_2), 26.64 (CH_2), 27.03 (CH_3), 38.69 (C), 39.72 (CH_2), 52.86 (CH_3), 58.54 (CH_2), 107.31 (C), 113.58 (C), 117.67 (C), 120.57 (CH), 124.34 (CH), 131.12 (C), 136.24 (C), 141.32 (C), 165.10 (C), 165.18 (C), 170.65 (C),

177.61 (C), 193.94 (CH); MS (EI) *m/z* (relative intensity) 446 (M^+, 3), 344 (55), 275 (76), 243 (100), 123 (73), 57 (53). Anal. calcd for $C_{25}H_{34}O_7$: C, 67.26; H, 7.62. Found: C, 67.28; H, 7.59.

5a More polar. IR (neat) 3508, 2970, 1736, 1687, 1143 cm^{-1}; ^1H NMR δ ppm 1.21 (s, 9H), 1.57 (s, 3H), 1.65 (s, 3H), 1.66 (s, 3H), 1.92–2.10 (m, 4H), 2.57 (dd, 1H, J_{gem} = 19.3 Hz, J¼7.4 Hz), 2.72–2.85 (m, 3H), 3.6–3.8 (bs, overlapping with two s, 1H, OH), 3.77 (s, 3H), 3.83 (s, 3H), 4.18 (br s, 1H), 4.79 (AB_{sys}, 2H, J = 16 Hz), 4.95–5.05 (two overlapping br t, 2H); ^{13}C δ ppm 16.13 (CH_3), 17.55 (CH_3), 25.50 (CH_3), 26.37 (CH_2), 27.02 (CH_3), 29.72 (CH_2), 33.38 (CH_2), 38.80 (CH_3), 39.81 (CH_2), 52.34 (CH_3), 52.67 (CH_3), 62.55 (C), 63.61 (CH_2), 68.76 (CH_2), 117.58 (CH_2), 123.93 (CH_2), 131.35 (C), 131.44 (C), 139.89 (C), 153.87 (C), 165.10 (C), 172.06 (C), 177.67 (C), 190.86 (C); MS (CI) 479 (M^++1, 100), 447 (8), 377 (9), 209 (9). Anal. Calcd for $C_{26}H_{38}O_8$: C, 65.27; H, 7.94. Found: C, 65.41; H, 7.85.

5b Less polar. IR 3500, 2971, 1736, 1144 cm^{-1}; ^1H NMR δ ppm 1.20 (s, 9H), 1.55–1.70 (br s overlapping with two s, 1H, OH), 1.58 (s, 3H), 1.62 (s, 3H), 1.66 (s, 3H), 1.90–2.15 m (4H), 2.56 (dd, 1H, J = 4.7 Hz, J_{gem} = 19.6 Hz), 2.56–2.80 (m, 2H), 2.92 (dd, 1H, J_{gem} = 19.6 Hz, J = 4.1 Hz), 3.70 (s, 3H), 3.82 (s, 3H), 4.61 (t, 1H, J = 4.4 Hz), 4.78 (AB_{sys}, 2H, J = 16 Hz), 5.03 (br t, 1H, J = 6.8 Hz), 5.25 (t, 1H, J = 7.5 Hz); ^{13}C NMR δ ppm 16.03 (CH_3), 17.71 (CH_3), 25.67 (CH_3), 26.42 (CH_2), 27.12 (CH_3), 28.46 (CH_2), 33.33 (CH_2), 38.90 (C), 39.92 (CH_2), 52.54 (CH_3), 52.68 (CH_3), 62.13 (C), 63.76 (CH_2), 69.33 (CH), 118.19 (CH), 123.93 (CH), 131.24 (C), 131.81 (C), 139.68 (C), 153.12 (C), 165.25 (C), 169.86 (C), 177.82 (C), 191.11 (C); MS (CI) m/z (relative intensity) 479 (M^++1, 100), 447 (7), 360 (16), 103 (16). Anal. calcd for $C_{26}H_{38}O_8$: C, 65.27; H, 7.94. Found: C, 65.43; H, 7.82.

The Mechanism

The initial step involves a Michael addition, where the enolate of 2-geranyl dimethyl 1,3-acetonedicarboxylate (**2**) reacts with the alkynal (**3**). This generates a vinyl carbanion intermediate (**A**), which is quickly protonated by the more acidic methine proton, leading to the formation of enolate (**B**). Enolate (**B**) serves as the precursor for the desired resorcinol (**4**). Despite its apparent stability, the sterically congested nature of this intermediate facilitates its isomerization to enolate (**C**), which acts as the precursor for the ketoalcohols (**5a** and **5b**). Alternatively, enolate (**C**) could also be formed directly from the vinyl carbanion (**A**) through protonation by the more acidic methine proton from the opposite side.

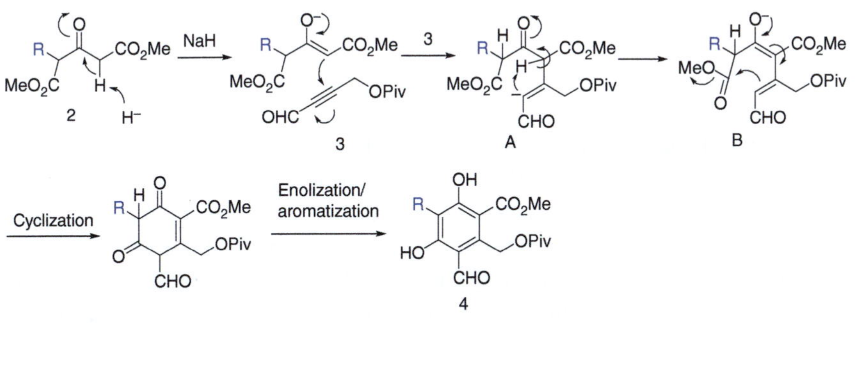

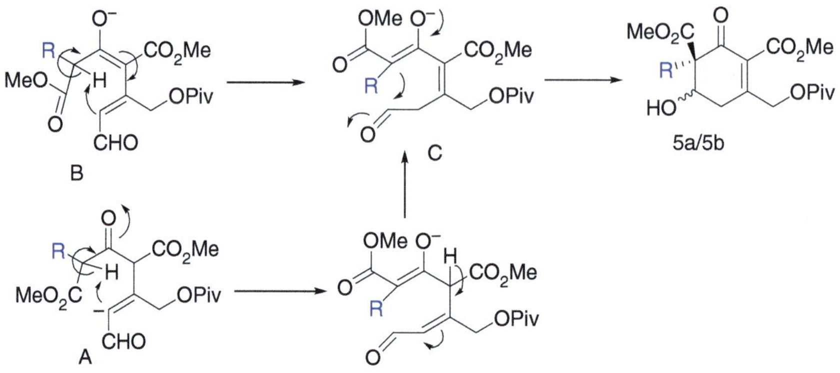

Problem 30: Vicarious Nucleophilic Substitution of Hydrogen in Aromatic Nitro Compounds

Golinski and Makosza reported a "vicarious nucleophilic substitution of hydrogen in aromatic nitro compounds [83, 84]." An *o*- or *p*-hydrogen in nitro aromatic compounds can be "substituted" by chloromethyl sulfones to form corresponding methylsulfone derivatives. For example, compound **3** was prepared in 67% from **1** and **2**. The scope of the reaction was explored by the same group and summarized in a review [85].

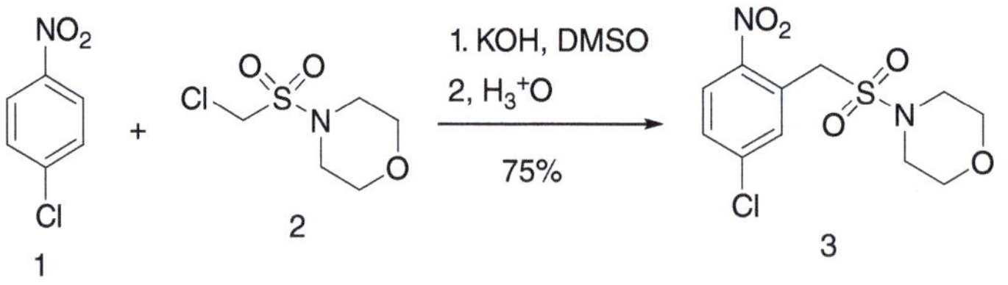

This reaction was applied to the synthesis of a series of glucan synthase (GS) inhibitors (**4**), which are potential arsenal antifungal drugs [86].

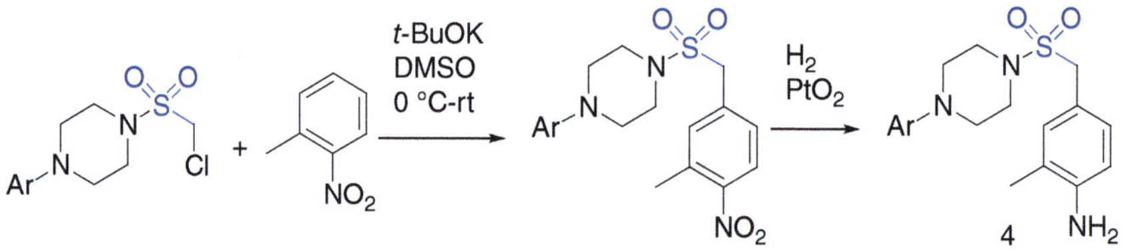

Please provide a mechanism for the formation of **3** from **1** and **2**.

Experimental Procedure (from *J. Org. Chem.*, **1984**, *49*, 1488)

Reaction of Nitroarenes with α-Halo Sulfonyl Compounds. General Procedure. Procedure A. To a stirred solution of nitroarene (0.01 mol) and 1-chloroalkyl phenyl sulfone or 1-chloroalkyl 4-morpholinyl sulfone (0.01 mol) in Me_2SO (15 mL) was added powdered KOH (4 g) and the reaction carried out at room temperature for 1 h. The mixture was poured into 2% HCl (100 mL) and extracted with chloroform, and the extract was dried over $MgSO_4$. A small portion of the extract was examined by ^{1}H NMR or by GLC to estimate the proportion of isomers in the crude mixture. The combined extracts were evaporated and the products were isolated and separated by column chromatography using chloroform as eluent. The products were purified by recrystallization.

Procedure C. Reaction and workup were completed as in **Procedure A**. After evaporation of the solvent, the crude product was purified by recrystallization. (This procedure was used when only one isomer was formed.)

Reaction of 4-Chloronitrobenzene (1) with 2. Reaction by **Procedure C** gave 5-chloro-2-nitrobenzyl 4-morpholinyl sulfone (**3**): yield 75%; mp 172–173 °C (AcOH). ^{1}H NMR 3.1 (m, 4H), 3.3–3.5 (m, 4H), 4.71 (s, 2H), 7.5–8.1 (m, 3H).

The Mechanism

The mechanism was proposed by Makosza [83, 85]. In this reaction, the Cl in compound **2** seems replaced by the nitroarenes at *ortho* and *para* to the nitro group. But in fact, the first stage is that σ-adduct formed from **2** and **1**. Then the σ-adduct is converted into the product by base-induced elimination of hydrogen chloride. Finally, acidic aqueous workup releases the product **3**.

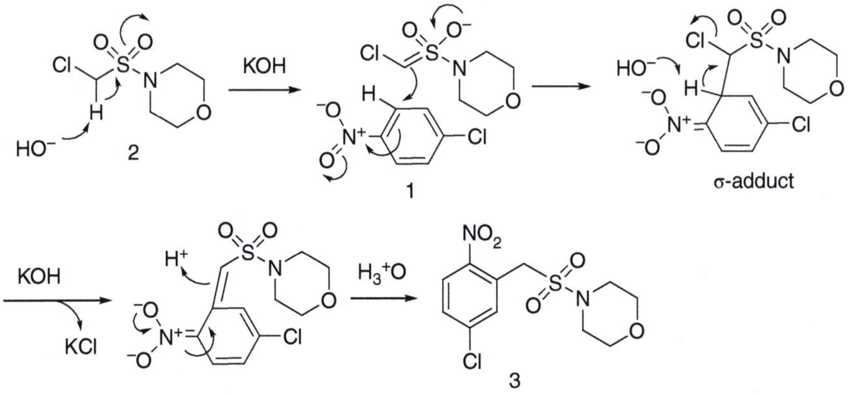

Problem 31: Synthesis of Tetralone from 1-Naphthol and 1,2-Dichlorobenzene

Sertraline hydrochloride (**1**) is an important pharmaceutical agent for the treatment of depression as well as dependency and other anxiety-related disorders [87]. The reported synthesis of **1** was starting from the tetralone **2**, which was prepared from 1-naphthol (**3**) and 1,2-dichlorobenzene (**4**) catalyzed by $AlCl_3$ [88–90].

Please propose a mechanism for the preparation of **2**.

Experimental Procedure (from *J. Org. Chem.*, **2011**, *76*, 10011)

4-(3,4-Dichlorophenyl)-3,4-dihydronaphthalen-1(2H)-one (2). To a stirred solution of 1-naphthol (5.0 g, 34.68 mmol), 1,2-dichlorobenzene (32 mL) and anhydrous $AlCl_3$ (11.56 g, 86.70 mmol) were added. The reaction mixture was stirred at 100 °C for 1.5 h under N_2. The reaction mixture was then cooled to room temperature and poured into ice and 1 N HCl (15 mL). The aqueous layer was extracted with CH_2Cl_2

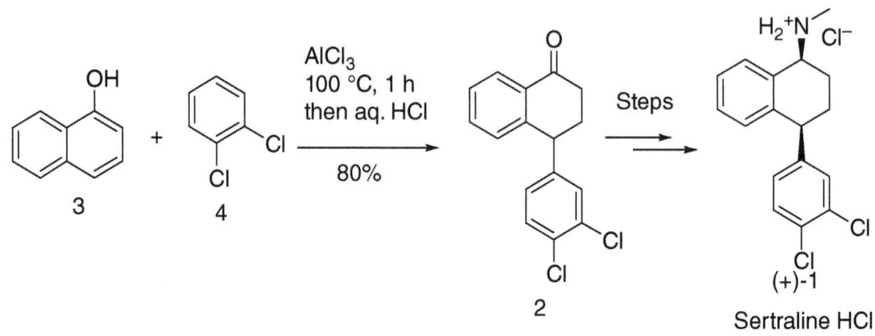

(2 × 100 mL). The organic layer was washed with H_2O, stirred with celite (6.7 g) and activated carbon (5 g), then filtered. The solvent was concentrated *in vacuo*. The residue was purified by flash column chromatography (*n*-hexane/EtOAc = 10/1) to afford 8.2 g (82%) of **2** as a white solid: $R_f = 0.13$ (*n*-hexane/EtOAc = 10/1); mp 99–101 °C; IR (KBr) υ 3422, 2859, 1675, 1591, 1470, 1286, 1203, 1029 cm^{-1}; ^1H NMR (300 MHz, $CDCl_3$) δ 2.20–2.32 (m, 1H), 2.42–2.52 (m, 1H), 2.58–2.77 (m, 2H), 4.28 (dd, $J = 8.0$ and 4.6 Hz, 1H), 6.95 (dd, $J = 8.4$, 2.0 Hz, 2H), 7.23 (d, $J = 1.9$ Hz, 1H), 7.39 (d, $J = 8.4$ Hz, 1H), 7.40 (t, $J = 7.2$ Hz, 1H), 7.48 (dt, $J = 7.2$, 1.5 Hz, 1H), 8.13 (dd, $J = 8.0$ and 1.9 Hz, 1H); ^{13}C NMR (125 MHz, $CDCl_3$) δ 31.6, 36.5, 44.5, 127.3, 127.5, 127.9, 129.2, 130.5, 130.6, 130.9, 132.6, 132.7, 133.8, 144.0, 144.8, 197.4; LC/MS (ESI) m/z 291.0 $(M + H)^+$; HRMS (EI) calcd for $C_{16}H_{12}Cl_2O$ $(M)^+$ 290.0265, found 290.0264.

The Mechanism

This reaction can be regarded as a Friedel-Crafts alkylation. Due to the electron-withdrawing property of the chlorine and activated by $AlCl_3$, the 1,2-dichlorobenzene can be an electrophile and attacked by 1-naphthol to give species **A**. After a series of isomerization-aromatization (H-shifts), the product **2** is produced.

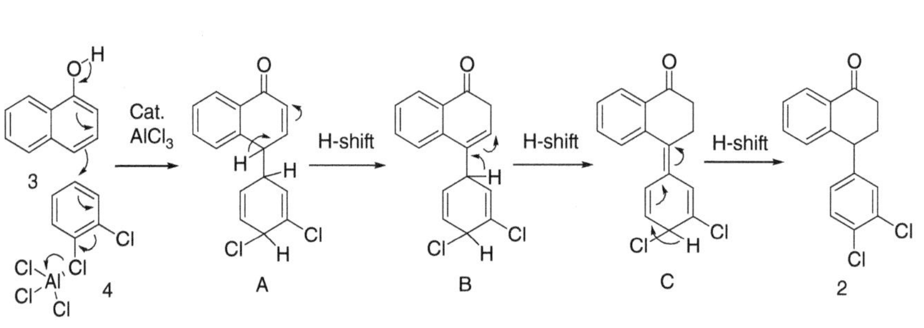

4.5 Formation of Nonaromatic Heterocycles

4.5.1 Lactone

Problem 32: Preparation of Thiolactone from Lactone

Biotin (**6**), also referred to as vitamin B_7 or vitamin H, is a member of the B-vitamin family. It plays a crucial role in various metabolic processes in humans and other organisms, particularly in the metabolism of fats, carbohydrates, and amino acids. A review published in 1997 outlined the synthesis of biotin [91]. Given its commercial significance, the production methods for biotin continue to be documented in patents and scientific articles. In one such synthetic process, thiolactone **2** served as a key intermediate. This compound was synthesized by heating lactone **1** with potassium thioacetate [92]. Alongside the target product **2**, several side products, namely **3**, **4**, and **5**, were also isolated and identified as impurities.

Please provide mechanism for the conversion of **1** to **2** and the formation of **3** and **4**.

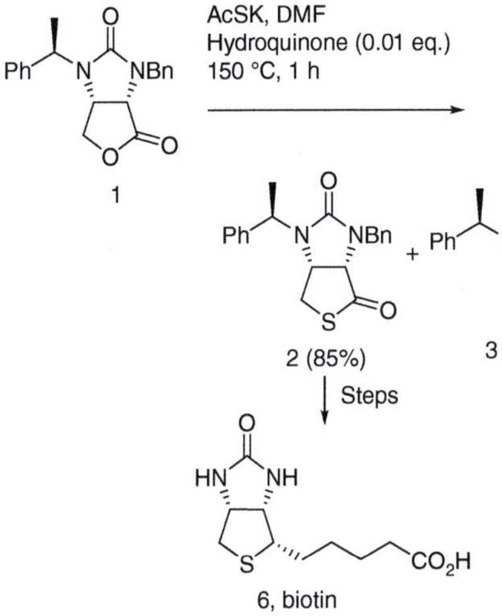

Experimental Procedure (from *Org. Process Res. Dev.*, **2003**, *7*, 272)
(3a*S*,6a*R*)-1-[(*R*)-1-Phenylethyl]-3-benzyltetrahydro-1H-thieno[3,4-*d*]imidazol-2,4-dione (2). A mixture of **1** (100 g, 0.29 mol), hydroquinone (0.32 g, 0.0028 mol), and potassium thioacetate (44.08 g, 0.39 mol) in 130 mL of *N,N*-dimethylacetamide was stirred in a 500-mL double-walled glass reactor. The system was nitrogen-purged three times before being heated to 150 °C under nitrogen for exactly 1 h. The suspension became a clear-brown solution at ca. 135 °C, and beige crystals were formed when the reaction reached ca. 150 °C. The suspension was cooled to 100 °C followed by the addition of glacial acetic acid (1.77 mL, 0.03 mol). The suspension was cooled to 55 °C over 30 min, and 432 mL of water was added. The suspension was stirred at 55 °C for 1 h at room temperature. Filtration followed by a water wash (240 mL) afforded crude **2** (99.9 g, 0.28 mol) as a light-brown/beige solid material. The crude **2** was then dissolved in 2-propanol (675 mL) at reflux. This solution was then cooled to 0–5 °C over a period of 3–4 h. The product was isolated by filtration and dried in a vacuum oven at 50 °C until the loss on drying was <0.5%. **2** was obtained as a beige solid (92.4 g, 0.26 mol) in 88% yield and with an assay of 99.1% (HPLC).

3, **4**, and **5**: **Side Product Isolation.** Approximately 1 L of plant 2-propanol recrystallization mother liquor was concentrated at the rotary evaporator to an oil, and a portion of this was separated on a silica gel column using ethyl acetate-hexane as eluent. From this chromatography, **3**, **4**, and **5** were isolated and subsequently identified.

3: (1-benzyl-4-methyl-3-(1R-phenethyl)-1,3-dihydroimidazol-2-one, $C_{19}H_{20}N_2O$ [292.38 g/mol]): ^{1}H NMR 1.62 (*s*, 3H), 1.80 (*d*, 3H, $J = 7.1$), 4.63 (*s*, 2H), 5.49 (*q*, 1H, $J = 7.1$), 5.61 (*s*, 1H), 7.1–7.5 (*m*, 10H). MS (CI-methane) 293 (M^+, 100), 217 (12), 189 (48), 105 (43), 91 (36).

4: (1-benzyl-3-(1R-phenethyl)-1,4-dihydro-3H-thieno-[3,4-d]imidazol-2,6-dione, $C_{20}H_{18}N_2O_2S$ [350.44 g/mol]): ^{1}H NMR 1.71 (*d*, 3H, $J = 7.1$), 3.12 (*d*, 1H, $J = 17.4$), 3.71 (*d*, 1H, $J = 17.4$), 4.95 (*s*, 2H), 5.61 (*q*, 1H, $J = 7.1$), 7.27–7.40 (*m*, 8H), 7.49 (*dd*, 2H, $J = 8, 1.5$). MS (CI-methane) 351 (M^+, 100), 247 (34), 105 (98), 91 (55).

The Mechanism

According to soft-hard acid-base principle, AcSK is classified as a soft base. The thioacetate would attack at the soft acid C-6 of **1** rather than at C-4. This reaction can be regarded as trans-esterification. Thio anion in AcSK is a much stronger nucleophile than oxygen anion in AcOK.

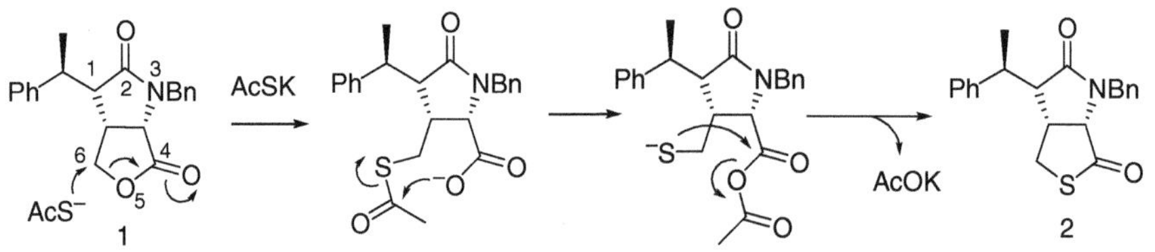

The formation of compound **3** is postulated as 1,2-hydrogen shift and decarboxylation side product.

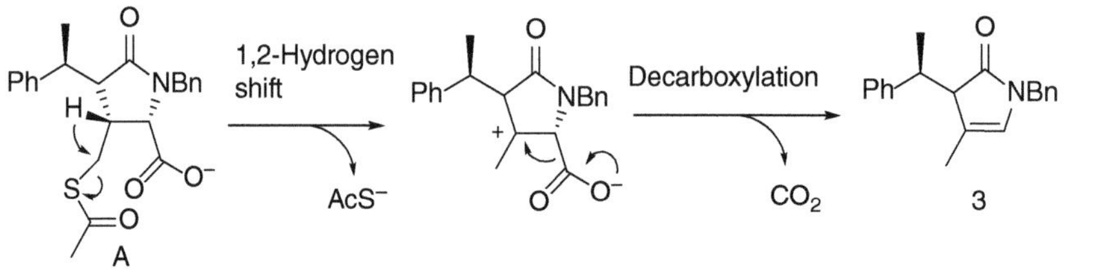

Compound **4** likely originates from its own oxidation by the oxygen present in the reaction system. The percentage of **4** was reduced from 2.0% to 0.4% by diminishing the oxygen level from air atmosphere to <0.3 ppm via argon protection [92].

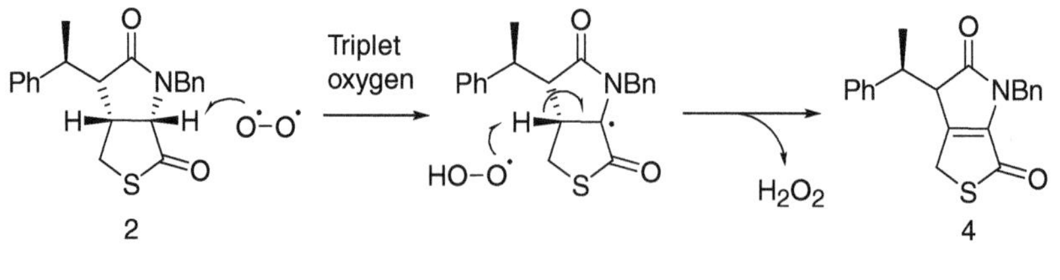

Problem 33: Preparation of D-Erythronolactone from Erythorbic Acid

Nelfinavir **1** is one of the most prescribed therapeutic agents to suppress the AIDS epidemic. (2*S*,3*R*)-3-(*N*-Benzyloxycarbonyl)amino-1-chloro-4-phenylthiobutan-2-ol (**2**) is a central building block of nelfinavir [93]. An enantioselective synthesis of **2** is starting from erythorbic acid (**3**) and the first step is the oxidation of **3** with hydrogen peroxide to give D-erythronolactone **4** [94, 95].

Please suggest a mechanism for the reaction from **3** to **4**.

HO HO H O O HO OH 3 — 1. 30% H_2O_2 Na_2CO_3 2. HCl → HO OH O O 4 — Steps → PhS NHCbz Cl OH 2 ⟸ HO O NH PhS OH N O NH H H 1 (Nelfinavir)

Experimental Procedure (from *J. Org. Chem.*, **1991**, *56*, 6225)

Dihydro-3(R),4(R)-dihydroxy-2(3H)-furanone (4). According to the procedure of Cohen, D-isoascorbic acid (704 g) in water (10 L, 4 °C) was sequentially treated with sodium carbonate (848 g over 30 min; conversion to sodium enolate), hydrogen peroxide (CAUTION, 30%, 880 mL over 1 h; exotherm from 5 to 35 °C; warmed to 40 °C for 30 min), and Darco G-60 (160 g in portions, 0–4 °C, then 25 °C). After observing a negative peroxide test, the mixture was filtered through Celite (2-L boiling water wash), treated with 6 N hydrochloric acid (2.1 L, pH 1.5), and concentrated to dryness under vacuum (45–50 °C). The solid was extracted with boiling ethyl acetate (5 × 3 L portions). These combined extracts were concentrated *in vacuo*, treated with diethyl ether (2 L), filtered, affording D-erythronohctone (436 g, 94%), which was suitable for further use: mp 101–103 °C $[\alpha]^{25}_{D}$ −73° (c 1.00, H_2O).

The Mechanism

One possible mechanism involves the free radical pathway. The initial step likely involves the oxidation of compound **3** by hydrogen peroxide through a free radical process, resulting in the formation of diketone **A**. Subsequently, the addition of hydrogen peroxide to the C-3 carbonyl group yields hydroperoxyl alcohol **B**. A hydroxyl radical then attacks the C-4 carbonyl, generating intermediate **C**, which undergoes ring cleavage with the loss of a hydroxyl radical. At this stage, the ester may be hydrolyzed to produce oxalic salt and the sodium salt **D**. Following acidification, cyclization occurs, leading to the formation of D-erythronolactone **4**.

Another plausible mechanism is the oxidation of the di-enol carbons by hydrogen peroxide and cleavage of the C3-C4 bond (next page).

Problem 34: Synthesis of L-lyxonolactone-2,3-*O*-isopropylidene

L-lyxonolactone-2,3-*O*-isopropylidene (**1**) is a key intermediate for the synthesis of UT-231B (iminosugar, **2**) [96], an analog of deoxynojirimycin. Iminosugar **2** is currently in clinical trials as an antiviral agent for the treatment of hepatitis C.

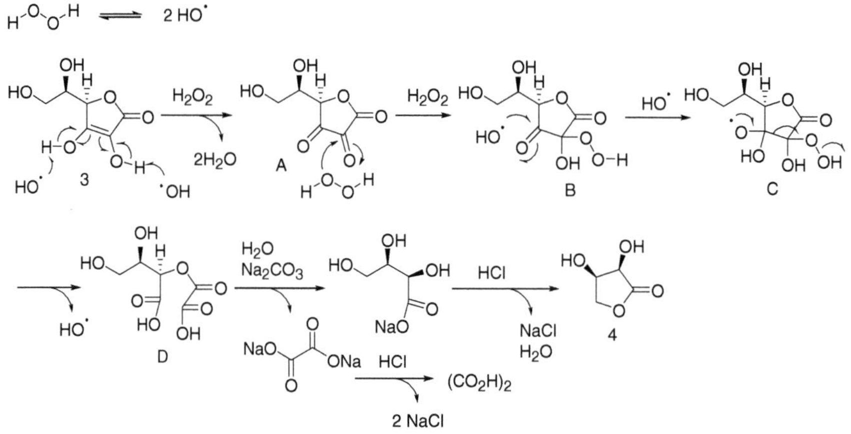

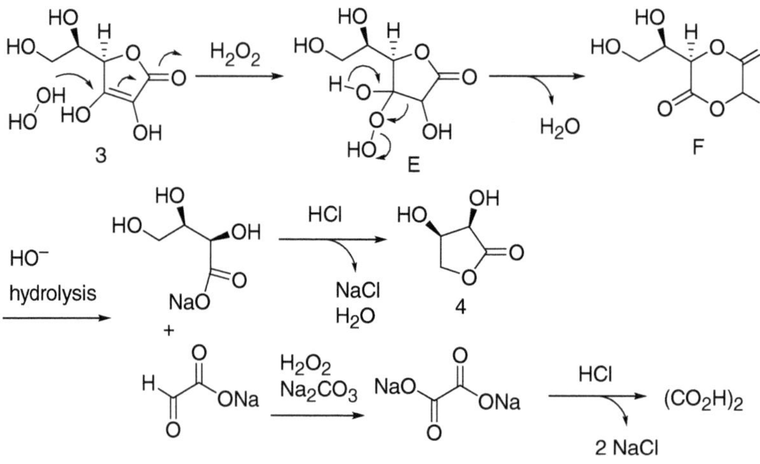

One of the reported synthesis of **1** is starting from D-ribose **3** as shown below [97]. The (*R*-) stereochemistry at C-4 in compound **5** is converted to (*S*-) form in **2** via mesylation and hydrolysis of the C-5 primary alcohol. Please suggest a mechanism for the reaction from **6** to **1**.

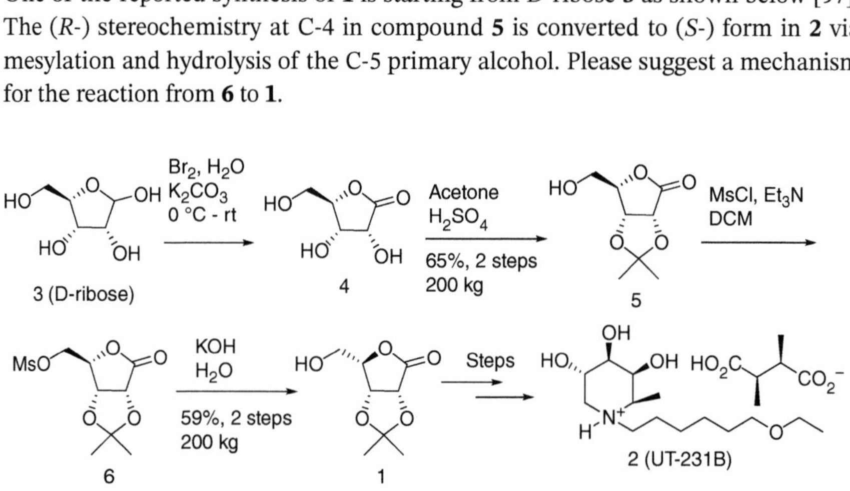

Experimental Procedure (from *Org. Process Res. Dev.*, **2006**, *10*, 484)

L-Lyxonolactone-2,3-*O*-isopropylidene (1). To D-ribonolactone-2,3-*O*-isopropylidene (**5**) (50 kg, 266 mol, 1.0 equiv) was added DCM (750 L) and triethylamine (29.8 kg, 41.0 L, 294 mol, 1.1 equiv) under nitrogen. This mixture was stirred at ambient temperature until it became clear, and then it was cooled to −20 °C. To this solution was slowly added methanesulfonyl chloride (33.3 kg, 22.5 L, 291 mol, 1.1 equiv), and the solution was stirred for 1 h at −20 °C. The solution was allowed to attain ambient temperature, and it was maintained at this temperature for 8 h. The reaction was quenched with water (400 L), and the organic layer was separated. The aqueous layer was extracted with DCM (2 × 200 L). The organic layers were combined, washed with water (2 × 200 L), and concentrated *in vacuo* below 30 °C to obtain the semisolid mesylate (**6**). To this crude mesylate was added a solution of KOH (36.5 kg, 651 mol, 2.44 equiv) in 245 L of water, maintaining the temperature below 30 °C. This solution was stirred for 4 h at the same temperature and then adjusted to pH 2.5–3.0 by adding 3 M hydrochloric acid (33.5 L). The acidic solution was concentrated *in vacuo* maintaining the temperature below 45 °C to afford a solid mass. The solid mass was triturated with acetone (300 L) and heated to reflux. The acetone was decanted, dried over sodium sulfate (35 kg), and filtered. The clear filtrate was concentrated *in vacuo* below 35 °C to yield the crude product. The product was crystallized from 2-propanol (100 L) to afford white, crystalline L-lyxonolactone-2,3-*O*-isopropylidene (**1**), 17.1 kg (43.1%, on a 200-kg scale the yield improved to 59%); mp 98–99 °C (lit mp 92–93 °C), $[R]_D^{25}$ −89° (*c* = 1.0, acetone), [lit $[R]_D^{20}$ −85.6° (*c* = 1.0, acetone)]. IR (KBr):

3423, 1778 cm^{-1}. ^{1}H NMR (300 MHz, $CDCl_3$): δ 1.36 (s, 3H, CH_3), 1.44 (s, 3H, CH_3), 2.72 (brs, 1H, OH), 3.75–3.77 (d, 1H, C 5′H), 3.91–4.01 (m, 2H, C-5H and C-5′H), 4.60–4.64 (m, 1H, C-4H), 4.83–4.89 (m, 2H, C-3H and C-2H). Anal. Calcd for $C_8H_{12}O_5$: C, 51.06; H, 6.38. Found: C, 50.96; H, 6.44.

The Mechanism

The process starts from a basic hydrolysis of the ester to form an intermediary epoxide **B** via anion **A**. The configuration at C-4 is retained at this stage. Subsequently intramolecular ring opening of the epoxide by the carboxylate nucleophile in a favorable 5-endo-*tet* process [98] proceeds with inversion of configuration to afford **1**.

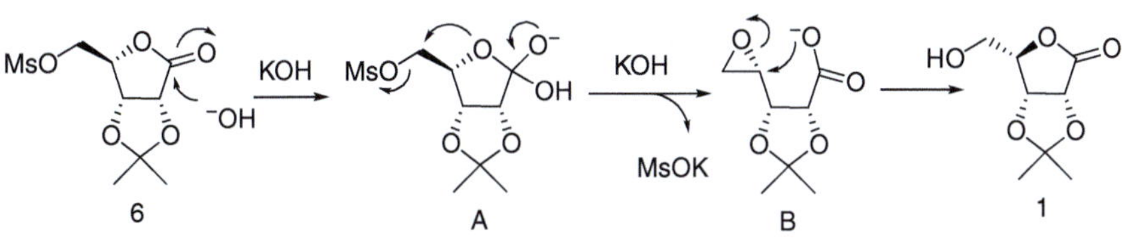

MsO ... 6 —KOH, $^-$OH→ A —KOH, −MsOK→ B → 1

Problem 35: Thermal Rearrangement of Ozonides

Ozonides (1,2,4-trioxolanes) are a class of cyclic peroxides that are structurally similar to 1,2,4-trioxanes. Arterolane (**OZ277**) [99], which features an ozonide core structure, is a recently approved antimalarial drug in several countries.

In 1993, our research group reported the thermal rearrangement of a stable steroidal ozonide (1,2,4-trioxolane) **1** to form lactone **2** [100]. This reaction was discovered serendipitously during a melting point determination. Ozonide **1** has a melting point of 98–99 °C. While performing the melting point determination using a hot plate, I momentarily stepped away to attend to another task. Upon returning, I realized the temperature had already reached 150 °C, and the crystalline sample had melted. Rather than discarding the molten material, I recovered it and analyzed it using TLC, comparing it with the original crystalline sample. Surprisingly, the TLC analysis revealed no trace of the starting ozonide **1**; instead, a new, more polar major spot was observed. The thermal reaction was subsequently repeated and scaled up to obtain sufficient quantities of the pure product, which was ultimately identified as lactone aldehyde **2**.

Please suggest a mechanism for the rearrangement from **1** to **2**.

150 °C
30 min
(80%)
Arterolane (OZ277)

Experimental Procedure (from *J. Chem. Res.*, **1993**, 30)
^{1}H and ^{13}C NMR spectra were measured on a Bruker AM-400 spectrometer using $CDCl_3$ as solvent.

3β-Acetoxy-15-oxo-13,14;14,15-disecocholestano-14,13-lactone (2). Crystalline **1** (100 mg) was heated under nitrogen at atmospheric pressure in an oil bath in which the temperature was controlled at 150 ± 2 °C for 20 min. The crude product was purified by chromatography (8 g silica gel, eluent: 20% diethyl ether in petroleum ether) to yield **2** as a viscous oil (80 mg, 80%) (Found: C, 72.9; H, 10.1. C29H48O5 requires C, 73.07; H, 10.15%). ν_{max}/cm^{-1} (film) 1734 br vs, 1245, 1030. δ_H 0.837 (3H, s, 19-H), 0.856, 0.859 (6H, 2d, *J* 6.6 Hz, 26, 27-H), 0.958 (3H, d, *J* 6.9 Hz, 21-H), 1.291 (3H, s, 18-H), 2.024 (3H, s, Ac-H), 2.315 (1H, ddd, *J* 16.0, 5.7, and 2.6 Hz, 16-H), 2.474 (1H, td, *J* 11.6 and 4.3 Hz, 8β-H), 2.718 (1H, ddd, *J* 16.0, 7.8, and 3.2 Hz, 16-H), 4.683 (1H, m, 3α-H), 9.813 (1H, dd, *J* 2.7 and 2.9 Hz, 15-H). δ_C 37.07 (C-1), 27.35 (C-2), 72.98 (C-3), 33.57 (C-4), 43.72 (C-5), 28.46 (C-6), 32.64 (C-7), 45.57 (C-8), 49.14 (C-9), 36.90 (C-10), 21.55 (C-11), 27.19 (C-12), 85.34 (C-13), 178.85 (C-14), 203.29 (C-15), 36.32 (C-16), 50.29 (C-17), 22.10 (C-18), 11.75 (C-19), 32.64 (C-20), 21.33 (C-21), 39.10 (C-22), 25.97 (C-23), 40.43 (C-24), 27.91 (C-25), 22.49 (C-26), 22.64 (C-27), 21.02 ($Ac\text{-}CH_3$), 170.48 (Ac-C=O). [^{13}C NMR data for 3β-acetoxyandrostane: 21.3 ($Ac\text{-}CH_3$), 169.9 (Ac-C=O)]. *m/z* 476 (2%, M^+), 458 (8), 432 (16), 363 (23), 307 (100).

The Mechanism

This thermal rearrangement should be similar to the mechanism of Baeyer–Villiger oxidation. The most substituted alkyl group migrates to the peroxide oxygen and thus the 1,2,4-trioxolane ring is cleaved to give the product **2**.

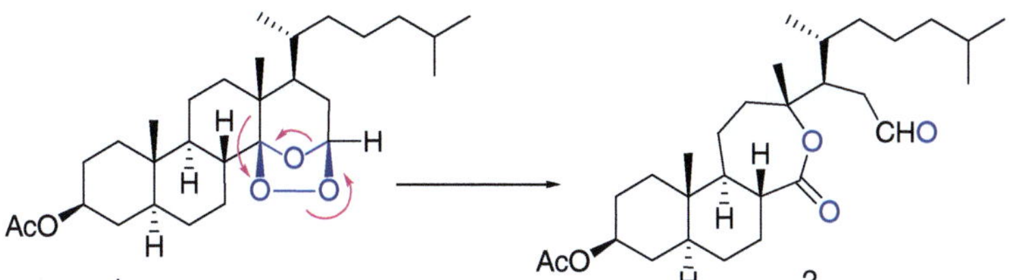

To demonstrate that the migrating group maintains its stereochemistry, we conducted the following reactions [101]. Heating ozonide **3** (neat) yielded lactone **4**. Reduction of **3** using Zn-HOAc produced the 1,5-dicarbonyl compound **5**. Subsequent Baeyer-Villiger oxidation of **5** proceeded with both regio and chemoselectivity, yielding lactone **4**, which was identical to the product obtained from the thermal rearrangement of **3**.

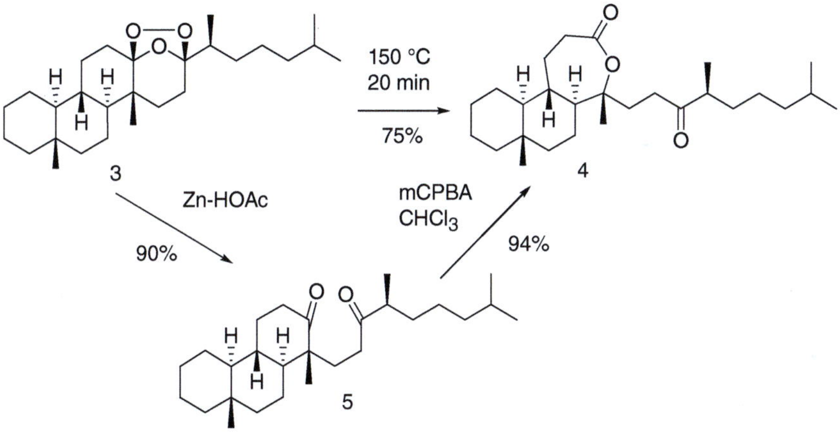

4.5.2 Hydrofuran

Problem 36: Preparation of 2,2-Dimethyl-5-phenylfuran-3(2H)-one by a Rearrangement

In 1958, Parker, Raphael, and Wilkinson reported an unusual rearrangement of 4-hydroxy-4-methyl-1-phenylpent-2-yn-1-one (**1**) to 2,2-dimethyl-5-phenylfuran-3(2H)-one (**2**) in the presence of diethylamine [102]. Recently, that reaction was applied to the synthesis of polmacoxib [103], which was approved in South Korea for the treatment of colorectal cancer. The mechanism of the reaction was discussed in the literature [102].

Please suggest your own mechanism for the reaction.

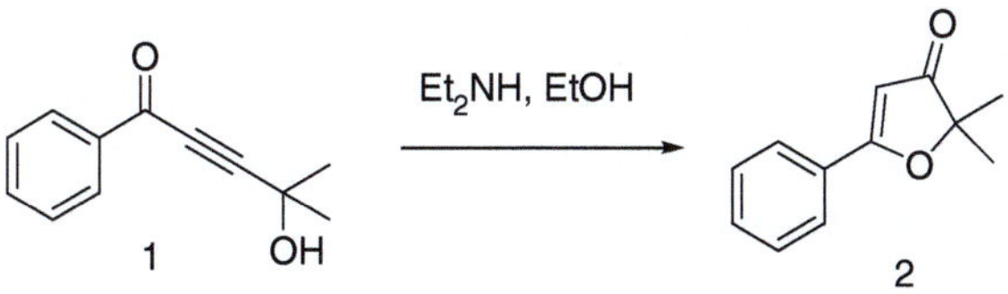

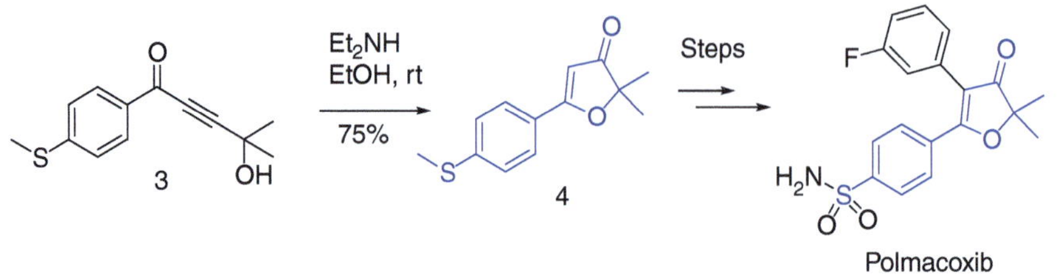

Experimental Procedure (from *J. Med. Chem.*, **2004**, *47*, 792)

Preparation of 2,2-Dimethyl-5-{4-(methylthio)-phenyl}-3(2*H*)-furanone (4). To a stirred solution of **3** (120 mg) in 20 mL of ethanol was added dropwise diethylamine (0.08 mL) diluted in 7 mL of ethanol over 5 min at room temperature. The reaction solution was stirred for another 1 h and then the solvent was removed *in vacuo*. The resulting residue was diluted with 50 mL of water and then extracted with DCM (30 mL × 3). The organic layer was concentrated *in vacuo* and subjected to column chromatographic separation (hexane/ethyl acetate) 4:1) to give 90 mg of **4** as a solid. mp: 107–109 °C. NMR: δ 1.48 (s, 6H), 2.54 (s, 3H), 5.91 (s, 1H), 7.30 (d, $J = 8.4$ Hz, 2H), 7.72 (d, $J = 8.4$ Hz, 2H). IR (cm^{-1}): 1676, 1579, 1485, 1376, 1174, 1095, 1050, 809.

Answer

Parker et al [102]. explained that the mechanism was initially a Michael addition of diethylamine to the ynone (**1**) to form adduct **A**. Cyclization of **A** provided hemiketal **B**. Subsequently, an oxotropic rearrangement led to dihydrofuran **C**. Finally, extrusion of diethylamine generated the product **2**.

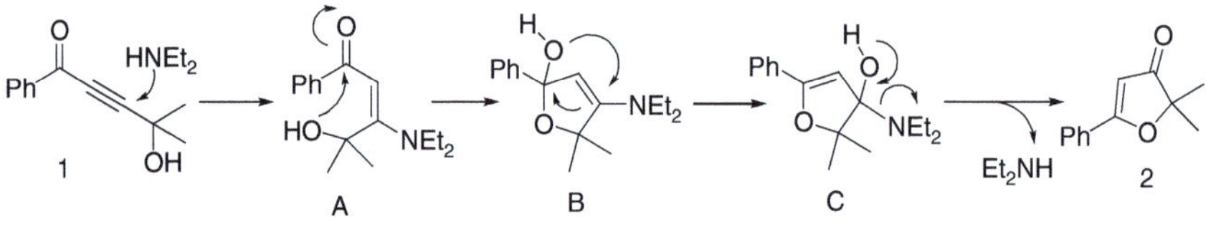

However, at the Michael addition stage, two geometric isomers, *E*-isomer **A** and *Z*-isomer **A′** could be produced. The *Z*-isomer **A′** could not cyclize to form such kind of **B′**. Although **A′** could be converted to **A** through tautomerization, another mechanism via the epoxide **D** or **E** is also possible.

Interconversion of the (*Z*)- and (*E*)-isomers (next page).

Problem 37: Decarboxylative Elimination of *β*-Hydroxycarboxylic Acid

Rovafovir etalafenamide (**GS-9131, 1**) is a phosphonamidate prodrug of the nucleotide reverse transcriptase inhibitor (NtRTI). **GS-9148** (**2**) is under investigation for the treatment of HIV-1 infection by Gilead. The process chemistry of **1** involves a key decarboxylative elimination of *β*-hydroxycarboxylic acid **3** along with protection of the adenine as an iminophosphorane to yield the 2,3-dihydrofuran **4** [104, 105].

Please suggest a mechanism for the reaction from **3** to **4**.

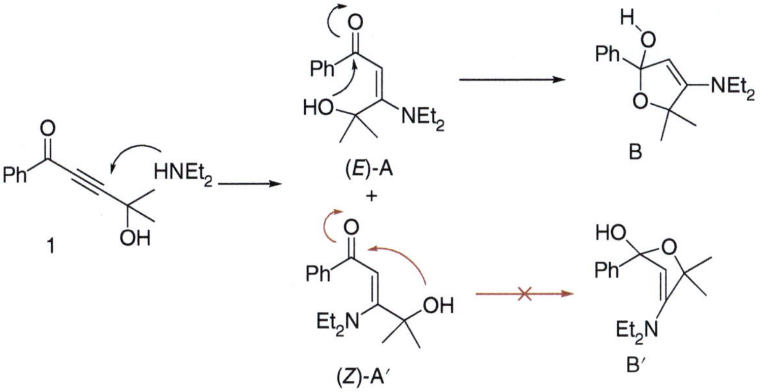

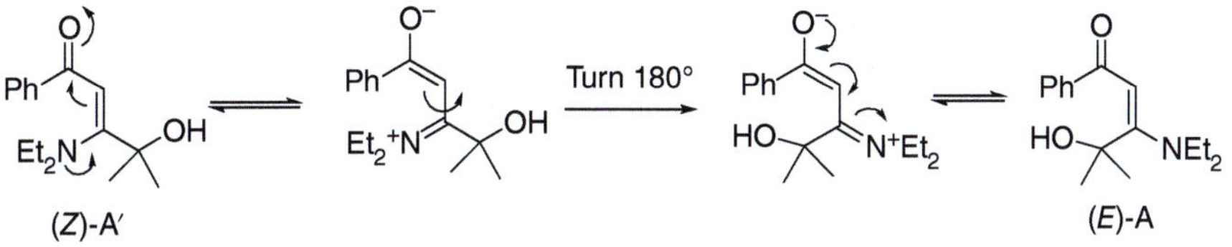

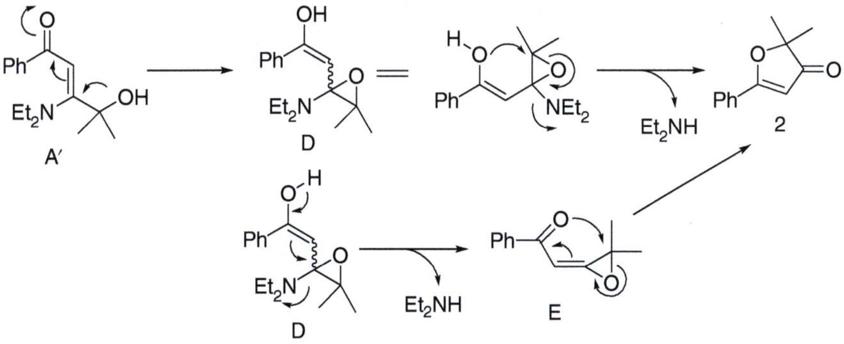

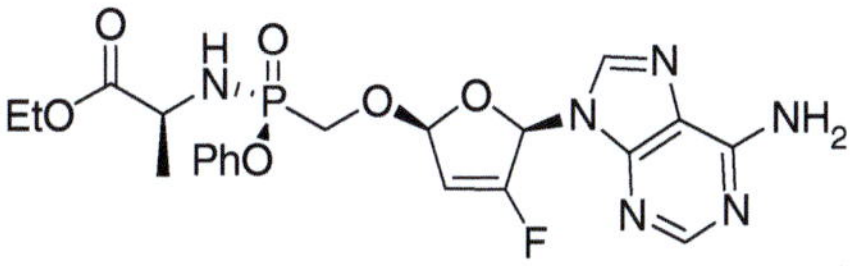

1   HNEt$_2$ → (E)-A + (Z)-A′

(E)-A → B

(Z)-A′ ✗→ B′

(Z)-A′ ⇌ Turn 180° ⇌ (E)-A

A′ → D = → 2 (−Et$_2$NH)

D → E (−Et$_2$NH) → 2

1, Rovafovir etalafenamide (GS-9131)

2, GS-9148

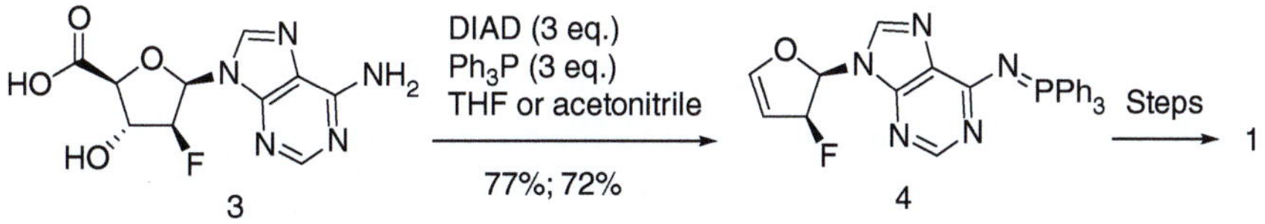

3 → DIAD (3 eq.), Ph$_3$P (3 eq.), THF or acetonitrile; 77%; 72% → 4 → Steps → 1

Experimental Procedure (from *Org. Process Res. Dev.*, **2021**, *25*, 1215)
Synthesis of *N*-(9-((2R,3S)-3-Fluoro-2,3-dihydrofuran-2-yl)-9H-purin-6-yl)-1,1,1-triphenyl-λ^5-phosphanimine (4). A reactor was charged with β-hydroxy-acid **3** (70.0 kg, 97.4 wt % purity, 240.7 mol, scaling factor), triphenylphosphine (2.7 kg/kg, 189.0 kg, 720.6 mol, 3.0 equiv), and acetonitrile (3.2 kg/kg, 224 kg, 285 L). To this mixture was slowly charged diisopropyl azodicarboxylate (2.1 kg/kg, 147.0 kg, 727.0 mol, 3.0 equiv) while maintaining the internal temperature of the reaction mixture below 30 °C. The mixture was agitated at 22 °C for 1 h, after which seed crystals of **4** (0.001 kg/kg, 0.07 kg) were charged and agitation was continued until the reaction was complete. After aging an additional 4 h, isopropyl acetate (10 kg/kg, 700 kg, 805 L) was charged and the slurry was concentrated under reduced pressure (45 °C jacket temperature) to ca. 560 L (8 L/kg). The temperature was adjusted to 37 °C, after which *n*-heptane (4.2 kg/kg, 294 kg, 430 L) was slowly charged to the slurry. The temperature was adjusted to 15 °C over 2 h, the contents aged for an additional 5 h at 15 °C, and then filtered into an agitated filter dryer. After deliquoring the filter cake, it was washed with pre-heated MTBE (3.0 kg/kg, 210 kg, 284 L) at 40 °C and deliquored. The wet filter cake was transferred back to the reactor and reslurried with MTBE (22 kg/kg, 1540 kg, 2081 L) at 40 °C for 3 h. The slurry was again filtered and after deliquoring the filter cake, it was washed with preheated MTBE (3.0 kg/kg, 210 kg, 284 L) at 40 °C and deliquored. The filter cake was again transferred back to the reactor and reslurried with MTBE (22 kg/kg, 1540 kg, 2081 L) at 40 °C for 3 h. The slurry was filtered a final time in an agitated filter dryer, rinsed with MTBE (1.1 kg/kg, 77 kg, 104 L), and thoroughly deliquored. The wet cake was dried under vacuum at 40 °C to yield 90.8 kg of **4** as a tan, crystalline solid. The potency-adjusted yield was 72.1%; 92.0 wt % purity. ^{1}H NMR (400 MHz, DMSO-d_6) δ 8.09 (d, J = 3.5 Hz, 1H), 8.02 (s, 1H), 7.90–7.83 (m, 6H), 7.66–7.60 (m, 3H), 7.59–7.52 (m, 6H), 7.26 (dd, 1H), 6.75 (dd, J = 28.6 and 5.9 Hz, 1H), 5.82 (ddd, J = 60.1, 5.9, and 2.6 Hz, 1H), 5.64 (dd, J = 2.2 and 2.2 Hz, 1H). ^{13}C{^{1}H} NMR (101 MHz, DMSO-d_6) δ 160.4 (d, J = 6.3 Hz), 153.5 (d, J = 12.1 Hz), 151.8, 149.9 (d, J = 5.1 Hz), 138.8 (d, J = 6.5 Hz), 132.7 (d, J = 9.9 Hz), 132.3 (d, J = 2.9 Hz), 128.7 (d, J = 12.2 Hz), 128.4 (d, J = 99.9 Hz), 125.1 (d, J = 24.3 Hz), 101.1 (d, J = 16.6 Hz), 90.4 (d, J = 192.9 Hz), 83.6 (d, J = 14.5 Hz). ^{19}F NMR (376 MHz, DMSO-d_6) δ −167.4 (dddd, J = 60.4, 28.7, 4.0, 4.0 Hz). ^{31}P{^{1}H} NMR (162 MHz, DMSO-d_6) δ 16.3. HRMS (ESI/TOF) m/z: $[M + H]^+$ Calcd for $C_{27}H_{22}FN_5OP$ 482.1541; Found 482.1554.

The Mechanism
This is a Mitsunobu type reaction. The initial steps follow the mechanism of Mitsunobu reaction. However, once the intermediate **A** is generated, decarboxylation occurs to give **B** and triphenylphosphine oxide, rather than cyclized to the unstable four-membered ring compound **C**. Since N—P bond energy (617 KJ/mol) is greater than O—P bond energy (597 KJ/mol), further reaction of **B** with $Ph_3P{=}O$ provides **4**.

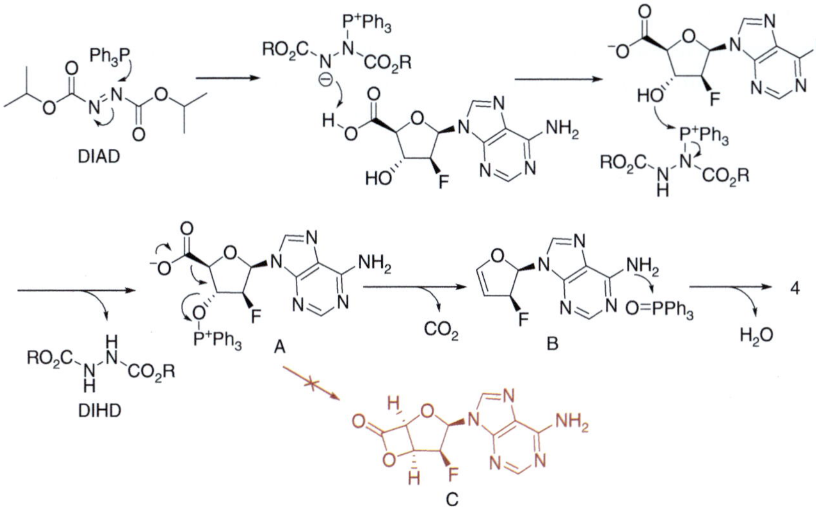

4.5.3 Tetrahydropyran

Problem 38: Construction of Fosdenopterin Core by Viscontini Reaction
Fosdenopterin is a medication used to reduce the risk of death due to a rare genetic disease known as molybdenum cofactor deficiency type A. The synthesis of fosdenopterin was published [106]. The core unit **3** was prepared from compounds **1** and **2** as shown below.

Please suggest a mechanism for the reaction.

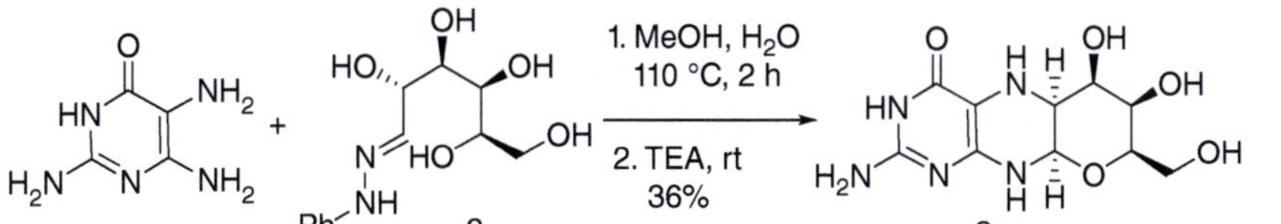

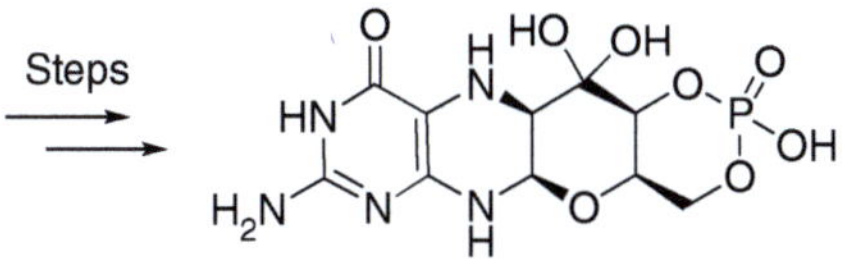

Experimental Procedure (from *J. Med. Chem.*, **2013**, *56*, 1730)
(5aS,6R,7R,8R,9aR)-2-Amino-6,7-dihydroxy-8-(hydroxymethyl)-3H,4H,5H,5aH,6H,7H,8H,9aH,10H-pyrano[3,2-g]pteridin-4-one (3). 2,5,6-Triamino-3,4-dihydropyrimidin-4-one dihydrochloride (**1**, 10.0 g, 46.72 mmol), D-galactose phenylhydrazone (**2**, 15.78 g, 58.38 mmol), and 2-mercaptoethanol (1 mL) were stirred and heated under reflux (bath temp 110 °C) in a 1:1 mixture of MeOH/H_2O (400 mL) for 2 h. After cooling to room temperature, diethyl ether (500 mL) was

added, the flask shaken, and the diethyl ether layer decanted off and discarded. The process was repeated with two further portions of diethyl ether (500 mL), and then the remaining volatiles were evaporated. Methanol (40 mL), H_2O (40 mL), and triethylamine (39.4 mL, 280 mmol) were successively added, and after 5 min the yellow solid was filtered off, washed with a little MeOH, and dried to give **3** (5.05 g, 36%) of suitable purity for further use. An analytical portion was recrystallized from boiling H_2O: mp 226 °C (decomp). $[\alpha]_D^{20}$ +135.6 (c 1.13, DMSO). ^{1}H NMR (DMSO-d_6): δ 10.19 (bs, exchanged D_2O, 1H), 7.29 (d, J = 5.0 Hz, slowly exchanged D_2O, 1H), 5.90 (s, exchanged D_2O, 2H), 5.33 (d, J = 5.1 Hz, exchanged D_2O, 1H), 4.66 (dt, J = 5.0 and 1.3 Hz, 1H), 4.59 (t, J = 5.6 Hz, exchanged D_2O, 1H), 4.39 (d, J = 10.3 Hz, exchanged D_2O, 1H), 3.80 (bt, J = 1.8 Hz, exchanged D_2O, 1H), 3.70 (m, 1H), 3.58 (dd, J = 10.3, 3.0 Hz, 1H), 3.53 (dt, J = 10.7 and 6.4 Hz, 1H), 3.43 (ddd, J = 11.2, 5.9, and 5.9 Hz, 1H), 3.35 (t, J = 6.3 Hz, 1H), 3.04 (bm, 1H). ^{13}C NMR (DMSO-d_6): δ 156.1 (C), 150.2 (C), 148.2 (C), 98.8 (C), 79.3 (CH), 76.3 (CH), 68.8 (CH), 68.4 (CH), 60.4 (CH_2), 53.7 (CH). ESI-HRMS calcd for $C_{10}H_{15}N_5NaO_5^+$, $(M + Na)^+$, 308.0965, found 308.0968. Anal. Calcd for $C_{10}H_{15}N_5O_5$ H_2O, 39.60 C, 5.65 H, 23.09 N, found 39.64 C, 5.71 H, 22.83 N. X-ray analysis of **3** confirmed the product as a monohydrate (Supporting Information).

The Mechanism

The hydrazone **2** is isomerized to the ketone **A** via Amadori rearrangement. Then regioselective Viscontini reaction occurs, the more basic 5-NH_2 group in **1** attacks the carbonyl in **A**. After dehydration, cyclization and elimination of $PhNHNH_2$, the product **3** was produced.

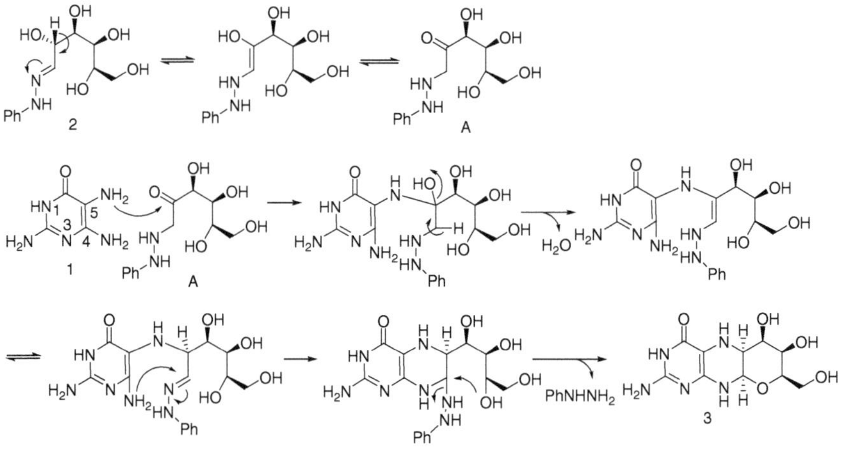

4.5.4 Pyrrolidine

Problem 39: Formation of 2,5-Dibenzyltetrahydropyrrolo[3,4-c]pyrrole-1,3-dione

RO5114436 is a CCR5 receptor antagonist which was developed by Roche [107]. 2,5-Dibenzyltetrahydropyrrolo[3,4-*c*]pyrrole-1,3-dione (**3**) was a key building block which was prepared from *N*-benzylmaleimide (**1**) and benzylglycine (**2**) in kilogram scale as shown in the following reaction.

Please propose a mechanism for the formation of **3**.

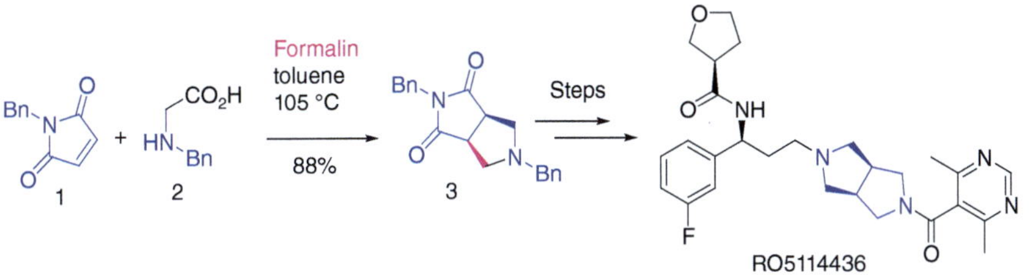

By using paraformaldehyde instead of formalin, lower yield (75%) [108] was resulted and the process was problematic on kilogram scale [107].

Experimental Procedure (from *Org. Process Res. Dev.*, **2010**, *14*, 592)

2,5-Dibenzyltetrahydropyrrolo[3,4-c]pyrrole-1,3-dione (3). Toluene (42 kg) was added to a mixture of *N*-benzylglycine (**2**, 2.16 kg, 13.11 mol) and *N*-benzylmaleimide (**1**, 1.76 kg, 9.38 mol) in a 120-L glass-lined reactor, and the resulting solution was heated to just under reflux (~105 °C). Formaldehyde/water (1.06 kg, 37% w/w, 13.1 mol) was added at a rate of 300 mL/h while collecting toluene/water distillate. After the completion of the reaction, the reaction mixture was cooled, and the solvent was distilled under reduced pressure (30–125 Torr) at 50 °C. MeOH (28 L) was added and then distilled off, the mixture was cooled to ambient temperature, and more MeOH (22.4 L) was added. Water (15 L) was slowly added to the solution over 2 h, and the slurry was stirred at ambient temperature overnight and then filtered. The solid was washed with MeOH/water (3 × 4 L, 60:40 v/v) and dried at 50 °C under reduced pressure (30–50 Torr) to give **3** (2.65 kg, 88% yield, 99% purity by HPLC area) as a light gray solid. ^{1}H NMR (300 MHz, $CDCl_3$) 2.13–2.58 (m, 2H), 3.07–3.25 (m, 2H), 3.30 (d, J = 10.2 Hz, 2H), 3.55 (s, 2H), 4.69 (s, 2H), 6.90–7.67 (m, 10H). MS m/z 321.1 $[M + H]^+$.

The Mechanism

This is a [3 + 2] cycloaddition, which can also be regarded as a special Mannich reaction. The reaction starts with the formation of an iminium ion **A** from benzylglycine (**2**) and formaldehyde. Then a [3 + 2] cycloaddition occurs between *N*-benzylmaleimide (**1**) and iminium ion **A** to give the adduct **B**. After decarboxylation, the product **3** is provided (next page).

Problem 40: Synthesis of *trans*-3-substituted Proline Derivatives

Schmalz's group reported a synthesis of *trans*-3-substituted proline derivatives **3** [109]. The first step is a one-pot reaction of proline methyl ester **1** with NBS followed by addition of Cbz-Cl to provide **2**. The proline derivatives **3** was demonstrated in the development of potent hepatitis C virus protease inhibitors [110] and farnesyl transferase inhibitors [111].

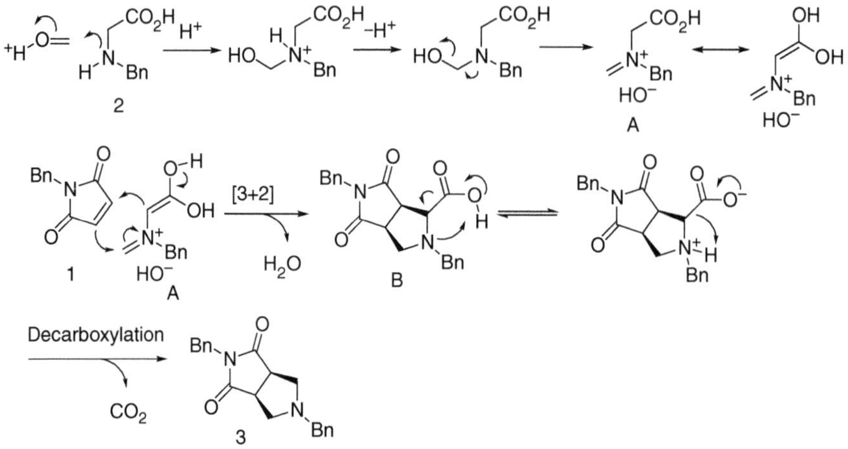

Please provide a mechanism for the reaction from **1** to **2**.

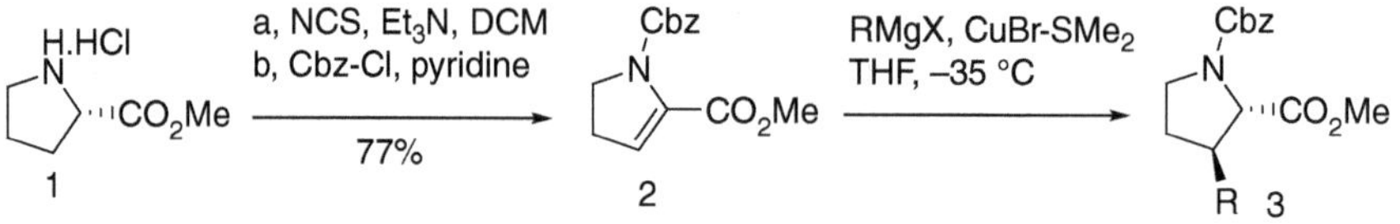

Experimental Procedure (from *Org. Lett.*, **2011**, *13*, 216, Supporting Information)

Synthesis of Methyl *N*-benzyloxycarbonyl-4,5-dihydropyrrole-2-carboxylate (2). To a solution of 43.28 g (264.1 mmol) of hydrochloride **1** in 650 mL of CH_2Cl_2 ([**1**] = 0.4 mol/L) were added successively under vigorous stirring at 0 °C 80 mL (2.2 eq., 575.1 mmol) of NEt_3 and 38.40 g (1.1 eq., 287.5 mmol) of NCS in small portions.[I] After stirring for 2.5 h at room temperature, 49 mL of pyridine (2.3 eq., 601.2 mmol) was added slowly within 1 h. The mixture was cooled to at −20 °C and 82 mL of Cbz-Cl (2.2 eq., 575.1 mmol) was added.[II] The stirred mixture (a suspension resulting from precipitated Et_3NHCl and pyridine-HCl) was allowed to slowly warm to room temperature (24 h) before it was washed twice with 150 mL of 1 N HCl, once with 150 mL of saturated $NaHCO_3$ and brine. The organic phase was dried over $MgSO_4$ and concentrated *in vacuo*. Finally, the crude product (98 g) was subjected to column chromatography on silica gel (EtOAc/c-hex = 2:5) to give 53.193 g (203.6 mmol, 77%) of **2** as a pale yellow oil. The succinimide derivative **C** was isolated as a side product. R_f = 0.27 (SiO_2, EtOAc/c-hex = 1:3); **^{1}H NMR** (300 MHz, $CDCl_3$): δ [ppm] = 7.33–7.27 (m, 5H, H-4′), 5.81 (t, 1H, H-3, 3J = 2.9(H-4) Hz), 5.11(s, 2H, H-2′), 3.96 (t, 2H, H-5, ^3J = 9.1 (H-4) Hz), 3.62 (s, 3H, H-2″), 2.63 (td, 2H, H-4, ^3J = 9.2 (H-5)/2.9(H-3) Hz); **^{13}C NMR** (100 MHz, $CDCl_3$): δ [ppm] = 162.4 (s, C-1″), 153.5 (s, C-1′), 136.2 (s, C-2), 135.8 (s, C-3′), 128.5/128.4/128.3/128.2/128.0 (d, C-4′), 120.1 (d, C-3), 67.7 (t, C-2′), 52.1 (q, C-2″), 48.4 (t, Cs-5), 28.6 (t, C-4); **IR** (ATR): ν [cm^{-1}] = 3090(w), 3063 (w), 3031 (w), 2952 (m), 2895 (m), 2853 (w), 1734 (vs), 1698 (vs), 1628 (s), 1585 (w), 1497(m), 1438 (s), 1405 (vs), 1317 (vs), 1278 (s), 1243 (vs), 1194 (vs), 1161 (vs), 1026 (s), 915 (m), 871 (m), 747 (s), 694 (s); **GC-MS** (EI, 70 eV): m/z (%) = 261 (1), 217 (5), 207 (5), 202 (5), 158 (1), 140 (5), 126 (5),

94 (<5,), 92 (10), 91 (100), 77 (5), 67 (5), 65 (15), 59 (<5), 51 (5); **Analytical data of C:** R_f = 0.33 (SiO_2, EtOAc/c-hex = 1:2); ^{1}H NMR (300 MHz, $CDCl_3$): δ [ppm] = 7.44–7.40 (m, 2H, H-4′), 7.39–7.29 (m, 2H, H-4′), 5.35 (s, 1H, H-2′), 2.74 (s, 4H, H-3); ^{13}C NMR (75 MHz, $CDCl_3$): δ [ppm] = 172.4 (s, C-2), 148.1 (s, C-1′), 133.9 (s, C-3′), 128.8/128.6/128.3 (d, C-4′), 69.9 (t, C-2′), 28.5 (t, C-3); **IR** (ATR): ν [cm^{-1}]= 3062 (w), 3034 (m), 2943 (m), 2917 (w), 2892 (w), 2258 (w), 2078 (w), 1970 (w), 1893 (w), 1807 (vs), 1778 (vs), 1698 (vs), 1585 (w), 1497 (m), 1454 (s), 1425 (s), 1378 (s), 1325 (vs), 1269 (vs), 1204 (vs), 1160 (vs), 1082 (s), 1000 (vs), 945 (m), 903 (s), 857 (w), 814 (s) m 746 (s), 694 (s), 662 (s).

[I]: Because of its low solubility in CH_2Cl_2, NCS was added as a solid. Too fast addition leads to decomposition of NCS (Cl_2 evolution).

[II]: Cbz-Cl also decomposes if added too fast. This is indicated through gas evolution (CO_2). To prevent decomposition, Cbz-Cl is added very slowly and the reaction mixture is brought slowly to room temperature.

The Mechanism

The beginning of the reaction is the *N*-chlorination of proline methyl ester **1** with NCS to give intermediate **A** and succinimide. Then elimination of HCl from **A** in the presence of triethylamine provides dihydropyrrole **B**. Final acylation of the dihydropyrrole by Cbz-Cl yields product **2**. The by-product succinimide also reacts with Cbz-Cl to generate the by-product **C**.

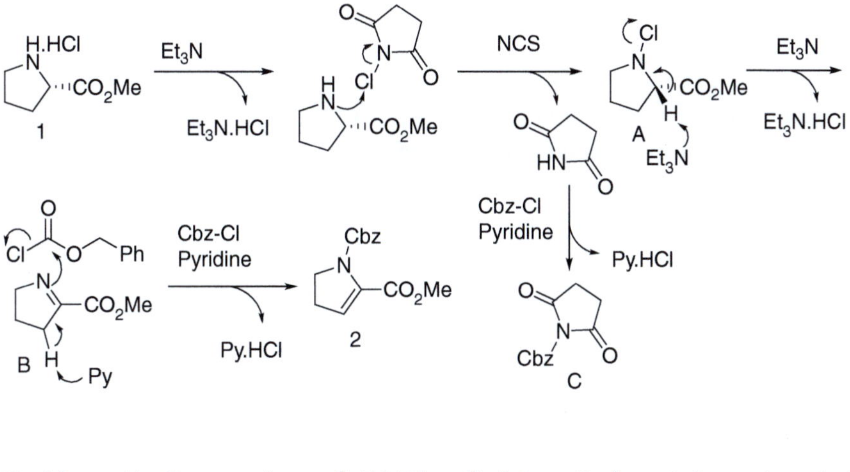

Problem 41: Preparation of 5,5-Dimethyl-3-methylene-1-(prop-1-en-2-yl) pyrrolidin-2-one

5,5-Dimethyl-3-methylenepyrrolidin-2-one (**3**) is used as a building block for the synthesis of Elexacaftor [112]. The Vertex patent applications include an example that described a 114-kg scale preparation of **3** [112]. In 1980, at almost the same time, Lai's group [113] and Lind's group [114] reported a rearrangement of 2,2,6,6-tetramethylpiperidin-4-one (**1**) to 5,5-dimethyl-3-methylene-1-(prop-1-en-2-yl)pyrrolidin-2-one (**2**) and **3**. Both groups proposed the mechanism for the products formation.

Without checking the references, could you please suggest the reaction mechanism?

1
$BnN^+Et_3Cl^-$, $CHCl_3$
50% NaOH (aq.)
5 °C
2
3
Elexacaftor

Experimental Procedure (from *J. Org. Chem.*, **1980**, *45*, 1513)

***N*-Isopropenyl-5,5-dimethyl-3-methylene-2-pyrrolidinone (2).** 2,2,6,6-Tetramethyl-4-piperidone hydrate (5.20 g, 30 mmol), chloroform (11.94 g, 100 mmol), and 18-crown-6 (0.40 g, 1.5 mmol) were placed in a 100-mL 3-neck flask immersed in a refrigerated circulating bath. The temperature was kept below 5 °C while 50% aqueous NaOH (24 g, 300 mmol) was added dropwise in 25 min. The solution was stirred at 5 °C for 7 h after the addition and then water was added until all solids dissolved. The two layers were separated and the aqueous layer was extracted with two 25-mL portions of $CHCl_3$. The combined organic layers were washed with one 10-mL portion of H_2O, dried, and concentrated under vacuum, 15 mL of hexane was added, the mixture was stirred, and the small amount of solid which formed was filtered off. The filtrate was concentrated and distilled to give 3.5 g (71%) of a clear oil at 63–7 °C (0.2 mm): IR (neat) 1680,1655,1640 cm^{-1}; ^1H NMR δ 1.35 (s, 6H), 2.01 (d, 3H), 2.70 (t, 2H), 4.89 (s, 1H), 5.20 (q, 1H), 5.32 (dt, 1H), 6.00 (dt, 1H); ^{13}C NMR δ 22.04 (q), 28.15 (q), 29.48 (t), 42.12 (s), 114.60 (t), 115.51 (t), 139.47 (s), 140.18 (s), 166.35 (s); mass spectrum, *m/e* 165 (M^+). Anal. Calcd for $C_{10}H_{15}NO$: C, 72.69; H, 9.15; N, 8.48. Found: C, 71.35; H, 8.93; N, 8.42.

5,5-Dimethyl-3-methylene-2-pyrrolidone (3). The procedure was as above except that 2.55 g (30 mmol) of piperidine was mixed with the reactants before the addition of 50% NaOH. The obtained crude product was stirred with 15 mL of hexanes to yield 1.95 g (52%) of a slightly yellowish solid after filtration. Recrystallization from heptane-toluene afforded colorless crystals: mp 139–142 °C; IR (KBr) 3180,1680,1645 cm^{-1}; ^1H NMR δ 1.32 (s, 6H), 2.61 (t, 2H), 5.33 (m, 1H), 5.97, (t, 1H); ^{13}C NMR δ 29.67 (q), 41.83 (t), 53.92 (s), 115.64 (t), 141.09 (s), 169.89 (s); mass spectrum, *m/e* 125 (M^+). Anal. Calcd for $C_7H_{11}NO$: C, 67.17; H, 8.86; N, 11.19. Found: C, 66.89; H, 8.75; N, 11.12.

(from WO/2019/028228 A1)

i. Under a nitrogen atmosphere, 2,2,6,6-tetramethylpiperidin-4-one (257.4 kg, 1658.0 mol, 1.00 eq), tri-butyl methyl ammonium chloride (14.86 kg, 63.0 mol, 0.038 eq), chloroform (346.5 kg, 2901.5 mol, 1.75 eq), and DCM (683.3 kg) were added to a 5000 L enamel reactor. The reaction was stirred at 85 rpm and cooled to 15~17 °C. The solution of 50 wt% sodium hydroxide (1061.4 kg, 13264.0 mol, 8.00 eq) was added dropwise over 40 h while maintaining the temperature between 15 and 25 °C. The reaction mixture was stirred and monitored by GC.

ii. The suspension was diluted with DCM (683.3 kg) and water (1544.4 kg). The organic phase was separated. The aqueous phase was extracted with DCM (683.3 kg). The organic phases were combined, cooled to 10 °C and then 3 M hydrochloric acid (867.8 kg, 2559.0 mol, 1.5 eq) was added. The mixture was stirred at 10~15 °C for 2 h. The organic phase was separated. The aqueous phase was extracted with DCM (683.3 kg × 2). The organic phases were combined, dried over Na_2SO_4 (145.0 kg) for 6 h. The solid was filtered off and washed with DCM (120.0 kg). The filtrate was stirred with active charcoal (55 kg) for 6 h. The resulting mixture was filtered and the filtrate was concentrated under reduced pressure (30–40 °C), −0.1 MPa). Then isopropyl acetate (338 kg) was added and the mixture was heated to 87~91 °C, stirring for 1 h. Then the solution was cooled to 15 °C in 18 h and stirred for 1 h at 15 °C. The solid collected by filtration was washed with 50% isopropyl acetate/hexane (80.0 kg × 2) and dried overnight in the vacuum oven at 50 °C to afford 5,5-dimethyl-3-methylene-2-one as an off-white solid, 55% yield (114.2 kg).

The Mechanism

Lind postulated [114] that the formation of epoxide **4** involved the addition of dichlorocarbene to the carbonyl group, whereas Lai later demonstrated that the reactive species was actually the trichloromethide ion [113]. This conclusion was supported by the absence of N-formylpiperidine upon the addition of piperidine.

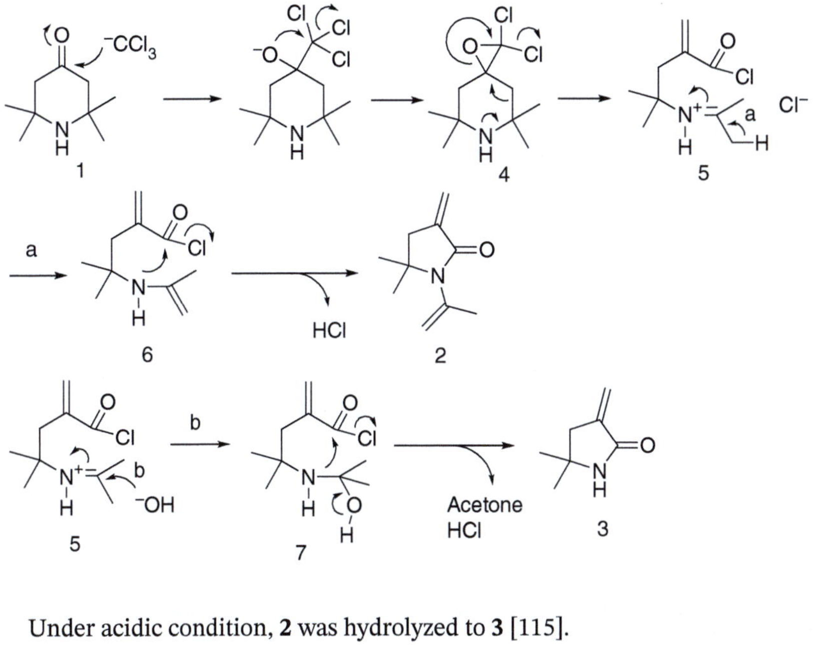

Under acidic condition, **2** was hydrolyzed to **3** [115].

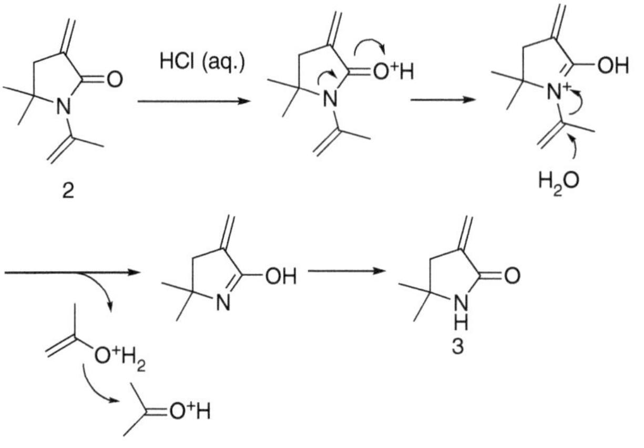

Problem 42: Synthesis of Heteroaromatic-Fused Pyrrolidines via Cyclopropane Ring-Opening Reaction

A one-step synthesis of heteroaromatic-fused pyrrolidines via cyclopropane ring-opening reaction was developed by Japanese chemists [116] and the method was applied to the synthesis of the protein kinase C-beta inhibitor JTT-010 [117]. The process is presented by the following reaction.

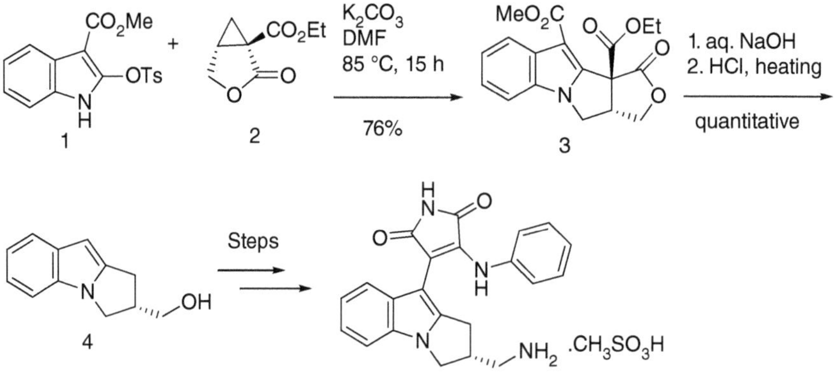

Please suggest mechanisms for the transformations from **1**, **2** to **3** and **3** to **4**.

Experimental Procedure (from *Org. Lett.*, **2007**, *9*, 3331, Supporting Information) **(3a*R*,10a*S*)-3a-Ethyl4-methyl3-oxo-3a,10a-dihydro-1*H*,10*H*-furo[3,4-c]pyrrolo[1,2-a]indole-3a,4-dicarboxylate (3).** To a solution of unpurified **2** (obtained from 609 mmol (*R*)-epichlorohydrin in 57% yield) in DMF (300 mL), **1** (100 g, 290 mmol) and K_2CO_3 (48.0 g, 347 mmol) were added, and the mixture was stirred for 18 h at 75–85 °C. After cooling to 40 °C, toluene (150 mL) and heptane (150 mL) were successively added. To the resulting slurry, water (300 mL) was added dropwise, and the mixture was stirred for 1 h at 0–5 °C. The precipitated solid was collected by filtration and rinsed with water (500 mL). The obtained wet cake was suspended in MeOH (500 mL) and stirred for 1 h at 15–25 °C. The precipitates were collected by filtration, rinsed with MeOH (300 mL), and dried *in vacuo* to provide **3** (68.2 g; yield 68%, purity 94%) as an off-white solid. The product was purified by preparative TLC (CHC13/EtOAc, 5:1) and further recrystallized from hexane/EtOAc to give analytically pure **3** as an off-white solid. Mp 205.0–206.0 °C; $[\alpha]_D^{20}$ −150.3° (c 0.48, CHC13); IR 1778, 1733, 1697, 1559, 1299, 1283, 1173, 1158, 1035, 981, 750, 743 cm^{-1}; ^1H NMR (400 MHz, CDC13) δ 8.23–8.19 (1H, m), 7.32–7.27 (3H, m), 4.66 (1H, dd, J = 9.8 and 6.6 Hz), 4.64 (1H, dd, J = 10.2 and 8.2 Hz), 4.42–4.25 (3H, m), 4.20–4.13 (1H, m), 4.10 (1H, dd, J = 10.3 and 5.9 Hz), 3.90 (3H, s), 1.30 (3H, t, J = 7.1 Hz); ^{13}C NMR (100 MHz, CDC13) δ 168.02, 167.12, 164.08, 142.43, 132.42, 130.84, 123.41, 122.64, 122.55, 109.98, 102.63, 68.63, 62.97, 60.49, 50.93, 50.91, 48.61, 14.09; Anal. Calcd for C18H17NO6: C, 62.97; H, 4.99; N, 4.08. Found: C, 62.84; H, 4.94; N, 3.92.

(*S*)-2,3-Dihydro-1*H*-pyrrolo[1,2-a]indol-2-ylmethanol (4). A 6 M aqueous NaOH solution (420 mL) was added to a suspension of **3** (106 g, purity 98%, 301 mmol) in EtOH (433 mL), and the mixture was heated to reflux for 3.5 h. The mixture was acidified by dropwise addition of concentrated HCl (220 mL) and then stirred for 1 h under reflux. A similar basifying–acidifying procedure was repeated twice (refluxing the mixture with successive addition of NaOH (55 mL), HCl (25 mL), NaOH (30 mL), and HCl (20 mL) every 30 min). The reaction mixture was cooled to room temperature, neutralized with a 6 M aqueous NaOH solution (10 mL), and concentrated to a weight of 930 g *in vacuo*. The organic layer of the obtained residue was separated with EtOAc (415 mL), and the aqueous layer was extracted with EtOAc (105 mL). The combined organic layers were dried over anhydrous Na_2SO_4 and then concentrated *in vacuo* to a weight of 63 g. The obtained residue was azeotroped with toluene (150 mL × 2) to afford ethanol-free **4** (60.0 g, containing toluene) as a pale brown solid. No further purification was attempted on this compound, which was used directly in the next step. Purification by preparative TLC (hexane/EtOAc, 1:1) followed by recrystallization from hexane/EtOAc gave analytically pure **4** as an off-white solid. Mp 80.3–81.5 °C; $[\alpha]^{20}{}_D$ +4.51° (c 0.40, $CHCl_3$); IR 3260, 1457, 1044, 1013, 764, 740 cm^{-1}; ^1H NMR (400 MHz, $CDCl_3$) δ7.54 (1H, d, J = 7.7 Hz), 7.23 (1H, dd, J = 8.1, 0.9 Hz), 7.11 (1H, td, J = 7.6 and 1.6 Hz), 7.05 (1H, td, J = 7.4 and 1.2 Hz), 6.15 (1H, d, J = 0.9 Hz), 4.21 (1H, dd, J = 10.2 and 7.4 Hz), 3.95 (1H, dd, J = 10.2 and 5.1 Hz), 3.82 (1H, dd, J = 10.2 and 6.3 Hz), 3.74 (1H, dd, J = 10.1 and 7.5 Hz), 3.28–3.12 (2H, m), 2.82 (1H, ddd, J = 15.4, 9.0, and 1.0 Hz); ^{13}C NMR (100 MHz, $CDCl_3$) δ 143.26, 133.11, 132.77, 120.34, 120.29, 119.22, 109.35, 92.83, 65.11, 46.34, 44.33, 27.37; Anal. Calcd for $C_{12}H_{13}NO$: C, 76.98; H, 7.00; N, 7.48. Found: C, 76.85; H, 7.04; N, 7.33. The enantiomeric excess was determined to be >99% by HPLC (Daicel CHIRALCEL OD-H 250 mm × 4.6 mm i.d.; Hexane/EtOH = 99/1–97/3 (10 min), then 97/3 (35 min); flow rate = 1.0 mL/min; temp = 25 °C; t_R = 36.5 min (*R*) or 41.3 min (*S*)).

The Mechanism

Under basic conditions (K_2CO_3), the NH group in compound **1** is deprotonated, forming a nucleophile that attacks the electron-deficient cyclopropane compound **2** to generate carbanion **3a**. This carbanion is stabilized by the electron-withdrawing diester groups and subsequently attacks the carbon bearing the leaving group Ots, completing the insertion of the heteroaromatic N(1)-C(2) unit to form compound **3**.

After the tri-ester hydrolysis by aqueous NaOH, the salt is acidified by HCl to release the tri-acid. Upon heating, decarboxylation occurs to give the product **4**. The triple decarboxylation may be stepwise and the last decarboxylation of indole-3-carboxylic acid should be catalyzed by acid [118].

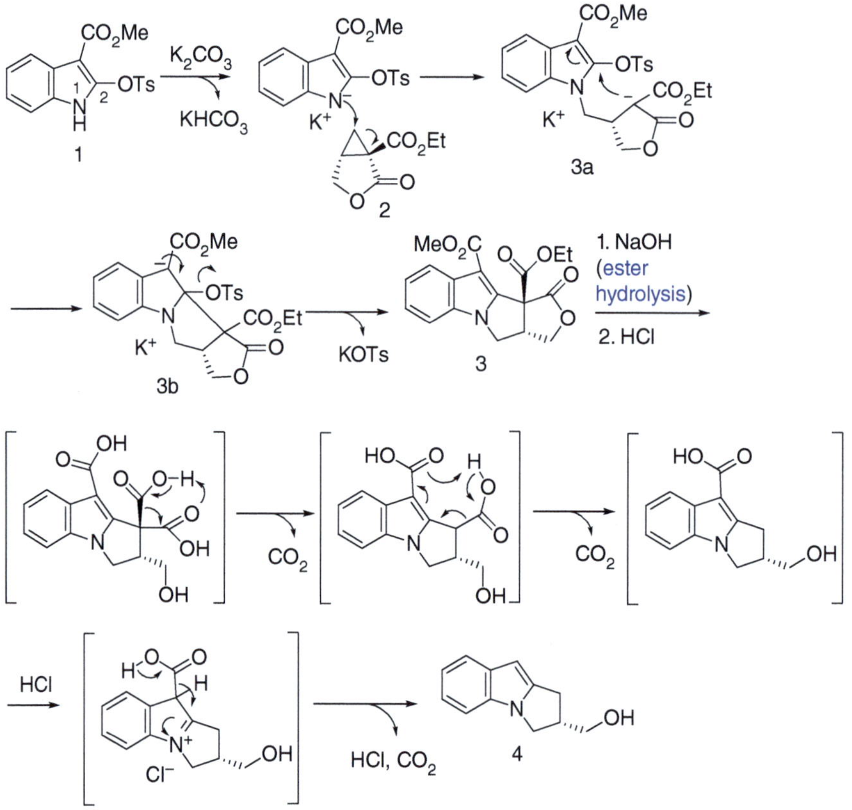

Problem 43: Synthesis of Spiro[3*H*-indole-3,2′-pyrrolidin]-2(1*H*)-one

BI-0252 (**1**), a novel spiro[3*H*-indole-3,2′-pyrrolidin]-2(1*H*)-one compound, was identified as an orally active inhibitor of the MDM2-p53 interaction [119]. The construction of the core spiro[3*H*-indole-3,2′-pyrrolidin]-2(1*H*)-one (**2**) was based on a three-component [3 + 2] cycloaddition [120]. The representative reaction is shown as below [119]. Treatment of 1-(2-fluoro-3-chlorophenyl)-2-nitroethene (**3**), 6-chloroisatin (**4**), and L-homoserine (**5**) in refluxing methanol provided rac-**2** in 46% yield [119].

Please suggest a mechanism for this transformation.

Experimental Procedure (from *J. Med. Chem.*, **2016**, *59*, 10147)

rac-(3S,3′S,4′S,5′S)-6-Chloro-3′-(3-chloro-2-fluorophenyl)-5′-(2-hydroxyethyl)-4′-nitro-1,2-dihydrospiro[indole-3,2′-pyrrolidine]-2-one (rac-2). 6-Chloroisatin (**4**, 8.0 g, 44.05 mmol), 1-(2-fluoro-3-chlorophenyl)-2-nitroethene (**3**, 8.9 g, 44.05 mmol), and L-homoserine (**5**, 5.24 g, 44.05 mmol) were stirred under reflux in MeOH (10 mL) for 18 h. The reaction mixture was concentrated *in vacuo*, and the residue titurated in acetonitrile. The solid was filtered and dried to give **rac-2**, which was used in the next step without further purification. **rac-2**, 8.9 g, 20.2 mmol, 46%. ^{1}HNMR (500 MHz, DMSO-d_6) δ ppm 1.49–1.63 (m, 2H) 3.44–3.52 (m, 1H) 3.52–3.61 (m, 1H) 4.09 (d, J = 7.25 Hz, 1H) 4.52–4.63 (m, 2H) 4.74 (d, J = 8.51 Hz, 1H) 6.06 (t, J = 8.51 Hz, 1H) 6.66 (s, 1H) 7.11 (dd, J = 7.88 and 0.95 Hz, 1H) 7.17 (t, J = 7.88 Hz, 1H) 7.44 (br t, J = 6.94 Hz, 2H) 7.56 (d, J = 7.88 Hz, 1H) 10.32 (s, 1H). LC/MS: $[M + H]^+ = 440$; $t_R = 1.21$ min.

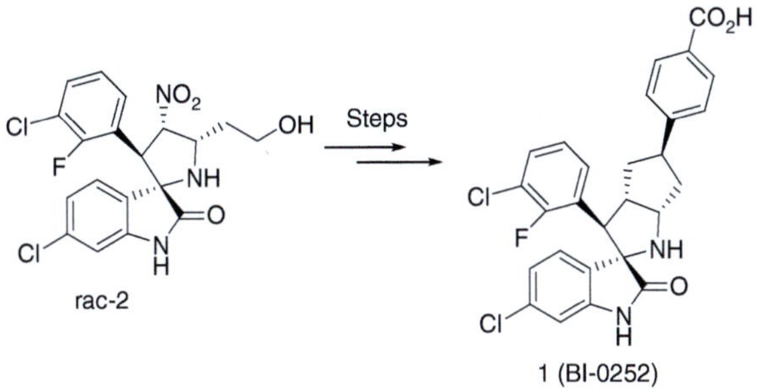

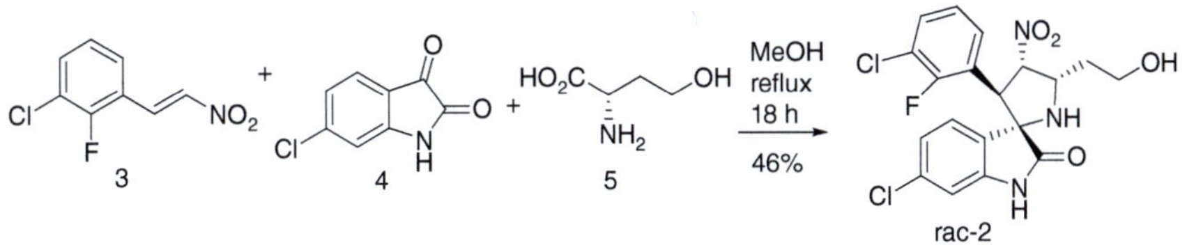

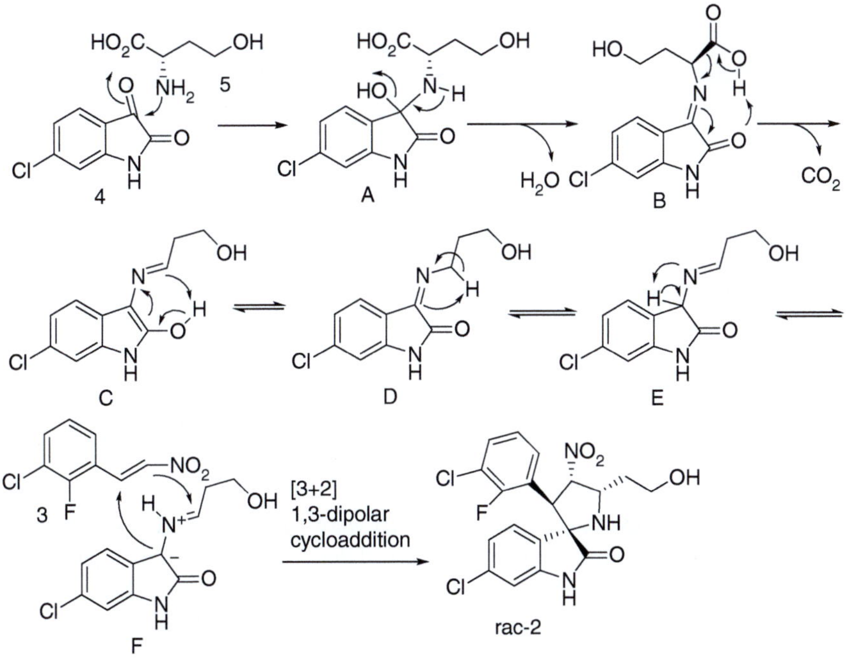

The Mechanism

The amino group in L-homoserine (**5**) adds to the carbonyl in 6-chloroisatin (**4**) to give adduct **A**. After H_2O is eliminated, the imine **B** is generated. Decarboxylation then occurs, yielding intermediate **C**, which can potentially resonate with its tautomers **D**, **E**, and azomethine ylide **F**. Finally 1,3-dipolar cycloaddition between the azomethine ylide **F** and 1-(2-fluoro-3-chlorophenyl)-2-nitroethene (**3**) leads to product rac-**2**.

4.5.5 Piperidine

Problem 44: Synthesis of (2*S*,4*R*)-4-Hydroxypipecolic Acid from Homoallyl Alcohol

In 1996, scientists from Bio-Mega/Boehringer Ingelheim reported the synthesis of (2*S*,4*R*)-4-hydroxypipecolic acid (**3**) [121]. The key intermediate was the lactone (*2S, 4R*)-**2**, which was prepared from homoallyl alcohol **1** by the following reaction sequence. Compound **3** was a key building block for synthesizing of palinavir, a potent inhibitor of the HIV-1 and type 2 (HIV-2) proteases [122].

Please provide a mechanism for the reactions from **1** to **2**.

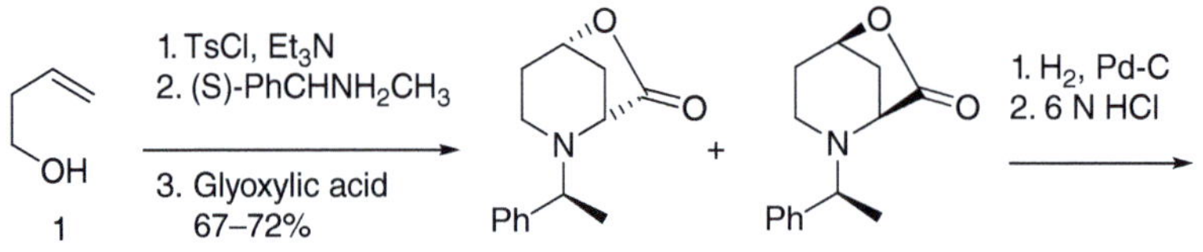

OH
1
1. TsCl, Et_3N
2. (S)-$PhCHNH_2CH_3$
3. Glyoxylic acid
67–72%
O N O Ph
+
O N O Ph
1. H_2, Pd-C
2. 6 N HCl
(2R, 4S)-2 (40 : 60) (2S, 4R)-2

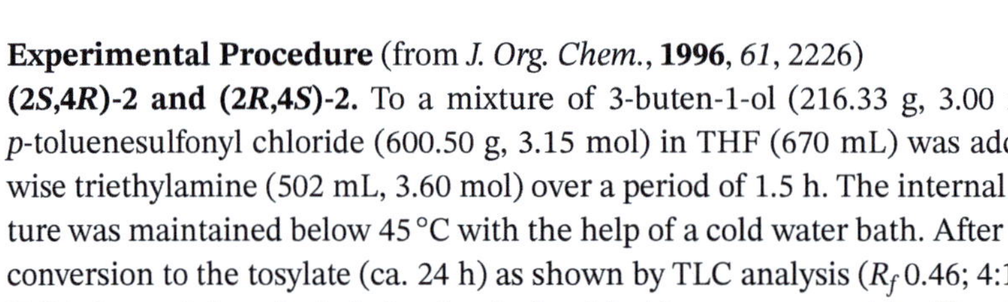

OH
N H CO$_2$H
(2S, 4R)-3
Steps
N H N O O N H OH N O H N O N
Palinavir

Experimental Procedure (from *J. Org. Chem.*, **1996**, *61*, 2226)

(2*S*,4*R*)-2 and (2*R*,4*S*)-2. To a mixture of 3-buten-1-ol (216.33 g, 3.00 mol) and *p*-toluenesulfonyl chloride (600.50 g, 3.15 mol) in THF (670 mL) was added dropwise triethylamine (502 mL, 3.60 mol) over a period of 1.5 h. The internal temperature was maintained below 45 °C with the help of a cold water bath. After complete conversion to the tosylate (ca. 24 h) as shown by TLC analysis (R_f 0.46; 4:1 hexane/EtOAc), precipitated triethylamine hydrochloride was removed by filtration using THF (2 L) for washings. Filtrate and washings were combined and concentrated under reduced pressure to a volume of ca. 2 L. Triethylamine (500 mL, 3.60 mol) was added to the solution followed by (*S*)-*R*-methylbenzylamine (465 mL, 3.60 mol, dropwise addition over 1 h). The reaction mixture was stirred 42 h at 70 °C after which point TLC analysis indicated complete disappearance of the tosylate. After cooling to room temperature, NaOH (150 g, 3.75 mol) in H_2O (1.5 L) was added, the organic layer separated, and the aqueous phase extracted with ether (2 × 1 L). The combined organic phases were concentrated under reduced pressure to a volume of ca. 1 L. The above solution was diluted with THF (500 mL) and warmed to 35 °C. Glyoxylic acid (50% w/w aqueous solution, 500 g, 4.5 mol) was added dropwise, allowing the internal temperature to reach 60 °C at which point the heating source was removed. After completion of addition (1.5 h), the reaction was heated to 60–65 °C for 7 h and stirred an additional 9 h at room temperature. TLC analysis indicated complete conversion to a mixture of (2*R*,4*S*)-**2** (R_f 0.87; EtOAc, minor diastereomer) and (2*S*,4*R*)-**2** (R_f 0.80; EtOAc, major diastereomer). Water (500 mL) and brine (500 mL) were added, and the mixture was basified to pH 8–9 with 5 N NaOH (200 mL). EtOAc (1 L) was added, the organic layer separated, and the aqueous phase extracted with EtOAc (2 ×1 L). The combined organic phases were washed with saturated $NaHCO_3$ (750 mL), water (750 mL), and brine (750 mL). After drying ($MgSO_4$), volatiles were removed under vacuum to give the diastereomeric lactones as a brown solid (466.6 g, 67% yield). ^{1}H NMR analysis showed a 60:40 ratio of (2*S*,4*R*)-**2** and (2*R*,4*S*)-**2**, and approximately 4% of *N,N*-dialkylated-(*S*)-*R*-methylbenzylamine. Analytical samples of each component were obtained by flash chromatography on silica gel using 25% EtOAc/hexane as eluant.

(2*R*,4*S*)-2 (minor diastereomer): R_f 0.73 (1:1 hexane/EtOAc). Mp 48–49 °C. $[R]_D^{25}$ −16.2° (*c* 1.05, $CHCl_3$). IR (KBr) ν 1770 cm^{-1}. ^{1}H NMR ($CDCl_3$) δ 7.35–7.2 (m, 5H), 4.79 (t, *J* = 5.2 Hz, 1H), 3.80 (d, *J* = 10.3 Hz, 1H), 3.58 (q, *J* = 6.7 Hz, 1H), 2.73 (dd, *J* = 12.4 and 6.7 Hz, 1H), 2.35–2.25 (m, 2H), 1.95 (d, *J*) 11.8 Hz, 1H), 1.91–1.84 (m, 1H), 1.79–1.69 (m, 1H), 1.44 (d, *J* = 6.7 Hz, 3H). ^{13}C NMR ($CDCl_3$) δ 174.0, 144.5, 128.3, 126.9, 76.6, 62.8, 55.2, 46.0, 37.5, 29.1, 21.0. MS *m/z* 232 (MH^+). Anal. Calcd for $C_{14}H_{17}NO_2$: C, 72.70; H, 7.41; N, 6.06. Found: C, 72.65; H, 7.46; N, 5.98. RP-HPLC (Symmetry C8; 10–80% CH_3CN in 20 mM Na_2HPO_4 (pH 7.4) in 25 min; 0.8 mL/min): t_R 20.7 min (>99.5% de).

(2*S*,4*R*)-2 (major diastereomer): R_f 0.62 (1:1 hexane/EtOAc). Mp 83–84 °C. $[R]^{25}_D$ −94.7° (*c* 1.01, $CHCl_3$). IR (KBr) ν 1765 cm^{-1}. ^{1}H NMR ($CDCl_3$) δ 7.42–7.37 (m, 2H), 7.35–7.29 (m, 2H), 7.28–7.22 (m, 1H), 4.77 (t, *J* = 5.1 Hz, 1H), 3.70 (q, *J* = 6.7 Hz, 1H), 3.33 (dd, *J* = 11.8 and 6.7 Hz, 1H), 3.18 (d, *J* = 5.1 Hz. 1H), 2.48 (dt, *J* = 11.8, 5.1 Hz, 1H), 2.12–2.02 (m, 2H), 1.95–1.86 (m, 1H), 1.82 (d, *J* = 11.8 Hz, 1H), 1.33 (d, *J* = 6.7 Hz, 3H). ^{13}C NMR ($CDCl_3$) δ 174.0, 144.5, 128.5, 127.4, 127.1, 76.4, 62.3, 57.1, 44.0, 37.2, 29.2, 21.3. MS *m/z* 232 (MH^+). Anal. Calcd for $C_{14}H_{17}NO_2$: C, 72.70; H, 7.41; N, 6.06. Found: C, 72.85; H, 7.50; N, 6.00. RP-HPLC (Symmetry C8; 10–80% CH_3CN in 20 mM Na_2HPO_4 (pH 7.4) in 25 min; 0.8 mL/min): t_R 19.6 min (>99.5% de).

The Mechanism

The first two steps are the tosylation of **1** followed by the alkylation of the chiral amine to give **B**.

OH 1 — TsCl, Et_3N → ($Et_3N^+HCl^-$) OTs A — NH_2 Ph → (TsOH; Et_3N → $Et_3N^+HTsO^-$) NH Ph B

The secondary amine **B** then reacts with the aldehyde group in glyoxylic acid to form iminium salt **D,** which, in the presence of water, undergoes cyclization to yield the 4-hydroxyl-2-carboxylic piperidine **E**. Finally, lactonization occurs to provide the product **2**. Actually, that kind of synthesis was earlier reported by Hays [123].

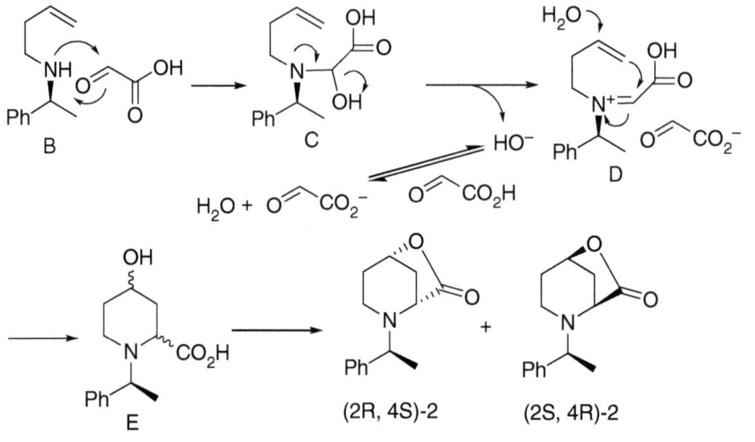

The synthesis was fundamentally based on the iminium chemistry developed by Grieco's group [124].

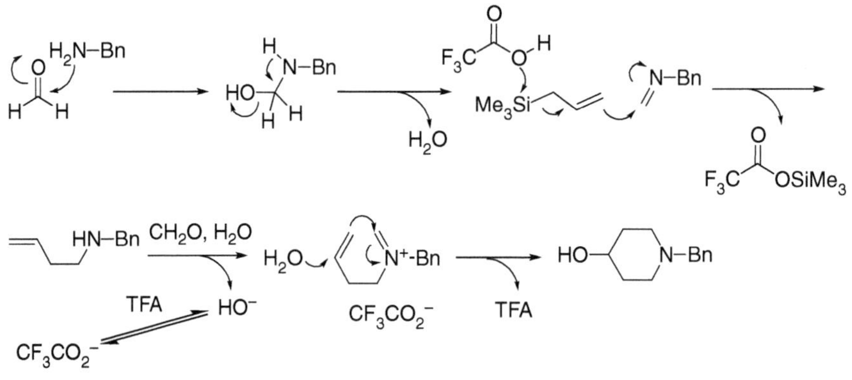

It is interesting that under the reaction condition, (*E*)-(cyclooct-1-en-1-ylmethyl) trimethylsilane gave 2-benzyl-1,2,3,5,6,7,8,9,10,10a-decahydrocycloocta[c]pyridine [124].

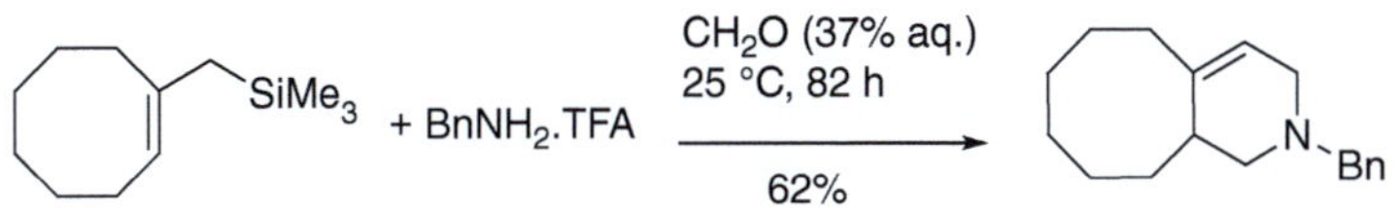

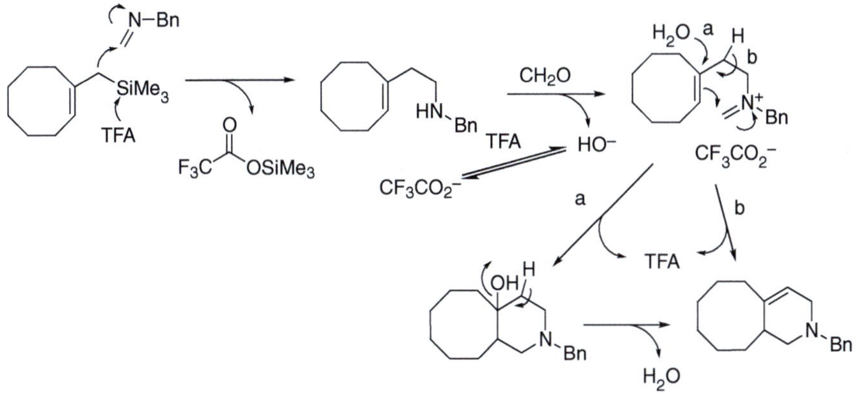

Problem 45: Preparation of a Spiropiperidine

Spiroamine **2** was prepared by Pfizer chemists in 36 kg scale from **1** as shown in the following reactions [125]. Compound **2** was a key building block for constructing a series of acetyl-CoA carboxylase (ACC) inhibitors, such as **3**.

Please provide stepwise reaction mechanisms from **1** to **2**.

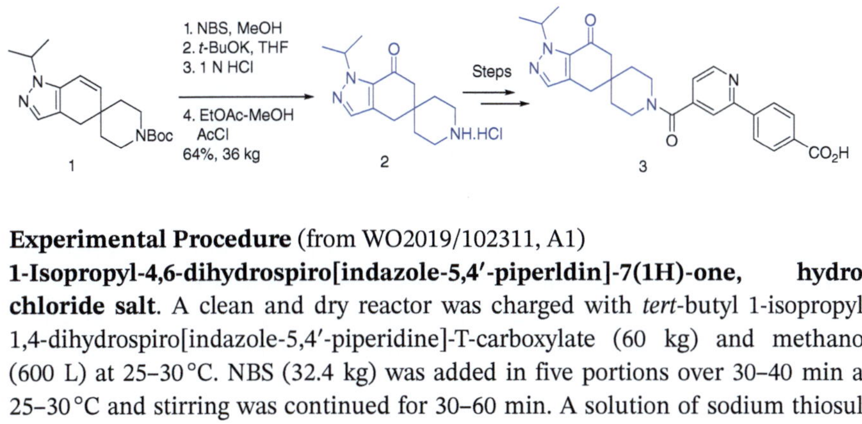

1. NBS, MeOH
2. *t*-BuOK, THF
3. 1 N HCl
4. EtOAc-MeOH AcCl
64%, 36 kg

Experimental Procedure (from WO2019/102311, A1)

1-Isopropyl-4,6-dihydrospiro[indazole-5,4′-piperldin]-7(1H)-one, hydrochloride salt. A clean and dry reactor was charged with *tert*-butyl 1-isopropyl-1,4-dihydrospiro[indazole-5,4′-piperidine]-T-carboxylate (60 kg) and methanol (600 L) at 25–30 °C. NBS (32.4 kg) was added in five portions over 30–40 min at 25–30 °C and stirring was continued for 30–60 min. A solution of sodium thiosulfate pentahydrate (5.4 kg) in water (102 L) was slowly added, maintaining internal temperature below 30 °C. The mixture was stirred for 20–30 min, then the solvent was evaporated under reduced pressure at below 45 °C. The residue was cooled down to 25–30 °C and 2-methyltetrahydrofuran (420 L) was charged in the reactor, along with water (90 L). The mixture was stirred for 15–20 min, then the layers were separated, the aqueous layer was further extracted with 2-methyltetrahydrofuran (120 L). Combined organic extracts were treated for 15–20 min at 25–30 °C with a solution of sodium hydroxide (4.8 kg) in water (120 L). Layers were separated and the organic layer was washed with water (120 L), followed by a solution of sodium chloride (12 kg) in water (120 L), and then dried over sodium sulfate (6 kg). After filtration, the cake was washed with 2-methyltetrahydrofuran (30 L) and combined filtrate were charged back into the reactor. The solvent was completely distilled at below 45 °C under reduced pressure and the residue was solubilized in THF (201 L). In another clean and dry reactor was charged potassium *tert*-butoxide (60.6 kg) and THF (360 L) at 25–30 °C. To that mixture was slowly added the solution of the residue in THF, maintaining a temperature below 30 °C. The reaction mixture was then warmed up to 60–65 °C and kept at this temperature for 1–2 h. Upon completion, the mixture was cooled to 0–10 °C, and slowly quenched with a solution of hydrochloric acid (1 N, 196 L), maintaining internal temperature below 10 °C. The reaction mixture was allowed to warm up to 25–30 °C, and ethyl acetate (798 L) was charged. After stirring for 15–20 min, the layers were separated, and the aqueous layer was further extracted with ethyl acetate (160 L). Combined organic layers were washed with water (160 L), dried over sodium sulfate (8 kg), filtered, and the cake was washed with ethyl acetate (300 L). The solvents were entirely distilled under reduced pressure at below 45 °C, and ethyl acetate (540 L) was charged into the reactor at 25–30 °C, followed by methanol (156 L). The mixture was cooled to 0–5 °C, at which point acetyl chloride (79.8 kg) was slowly added, maintaining the temperature in the specified range. The mixture was then allowed to warm up to 20–25 °C and was kept at this temperature for 4–5 h with stirring. The resulting slurry was filtered and the solids were washed with ethyl acetate (120 L), then dried at 40–45 °C for 8–10 h to furnish the desired crude product (33.5 kg, 65%). A final purification step was performed by solubilizing this crude solid (56.8 kg) in methanol (454.4 L) in a clean a dried reactor at 25–30 °C. The solution was stirred for 30–45 min, then passed through a 0.2 micron cartridge filter into a clean and dry reactor at 25–30 °C. Methanol was distilled under reduced pressure at below 50 °C until ~1vol solvent remains. The reaction mixture was cooled to 25–30 °C and fresh acetonitrile (113.6 L) was charged through a 0.2 micron cartridge filter. The solvents were distilled under reduced pressure at below 50 °C until ~1vol solvent remains. The reaction mixture was cooled to 25–30 °C and fresh acetonitrile (190 L) was charged into the reactor through a 0.2 micron cartridge filter. The mixture was warmed up to 65–70 °C and stirred for 45 min, then cooled down to 25–30 °C and stirred for 1 h. The resulting slurry was filtered, and the cake was washed with chilled (15 °C) acetonitrile (56.8 L). The solids were dried under reduced pressure at 40–50 °C for 8 h to afford Intermediate **2** (36.4 kg, 64%). ^{1}H NMR (400 MHz, CD_3OD) δ ppm 7.43 (s, 1H), 5.32–5.42 (m, 1H), 3.15–3.25 (m, 4H), 2.89 (s, 2H), 2.64 (s, 2H), 1.69–1.90 (m, 4H), 1.37–1.45 (m, 6H); ESI $[M + H]^+ = 248$.

The Mechanism

Step 1. The first step is a cobromination [126] of the carbon-carbon double bond to form compound **A**.

Step 2. Elimination of HBr from **A** by *t*-BuOK produces the vinyl ether **B**.

Step 3. Hydrolysis of the vinyl ether catalyzed by HCl gives the ketone **C**. Boc group might be partially deprotected at this stage.

Step 4. The final step is the deprotection of Boc by HCl (generated from the reaction of AcCl with MeOH).

1 → NBS → MeOH → A + HN(succinimide)

A → t-BuOK (− t-BuOH + KBr) → B

B → H^+ → H_2O (− MeOH + H^+) → H^+ → C

C → HCl (− isobutylene) → Cl^-, H^+

→ (− CO_2) → NH → HCl → 2 (NH.HCl)

Problem 46: Dealkylation of Tertiary Amine with DEAD

Thebaine **1** is a controlled drug. When it was treated with DEAD in refluxing acetonitrile, the demethylated northebaine **2** was obtained in high yield [127]. Other demethylated alkaloids were prepared by this method [128]. This demethylation method was earlier reported by Smissman and Makriyannis [129]. Though this demethylation is important in alkaloid chemistry [130], the mechanism has not been clearly described.

Please suggest a mechanism for this reaction.

Thebaine → 1. DEAD, AcCN reflux, 1.5 h; 2. Py.HCl; 88% → Northebaine

Experimental Procedure (from *J. Med. Chem.*, **1980**, *23*, 698)
Northebaine hydrochloride was prepared from **thebaine** (Merck, Sharp, and Dohme) in 88% yield by a modification of DEAD demethylation (Eli Lilly and Co., Netherlands Appl. 6515815 (1966); *Chem. Abstr.*, 65, 15441d (1966)). A solution of 25 mL of DEAD in 50 mL of MeCN was added over 0.5 h to a stirred, refluxing solution of 47 g of **thebaine** in 250 mL of MeCN. Refluxing was continued an additional 1.5 h; pyridine hydrochloride (30 g) was added and the mixture was allowed to cool to room temperature. The solid was collected, washed with MeOH, and dried at room temperature; concentration of the mother liquor under reduced pressure and trituration of the residue with MeOH gave a second and third crop. The crude **northebaine hydrochloride** (44 g, mp 265–270 °C, lit.[12] mp 270–272 °C) was used without further purification for reaction with cyclopropylcarbonyl chloride, followed by $LiAlH_4$ reduction of the amide to give *N*-(cyclopropylmethyl)northebaine.

(from *J. Org. Chem.*, **1973**, *38*, 1652)

Demethylation of *N*-Methylpiperidine (Preparative Method). To a solution of *N*-methylpiperidine (2.48 g, 25.00 mmol) in C_6H_6 (20 mL, Na dry) was slowly added a solution of diethyl azodicarboxylate (6.53 g, 37.50 mmol) in C_6H_6 (20 mL, Na dry) and the mixture was refluxed for 30 min. The solvent and the unreacted *N*-methylpiperidine (0.25 g, 2.52 mmol, 10%) were removed under reduced pressure and the residue was dissolved in a mixture of 4 **N** HCl (25 mL) and EtOH (10 mL). The solution was refluxed for 2 hr and taken to dryness under reduced pressure. The residue was triturated with 10 **N** NaOH solution (1 mL) and extracted with Et_2O (3 × 25 mL). The combined extracts were dried ($MgSO_4$) and distilled to give piperidine (1.3 g, 15.27 mmol, 61%).

The Mechanism

Thus, the following mechanism is based on the procedure described in *J. Org. Chem.*, **1973**, *38*, 1652 [129].

The *N*-methylpiperidine as a nucleophile attacks the N=N double bond in DEAD to form intermediate **A**. Intramolecular H shift provides **B** and subsequently **B**

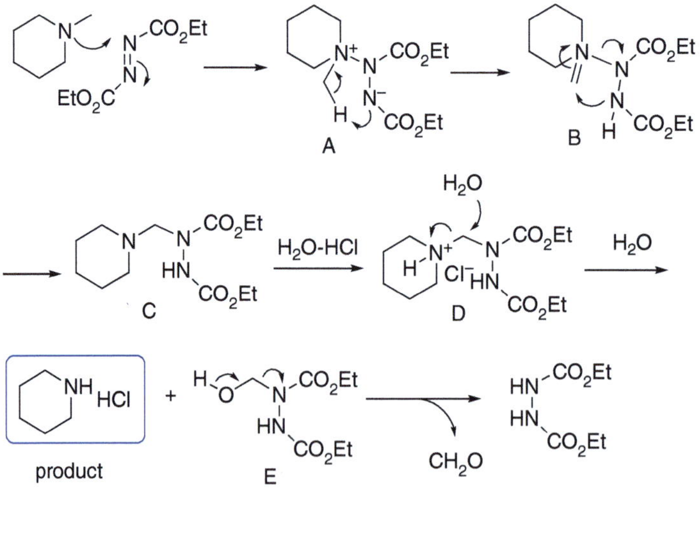

EtO_2C CO_2Et A B H_2O C H_2O-HCl D H_2O product E CH_2O

rearranges to methylenediamine **C**. Since the N in the tertiary amine is the most basic among the three nitrogens, after being treated with aqueous HCl, it will quickly get protonated to generate the intermediate **D**. Thus, the methylene is activated and readily attacked by H_2O to provide the demethylated product and the hydroxylmethyl dimethylhydrazine dicarboxylate **E**. The by-product **E** may further decompose to formaldehyde and dimethylhydrazine dicarboxylate.

Problem 47: A Novel Synthesis of δ-Keto-α-amino Acids

In a synthesis of Relebactam [131], chemists from Merck reported the preparation of the key intermediate **4** from (*S*)-1-(*tert*-butoxycarbonyl)-5-oxopyrrolidine-2-carboxylic acid (**1**) as illustrated in the following reactions.

Please provide mechanism for the transformation from **1** to **4**.

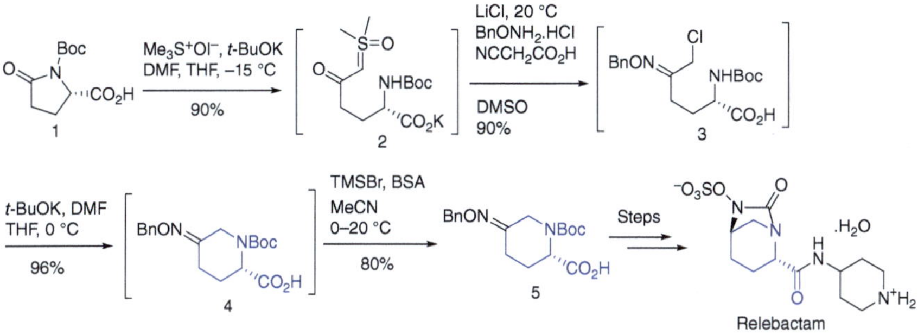

Experimental Procedure (from *J. Org. Chem.*, **2020**, *85*, 994)

Preparation of (S)-5-((Benzyloxy)imino)piperidine-2-carboxylic acid (5). To a mixture of (*S*)-*N*-Boc-pyroglutamic acid (**1**, 40.0 g, 174.4 mmol) and Me_3SOI (49.9 g, 226.7 mmol) in DMF (200 mL) at −20 °C was added a solution of KO*t*Bu (20 wt% in THF, 231 mL, 382.5 mmol) under good agitation over 10 h maintained at −20 to −10 °C. Upon completion of addition, the reaction mixture was aged at −20 to −10 °C for additional several hours until completion of the reaction (99% conversion) to the sulfur ylide **2**; M.S. (ESI): 322 $(M + H)^+$. DMSO (180 mL) and THF (180 mL) were added while maintaining an internal temperature of ≤−5 °C. Cyanoacetic acid (44.5 g, 523.1 mmol) followed by LiCl (37.0 g, 872.8 mmol) and $BnONH_2$·HCl (27.9 g, 174.2 mmol) was then added in portions while maintaining an internal temperature of ≤0 °C. The batch was warmed to 25 °C over 2–3 h and agitated for an additional 20 h at 25 °C. HPLC analysis confirmed the consumption of **2** and formation of **3**. The reaction mixture was cooled to 15 °C. Toluene (240 mL) followed by water (360 mL) was added to quench the reaction while maintaining an internal temperature of ≤25 °C. To the separated organic phase was added a saturated $NaHCO_3$ aqueous solution (~80 mL) slowly until the pH in the aqueous layer was stable at ~5.4. The organic phase containing **3** (81% assay yield) was separated and azeotropically dried in vacuum below 35 °C to a volume of ~140 mL, which was then directly used for the subsequent step without further purification. Crude sample **3** was confirmed by mass spectroscopy and ^{1}H NMR. M.S. (ESI): 385 $(M + H)^+$, 407 $(M+Na)^+$; ^{1}H NMR (DMSO-d_6) δ 13.21 (brs, 1H), 7.38–7.18 (m, 5H), 5.09 (brs, 2H),

4.30–4.24 (m, 4H), 3.94–3.83 (m, 1H), 2.51–2.34 (m, 2H), 2.02–1.93 (m, 1H), 1.83–1.72 (m, 1H), 1.72 (s, 9H).

To a solution of chlorooxime **3** (54.0 g by assay, 140.3 mmol) in toluene (190 mL) was added DMF (27 mL). The batch was cooled to 0 °C. KO*t*Bu (20 wt % in THF, 211 mL, 349.4 mmol) was added dropwise over 3 h while maintaining the internal temperature below 5 °C. After an additional 0.5 h aging at 0–5 °C, water (216 mL) was added dropwise while maintaining the internal temperature below 5 °C. MTBE (216 mL) was added to the separated aqueous phase. The batch was then pH-adjusted to ~4 with 6 M HCl while maintaining the internal temperature below 5 °C. The separated organic phase containing **4** (96% assay yield) was azeotropically dried and solvent switched to acetonitrile in a vacuum below 35 °C to a final volume of ~240 mL, which was directly used for the subsequent step without further purification. A crude sample of **4** was confirmed by mass spectroscopy and ^{1}H NMR. M.S. (ESI): 349 $(M + H)^+$, 371 $(M + Na)^+$; NMR showed that it is a mixture of *E* and *Z* oximes and Boc conformers; ^{1}H NMR (DMSO-d_6) δ 7.45–7.20 (m, 5H), 5.03 (s, 2H), 4.39–4.19 (m, 1H), 4.18–4.07 (m, 1H), 3.92–3.79 (m, 1H), 2.70–2.47 (m, 1H), 2.43–2.19 (m, 1H), 2.08–2.89 (m, 2H), 1.41– 1.31 (m, 9H). ^{13}C{^{1}H} NMR (DMSO-d_6) δ 173.0 (3C), 154.8 (4C), 154.2 (4C), 137.9 (3C), 128.3 (3C), 127.8 (3C), 127.6 (3C), 79.7 (3C), 74.8 (4C), 62.9, 54.2 (4C), 27.9 (4C), 24.6 (4C), 22.1 (4C). HRMS (ESI) calcd for C18H25N2O5 [M + H]+, 349.1763; found, 349.1751. An analytical sample of **4** as the cyclohexylamine salt was also prepared and characterized by NMR: ^{1}H NMR (DMSO-d_6) δ 8.13 (br s, 3H), 7.38–7.26 (m, 5H), 5.07–4.97 (m, 2H), 4.97–4.82 (m, 1H), 4.37–4.11 (m, 1H), 4.11–3.89 (m, 1H), 3.82–3.60 (m, 1H), 2.91–2.65 (m, 2H), 2.25–1.91 (m, 2H), 1.91–1.80 (m, 2H), 1.78–1.61 (m, 3H), 1.61–1.50 (1H), 1.42–129 (m, 9H), 1.29–1.15 (4H), 1.14–0.99 (m, 1H); ^{13}C{^{1}H} NMR (DMSO-d_6) δ 173.2 (4C), 155.6 (4C), 154.6 (4C), 138.1 (4C), 128.2 (2C), 127.7 (2C), 127.5 (4C), 78.4 (4C), 74.6 (4C), 55.8 (4C), 53.6, 49.0, 43.7 (2C), 31.1, 28.1 (3C), 25.8 (3C), 24.8 (2C), 21.2 (2C).

To a mixture of Boc oxime acid **4** (45.0 g by assay, 129.1 mmol) in acetonitrile (315 mL) at 15–25 °C was added BSA (17.1 g, 84.0 mmol). The mixture was agitated at ambient temperature for 1 h and then at 50 °C for 1 h. The reaction solution was cooled to 0–5 °C. TMSBr (35.6 g, 232.5 mmol) was added dropwise at 0–5 °C over 1 h. After 16–20 h aging at 25 °C, HPLC analysis showed <2% of **4**; water (10.7 g) was slowly added at 15–25 °C. The mixture was aged for an additional 1 h. A solution of *n*-Bu_4NOAc (88.7 g, 294.2 mmol) in MeCN (300 mL) was added dropwise. The batch was seeded after ~50% of the *n*-Bu_4NOAc was added. The seeded batch was aged at 15–25 °C for 1 h and then agitated at 50 °C for an additional 2 h. The rest of the *n*-Bu_4NOAc acetonitrile solution was added dropwise at 50–55 °C over 3 h. The slurry was agitated at 50–55 °C for 4 h, then cooled to 20 °C, and agitated for additional several hours before filtration. The wet cake was displacement-washed with 10% H_2O in MeCN (135 mL). The wet cake was slurried in 10% H_2O in MeCN (450 mL) at 50–55 °C for 5 h, then cooled to 20 °C, and agitated for additional several hours before filtration. The wet cake was displacement-washed with 10% water in MeCN (135 mL). The wet cake was dried in a vacuum oven with N_2 sweep at 40–50 °C to afford 26.2 g of **5** (80% yield). ^{1}H NMR (DMSO-d_6 + traces of MeOH-d_4) δ 7.39–7.28 (m, 5H), 5.07–5.02 (m, 2H), 3.55–3.47 (m, 2H), 3.42–3.37 (m, 1H),

2.94–2.85 (m, 1H), 2.29–2.19 (m, 1H), 2.16–2.04 (m, 1H), 1.69–1.58 (m, 1H); $^{13}C\{^{1}H\}$NMR (DMSO-d$_6$ + traces of MeOH-d$_4$) δ 169.6, 152.9, 137.8, 128.3 (2C), 127.8 (2C), 127.7 (2C), 74.9 (2C), 57.2 (2C), 44.9, 25.0, 22.2. HRMS (ESI) calcd for $C_{13}H_{17}N_2O_3$ $[M + H]^+$, 249.1239; found, 249.1231.

The Mechanism

The method was based on the reaction developed by Baldwin et al. in 1993 [132].

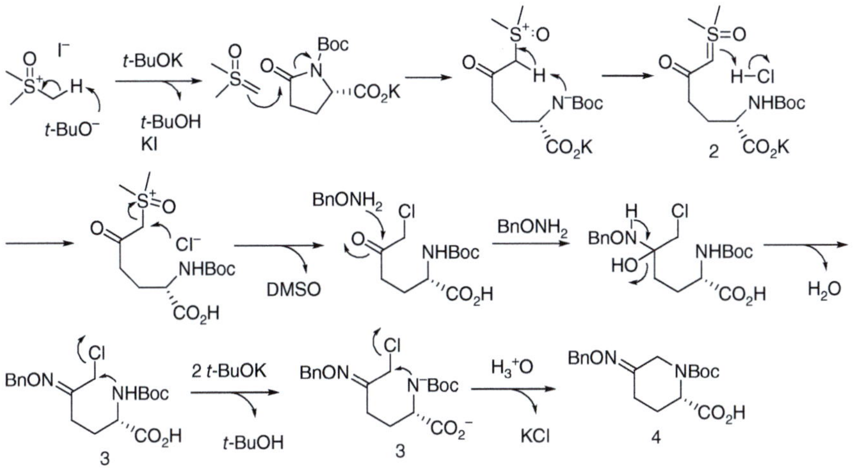

Problem 48: Construction of Morphine Core by Acid Promoted Cyclization

Samidorphan in combination with olanzapine was approved by the FDA in 2021. They are medications for the treatment of patients with schizophrenia and bipolar I disorder. In a synthesis of samidorphan [133, 134], treatment of compound **1** with 55% aqueous H_3PO_4 gave the morphine core structure **2** in 66% yield.

Please suggest a mechanism for this transformation.

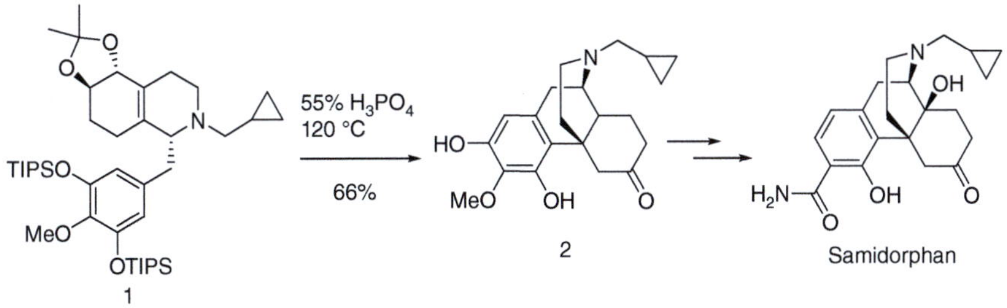

Experimental Procedure (from *Chem. Sci.*, **2019**, *10*, 535, supplementary information) **(4bS,9R)-11-(Cyclopropylmethyl)-2,4-dihydroxy-3-methoxy-8,8a,9,10-tetrahydro-5H9,4b-(epiminoethano)phenanthren-6(7H)-one (2)**: In a glovebox, to a 75-mL pressure tube vial containing the tetrahydropyridine **1** (1.59 g, 2.18 mmol, 1.00 equiv) equipped with a stir bar was added a mixture of H_3PO_4 (85% in H_2O, 14.4 mL) and HPLC grade H_2O (7.6 mL) [both sparged with N_2 for 30 min prior to use]. The tube was taken out of the glovebox and the reaction mixture was heated at 125 °C with vigorous stirring over 5 h. The reaction mixture was cooled to room temperature, transferred to a separatory funnel with H_2O (20 mL), and the aqueous layer was washed with $CHCl_3$ (3 × 50 mL) to discard the silyl by-products dissolved in the organic layer. To the remaining aqueous layer was added ice and NH_4OH was slowly added dropwise until the pH was confirmed to be pH 8 [because the reaction is exothermic, the NH_4OH should be added slowly with caution]. A solvent mixture of 5% MeOH in $CHCl_3$ was used to extract out the phenol (6 × 50 mL). The combined organic layer was concentrated under reduced pressure. The crude product was purified by silica gel chromatography (90:10:1 CH_2Cl_2:MeOH:NH_4OH) to provide **2** (517 mg, 66%) as a tan colored solid. Melting point 145–148 °C. $[\alpha]_D$ + 84.6°, ($CHCl_3$, c = 0.73). IR (neat) 3520, 3296, 2923, 2870, 1707, 1603, 1504, 1429, 1364, 1216 cm^{-1}. ^{1}H NMR (500 MHz, $CDCl_3$) δ 6.24 (s, 1H), 4.12 (dd, J = 13.5 and 1.8 Hz, 1H), 3.80 (s, 3H), 3.40–3.36 (m, 1H), 2.87–2.77 (m, 2H), 2.74–2.63 (m, 1H), 2.61–2.53 (m, 1H), 2.48–2.30 (m, 3H), 2.29–2.22 (m, 2H), 2.14–2.07 (m, 1H), 1.94–1.80 (m, 3H), 1.71–1.62 (m, 1H), 1.03–0.92 (m, 1H), 0.61–0.56 (m, 2H), 0.22 – 0.15 (m, 2H). ^{13}C NMR (126 MHz, $CDCl_3$) δ 211.4, 148.6, 147.1, 133.7, 115.6, 107.1, 61.4, 59.7, 55.1, 50.6, 45.2, 44.2, 41.4, 41.2, 38.1, 27.3, 24.9, 8.8, 4.4, 4.1 (two aromatic Cs overlapped). HRMS (ESI): Mass calcd for $C_{21}H_{28}NO_4{}^+$ $[M + H]^+$, 358.2013. Found $[M + H]^+$, 358.2016.

The Mechanism

At acidic condition, deprotection of both TIPS groups and the acetal group from **1** should occur to give **A**. Then neutral conversion of the diol to the keto group *via* allylic alcohol ionization and a hydride shift provide **B**. Finally, a Grewe cyclization [135] affords product **2**.

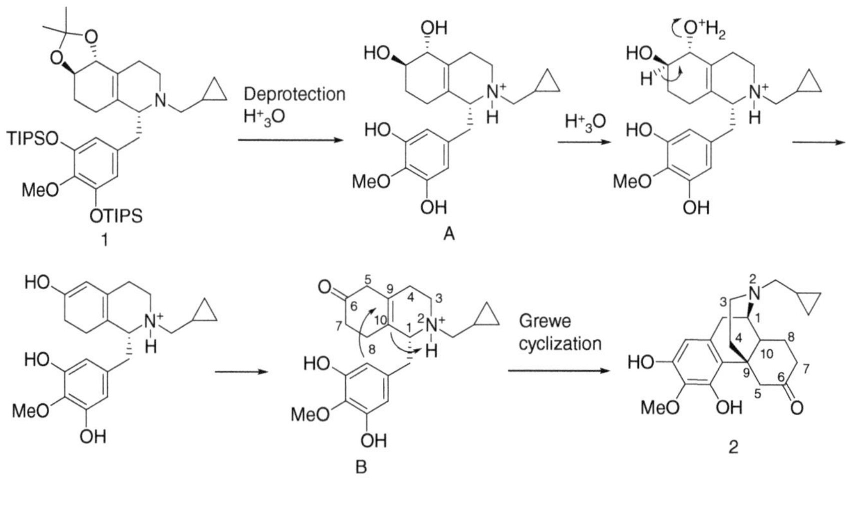

Problem 49: Preparation of 8a-Phenyloctahydroquinolin-4a(2*H*)-ol

Compound (-)-**4** and analogs were evaluated for their antagonistic effects on cGMP levels in mice, and **4** was found to exhibit in vivo potency [136]. The synthesis of (-)-**4** was started from 2,3,4,4a,5,6,7,8-octahydroquinoline (**1**) [136, 137]. The first two reaction steps led to intermediate **3** as shown below.

Please provide a mechanism for the transformations from **1** to **2** and from **2** to **3**.

1 → (O_2, EtOAc, rt, 3.5 h; 87%) → 2 → (PhLi (2.4 eq.), Et_2O, then H_2O; 64%) → 3 → (1. H_2SO_4-DCM; 2. H_2, PtO_2-MeOH; 3. Resolution) → 4

Experimental Procedure (from *J. Am. Chem. Soc.*, **1955**, *77*, 6595)

10-Hydroperoxy-$\Delta^{1(9)}$-octahydroquinoline (2). A solution of 2.8 g (0.02 mole) of **1** in 20 mL of ethyl acetate was magnetically stirred in a closed system containing oxygen. In 3.5 h, absorption had stopped with the uptake of 0.9 molar equivalent. When 50% had been absorbed, the hydroperoxide began to crystallize. (To obtain a crystalline hydroperoxide, it is essential that the octahydroquinoline be of high purity. The rate of oxidation is also highly sensitive to the purity of the base, freshly distilled samples absorbing 200–300 mL of oxygen within 10 min.) The flask was chilled in ice, the product filtered and washed with 25 mL of cold ethyl acetate. The dry solid weighed 2.65 g (87%) (small needles) m.p. 99–100 °C. In the solid state, the hydroperoxide is fairly stable. Samples kept at −5 °C for two months still gave strong positive tests with starch-iodide paper. *Anal.* Calcd. for $C_9H_{15}NO_2$: C, 63.88; H, 8.94; N, 8.28. Found: C, 63.78; H, 8.89; N, 8.05.

(from *J. Med. Chem.*, **1992**, *35*, 1634)

***cis*-4a-Hydroxy-8a-phenyldecahydroquinoline (3).** To a solution of 4a-hydroperoxy-$\Delta^{8,8a}$-octahydroquinoline (**2**, 20.8 g, 0.123)

mol) in 250 mL of ether was added 250 mL of 1.2 M phenyllithium in ether. The solution was refluxed under a nitrogen atmosphere for 2 h, cooled in ice water, and quenched in water. The aqueous solution was extracted with ether, and the combined organic extracts were washed with water and dried over sodium sulfate. After concentration, the crude oil was submitted to flash chromatography on silica gel using first 10% and then 33% ethyl acetate-hexane as eluent to give 18.2 g (64%) of compound **3**: mp 90.5–91.5 °C; IR (KBr) 3216 (b), 3051,3016,2913,2853,1489,1443, 1140,1034,995, 812,752,700, 542 cm^{-1}; ^1H NMR ($CDC1_3$) δ 7.94 (d, 2H, J = 8.0 Hz), 7.31 (t, 2H, J = 7.7 Hz), 7.19 (t, 1H, J = 7.3 Hz), 2.5–2.9 (b, 3H), 1.2–2.05 (b, 13H); ^{13}C NMR ($CDC1_3$) δ 146.3,128.7 (2 C), 127.9 (2 C), 125.9, 77.2, 73.4, 62.0, 41.1, 35.4 (2 C), 24.6, 21.6 (2 C); mass spectrum m/z 231 (M^+), 214, 202, 188,175,160, 154,145,132, 119, 104, 91, 86, 77, 55.

The Mechanism

The first step likely involves an ene-type imine oxidation to form hydroperoxide **2**. This is followed by nucleophilic addition of PhLi to the C=N bond, and final reduction of the hydroperoxy group yields product **3**. Alternatively, this transformation can be viewed as phenyllithium being oxidized by the hydroperoxide to phenol [138, 139]. Based on this mechanism, three equivalents of PhLi are required. Therefore, using more than three equivalents—rather than the 2.4 equivalents [138] reported—might improve the yield of product **3**.

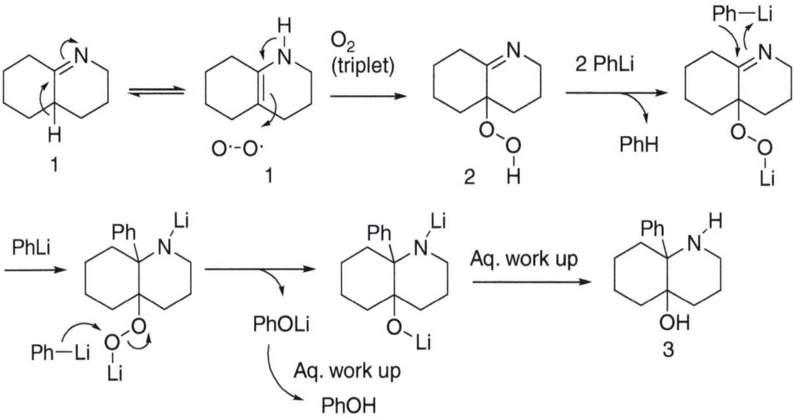

Problem 50: Preparation of an Octahydroquinolizine

MLN1251 (**1**) is a CCR5 antagonist for treatment of HIV. A multi-kilogram scale synthesis of **1** is reported by Millennium Pharmaceuticals [140]. As shown below, the key building block 1,3,4,6,7,8-hexahydro-2*H*-quinolizine (**3**) is prepared from ethyl 5-(2-oxopiperidin-1-yl)pentanoate (**2**) by heating with soda lime (a mixture of sodium hydroxide, calcium oxide, and water). The methodology was initially developed by Seiji Miyano [141]. Later on, the method was applied by John McIntosh [142] and R. W. Alder's group [143].

Please suggest a mechanism for the transformation from **2** to **3**.

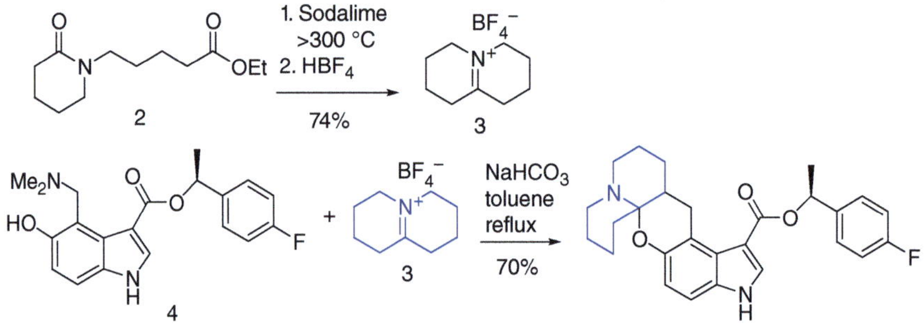

Experimental Procedure (from *Org. Process Res. Dev.*, **2007**, *11*, 241)

1,3,4,6,7,8-Hexahydro-2*H*-quinolizine Tetrafluoroboric Acid Salt (3). Ethyl 5-(2-oxopiperidin-1-yl)pentanoate (**2**, 394.0 g, 1.73 mol) and crushed soda lime (524.2 g) were charged to a 2-L, three-necked round-bottomed flask with thermometer, distillation apparatus, and heating mantle. The apparatus was vacuum flushed with nitrogen, and the mixture was heated and distilled to give two fractions. Fraction 1: approx. 150 mL, distillate temp below 110 °C (mainly ethanol and water distilled over). Fraction 2: approx. 300 mL, distillate temp 110–210 °C (up to 300 °C pot temperature, water and product distilled). Fraction 1 was taken up in TBME (400 mL), shaken, and separated. The organic layer was washed with brine (400 mL), added to Fraction 2, and diluted with additional TBME (400 mL). The solution was washed with brine (400 mL), the aqueous was extracted with TBME (400 mL), and the organic layers were combined, dried ($MgSO_4$, 84 g), and filtered. The solution was cooled while stirring in an ice water bath, and tetrafluoroboric acid (215 mL of a 54 wt % solution in diethyl ether, 1.56 mol) was added at a rate such that the temperature was maintained below 30 °C. The tetrafluoroborate salt precipitated, was collected by filtration, and dried under vacuum at ambient temperature for 16 h to give 287.35 g, 74%, of **3** as a light-peach solid, which was stored in a freezer. ^{1}H NMR (DMSO-d_6), 300 MHz) δ (ppm): 1.62 (m, 4H), 1.83 (m, 4H), 2.68 (m, 4H), 3.60 (m, 4H). ^{13}C NMR (DMSO-d_6, 75 MHz) δ (ppm): 186.0, 53.2, 31.8, 20.1, 16.3.

The Mechanism

The first step is the hydrolysis of the ester under the strong base at high temperature. This is followed by an intramolecular condensation between the acid (Na^+ and/or Ca^{2+} salt form) and the amide moiety to generate intermediate **A**. Subsequent decarboxylation from **A** yields the free amine quinolizine **B**. Treatment of the free amine with HBF_4 leads to the salt **3**.

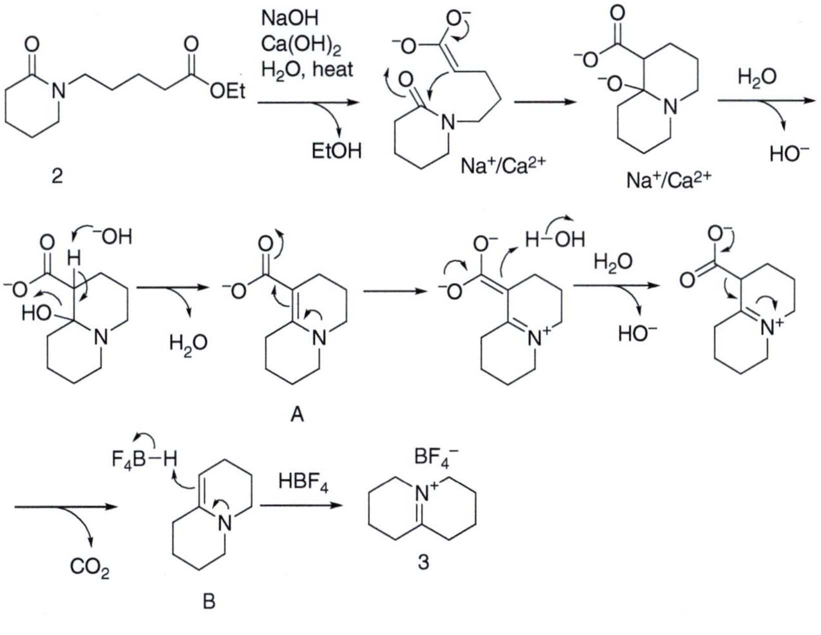

Problem 51: Preparation of a Chromeno[3,2-i]quinolizine

This is the second mechanism problem for the synthesis of the target compound mentioned in **Problem 50**. The final step synthesis of MLN1251 (**1**) is achieved by heating intermediates **2** and **3** in toluene in the presence of sodium bicarbonate [144].

Please suggest a mechanism for the reaction.

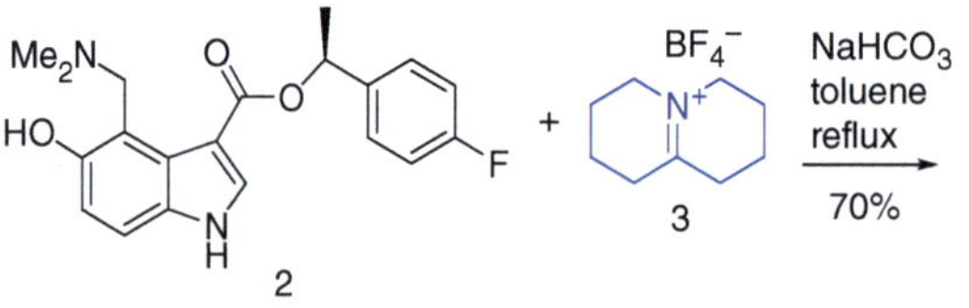

Me_2N, HO, O, O, N, H, 2, F, +, BF_4^-, N^+, 3, $NaHCO_3$, toluene, reflux, 70%, N, O, O, O, N, H, F, 1 (MLN1251)

Experimental Procedure (from *Org. Process Res. Dev.*, **2007**, *11*, 241)

MLN1251 (1). A 10-L round-bottomed flask with anchor stirrer, reflux condenser, and heating mantle was charged with toluene (1 L), saturated aqueous sodium hydrogen carbonate (1 L), 1,3,4,6,7,8-hexahydro-2*H*-quinolizine tetrafluoroboric acid salt (**3**, 104.8 g, 464 mmol), and 1-(4-fluorophenyl)-ethyl 4-dimethylaminomethyl-5-hydroxy-2-methyl-1*H*-indole-3-carboxylate hydrochloride (**2**, 100 g, 442 mmol), and the mixture was heated at reflux for 160 min. The mixture was cooled, and the layers were separated. The organic layer was dried ($MgSO_4$, 60 g), filtered, and concentrated. The residue was taken up in ethanol (300 mL) and heated to 60 °C. Water (100 mL) was added, and the mixture stirred while cooling. When the temperature was 42 °C, 300 mg of seeds were added. The mixture was stirred at ambient temperature for 2.5 days. The resulting precipitated solid was collected by filtration, washed with 4:1 ethanol/water (2 × 125 mL), and dried in a vacuum oven at 30 °C for 23 h to give 80.0 g, 70% of **1** as an off-white solid. ^1H, ^{19}F NMR consistent with structure. Ratio of diastereomers by ^{19}F NMR: 1.02:1.00. ^1H NMR (DMSO-d_6, 300 MHz) δ (ppm): 11.55 (br s, 1H), 7.45–7.53 (m, 2H), 7.16–7.24 (m, 2H), 7.03–7.07 (m, 1H), 6.59–6.63 (m, 1H), 5.94–6.03 (m, 1H), 3.10–3.32 (m, 2H), 2.86–2.94 (m, 1H), 2.63–2.71 (m, 2H), (2.54, 2.53; s, 3H), 2.35–2.47 (m, 2H), 1.83–1.92 (m, 1H), 1.09–1.76 (m, 12H). ^{13}C NMR (DMSO-d_6, 75 MHz) δ (ppm): 14.7, 19.4, 21.9, 22.2, 24.6, 25.0, 26.8, (30.0, 30.1), 30.5, 36.0, (48.8, 48.9), (70.3, 70.4), (86.4, 86.5), 104.4, (109.9, 109.9), (110.2, 110.3), 112.7, 115.0 (d, $^2J_{CF} = 21.3$ Hz), (125.8, 125.8), (128.0, 128.4; d, $^3J_{CF} = 8.2$ Hz), 129.5, (138.2, 138.5; d, $^4J_{CF} = 3.0$ Hz), (142.6, 142.7), (147.9, 148.0), (161.4, 161.5; d, $^1J_{CF} = 243$ Hz), (164.2, 164.2). MS 463.2 ES $[M + H]^+$. R_f 0.2 (1:1 EtOAc/hexanes). HRMS $[M + H]^+$ observed = 463.2389, calculated = 463.2397. Anal. Calcd for $C_{28}H_{31}FN_2O_3$: C, 72.70; H, 6.76; N, 6.06. Found: C, 72.94; H, 6.72; N, 6.07.

The Mechanism

The free enamine **A** is released by treatment of salt **3** by $NaHCO_3$. The dimethylamino group in compound **2** is activated by a neighboring hydroxyl group through hydrogen bonding, making it a good leaving group. As a nucleophile, the β-alkenic carbon of the free enamine **A** attacks the benzylic position in **2**, replacing the dimethylamino group to give an inner salt **B**, which should quickly cyclize to the final product **1**.

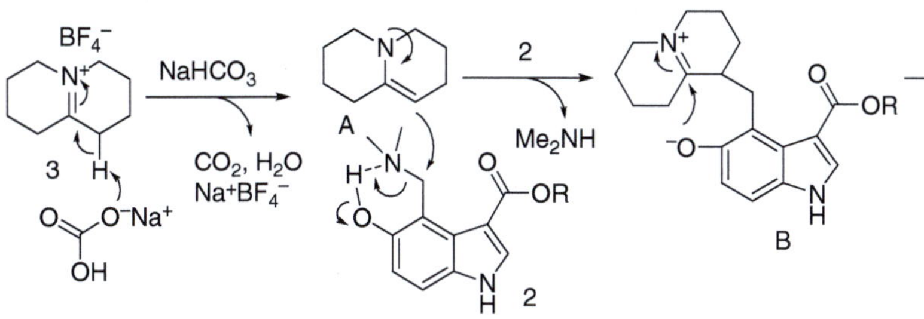

4.5.6 Imidazoline

Problem 52: Preparation of Phenytoins from Benzils

Some phenytoin derivatives, such as **3**, showed anti-inflammatory and analgesic activities [145]. Mahmoodi and Emadi reported a one-pot synthesis of phenytoin (**2**) from benzil (**1**) and urea in refluxing aqueous NaOH and ethanol [146]. Please provide a mechanism for the reaction from **1** to **2**.

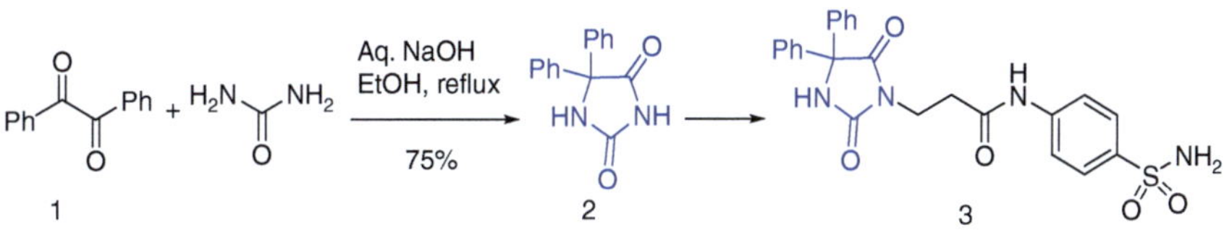

Experimental Procedure (from *Rus. J. Org. Chem.*, **2004**, *40*, 377)

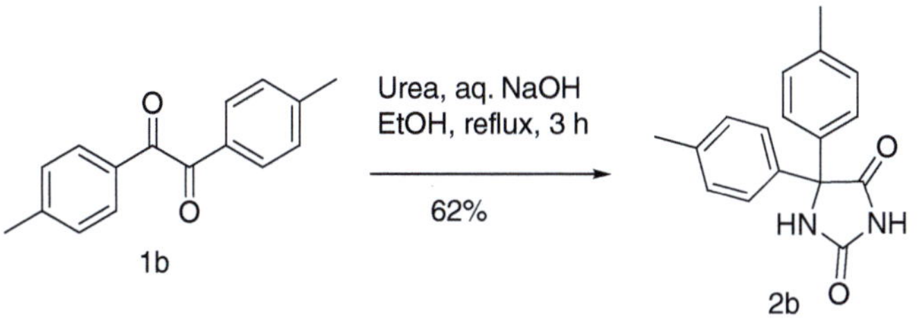

5,5-Di-*p*-tolylimidazolidine-2,4-dione (2b). A 100-ml round bottom flask equipped with a reflux condenser was charged with 5.95 g (25 mmol) of compound IIb, 3 g (50 mmol) of urea, 15 mL of 40% aqueous NaOH, and 70 mL of 96% ethanol. The mixture was heated for 3 h under reflux with stirring, cooled to room temperature, and poured into 120 mL of distilled water. The resulting mixture was vigorously stirred for 15 min, and the precipitate was filtered off, and recrystallized from 96% ethanol. Yield 2.41 g (62%), colorless crystals, mp 225 °C. IR spectrum (KBr), ν 3200 s, 1720 s cm^{-1}. ^1H (DMSO): 2.5 s (6H), 3.5 s (H_2O), 7.4 s (8H), 9.3 s (1H), 11.2 s (1H); DMSO–D_2O: 2.1 s (6H), 4 s (H_2O), 7.2 s (8H). ^{13}C (DMSO): 20.9, 75.4, 128.3, 129.7, 135.1, 140, 160, 170.6.

The Mechanism

The key step of the reaction is a carbanion rearrangement. Urea as nucleophile attacks benzil (**1**) to give intermediate **A**. Then cyclization occurs to provide **B**. At a strong basic condition, **B** is deprotonated, followed by phenyl migration to generate anions **D**/**E**/**F**. Final protonation of **D**/**E**/**F** leads to phenytoin (**2**).

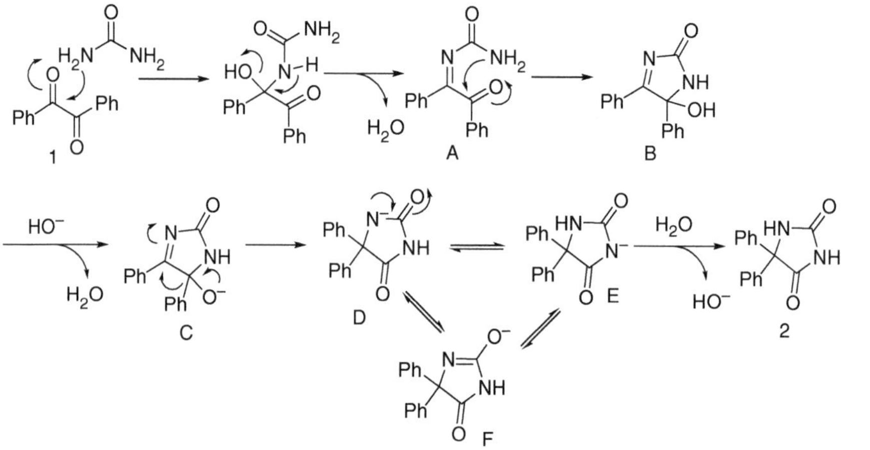

4.5.7 Oxazolidine

Problem 53: Preparation of a Chiral (*R*)-5-(Hydroxymethyl)-3-aryloxazolidin-2-one

Linezolid (**1**) is a synthetic antibiotic used for the treatment of serious infections caused by Gram-positive bacteria that are resistant to several other antibiotics. A key step for the synthesis of linezolid is the formation of the chiral oxazolidin-2-one (**2**), which is prepared from the Cbz protected aniline **3** and (*R*)-glycidyl butyrate **4** [147]. The reaction is shown as below. Please provide a mechanism for the formation of **2** from **3** and **4**.

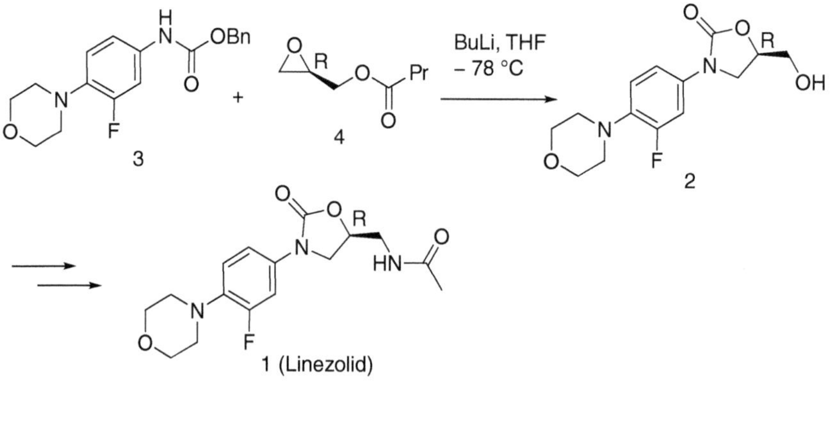

Experimental Procedure (from *J. Med. Chem.*, **1996**, *39*, 673)

(*R*)-[*N*-3-(3-Fluoro-4-morpholinylphenyl)-2-oxo-5-oxazolidinyl]methanol, 2. To a solution of 40.604 g (123.0 mmol) of **3** in THF (500 mL) under nitrogen at −78 °C was added *n*-butyllithium (77 mL, 1.6 M in hexane, 123.2 mmol) over 20 min via syringe. The solution was stirred at −78 °C for 35 min; then a THF solution (25 mL) of (*R*)-glycidyl butyrate (**4**, 17.8 mL, 125.7 mmol) was added in a dropwise fashion via addition funnel, over 30 min. After stirring at −78 °C for 1 h, the bath was removed and the reaction mixture was stirred at room temperature overnight. The reaction was then quenched with 20 mL of saturated ammonium chloride, ethyl acetate (300 mL), saturated ammonium chloride (400 mL), and water (300 mL) were added, the phases were separated, and the aqueous portion was extracted with ethyl acetate (3 × 300 mL). The combined organic portions were washed with saturated sodium chloride, dried ($MgSO_4$), and evaporated to give **2** (60.725 g) as a yellow solid; this was recrystallized from ethyl acetate and hexanes to give 30.814 g (85%) of **2** as a light gray amorphous solid, mp 112–114 °C. A second crop (2.431 g, 7%) of less pure material was obtained. ^{1}H NMR ($CDCl_3$, 400 MHz): δ 7.44 (dd, J = 14.4 Hz, J' = 2.4 Hz, 1H), 7.10 (dd, J = 8.4 Hz, J' = 1.6 Hz, 1H), 6.91 (t, J = 8.8 Hz, 1H), 4.73 (m, 8 lines, 1H), 3.96 (overlapping m, 3H), 3.86 (t, J = 4.8 Hz, 4H), 3.74 (m, 1H), 3.04 (t, J = 4.8 Hz, 4H), 2.93 (t, J = 6.4 Hz, 1H). IR (mineral oil mull, cm^{-1}): 3432, 3263, 2925, 1748, 1734, 1523, 1465, 1454, 1441, 1424, 1231. MS: *m/z* 296 (100, M^+), 238 (41.6), 151 (14.2). Exact mass: calcd for $C_{14}H_{17}FN_2O_4$, 296.1172; found, 296.1180. $[a]_D^{20}$ − 50 ° (*c* 0.990, $CHCl_3$). Anal. ($C_{14}H_{17}FN_2O_4$) C, H, N.

The Mechanism

The carbamate **3** is deprotonated by BuLi to give the anion species **A**. As a nucleophile, **A** would attack the epoxide **4** from the less steric hindered side to generate the alkoxide **B**, which would readily attack the carbamate carbonyl to produce the cyclized oxazolidin-2-one **C**. Then transesterification occurs between **C** and the produced BnO^-, and after aqueous workup, to yield the product **2**. The by-product of this reaction is benzyl butyrate.

4.5.8 Tetrahydropyrimidine

Problem 54: Synthesis of Tetrahydropyrimidinol

Compound **1** is a $\alpha_v\beta_3$ integrin antagonist and is being studied for the treatment of cancer. Tetrahydropyrimidinol **2** is a key intermediate for the synthesis of **1**. The preparation of **2** is from **3** and 1,3-diaminopropan-2-ol (**4**) as shown below [148]. An impurity **5** is also generated during the reaction.

Please suggest a mechanism for the formation **2** and **5**.

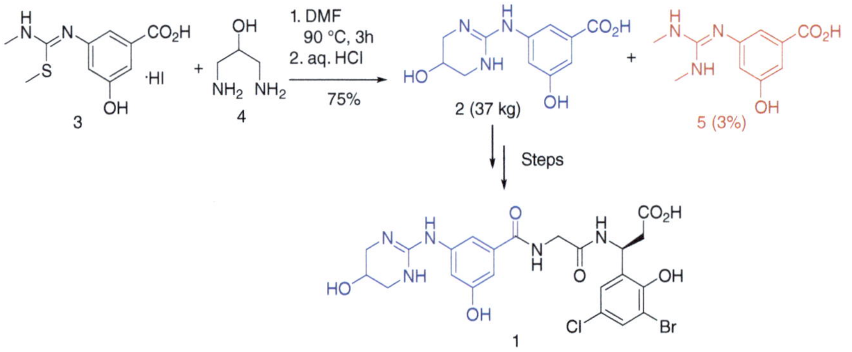

Experimental Procedure (from *Org. Process Res. Dev.*, **2004**, *8*, 571)

***N*-[2-(5-Hydroxy-4,6-tetrahydropyrimidine)]-3-amino-5-hydroxybenzoic Acid (2)**. To a separate 100-gal reactor vented to sequential bleach and sulfuric acid scrubbers was charged premelted (mp 40–45 °C) 1,3-diaminopropan-2-ol (**4**, 50.8 kg, 597 mol) and DMF (41.7 kg). The 1,3-diaminopropan-2-ol/DMF solution was transferred to the reactor containing **3** at such a rate to keep the contents below 50 °C. Upon complete addition, the contents were heated to 90 °C. Vigorous off-gassing began at approximately 60 °C. Once the reactor contents reached 90–95 °C, it was held within this range for 3 h.

The reactor jacket was cooled to 40 °C. Into a separate 500-gal reactor was charged 181 kg of water. With agitation, the solution containing **2** was transferred into the 500-gal reactor. The contents were then cooled to 10 °C, and the pH was carefully adjusted to pH 6 with 37% aqueous HCl (27.7 kg). The resulting precipitate was isolated via a variable speed 40-in. Tolhurst centrifuge. The damp cake was washed with water (18 kg) and acetonitrile (14 kg) and allowed to spin dry. The product was transferred to a vacuum tray dyer and dried overnight at 70–75 °C at 25–28 in. Hg to give 37 kg (74.9%) of **2** with an HPLC assay of 96.4%: TLC: $R_f = 0.53$ ($CHCl_3$/MeOH/H_2O, 25:25:5); ^{1}H NMR (D_2O, 90 MHz) δ 7.42 (s, 2H), 7.01 (t, 1H), 4.41 (t, 1H), 3.46 (br s, 4H). This product was identical to an authentic sample as determined by ^{1}H NMR, ^{13}C NMR, and HPLC analysis.

The Mechanism

The mechanism can be interpreted as a substitution of both the methylthio and methylamino groups by 1,3-diamine (**4**). The whole process should be reversible. However, due to the volatility of methanethiol and methylamine, most of the two by-products are evaporated from the reaction system.

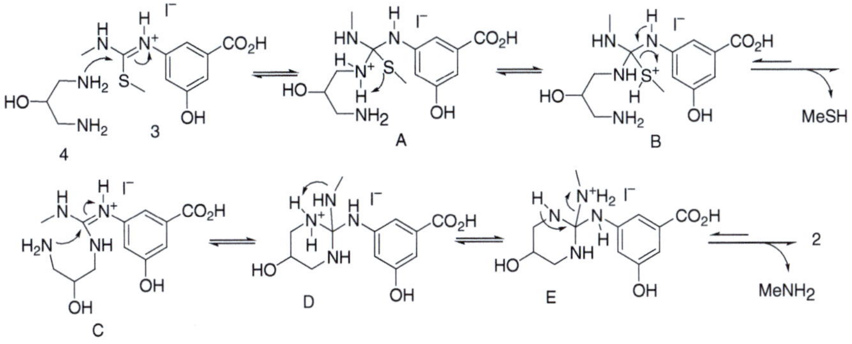

Once methylamine is generated, it would be a nucleophile and attacks the unreacted starting material **3** to form the impurity **5**. Indeed, it was found that the level of **5** could be diminished from 3% to less than 0.5% by either adjusting the mode of reagent interaction or by physically facilitating gaseous methylamine expulsion. For example, the mode of reagent addition was changed by either adding the crude solution of **3** to a hot solution of 1,3-diaminopropan-2-ol in DMF (inverse addition) or by increasing the rate of 1,3-diaminopropan-2-ol addition to the crude solution of **3**.

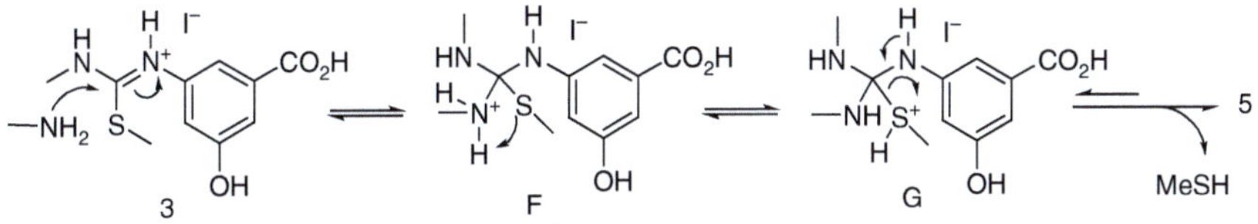

4.5.9 Diazepine

Problem 55: Synthesis of 2,3,4,5-Tetrahydro-1H-benzo[e][1,4]diazepine
SB-214857-A (lotrafiban, **1**) is a potent nonpeptidic glycoprotein IIb/IIIa antagonist and inhibits platelet aggregation [149]. In a process research, the key intermediate **8** was prepared from 2-nitrobenzyl alcohol (**2**) by the following telescope reactions [150].

Please suggest a mechanism for the transformations from intermediate **5** to intermediate **8**.

Experimental Procedure (from *Org. Process Res. Dev.*, **2003**, *7*, 655)
2,3,4,5-Tetrahydro-4-methyl-3-oxo-1*H*-1,4-benzodiazepine-2-acetic Acid Methyl Ester (9) from Nitrobenzyl Alcohol (2). A solution of methanesulfonyl chloride (87 kg, 795 mol) in THF (35 kg) was added to a solution of 2-nitrobenzyl alcohol **2** (100 kg, 653 mol) and triethylamine (71.5 kg, 706 mol) in THF (620 kg) over 2 h, allowing the temperature to rise from 17 °C to a maximum of 34 °C. The head tank and transfer lines were washed through with THF (10 kg). The reaction mixture was stirred at 32 °C for 1 h. The mixture was added to 40% w/w methylamine in water (620 kg, 7985 mol) over 45 min, allowing the temperature to rise from 20 to a maximum of 33 °C. The transfer line was washed through with THF (89 kg), and stirring continued at 32 °C for 1.5 h. The phases were allowed to separate for 15 min. The phases were separated, and the aqueous phase was extracted with toluene (256 kg) and then discarded. The organic

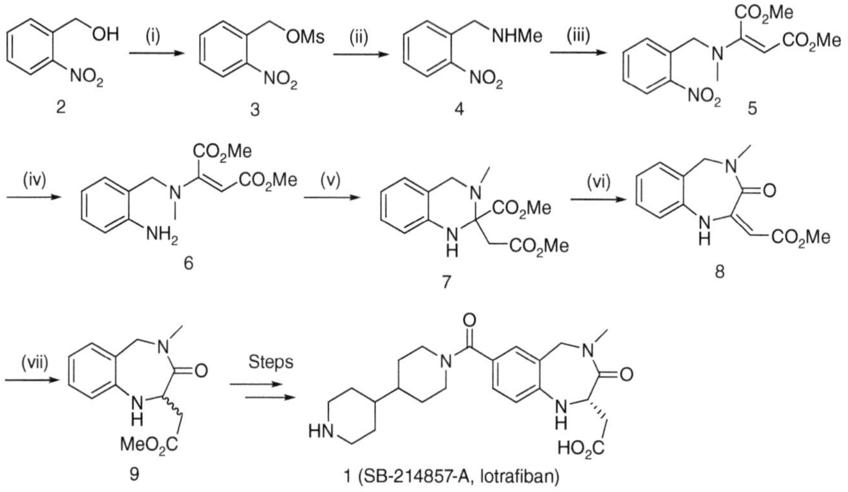

Reagents: (i) $MeSO_2Cl$, Et_3N, THF; (ii) aq $MeNH_2$; (iii) dimethyl acetylenedi-carboxylate (DMAD), EtOAc; (iv) cyclohexene, Pd-C, or H_2, Raney Ni; (v) AcOH, MeOH; (vi) DBU, or NaOMe, MeOH; (vii) NH_4HCO_2 or Pd-C, MeOH, H_2O.

phases were combined, and solvent (900 L) was distilled off under vacuum, maximum base temperature 22 °C. The resulting concentrate was cooled to 8 °C (pH 10.3) and then pH-adjusted to 4.3 by the addition of 12% v/v aqueous sulfuric acid solution (177 L), taken from a bulk solution of concentrated sulfuric acid (340 kg) and demineralized water (1500 L). The temperature was allowed to rise to a maximum of 17 °C. The mixture was stirred for 5 min, and the phases were allowed to separate for 15 min. The phases were separated, and the organic phase was discarded. The aqueous phase was diluted with demineralized water (150 L) and then basified to pH 11.7 with 50% w/w sodium hydroxide solution (110 kg). Compound **4** was extracted into ethyl acetate (270 kg), and the mixture was warmed to 27 °C. The aqueous phase was separated off and extracted with further ethyl acetate (135 kg), maintaining the temperature at 25–30 °C, and then discarded. The organic phases were combined.

A solution of dimethylacetylene dicarboxylate (86 kg, 605 mol) in ethyl acetate (90 kg) was added over 1 h with cooling applied, allowing the temperature to rise to a maximum of 29 °C. The head tank and transfer lines were washed through with ethyl acetate (9.0 kg). The reaction mixture was stirred at 23–24 °C for 40 min. The mixture was cooled to 4 °C, and demineralized water (50 L) was charged, followed by ethyl acetate (180 kg). The pH was adjusted from 10.9 to 2.6 by the careful addition of 12% v/v aqueous sulfuric acid (45 L), maintaining the temperature between 0 and 5 °C. The mixture was stirred for 5 min, and the phases were allowed to separate for 15 min. The aqueous phase was separated off and discarded, and the organic phase washed with an aqueous solution of sulfuric acid and sodium sulfate (135 kg), taken from a bulk solution of sodium sulfate (50 kg), concentrated sulfuric acid (9.0 kg), and demineralized water (650 L), maintaining the temperature between 0 and 5 °C. The head tank and transfer lines were washed through with demineralized water (10 L). The mixture was stirred for 5 min, and the phases were allowed to separate for 15 min. The aqueous phase was separated off and discarded and the organic phase washed twice with 10% w/w sodium bicarbonate solution (2 × 125 kg), allowing the temperature to rise to 7 °C. After addition of the second bicarbonate wash, the head tank and transfer lines were washed through with demineralized water (10 L).

Ten percent palladium on charcoal (Pd—C), 59% water wet (30 kg, source Johnson Matthey, 87 L) was charged and the mixture heated to 66 °C. Cyclohexene (240 kg, 2921 mol) was added over 35 min, maintaining reflux. The head tank and transfer lines were washed through with ethyl acetate (9 kg). The reaction mixture was stirred at reflux (68 °C) for 4.5 h, and then the water was separated off by Dean-Stark distillation over a further 4 h 20 min. The total volume of water collected was 54 L.

The mixture was cooled to 25 °C and filtered, washing the filters and lines through with ethyl acetate (89 kg). A mixture of ethyl acetate and unreacted cyclohexene was distilled off at atmospheric pressure (950 L), the temperature rising to a maximum of 82 °C. The remaining contents of the vessel were cooled to 62 °C, and methanol (550 kg) was added. The distillation was continued at atmospheric pressure until a further 200 L of solvent had been distilled out (maximum temperature 65 °C). Glacial acetic acid (2.1 kg, 35 mol) was charged, washing through with methanol (8 kg), and the mixture stirred at reflux for 1.5 h.

A 30% solution of sodium methoxide in methanol (19.7 kg, 109 mol) was added, maintaining reflux, and the head tank and lines were washed through with methanol (8 kg). The mixture was stirred at reflux for 2 h. The mixture was cooled to 33 °C and the pH (10.0) adjusted to 7.7 by the addition of glacial acetic acid (5.1 kg, 85 mol). The mixture was stirred at 30–35 °C for 33 min (final pH 7.5). Ten percent Pd-C, 59% water wet (15 kg, source Johnson Matthey 87 L) was charged, and a solution of sodium borohydride in demineralized water containing sodium hydroxide [sodium borohydride (20.0 kg, 528.7 mol) in demineralized water (265 L) containing sodium hydroxide (0.12 kg)] was added over 2 h 45 min, allowing the temperature to rise from 33 °C and maintaining at 40–45 °C. The head tank and transfer lines were washed through with demineralized water (5 L). The mixture was stirred at 40–42 °C for 55 min.

The pH was adjusted from 9.6 to 7.0 by the addition of glacial acetic acid (25 kg, 416 mol). The head tank and transfer lines were washed through with demineralized water (10 L). DCM (930 kg) was added and the slurry (35 °C) filtered to a clean vessel. The filters and lines were washed through with DCM (266 kg) at 30 °C. A mixture of DCM and methanol was distilled off at atmospheric pressure to a base temperature of 50 °C (volume of distillate 800 L). The distillation was continued under vacuum to a base temperature of 53 °C (volume of distillate 511 L). Demineralized water (500 L) was added to the concentrate and the mixture cooled to 20 °C to crystallize the product. After stirring the slurry for 1 h at 17–2 °C, the product was filtered off in a filter drier, washed with demineralized water (200 L), and dried at 35–40 °C under vacuum for 14 h to give **9** as a white solid, 117 kg 70% from 2-nitrobenzyl alcohol, **9**: mp 174–176 °C; ^{1}H NMR (400 MHz, $CDCl_3$) δ 7.07 (td, 1H, J = 7.7 and 1.4 Hz), 6.93 (dd, 1H, J = 7.5 and 1.1 Hz), 6.67 (td, 1H, J = 7.4 and 1.0 Hz), 6.57 (dd, 1H, J = 8.1 and 0.9 Hz), 5.42 (d, 1H, J = 16.3 Hz), 5.00 (q, 1H, J = 6.1 Hz), 4.09 (d, 1H, J = 5.4 Hz), 3.73 (s, 3H), 3.72 (d, 1H, J = 16.4 Hz), 3.07 (s, 3H), 2.98 (dd, 1H, J = 16.0 and 6.8 Hz), 2.65 (dd, 1H, J = 16.0 and 6.5 Hz); ^{13}C NMR (100 MHz, $CDCl_3$) δ 171.9, 169.6, 145.1, 129.5, 129.0, 119.8, 118.2, 117.5, 53.3, 51.9, 51.8, 35.9, 34.6; IR (neat, cm^{-1}) 1728, 1650; LRMS (CI + ve) m/z 249 (M^+ + H).

The Mechanism

The first reaction is a selective hydrogenation of the nitro to give the aniline **6**. Then Michael addition (cyclization) of the amino to the C=C double bond yields the intermediate **7**. At the basic condition, the heterocycle can be opened to the diamino compound **A** which in turn cyclizes via amidation to produce the seven-membered product **8**. The α,β-unsaturated double bond of the ester should be *Z*-form due to the intra molecule hydrogen bond.

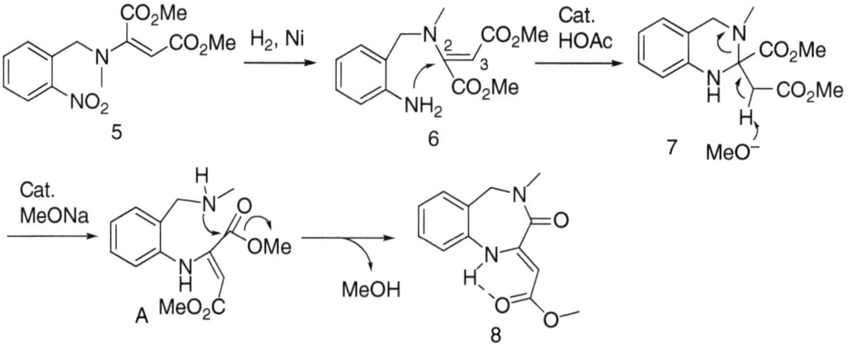

Although it is less likely, another possibility is, at the second stage, the amino adding to the C-3 of the double bond in **6** to form **B**. Subsequently, the reaction follows ring-opening-ring-close process to lead to product **8**.

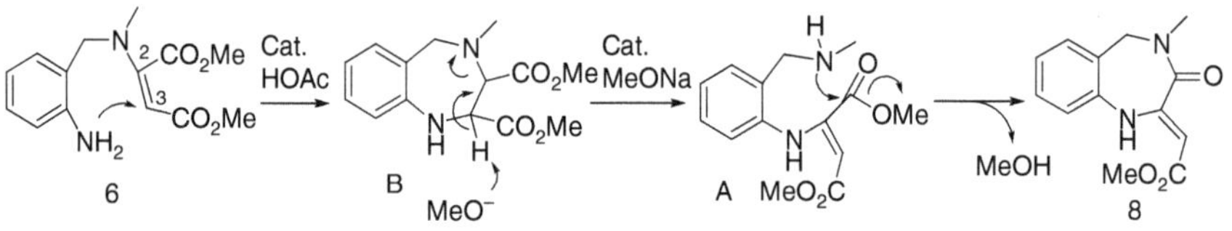

4.5.10 Trioxane

Problem 56: Formation of 1,2,4-Trioxanes from Cyclic Peroxides and Carbonyl Compounds

1,2,4-Trioxane core structure in artemisinin is proven to be the active site for its anti-malaria activity. In 1983, Jefford's group reported that when cyclic peroxide **2** was treated with acetaldehyde in the presence of Amberlyst-15, a 1,2,4-trioxane **4** was obtained in 13% yield as a pair of epimers. Meanwhile, 1,4-dimethoxynaphthalene (**1**) (24%) and naphthalene-1,4-dione (**3**) (37%) were also obtained [151].

More stable 1,4-endoperoxide **7** was subjected to the same reaction with acetaldehyde, the epimeric mixture of cis-fused 1,2,4-trioxanes **8** were obtained in quantitative yield. Ketones proved less reactive with **7.** Catalyzed by trimethylsilyl trifluoromethanesulfonate (TMSOTf) in DCM, the yield of product **9** was improved to 83% [152].

Please provide mechanism for the formation of products **1**, **3**, and **4**.

Artemisinin (Qinghaosu)

$^1O_2$; CH_3CHO, Amberlyst-15, THF; 1 (24%), 3 (37%), 4 (13%)

Amberlyst-15, THF; 8, R=H, 100%; 9, R=Me, 30%; 10, R=CO_2Me, 31%

Experimental Procedure (from *J. Chem. Soc., Chem. Commun.*, **1984**, 523)
The peroxide (0.2 mmol) is mixed with an excess of carbonyl compound (200 μL for acetaldehyde and acetone and 60 μL for cyclopentanone) in dry DCM (2 mL). To this solution at −78 °C, TMSOTf (0.3–8.4 equiv) is added with stirring under nitrogen. Once the reaction is complete (*ca.* 1.5 h), a portion of trimethylamine (half the volume of TMSOTf used) is added and the resulting mixture is diluted with DCM (20 mL). The organic layer is washed with water (3 times) and dried ($MgSO_4$). After evaporation of solvent, the 1,2,4-trioxane is preparatively isolated by layer chromatography over silica gel (thickness 2 mm).

The Mechanism

Due to the thermal instability of **2**, a retro Diels-Alder reaction gives product **1** and singlet oxygen. Acid-catalyzed hydrolysis of **2** provides the quinone **3**. Protonation of **2** has generated the hydroperoxide **5** as the crucial first step for the formation

H^+; H_2O; H_3^+O, 2MeOH, H_2O_2; Retro D-A reaction, $^1O_2$; CH_3CHO; H_3^+O, MeOH

of the trioxane **6**. Normally, its hydrolysis is easy, giving 1,4-naphthoquinone (**3**). However, **5** has the chance to nucleophilically attack the acetaldehyde molecule, which in turn cyclizes to give the 1,2,4-trioxane ring, which is cis-fused for reasons of geometry control. The resulting product **6** is hydrolyzed in situ to **4**, which consists of a pair of epimers.

Problem 57: Synthesis of 1,2,4-Trioxanes from 4-Alkylphenols and Aldehydes
The 1,2,4-trioxane moiety appears to be a key pharmacophore in combating malaria [153]. M. W. George's group reported a continue flow synthesis of 1,2,4-trioxanes (such as **4**) from 4-alkylphenols (such as **1**) and aldehydes (such as **3**) by photo-generated singlet oxygen via the peroxyquinol intermediate **2** [154]. An example is shown below. Please suggest a mechanism for the transformation from **1** to **4**.

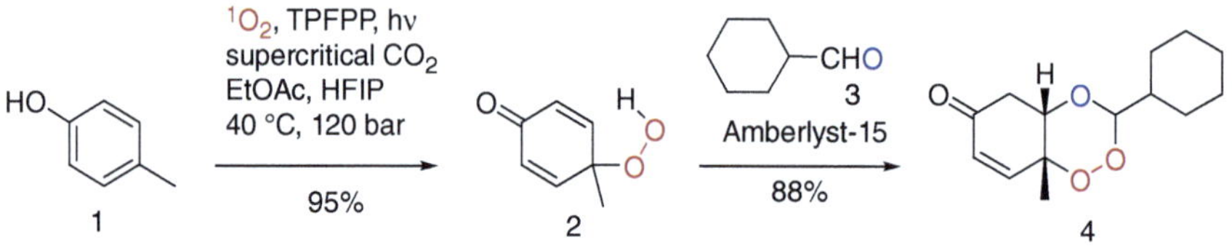

Experimental Procedure (from *Org. Process Res. Dev.*, **2021**, *25*, 1873, Supporting Information)

Procedure for the Synthesis of *para*-peroxyquinol Derivatives: The desired concentration of starting material was pumped into the rig by organic pump 1 at the desired flow rate; a second pump delivered more solvent after the photo reactor to avoid the blockage in the bicyclic peroxy radical (BPR). The para-peroxyquinol derivatives crude could be collected after BPR, ready to be isolated using flash chromatography and identified.

Procedure for the Synthesis of 1,2,4-Trioxane Derivatives: As described in the procedure for the synthesis of *para*-peroxyquinol derivatives, instead of pumping the pure solvent from the second pump, the desired aldehyde dissolved in the solvent was pumped at the same flow rate as organic pump 1.

Procedure in Batch Experiments. In a round bottle flask, 1 mmol of *para*-cresol (**1**) was dissolved in the desired solvent to prepare 0.067 M of the solution. To this solution, 0.05–0.1 mol% of TPFPP to **1** was added. Oxygen was provided from a balloon loaded with pure oxygen at atmospheric pressure. The mixture was irradiated by white light-emitting diode (LED) light at room temperature for a time period of 4–5 h. The crude was purified using cyclohexane/EtOAc to eluent products using automatic flash chromatography after removing the solvent. For the reaction with co-catalyst, 0.5 equivalent co-catalyst was added into the solution before irradiation.

4-Hydroperoxy-4-methylcyclohexa-2,5-dien-1-one (2). The product was purified using flash chromatography column and eluted with a mixture of EtOAc/cyclohexane (1:1). White solid. ^{1}H NMR (400 MHz, $CDCl_3$) δ 8.34 (s, 1H), 6.94 – 6.87 (m, 2H), 6.33 – 6.26 (m, 2H), 1.42 (s, 3H). ^{13}C NMR (101 MHz, $CDCl_3$) δ 185.5, 149.7, 130.5, 78.7, 22.9. HRMS (ESI, *m/z*): calc for $C_7H_8O_3$ $[M]^+$: 140.0473, found: 140.0472.

3-Cyclohexyl-8a-methyl-4a,8a-dihydrobenzo[e][1,2,4]trioxin-6(5H)-one (4). The product was purified using flash chromatography column and eluted with a mixture of EtOAc/cyclohexane (1:4). White solid. ^{1}H NMR (400 MHz, $CDCl_3$) δ 6.84 (dd, $J = 10.4$ and 2.8 Hz, 1H), 6.07 (dd, $J = 10.3$ and =, 0.9 Hz, 1H), 5.02 (d, $J = 5.4$ Hz, 1H), 4.14 (q, $J = 3.0$ Hz, 1H), 2.70 (t, $J = 2.9$ Hz, 2H), 1.74 – 1.58 (m, 6H), 1.33 (s, 3H), 1.22 – 0.98 (m, 5H). ^{13}C NMR (101 MHz, $CDCl_3$) δ 195.3, 151.3, 129.7, 106.7, 78.0, 76.5, 41.2, 40.5, 27.1, 26.3, 25.8 (2×C), 20.7.

The Mechanism

4-Methylphenol (**1**) undergoes regioselective oxidation at the 4-position by singlet oxygen, yielding the peroxyquinol intermediate **2**. Upon the addition of aldehyde **3** to the system, it is activated by the solid acid catalyst, amberlyst-15. The hydroperoxyl group in **2** acts as a nucleophile, attacking the aldehyde group of **3** to form the adduct **A**. Subsequently, an intramolecular Michael addition takes place between the newly generated hydroxyl group and the quinone moiety in **A**, leading to the formation of the cyclized 1,2,4-trioxane product **4**.

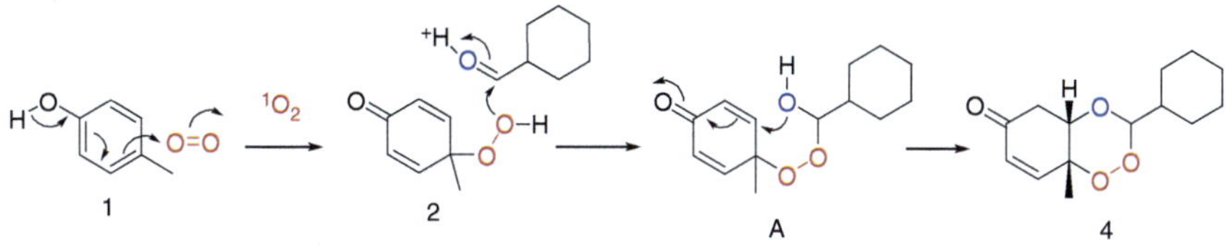

Problem 58: A Key Step for Industrial-Scale Synthesis of Artemisinin (Qinghaosu)

Artemisinin **2** (Qinghaosu) and its derivatives, including dihydroartemisinin, artemether, and artesunate, are critical components of modern malaria treatment. This natural compound was discovered by Chinese scientists, most notably Tu Youyou, who was jointly awarded the 2015 Nobel Prize in Physiology or Medicine alongside William C. Campbell and Satoshi Ōmura for her groundbreaking work. Artemisinin is industrially extracted from the plant Artemisia annua, commonly known as sweet wormwood. In 2014, researchers at Sanofi developed a synthetic route for the industrial-scale production of artemisinin [155]. The key steps in this process involve the Schenck ene reaction and Hock cleavage, as illustrated below.

Please propose a detailed mechanism for the conversion of compound **1** to compound **2**.

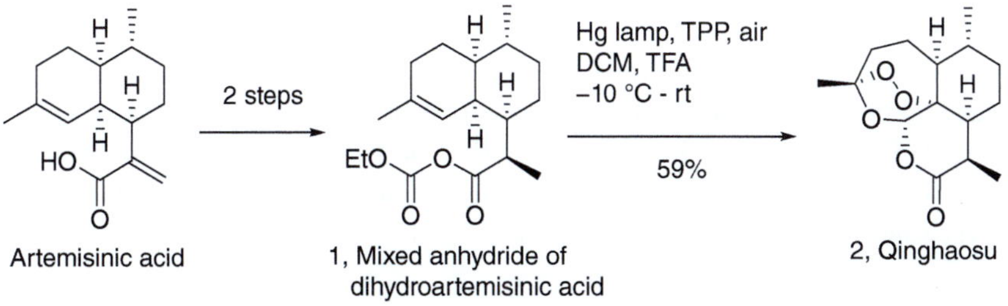

Artemisinic acid

2 steps

1, Mixed anhydride of dihydroartemisinic acid

Hg lamp, TPP, air
DCM, TFA
−10 °C - rt
59%

2, Qinghaosu

Experimental Procedure (from *Org. Process Res. Dev.*, **2014**, *18*, 417)
The process started with 600 kg of artemisinic acid in **Step 1**.

Step 3: Photooxidation of Mixed Anhydride (dihydroartemisinic acid ester mixed anhydride (DHAEMC)) to Artemisinin. To the solution of DHAEMC were added 2570 kg DCM and 300 g tetraphenylporphyrin before the solution was exposed to light irradiation using photoreactors containing mercury vapor lamps and ambient air bubbling at about −10 to −15 °C. In the beginning of the irradiation, 132 kg TFA was added to the reaction mixture. The reaction was monitored by HPLC. As soon as the reaction was completed, the solution was treated twice with 720 L aqueous solution of sodium bicarbonate and was subsequently washed with 1440 kg water. The washed organic phase was treated with 30 kg activated charcoal and filtered. Before the crystallization step, the ADT24h of the reaction mixture was checked.

Step 4: Crystallization of Artemisinin. By the organic solution of artemisinin coming from the photooxidation, DCM was partially removed by distillation under vacuum, while 2700 kg n-heptane was added in order to maintain a constant volume in the reactor. During the solvent switch, artemisinin was crystallized. The n-heptane suspension of artemisinin was added with 180 kg ethanol, and then it was cooled at about 20 °C to accomplish the crystallization. The wet product was isolated by centrifugation, and washed twice with 1200 kg mixture of n-heptane/ethanol. The artemisinin was dried under reduced pressure at a temperature not exceeding 36 °C. About 370 kg of artemisinin as crystalline white powder was typically obtained for a batch size of 600 kg of artemisinic acid.

Analytical Data of Semisynthetic Artemisinin. Optical Rotation: $[\alpha]^{20}{}_{D} = +74{-}78$ [10 mg/mL in ethanol]. The melting point of the crystalline artemisinin was found to be about 159 °C. The theoretical mass of $[M + H]^+$ is 283.1545 amu. The high-resolution mass spectrum shows the $[M + H]^+$ at $m/z = 283.1557$ amu. This measured mass is consistent with the $[M + H]^+$ formula $C_{15}H_{22}O_5$ within a deviation of 4.2 ppm. (amu: atomic mass unit) Data acquisition of all NMR spectra was achieved using a spectrometer operating at a proton resonance frequency of 600 MHz and a carbon 13 resonance frequency of 150 MHz (see Table 1). The NMR spectra were recorded using a solution in deuterated dimethyl sulfoxide (DMSO-d_6). Calibration was performed using the residual solvent signal. ^{1}H NMR δ 0.93 (d), 1.08 (d), 1.15 (dq), 1.34 (m), 1.36 (s), 1.54 (m), 1.65 (m), 1.73 (dq), 1.79 (dt), 1.94 (m), 2.06 (m), 2.28 (m), 3.17 (m), 6.11 (s). ^{13}C NMR δ 12.30 (CH_3), 19.46 (CH_3), 22.36 (CH_2), 24.28 (CH_2), 24.85 (CH_3), 32.44 (CH), 33.07 (CH_2), 35.42 (CH_2), 36.02 (CH), 43.85 (CH), 49.43 (CH), 77.48 (C), 93.20 (CH), 104.62 (C), 171.34 (C=O).

The Mechanism

It is assumed that a regioselective Schenck ene reaction [156] between singlet oxygen and the double bond in **1** takes place to form intermediate **A**. Then a Hock cleavage [157] occurs to produce intermediate **C** via **B**. Subsequently, a free radical oxygenation of **C** occurs with triplet oxygen to generate intermediate **D**. Finally, cascade cyclization of **D** leads to Qinghaosu **2**.

$^1O_2$ (singlet oxygen)
Schenck ene reaction
H^+
Hock cleavage
H_2O
1
A
B
C
D
E
$-H^+$
$^3O_2$ (triplet oxygen)
(free radical reaction)
CO_2
EtOH
2, Qinghaosu

4.6 Formation of Monocyclic Aromatic Heterocycles

4.6.1 Oxathiolane

Problem 59: Synthesis of Substituted 1,3-Oxathiolane

Apricitabine (**5**, ATC, 2-(*R*)-hydroxymethyl-4-(*R*)-(cytosin-10-yl)-1,3-oxathiolane) is classified as a deoxycytidine analog nucleoside reverse transcriptase inhibitor and is in phase III clinical development for the treatment of HIV infection. In 2011, Australian chemists reported a three-step process of ATC as shown below [158].

Please suggest a mechanism for the reaction from **2** and **3** to **4**.

Experimental Procedure (from *Org. Process Res. Dev.*, **2011**, *15*, 763)

2-(R)-Benzoyloxymethyl-4-(R)-(N-benzoylcytosin-1-yl)-1,3-oxathiolane (4 - 4c). 2-(*R*,*S*)-Benzoyloxymethyl-1,3-oxathiolane-S-oxide (**2**, **2a**) (12.0 g, 0.050 mol) was dissolved in DCM in a 500-mL three-neck flask and this was cooled to −50 °C. To this was added triethylamine (15.3 mL, 0.110 mol), followed by iodotrimethylsilane (21.4 mL, 0.150 mol) via a dropping funnel, dropwise, at a rate so that the internal temperature was between −30 °C and −50 °C. The resulting light yellow solution was stirred for 45 min while maintaining the temperature between −40 and −50 °C. To the reaction mixture was added copper(II) chloride (1.3 g, 0.010 mol); after 5 more minutes was added *N*-benzoylcytosine (**3**) (10.1 g, 0.047 mol). The resulting mixture was stirred at −50 °C for 15 min and then was allowed to warm to 0 °C over 1 h. The reaction mixture was stirred at this temperature overnight. After overnight stirring, the reaction was stirred at room temperature for 1 h, cooled again in ice, and quenched with the addition of water (100 mL) followed by 5% ammonia (100 mL). This was stirred for 5 min, diluted with DCM (50 mL), and filtered through a Celite plug. The plug was washed with additional DCM (2 × 50 mL), and the combined filtrates were poured into a separating funnel. The organic layer was separated, washed with 2% phosphoric acid (2 × 60 mL) and again with 2.5% ammonia (2 × 100 mL). The combined aqueous layers were re-extracted with DCM (100 mL). The combined organic layers were dried with magnesium sulfate, filtered, and evaporated to give a light yellow/brown thick oil 17.8 g (86% recovery). This crude mixture consisted of *cis*-3 and *trans*-4 combined coupled product at 62% purity (NMR) and with a cis-/trans-isomer ratio 2.86:1 (NMR).

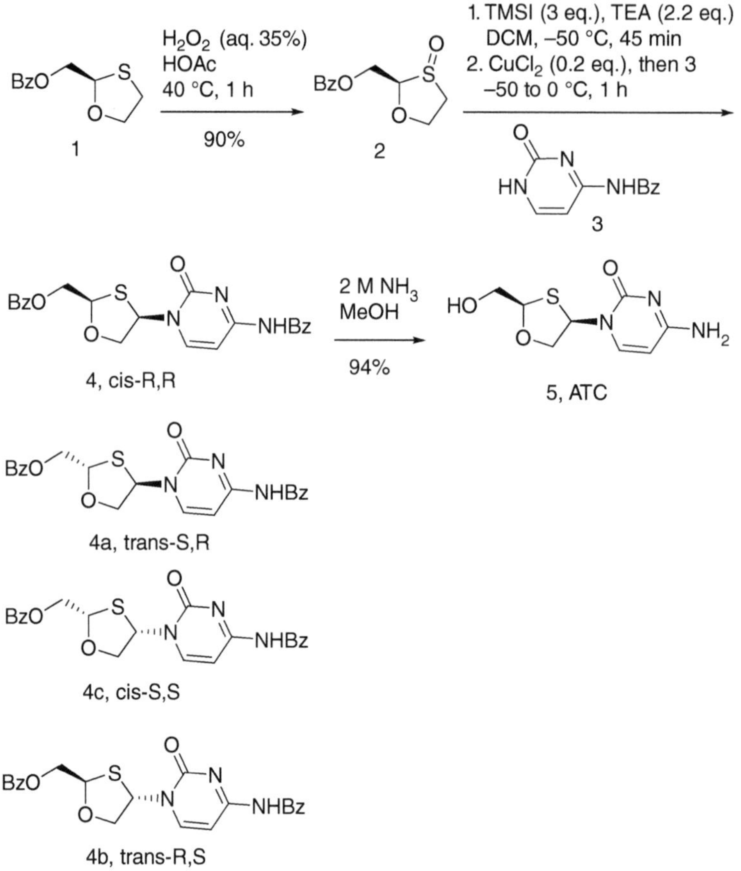

Recrystallization of the crude 2-(*R*,*S*)-benzoyloxymethyl- 4-(*R*,*S*)-(*N*-benzoylcytosin-1-yl)-1,3-oxathiolane (4, 4a, 4b, 4c) to give pure 2-(*R*)-benzoyloxymethyl-4-(*R*)-(*N*-benzoylcytosin-1-yl)-1,3-oxathiolane (4). To the crude (*R*,*S*)-benzoyloxymethyl-4-(*R*,*S*)-(*N*-benzoylcytosin-1-yl)-1,3-oxathiolane (**4, 4a, 4b, 4c**) (17.8 g) in a 500-ml round bottom flask was added 14.5 times (by volume) methanol (258 mL), and the mixture was refluxed until a clear solution could be seen. This was then allowed to cool to room temperature gradually and was left standing overnight. The resulting crystallized product was filtered, followed by washing with methanol (2 × 100 mL) and drying under vacuum. The resulting slightly colored, feathery, crystalline solid on analyzing with NMR showed that it is 99.1% cis-(2*R*,4*R*)-isomer **4**, with 0.9% of cis-(2*S*,4*S*)-isomer **4c**, isolated yield, 7.2 g (35% yield for the isomer). ^{1}H NMR ($CDCl_3$): δ 8.5 (br s, 1H),

8.25 (d, 1H), 8.0 (d, 2H), 7.8 (d, 2H), 7.6 (m, 2H), 7.45 (m, 4H) 7.3 (poorly resolved d, 1H), 6.6 (d, 1H) 5.5(t, 1H), 4.8 (m, 2H), 4.5 (d, 1H), 4.05 (dd, 1H). Chiral HPLC for cis-(2*R*,4*R*) isomer (**4**) R_t 12.14 min. Melting point 163–165 °C.

The Mechanism

The reaction proceeds through an initial activation of (**2**) undergoing a sila-Pummerer rearrangement [159]. The TMSI is activated by $CuCl_2$ and then attacked by **2** to form the active electrophile **A**. Addition of **3** to **A** provides the major isomer **4** and minor isomers **4a**, **4b**, and **4c**.

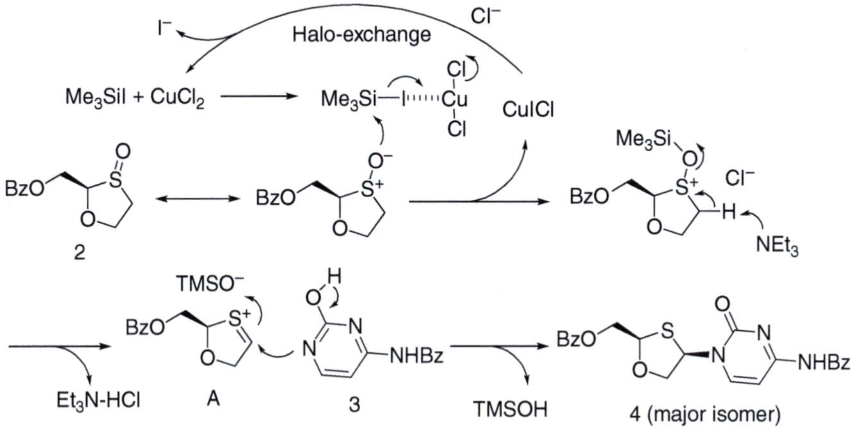

4.6.2 Pyrrole

Problem 60: Preparation of Pyrrole-2-Aldehyde from Pyrrole-2-*t*-butyl ester

Sunitinib (**3**) is an anticancer drug for the treatment of renal cell carcinoma. In a synthesis of **3**, pyrrole aldehyde (**2**) is an important intermediate. It was produced by the following process from **1** [160, 161].

Please provide a mechanism for this reaction.

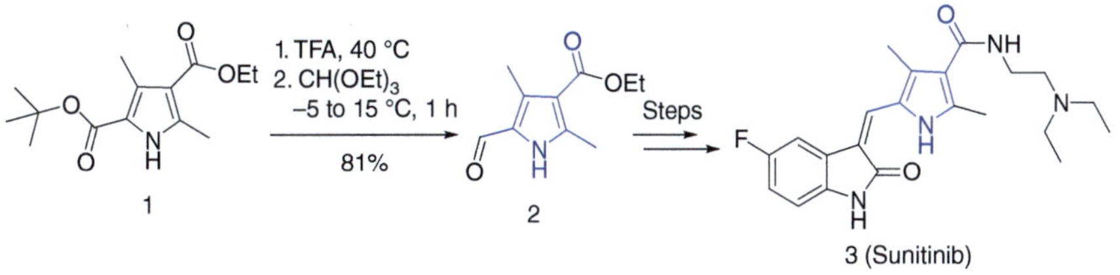

Experimental Procedure (from US2004/266843, A1)

3,5-Dimethyl-1*H*-pyrrole-2,4-dicarboxylic acid 2-*tert*-butyl ester 4-ethyl ester (**1**, 80.1 g, 0.3 mol) and 400 mL of TFA were stirred for 5 minutes in a 2-L 3-neck round bottom flask equipped with mechanical stirring and warmed to 40 °C in an oil bath. The mixture was then cooled to −5 °C and triethyl orthoformate (67.0 g, 0.45 mol) was added all at once. The temperature increased to 15 °C. The mixture was stirred for about 1 minute, removed from the cold bath and then stirred for 1 hour. The TFA was removed by rotary evaporation and the residue put in the refrigerator where it solidified. The solid was dissolved by warming and poured into 500 g of ice. The mixture was extracted with 800 mL of DCM to give a red solution and a brown precipitate, both of which were saved. The precipitate was isolated and washed with 150 mL of saturated sodium bicarbonate solution. The DCM phase was washed with 150 mL of sodium bicarbonate and both bicarbonate solutions discarded. The DCM solution was washed 3 times with 100 mL of water each time. The DCM solution was evaporated to dryness and the dark residue recrystallized twice from hot ethyl acetate after decolorizing with Darco to give golden yellow needles. The brown precipitate was recrystallized from 350 mL of hot ethyl acetate after decolorizing with Darco to give a yellow-red solid. All the recrystallized solids were combined and recrystallized from 500 mL of hot ethanol to give 37.4 g (63.9%) of 5-formyl-2,4-dimethyl-1H-pyrrole-3-carboxylic acid ethyl ester as yellow needles (mp 165.6–166.3 °C., lit. 163–164 °C.). The evaporated residues from the ethyl acetate and ethanol mother liquors were recrystallized from 500 mL of ethanol to give 10.1 g (17.1%) of a second crop of dirty yellow needles.

Analytical data of product **2** from *Chem. Pharm. Bull.*, **2007**, *55*, 1439.

White solid, yield: 63%; mp: 163–164 °C; TLC: $R_f = 0.49$ (silica gel GF_{254}; hexane/EtOAc, 5/2), UV; δ_H [400 MHz/$CDCl_3$/TMS]: 12.19 (s, 1H), 9.62 (s, 1H), 4.18 (2H, q, $J = 7.2$ Hz), 2.46 (s, 3H), 2.42 (s, 3H), 1.28 (3H, t, $J = 7.2$ Hz).

The Mechanism

The first stage is the selective hydrolysis of the *t*-butyl ester under acidic condition. Then decarboxylation occurs, followed by formylation of the pyrrole **A** with triethyl orthoformate, and finally hydrolysis of the acetal to provide the aldehyde **2**.

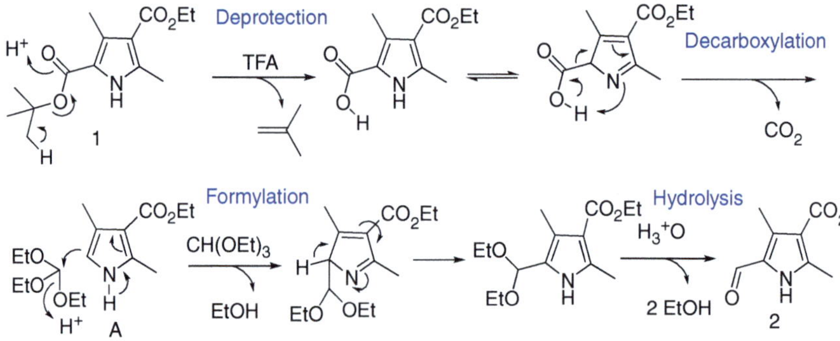

The decarboxylation might also occur via the following path.

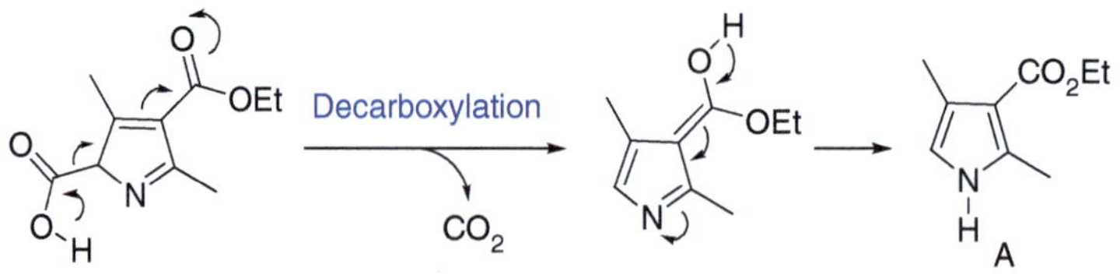

Problem 61: Synthesis of 2-Formylpyrroles from Pyridinium Iodide Salts

In 2020, Fu's group reported a synthesis of 2-formylpyrroles (**2**) from pyridinium iodide salts (**1**) as shown below [162]. Pyrroles are important building blocks for constructing some pharmaceutical molecules such as bradykinin receptor antagonists (**3**) [163]. Please provide a mechanism for this reaction.

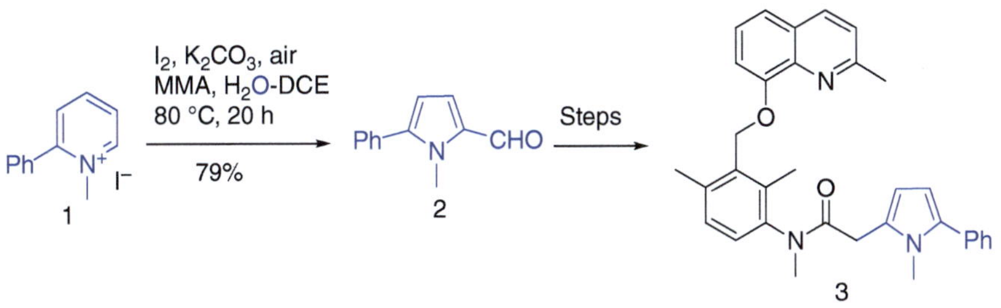

Experimental Procedure (from *Org. Lett.*, **2020**, *22*, 6107, Supporting Information)
An oven-dried round-bottomed flask equipped with magnetic stirring bar was sequentially charged with 1-methyl-2-phenyl-1-pyridinium iodide **1** (1 g, 3.37 mmol), I_2 (2.15 mmol, 538.72 mg), K_2CO_3 (13.47 mmol, 1.86 g), methyl methacrylate (3.37 mmol, 3.37 μL), H_2O (6.37 mL), and DCE (6.37 mL) in the air. The mixture was stirred at 80 °C (heated by oil bath) for 24 hours. After cooling to room temperature, the mixture was extracted with DCM, and the combined organic layers were washed with H_2O and brine, dried over anhydrous $MgSO_4$, concentrated by rotary evaporation. The crude material was purified by flash chromatography (petroleum ether/ethyl acetate) on silica gel to give 1-methyl-5-phenyl-1*H*-pyrrole-2-carbaldehyde (**2**) (66%). $R_f = 0.4$ [petroleum ether/ethyl acetate = 20:1 (v/v)]. Colorless viscous liquid. IR (KBr) 3294, 2956, 1660, 1463 cm^{-1}. ^1H NMR (400 MHz, Chloroform-d) δ 9.58 (s, 1H), 7.47 – 7.40 (m, 5H), 6.98 (d, $J = 4.1$ Hz, 1H), 6.31 (d, $J = 4.1$ Hz, 1H), 3.93 (s, 3H). ^{13}C NMR (101 MHz, $CDCl_3$) δ 179.7, 144.4, 133.1, 131.2, 129.3, 128.8, 128.7, 124.6, 110.8, 34.4. HRMS (ESI-TOF) m/z: $[M + H]^+$ Calcd for $C_{12}H_{12}NO$: 186.0913; Found: 186.0915.

The Mechanism

This process exemplifies an **ANRORC** (**A**ddition of the **N**ucleophile, **R**ing **O**pening, and **R**ing **C**losure) reaction, a specific type of substitution mechanism observed in nucleophilic attacks on ring systems. The term ANRORC refers to the sequential steps of nucleophile addition, ring opening, and subsequent ring closure [164]. In this case, partial oxidation of I^- by atmospheric oxygen generates I_2, which serves as the iodine source. The reaction begins with the nucleophilic attack of H_2O on the C6 position of the pyridinium salts (**1**), yielding intermediate **A**. Ring opening of **A** produces a conjugated enamine aldehyde (**B**), which is then oxidized by a molecule of iodine. This oxidation leads to the formation of species **C**, characterized by a three-membered iodonium ring. Species **C** exists in equilibrium with the iodonium enamine aldehyde (**D**). Alternatively, species **C** can be directly formed through the oxidation of **A** by I_2. The final product, *N*-methyl-5-phenyl-2-formylpyrrole (**2**), is obtained via intramolecular cyclization of **D**, followed by the elimination of HI through intermediate **E**.

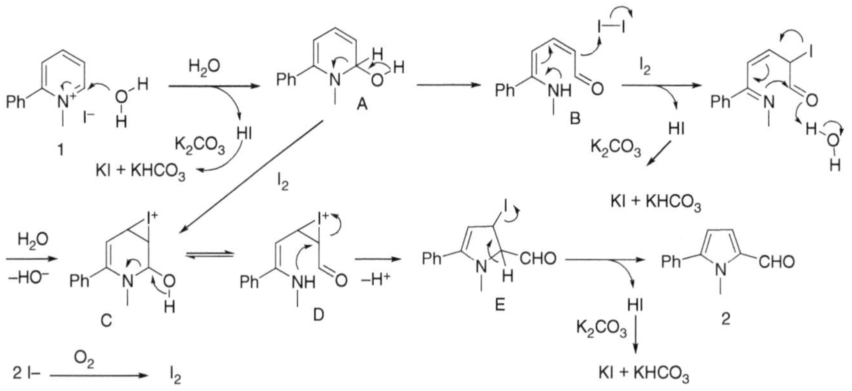

Another example of **ANRORC** reaction is the reaction of ammonia with 4*H*-pyran-4-one to form 4-hydroxypyridine [165].

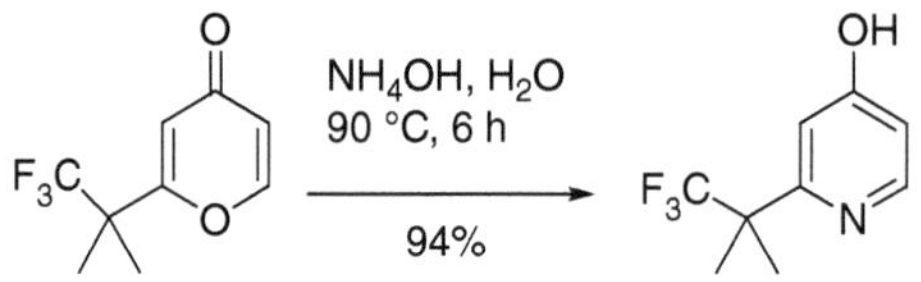

Problem 62: Preparation of 2-Aminopyrroles

Vemurafenib is a medication used for the treatment of late-stage melanoma. It is an inhibitor of the B-Raf enzyme. 5-Amino-1-(*tert*-butyl)-1*H*-pyrrole-3-carbonitrile (**2**) is a key building block for the manufacturing process of Vemurafenib [166]. In a patent application, compound **2** was prepared from succinonitrile **1**, ethyl formate, and *t*-butylamine through a one-pot process [166]. In a previous process report [167], the 2-aminopyrrole **2** was prepared in 6.7 kg scale and the intermediate **2A** was characterized. Compound **2** was also applied for the synthesis of 7-azaindolylcarboxy-endo-tropanamide (**DF 1012**), a selected candidate drug in a new class of nonnarcotic antitussive compounds [167].

Please provide a mechanism for the formation of **2** from **1**.

Experimental Procedure (from *Org. Process Res. Dev.*, **2003**, *7*, 209)

***N-tert*-Butylaminomethylenesuccinonitrile (2A)**. A 250-L steel reactor in nitrogen atmosphere was loaded with sodium methoxide (4 kg, 74 mol) and anhydrous toluene (60 L).The suspension was cooled to 5 °C under stirring, and a solution of ethyl formate (6.17 kg, 83.3 mol) and succinonitrile (**1**, 5.60 kg, 69.9 mol) in anhydrous toluene (60 L) was added dropwise. The addition rate was controlled to avoid the inner temperature exceeding 20 °C. Once the addition was done, the mixture was stirred for 2.5 h at room temperature, still under nitrogen atmosphere. *tert*-Butylamine (5.22 kg, 71.4 mol) and acetic acid (5.0 kg, 83.3 mol) were subsequently added to the reaction mixture, ensuring that the temperature did not exceed 40 °C. The mixture was refluxed for 2.5 h and then cooled to room temperature; an 18% NaCl solution (17 L) was added, and the two phases were separated. The aqueous layer was extracted twice with toluene (7.5 L), and the collected organic extracts were evaporated under reduced pressure to give **2A** (10.28 kg, 63.0 mol) as a brown solid (*E*/*Z* isomers mixture 2:1). Typically, the obtained solid was subsequently used without further purification. An analytically pure sample was recrystallized from acetone/diethyl ether: mp 118–119 °C; ^{1}H NMR ($CDCl_3$) δ 1.35 (s, 9H), 3.18 (s, 2H), 5.0 (bs, 1H), 6.88 (d, 1H, (*E*-isomer), J = 14.2 Hz), 6.99 (d, 1H, (*Z*-isomer), J = 14.5 Hz); IR (Nujol) 3300, 2170, 1650, 1210 cm^{-1}. Anal. Calcd for $C_9H_{13}N_3$: C, 66.23; H, 8.03; N, 25.74. Found: C, 65.98; H, 8.01; N, 25.81. ESI-MS m/z 164 $[M + H]^+$.

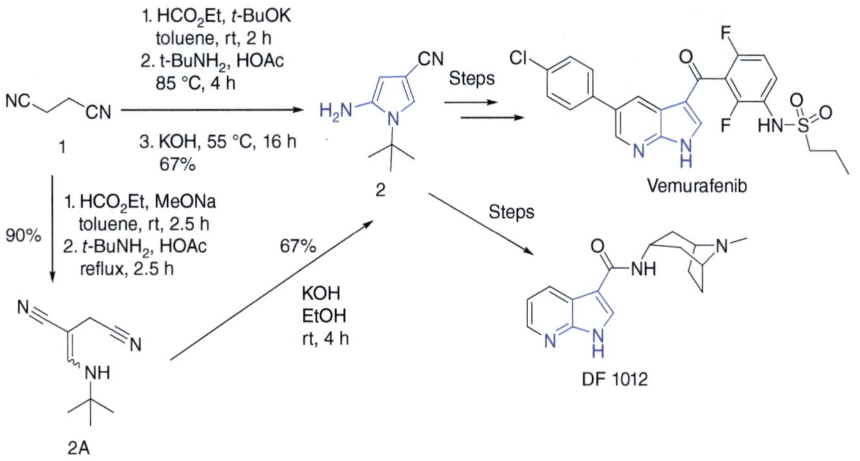

2-Amino-1-*tert*-butyl-4-cyanopyrrole (2). In a 250-L steel reactor, 85% KOH (7.35 kg, 111.3 mol) was dissolved in denatured ethanol (60 L) at 50 °C, and the opalescent solution was then cooled at room temperature. To the latter, a solution of 7 (10.00 kg, 61.3 mol) in denatured ethanol (17 L) was added while stirring. This mixture was continuously stirred for 4 h at room temperature and then concentrated under reduced pressure until a solid started to separate. The mass was then diluted with water (50 L) and stirred at room temperature for another hour. The precipitate was separated by centrifugation, washed with water (15 L), and desiccated at 50 °C to give **2** (6.73 kg, 41.2 mol) as brown solid. The yield based on succinonitrile was 60.6%. Typically, the solid was used in subsequent reactions without further purification. An analytically pure sample was recrystallized from isopropyl ether: mp 112–113 °C; ^{1}H NMR ($CDCl_3$) δ 1.60 (s, 9H), 3.15 (bs, 2H), 5.76 (d, 1H, J = 2.2 Hz), 6.96 (d, 1H, J = 2.2 Hz); IR (Nujol) 3380, 3320, 2190, 1640, 1200, 785, 715 cm^{-1}. Anal. Calcd for $C_9H_{13}N_3$: C, 66.23; H, 8.03; N, 25.74. Found: C, 66.06; H, 8.04; N, 25.76. ESI-MS m/z 164 $[M + H]^+$.

The Mechanism

The first step is an aldol condensation between succinonitrile **1** and ethyl formate under basic condition to form **A**.

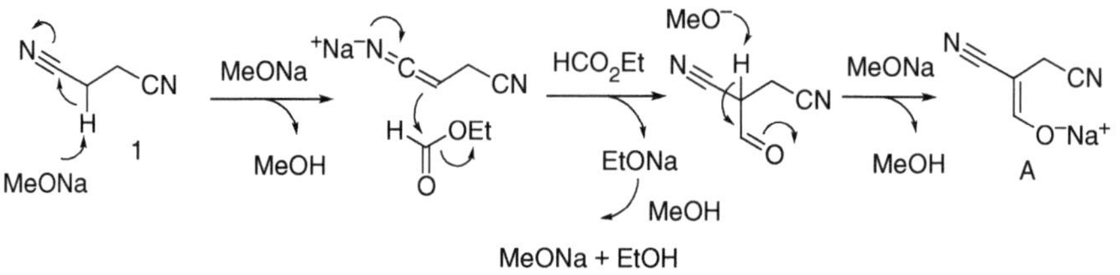

After neutralized with HOAc, the second step is the *t*-butyl amine condensed with the aldehyde form a Schiff base **B**, which is then isomerized to enamine **2A** under the weak acidic condition.

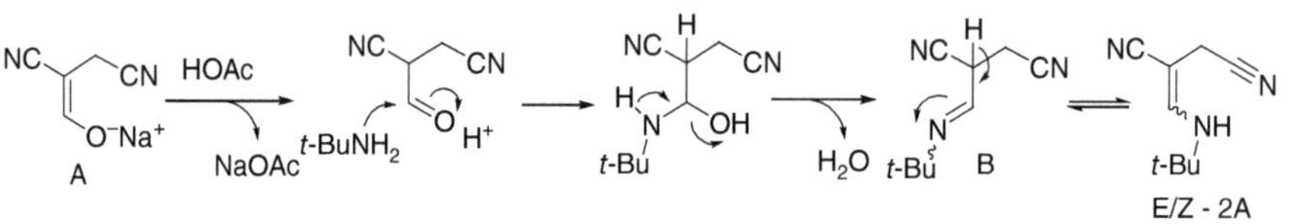

The final step is the cyclization of NH in intermediate **2A** with the nitrile group. After isomerization, product **2** is afforded.

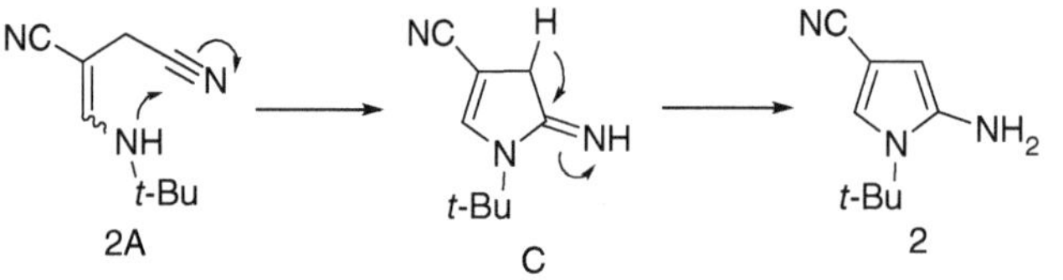

Problem 63: Synthesis of 3-(3,4-Dihydro-2*H*-pyrrol-5-yl)pyridine (Myosmine)

3-(3,4-Dihydro-2*H*-pyrrol-5-yl)pyridine (**4**, myosmine) is a key intermediate for the synthesis of nicotine (**5**). In a patent application [168], myosmine (**4**) was prepared from nicotinic acid ethyl ester (**1**) and *N*-vinylpyrrolidone (**2**) by a two-step procedure as shown below. That method for preparation of myosmine was first reported by Brandänge and Lindblom in 1976 [169]. Later on, the methodology was applied to synthesize a series of 2-substituted and 2,3-disubstituted pyrrolines [170, 171].

Please suggest a mechanism for the transformation.

Experimental Procedure (from CN114644614A, **2022**)

Add 20 kg of xylene, 5 kg ethyl nicotinate (**1**, that is, R = ethyl), 2.5 kg potassium *tert*-butoxide, and 4 kg *N*-vinylpyrrolidone. After the system was heated to 100 °C and reacted for 4 h, 3 kg of 15% hydrochloric acid was added dropwise at a temperature of 50 °C for 3 h. After the dropwise addition, the temperature of the system was raised to 90 °C for 2 h, and the temperature was cooled to 30 °C. The pH of the

CO_2Et 1 + 2 → 1. *t*-BuONa, xylene 100 °C, 4 h 2. HCl-H_2O, 90 °C, 2 h; 77% → 3 (NH_2)

THF-H_2O (3:1, w/w) NaOH (1.23 eq.) 40 °C; 98% → 4 (myosmine) → 2 steps → 5 (nicotine)

system was adjusted to 7 with 10% sodium hydroxide; then ethyl acetate was added for extraction twice, each with 20 kg. The combined organic phases were dried and concentrated under reduced pressure at 40 °C to obtain 4.11 kg of the compound of formula **3a**. The purity of the liquid phase was 97.3% and the yield was 76.7%, which was directly used in the next reaction. ^{1}H NMR ($CDCl_3$, 400 MHz) δ: 8.91 (d, J = 7.8 Hz, 1H), 8.51 (d, J = 8.0 Hz, 1H), 8.35 (d, J = 7.8 Hz, 1H), 7.68 (d, J = 7.8 Hz, 1H), 3.03 (t, J = 8.0 Hz, 2H), 2.81 (t, J = 8.0 Hz, 2H), 2.22–2.20 (m, 2H), 1.90–1.88 (br, 2H). LCMS Cacl: 164.21, Detec. M+1: 165.2.

To the system, 15 kg of THF, 5 kg of water, 5 kg of the compound of formula **3a** was added. The temperature of the system was maintained at 40 °C, then 1.5 kg of sodium hydroxide was added. The reaction mixture was stirred at 40 °C until the compound of formula **3a** was completely consumed. The mixture was extracted twice with 10 kg of DCM each time. The organic phases were combined and concentrated under reduced pressure at 30 °C to obtain 4.35 kg of myosmine. The liquid phase purity was 97.1%, and the yield: 97.8% was directly used in the next reaction. ^{1}H NMR ($CDCl_3$, 400 MHz) δ: 8.58 (d, J = 8.0 Hz, 1H), 8.45 (d, J = 7.8 Hz, 1H), 7.68 (d, J = 8.0 Hz, 1H), 7.24–7.20 (m, 1H), 4.13 (t, J = 4.0 Hz, 1H), 3.20–3.17 (m, 1H), 3.06–3.01 (m, 1H), 2.22–2.20 (m, 1H), 1.90–1.88 (m, 1H), 1.80–1.60 (m, 1H). LCMS Cacl: 146.19, Detec. M+1: 147.2.

The Mechanism

The first step is an ester condensation between the ester **1** and amide **2** to generate intermediate **A**. When the reaction system is adjusted to acidic, deprotection of the vinyl group from **A** occurs to give the cyclic amide **B**. Further hydrolysis of amide **B** provides carboxylic acid **C**, which should be decarboxylated simultaneously at acidic condition to yield the ketoamine **3**. Final cyclization and dehydration leads to myosmine **4** (next page).

Problem 64: Formation of Substituted Pyrrole by Cyclization of γ-ketonitrile

Esaxerenone is a novel, nonsteroidal, selective mineralocorticoid that was approved in Japan for the treatment of hypertension. In a synthesis of Esaxerenone, the formation of the core pyrrole **2** was achieved by the cyclization of γ-ketonitrile (**1**) by the reaction as shown below [172]. The method for synthesizing of 2-chloro-3-carboethoxy- or 2-chloro-3-cyano- 4,5-disubstituted and 5-substituted pyrroles was developed by Louise H. Foley in 1994 [173]. The methodology, with some modifications by medicinal chemists, has been applied in some drug discovery projects [172, 174, 175].

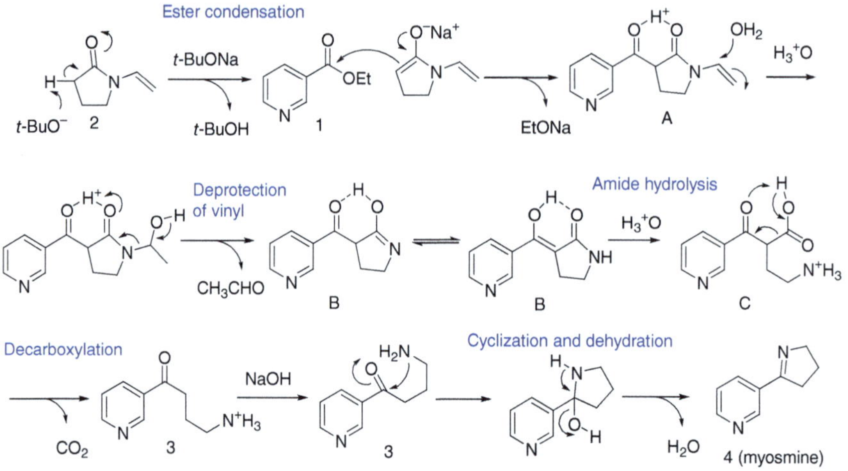

Please provide a mechanism for the transformation from **1** to **2**.

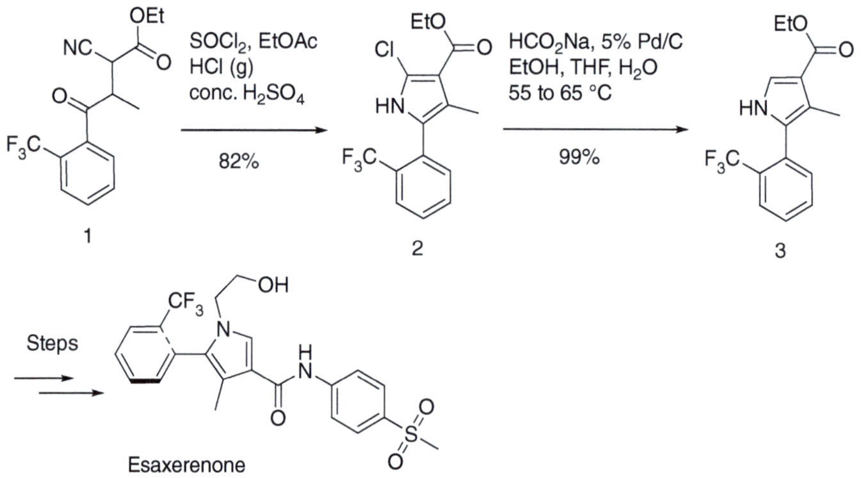

Experimental Procedure (from US **2016**/0096803 A1)

Ethyl 2-chloro-4-methyl-5-[2-(trifluoromethyl)phenyl]-1H-pyrrole-3-carboxylate (2). To the toluene solution (217 mL) of ethyl 2-cyano-3-methyl-4-oxo-4-[2-(trifluoromethyl)phenyl]butanoate (**1**, 112.2 g, 358 mmol) obtained by the production method of Example 2, ethyl acetate (362 mL) and thionyl chloride (42.59 g (358 mmol)) were added at 20 to 30 °C, followed by cooling to −10 to 5 °C. Then, hydrogen chloride gas (52.21 g (1432 mmol)) was blown into the mixture, and concentrated sulfuric acid (17.83 g (179 mmol)) was further added thereto, and the resulting mixture was heated and stirred at 15 to 30 °C. After the mixture was stirred for about 20 h, ethyl acetate (1086 mL) was added thereto, followed by heating to 30 to 40 °C, and water (362 mL) was added thereto, and then, the resulting mixture was subjected to liquid separation. To the organic layer obtained by liquid separation, water (362 mL) was added, followed by liquid separation, and then, a 5 w/v % aqueous sodium hydrogen carbonate solution (362 mL) was added thereto, followed by liquid separation. The organic layer was subsequently concentrated under reduced pressure. Toluene (579 mL) was then added, and the mixture was again concentrated under reduced pressure. Afterward, toluene (72 mL) was added, and the mixture was cooled to 0–5 °C. After the mixture was stirred for about 2 hours, the deposited crystal was filtered and washed with toluene (217 mL) and cooled to 0 to 5 °C. The obtained wet crystal product was dried under reduced pressure at 40 °C, whereby the title compound was obtained (97.55 g, yield: 82.1%). ^{1}H NMR (400 MHz, $CDCl_3$) δ: 1.38 (3H, t, J = 7.1 Hz), 2.11 (3H, s), 4.32 (2H, q, J = 7.1 Hz), 7.39 (1H, d, J = 7.3 Hz), 7.50–7.62 (2H, m), 7.77 (1H, d, J = 8.0 Hz), 8.31 (1H, br).

The Mechanism

The enol form of ketone **1** is chlorinated by $SOCl_2$ to yield vinyl chloride **A**. Subsequent addition of HCl to the nitrile produces imidoyl chloride, which isomerizes to 1-chloroenamine **B**. Cyclization of **B** yields 2-chloropyrrole **2**. In the final step, compound 2 undergoes transfer hydrogenolysis of **2** to afford the dechlorinated product **3**.

Another possible mechanism, as proposed by Foley [173], does not involve the formation of vinyl chloride. Instead, the imidoyl chloride **C** or chloroenamine **D** undergoes cyclization to yield dihydropyrrole **E**, which then undergoes dehydration to yield pyrrole **2**. However, this mechanism cannot explain the role of thionyl chloride.

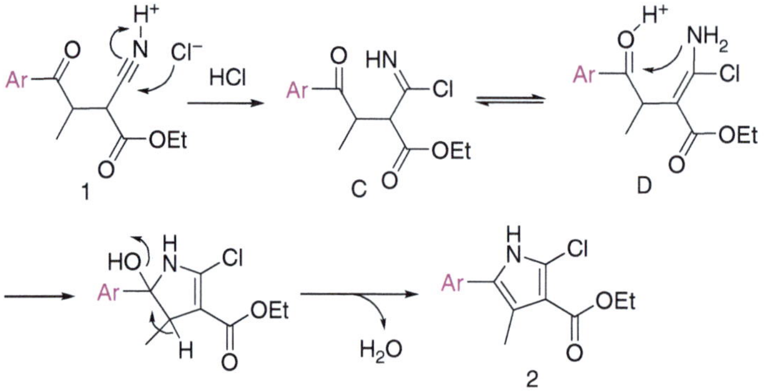

4.6.3 Thiophene

Problem 65: Synthesis of Trisubstituted Thiophene

AZD-7762 is a potent, ATP-competitive inhibitor of checkpoint kinase 1 (Chk1). It advanced to phase II clinical trials as part of a combination therapy for the treatment of various solid tumors. In a multi-kilogram scale synthesis of AZD-7762, the key intermediate aminothiophene **5** is synthesized in high yield from aminopiperidine **1** and chloro cinnamonitrile **4,** via intermediates **2** and **3**, by a telescope process [176]. The process strategy for **5** is based on a method developed by Hartmann [177] in 1984. The methodology was also applied to synthesize nonpeptide classical competitive inhibitor class of protein tyrosine phosphatase 1B [178].

Please provide a mechanism for the reaction from **3** and **4** to **5**.

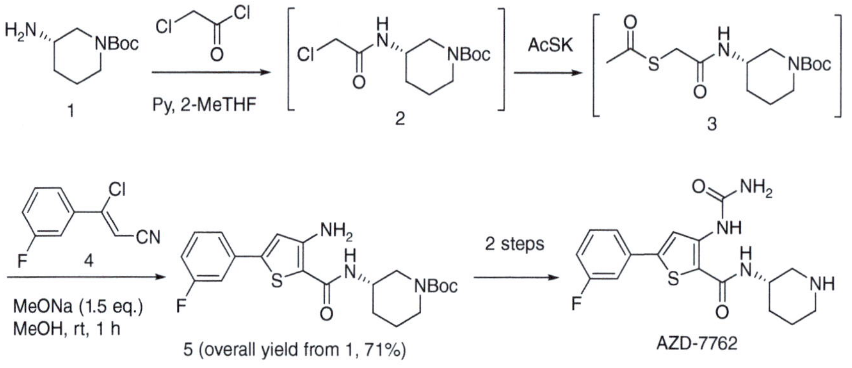

Experimental Procedure (from *Org. Process Res. Dev.*, **2017**, *21*, 310)

(S)-3-{[3′-Amino-5′-(3″-fluorophenyl)thiophene-2′-carbonyl]amino}-piperidine-1-carboxylic Acid *tert*-Butyl Ester (5). Piperidine **1** (120.0 g, 0.599 mol) was dissolved in 2-methyltetrahydrofuran (540 mL). Pyridine (58.14 mL, 0.719 mol) was added, followed by a limewash of 2-methyltetrahydrofuran (60 mL). Chloroacetyl chloride (55.32 mL, 0.689 mol) was added dropwise, maintaining the temperature at 21–25 °C, followed by a limewash of 2-methyltetrahydrofuran (60 mL). After 2.5 h at ambient temperature, the reaction mixture was sampled for conversion to **2** by GC before the addition of a 16% (w/w) aqueous solution of sodium chloride (360 mL). The mixture was stirred for 30 min before separating off the aqueous phase.

To the resulting organic phase was added a filtered solution of potassium thioacetate (102.65 g, 0.899 mol) in water (204 mL), followed by a limewash of water (36 mL), maintaining the temperature at 19–26 °C throughout. After stirring overnight at ambient temperature, the organic phase was sampled for conversion to **3** by HPLC before separating off the aqueous phase.

To the organic phase was added solid **4** (97.93 g, 0.539 mol) before dropwise addition of a solution of sodium methoxide in methanol (202 mL @ 25% w/w, 0.899 mol), maintaining the temperature at 21–24 °C. This was followed by a limewash of methanol (36 mL). After stirring for 1 h 50 min at ambient temperature, the reaction mixture was sampled by HPLC for conversion to **5** before heating to 33 °C, followed by dropwise addition of water (600 mL). After stirring for 10 min, the aqueous phase was separated off. To the organic phase was added isohexane (960 mL) dropwise before removing a small sample of the reaction mixture, allowing it to cool and crystallize before returning it to the bulk mixture to seed crystallization. Dropwise addition of a second portion of isohexane (480 mL), followed by cooling to 3 °C over 1 h, and stirring at this temperature overnight yielded crystallization of the product. Filtration, displacement, washing the solid with ice-cold *tert*-butyl acetate (240 mL), and 2 × ice-cold mixed solvent system of *tert*-butyl acetate and isohexane (1:1, 2 × 240 mL) and drying at 40 °C over 3 days afforded **5** as a pale yellow solid (176.70 g, 70% yield based on **1**). This process was operated in 3 batches of up to 4.8 kg input of **1** each, generating a total of 15.3 kg of **5**. The average yield was 71%.

^{1}H NMR (400 MHz, DMSO-d_6, 353 K): 7.49–7.32 (m, 3H), 7.19–7.12 (m, 1H), 7.01 (s, 1H), 6.91 (d, 1H), 6.29 (br.s, 2H), 3.91–3.64 (m, 3H), 2.96–2.77 (m, 2H), 1.92–1.77 (m, 1H), 1.74–1.30 (m, 12H). ^{13}C NMR (100 MHz, DMSO, 353 K): 164.3 (s), 163.1 (d, J = 245), 154.6 (s), 153.9 (s), 142.2 (d, J = 3 Hz), 136.1 (d, J = 8 Hz), 131.6 (d, J = 9 Hz), 121.9 (d, J = 3 Hz), 118.7 (s), 115.6 (d, J = 21 Hz), 112.4 (d, J = 23 Hz), 102.5 (s), 79.2 (s), 48.6 (s), 46.1 (s), 44.1 (s), 30.3 (s), 28.6 (s), 24.1 (s); LC-HRMS calcd for $[M + H]^+$ $C_{21}H_{27}FN_3O_3S$: 420.1752; found $[M + H]^+$: 420.1748.

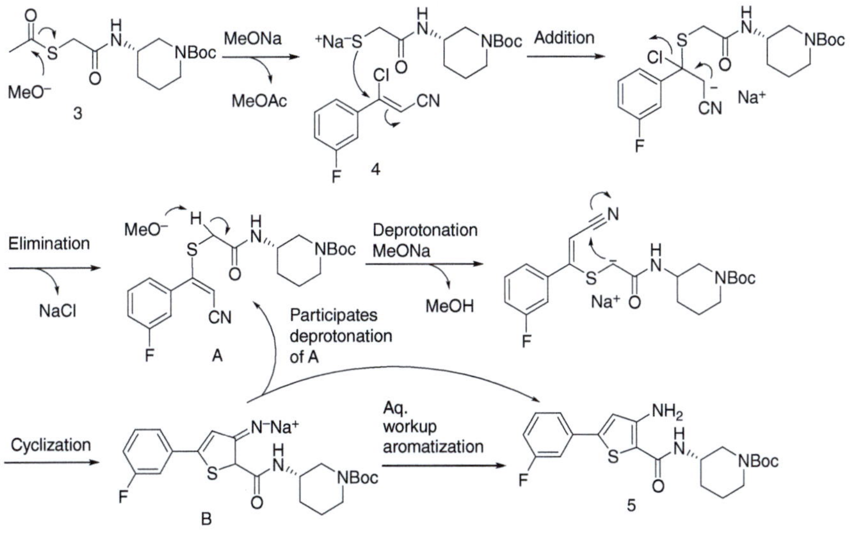

The Mechanism

Methanolysis of the thioacetate group by methoxide anion in **3** leads to a thiolate anion which takes part in an addition-elimination process to give cinnamonitrile **A**. Under basic conditions, the anionic cyclization is potentially autocatalytic, possibly because the proposed nitrile anion **B** is sufficiently basic to deprotonate **A**, thereby propagating the catalytic cycle. This rationale explains why only 1.5 equivalents of sodium methoxide were added (based on 100% of **1** converted to **3**). Final intramolecular cyclization and aqueous workup produces the product **5**.

4.6.4 Pyrrol-2-one

Problem 66: Preparation of Spiro 1*H*-Pyrrol-2(5H)-ones by One-Pot Multicomponent Reaction

1*H*-Pyrrol-2(5H)-ones **1** represent important heterocycles with a variety of pharmaceutical uses. For example, pulchellalactam (**2**) is a CD45 protein tyrosine phosphatase inhibitor [179]. Commeiras and coworkers reported new multicomponent synthesis of either γ-hydroxy-γ-butyrolactams **3** or various γ-lactam motifs **4–6** from (*Z*)-3-iodoacrylic acids or (*E*)-2,3-dihalogenoacrylic acids **7**, terminal alkynes **8**, and primary amines **9** in the presence of a copper(I) catalyst [180, 181].

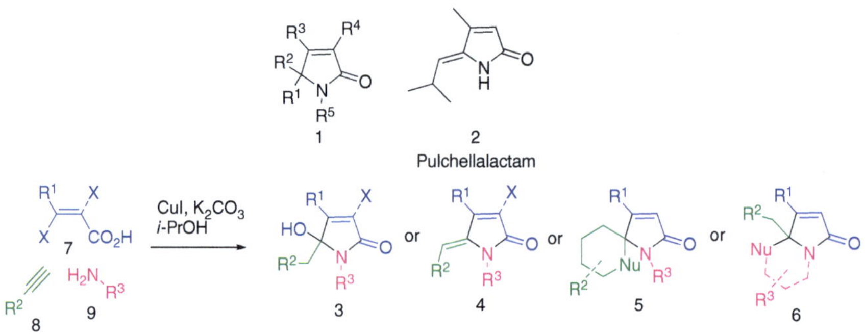

R3 R4 R2 R1 N R5 O 1
N H O 2 Pulchellalactam
R1 X X 7 CO2H R2 8 H2N R3 9 — CuI, K2CO3, i-PrOH → HO R1 X R2 N R3 O 3 or R1 X R2 N R3 O 4 or R1 R2 Nu N R3 O 5 or R1 R2 Nu N R3 O 6

The following is a representative example [181]. One-pot, two-step reaction of **7a**, **8a**, and **9a** under the specified conditions afforded the desired product **5a** along with a side product **10a**.

Please suggest mechanisms for the formation of **5a** and **10a**.

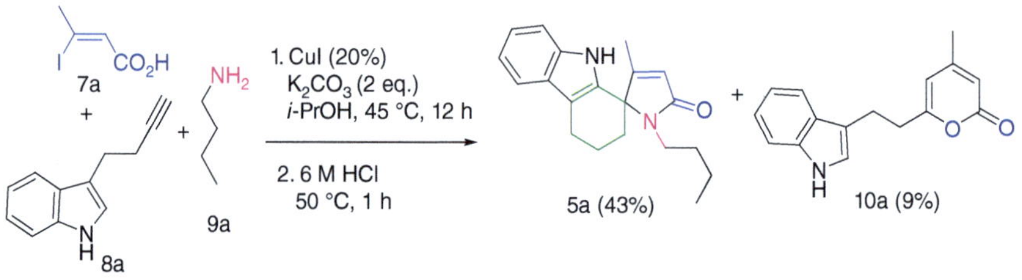

I CO2H 7a + 8a (N H) + NH2 9a — 1. CuI (20%), K2CO3 (2 eq.), i-PrOH, 45 °C, 12 h; 2. 6 M HCl, 50 °C, 1 h → NH N O 5a (43%) + N H O O 10a (9%)

Experimental Procedure (from *Org. Biomol. Chem.*, **2017**, *15*, 3304, supplementary information)
(*Z*)-3-Substituted-3-iodoprop-2-enoic acid derivative **7** (0.5 mmol, 1 equiv) was dissolved in *i*PrOH (1.75 mL) in oven-dried-Schlenk tube. K_2CO_3 (139 mg, 1.0 mmol, 2 equiv) was then added to the solution and the suspension was stirred for 10 min under argon. The mixture was then degassed at −78 °C for 2 × 10 min and the vessel was back filled with argon. After warming to room temperature, terminal alkyne **8** (1.0 mmol, 2 equiv), primary amine **9** (2.0 mmol, 4 equiv), and CuI (19 mg, 0.1 mmol, 0.2 equiv) were respectively added into the mixture. The mixture was then rapidly degassed and the vessel was back filled with argon. The sealed Schlenk tube was placed in the preheated oil bath (45 °C) and was stirred overnight. The reaction mixture was cooled to 0 °C, then 1.7 mL of a solution of hydrochloric acid (6 M, 20 equiv) was added dropwise. The reaction was then heated at 50 °C until the disappearance of the γ-hydroxybutyrolactam checked by TLC. The reaction mixture was cooled to 0 °C, then filtered through a pad of Celite®. The aqueous phase was extracted with ethyl acetate and the combined organic layers were washed with brine, dried over Na_2SO_4, and concentrated under reduced pressure. The crude product was then purified by flash chromatography on silica gel using petroleum ether: ethyl acetate as eluent.

1′-Butyl-3′-methyl-2,3,4,9-tetrahydrospiro[carbazole-1,2′-pyrrol]-5′(1′H)-one (5a). Purification: flash chromatography on silica gel (PE/EtOAc: from 8/2 to 6/4). Yield: 43% (66 mg). Physical appearance: white solid; m.p. (crystal): 105 °C; ^{1}H NMR ($CDCl_3$, 400 MHz): δ (ppm) 0.81 (3H, CH_3, t, J = 7.3 Hz), 1.15–1.25 (2H, CH_2, m), 1.30–1.41 (1H, CH_2, m), 1.53–1.63 (1H, CH_2, m), 1.81 (3H, CH_3, br d, J = 1.5 Hz), 2.01–2.17 (4H, 2 x CH_2, m), 2.74–2.81 (1H, CH_2, m), 2.88–2.98 (2H, 2 x CH_2, m), 3.29 (1H, CH_2, ddd, J = 15.9, 10.5 and 5.4 Hz), 5.91 (1H, CH, br q, J = 1.5 Hz), 7.12 (1H, HAr, br dd, J = 8.0 and 7.1 Hz), 7.20 (1H, HAr, br dd, J = 8.0 and 7.1 Hz), 7.26–7.30 (1H, HAr, m), 7.55 (1H, HAr, d, J = 8.0 Hz), 7.96 (1H, NH, br s); ^{13}C NMR ($CDCl_3$, 100 MHz): δ (ppm) 13.8 (CH_3), 14.2 (CH_3), 20.6 (CH_2), 20.8 (CH_2), 21.5 (CH_2), 31.5 (CH_2), 32.6 (CH_2), 41.0 (CH_2), 67.9 (C), 111.6 (CHAr),115.8 (CAr), 118.6 (CHAr),119.2 (CHAr), 122.0 (CH), 122.7 (CHAr), 127.0 (CAr), 129.1 (CAr),136.9 (CAr), 163.1 (C), 171.3 (C); IR (nujol): 3178, 2931, 1672, 1641, 1407, 1228, 779, 729 cm^{-1}; HRMS (ESI-MS) calcd for $C_{20}H_{25}N_2O[M + H]^+$ 309.1961, found 309.1962.

The Mechanism
A plausible mechanism involving a Sonogashira-type coupling without the use of palladium has been proposed in the literature [180, 182]. The process is thought to begin with the oxidative addition of copper(I) into the C—I bond of compound **A**, forming the copper(III) intermediate **B**. Subsequent steps likely include the π-coordination of alkyne 8a, followed by the formation of complex **D** through base-mediated deprotonation of the terminal acetylenic proton. Reductive elimination of intermediate **D** then yields intermediate E.

At this stage, the pathway diverges: a 6-endo-dig cyclization [183] facilitated by copper(I) could lead to the formation of side product **10a** (path b), while a 5-exo-dig cyclization [183] assisted by copper(I) would generate lactone **F** (path a) via protodemetalation. The primary amine **9a** then undergoes nucleophilic addition onto lactone **F**, producing hydroxylactam **3a** through intermediate **G**. Subsequent elimination of water from **3a** results in the formation of lactam **4a**.

The addition of excess HCl to the mixture ultimately provides the final product **5a** [181]. It is proposed that the mechanism involves the cyclization of **3a** via acyliminium chloride H to yield the spiro product **5a**.

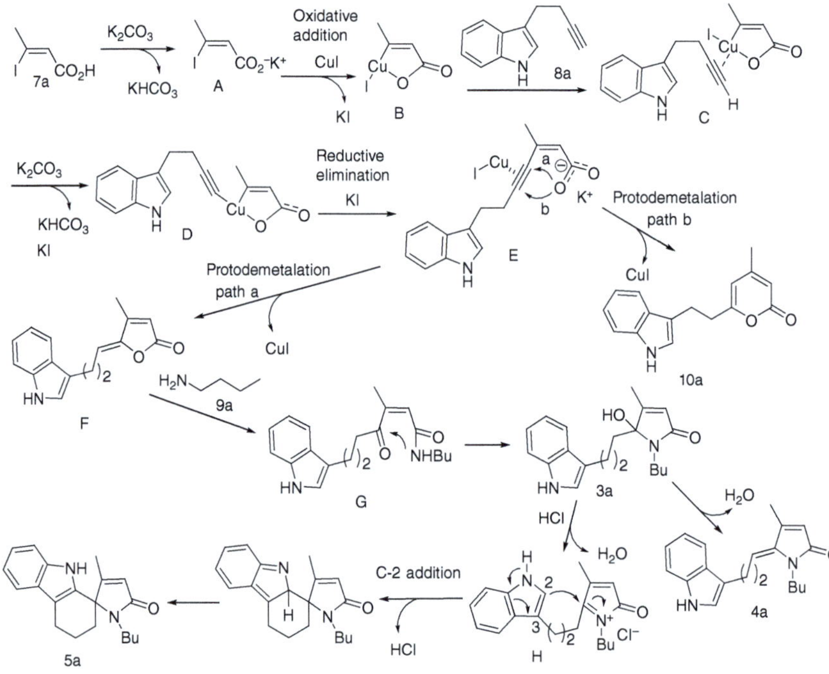

At the stage of intermediate **H**, there is a second possible cyclization, namely C-3 addition, to give the spiro compound intermediate **I**. The latter would undergo rearrangement to give **J**, followed by β-elimination to furnish the final product **5a**.

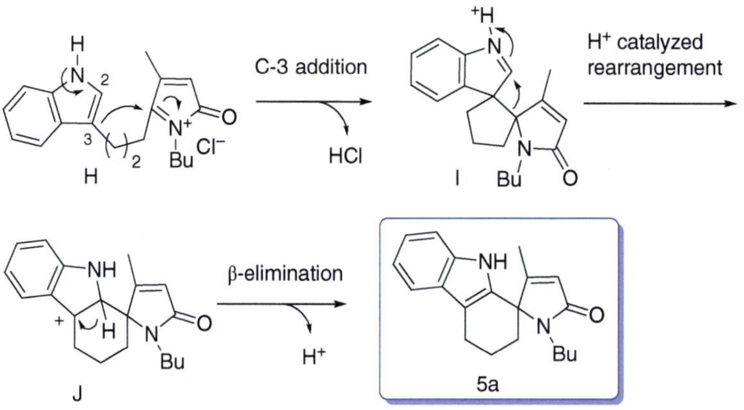

4.6.5 Pyridine

Problem 67: A General Preparation of Pyridines via the Annulation of Ketones

In 2001, chemists from Merck reported a general preparation of pyridines via the annulation of ketones [184]. A representative example is shown as the preparation of nicotinate **3**, which is a key building block for the synthesis of a c-Met kinase inhibitor **4** [185]. The methodology was also applied for the synthesis of COX-2 specific inhibitor Etoricoxib [186].

Please suggest a mechanism for this reaction.

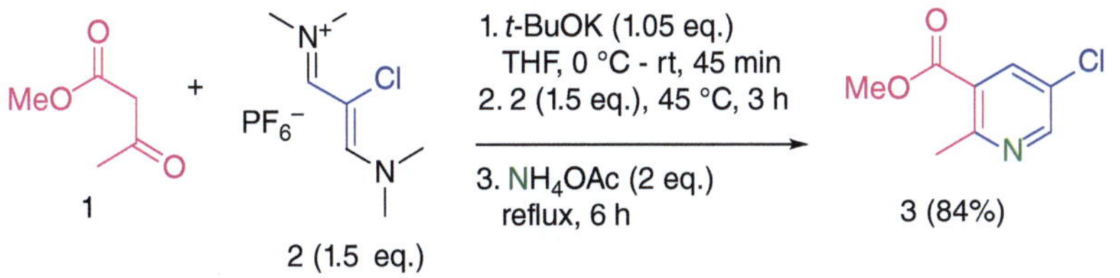

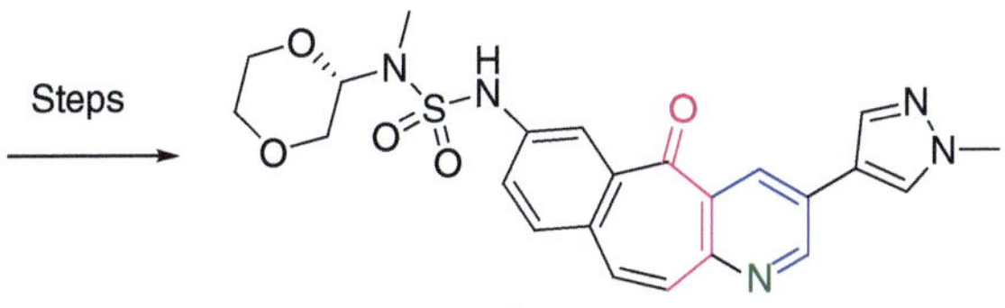

1. *t*-BuOK (1.05 eq.) THF, 0 °C - rt, 45 min
2. 2 (1.5 eq.), 45 °C, 3 h
3. NH_4OAc (2 eq.) reflux, 6 h

1 + 2 (1.5 eq.) → 3 (84%)

Steps → 4

1. *t*-BuOK, THF
2. HOAc
3. NH_4OAc reflux 97%

5 + 2 (1.5 eq.) → 6 (Etoricoxib)

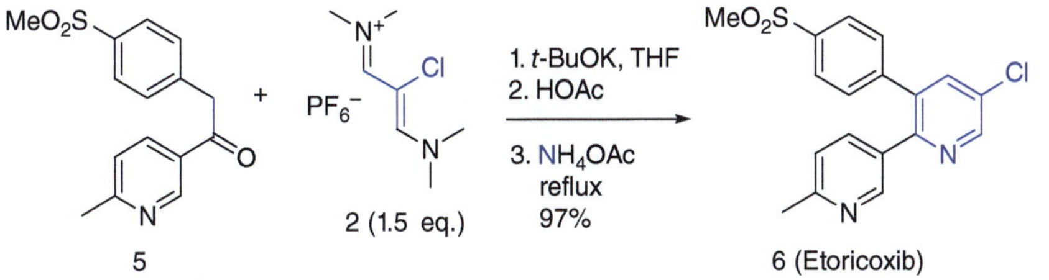

Experimental Procedure (from *J. Org. Chem.*, **2001**, *66*, 4194)

General Experimental Procedure for the Annulation of Ketones. To a suspension of the ketone (**1**, 25 mmol) in dry THF (50 mL) at 0 °C was added dropwise a 20 wt % solution of *t*-BuOK in THF (16.4 mL, 26.3 mmol, 1.05 equiv). The slurry was stirred at room temperature for 45 min, and the vinamidinium hexafluorophosphate salt (**2**, 11.5 g, 37.5 mmol, 1.5 equiv) was added in one portion. The resulting mixture was stirred at 45 °C for 3 h (in the presence of 1.0 equiv of DABCO when specified), and ammonium acetate (2 equiv) was added in one portion. The resulting dark solution was heated at reflux for 6 h and concentrated under reduced pressure. The residue was directly purified by chromatography on silica gel (CH_2Cl_2 or $CHCl_3$/MeOH was used as eluent). Alternatively, the residue after evaporation of the THF can be extracted with ethyl acetate and washed with water and saturated sodium chloride solution prior to the purification.

2-Methyl-5-chloro Methylnicotinate (3). Mp 186–187 °C. ^{1}H NMR (400 MHz, $CDCl_3$): δ 8.58 (d, $J = 2$ Hz, 1H), 8.18 (d, $J = 2$ Hz, 1H), 3.94 (s, 3H), 2.81 (3H). ^{13}C NMR (100 MHz, $CDCl_3$): δ 165.6, 157.9, 150.5, 137.7, 129.1, 125.9, 52.4, 24.1. IR (cm^{-1}): 3074, 3036, 3000, 2957, 2925, 2858, 1732. Anal. Calcd for $C_8H_8ClNO_2$: C, 51.77; H, 4.34; N, 7.55. Found: C, 51.98; H, 4.43; N, 7.44.

The Mechanism

The reaction mechanism, investigated by Merck scientists [184], can be succinctly summarized in the following scheme. The ketone must possess an active CH_2 or CH_3 group at the α-position, enabling the generation of enolate **A** using *t*-BuOK. The reaction of enolate **A** with the vinamidinium salt **2** can proceed through two distinct transition state models—1,2-addition or 1,4-addition—both leading to the same enamine intermediate **B**. This enamine **B** is stabilized by the presence of an electron-withdrawing substituent at the β-position (R^3), facilitating its partial tautomerization to form **C**. Typically, the adduct **B**/**C** undergoes elimination of dimethylamine to yield the corresponding dienamine **D**, which is stable and can be isolated and characterized. The cyclization of intermediate **D** in the presence of ammonium acetate proceeds through the formation of dienamine intermediate **E**, accompanied by the release of dimethylamine. Notably, there is no evidence suggesting that ammonia attacks the ketone moiety in **D** during the final cyclization step.

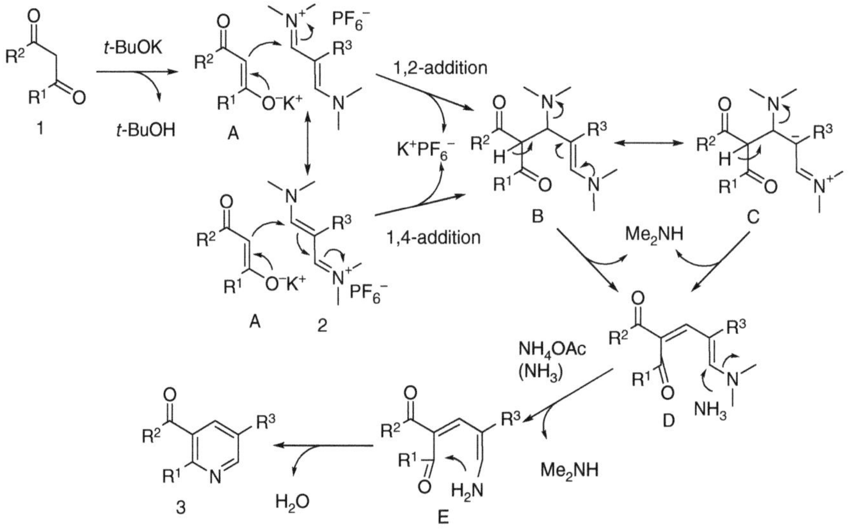

Problem 68: Conversion of Benzenes to Pyridines by *ipso*-selective Nitrene Internalization

The pyridine moiety serves as a significant isostere of benzene in medicinal chemistry research. However, the absence of direct carbon-to-nitrogen replacement techniques means that performing a nitrogen scan typically necessitates the bottom-up synthesis of each azine isomer. The synthetic approach for these isomers often differs considerably from that of their benzene counterparts. Specific nitrogen scans have played a crucial role in the discovery of several recently approved pharmaceuticals, including FDA-approved drugs such as imatinib, selpercatinib, sotorasib, and omidenepag, as well as flumatinib, which were approved by China's National Medical Products Administration.

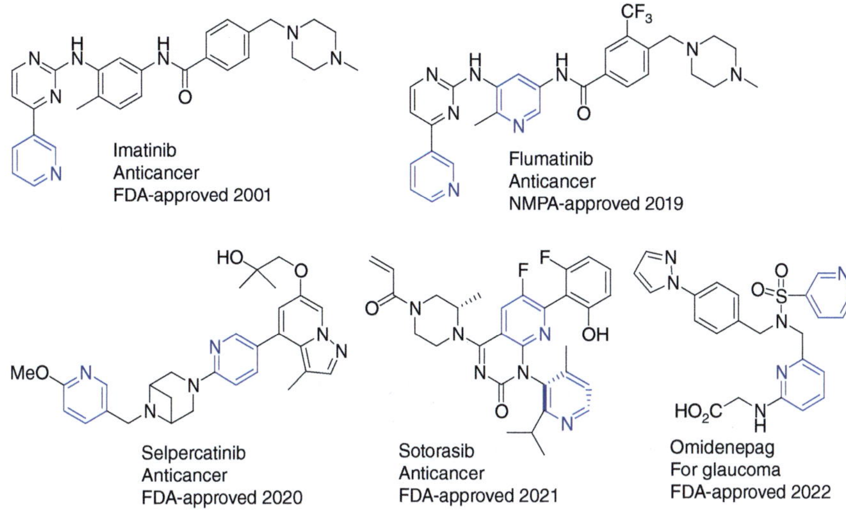

Imatinib
Anticancer
FDA-approved 2001

Flumatinib
Anticancer
NMPA-approved 2019

Selpercatinib
Anticancer
FDA-approved 2020

Sotorasib
Anticancer
FDA-approved 2021

Omidenepag
For glaucoma
FDA-approved 2022

A recent breakthrough by Levin's group at the University of Chicago introduced an innovative approach for aromatic nitrogen scanning through ipso-selective nitrene internalization [187]. In a representative example, 1-azido-3-(tert-butyl)benzene (**1**) and ethylaminoethanol (**2**) in acetonitrile were irradiated using a 427 nm LED. This was followed by treatment with 1,8-diazabicyclo(5.4.0)undec-7-ene (DBU) and *N*-bromocaprolactam (NBC) in dioxane at 80 °C, yielding the corresponding pyridine product **4** in 32% yield. This method was successfully applied to synthesize a potential 5α-reductase inhibitor, **7**, an estrone pyridine analog, demonstrating significantly improved efficiency compared to previously reported synthetic routes [188].

Please suggest a mechanism for the transformation from **1** to **4**.

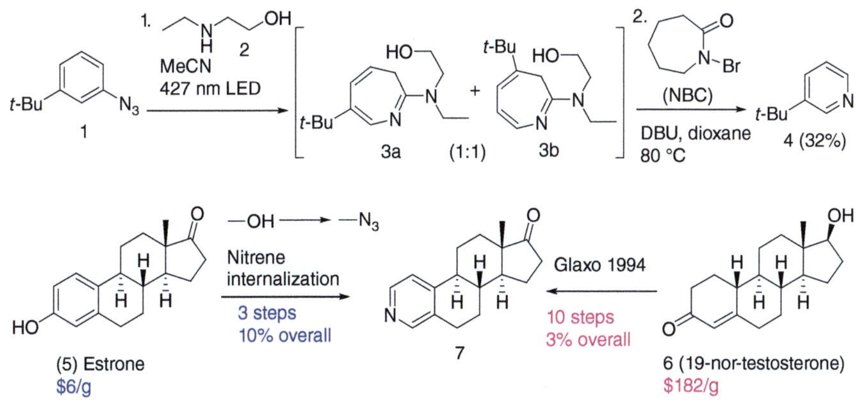

1. 2, MeCN, 427 nm LED → 3a + 3b (1:1); 2. (NBC), DBU, dioxane, 80 °C → 4 (32%)

(5) Estrone $6/g → (—OH → —$N_3$, Nitrene internalization, 3 steps, 10% overall) → 7 ← (Glaxo 1994, 10 steps, 3% overall) ← 6 (19-nor-testosterone) $182/g

Experimental Procedure (from *Science*, **2023**, *381*, 1474, Supplementary Materials)
General Procedure: A 2 dram vial was charged with 0.3 mmol of aryl azide (**1**), 0.3 mmol ethylaminoethanol (**2**), 3 mL MeCN, and a stir bar. The vial was sealed under a septum cap under air and subjected to photolysis using a 427 nm LED lamp while cooling to room temperature with a fan. After TLC indicated consumption of starting material, the solvent was removed under a stream of nitrogen. The atmosphere was then replaced with dry N_2 and then 3 mL dry, degassed dioxane was added. Then, 1.0 eq. (0.3 mmol) DBU was added, and the reaction was warmed to 80 °C. Separately, a solution of 115 mg (0.6 mmol, 2 eq.) of NBC was prepared in 1 mL dioxane (note: degassing of this latter solution was found experimentally not to significantly impact yields) and added dropwise to the reaction at 80 °C over the course of 2 minutes. The reaction was then concentrated under a stream of nitrogen and the residue purified by column chromatography on silica gel.

3-(*tert*-Butyl)-pyridine (4). Synthesized according to the general procedure from 1-azido-3-(tert-butyl) benzene (**1**, 52.5 mg, 0.3 mmol). The title compound was obtained as a yellow oil in 32% yield (12.9 mg) after purification by silica gel chromatography. ^{1}H NMR (400 MHz, $CDCl_3$) δ 8.70 – 8.65 (m, 1H), 8.43 (dd, J = 4.8 and 1.6 Hz, 1H), 7.69 (ddd, J = 8.1, 2.6, and 1.6 Hz, 1H), 7.23 (ddd, J = 8.0, 4.8, and 0.8 Hz, 1H), 1.35 (s, 9H). Spectra match those previously reported.

The Mechanism

Benzene ring enlargement by photolysis of phenylazide was reported by Doering and Odum in 1966 [189]. Under the irradiation of an LED, the starting azide **1** gives the intermediate nitrene **A**. Nitrene **A** then rearranges to form azepine **B** or **C**, both of which are electrophilic at the carbon where the azide was originally attached. Nucleophilic addition of amine **2** to azepines **B** or **C** provides the intermediate **3a** or **3b**. Cyclization of **3a** provides the spiro azepine oxazolidine **D**. The azepine **D** is then brominated by NBC to yield the bromide **E**. Elimination of one molecule of HBr by DBU from **E** affords azaheptatriene **F**. 3,3-Sigma tropic rearrangement of **F** provides the azanorcaradiene **G**. Due to the instability of the aziridine ring, it cleaves to the zwitterionic intermediate **H**. Aromatization of **H** by extrusion of carbene **I** leads to the pyridine product **4**.

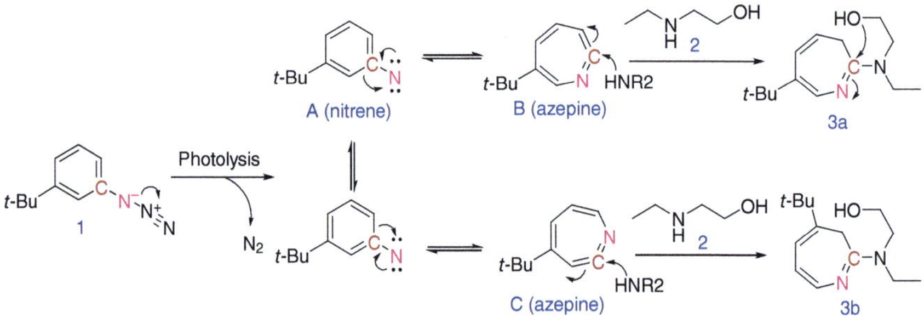

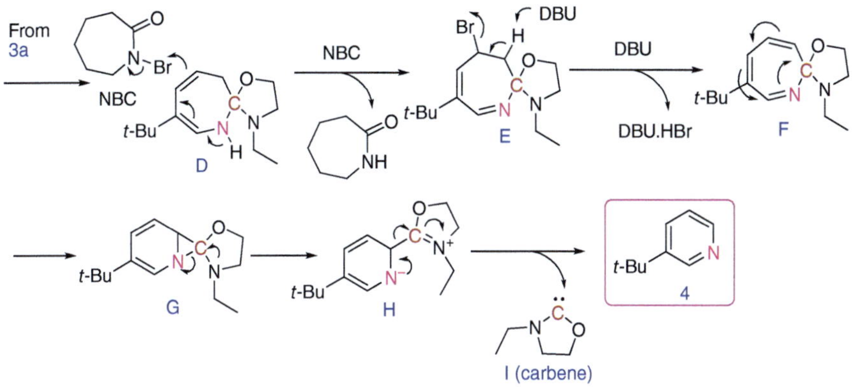

Following the similar manner, another intermediate **3b** could also undergo cyclization to give **J**. Due to the steric hindrance of the *t*-butyl group, the following bromination with NBC occurs at C-4 rather than at C-6. Thus, the corresponding product is still **4**.

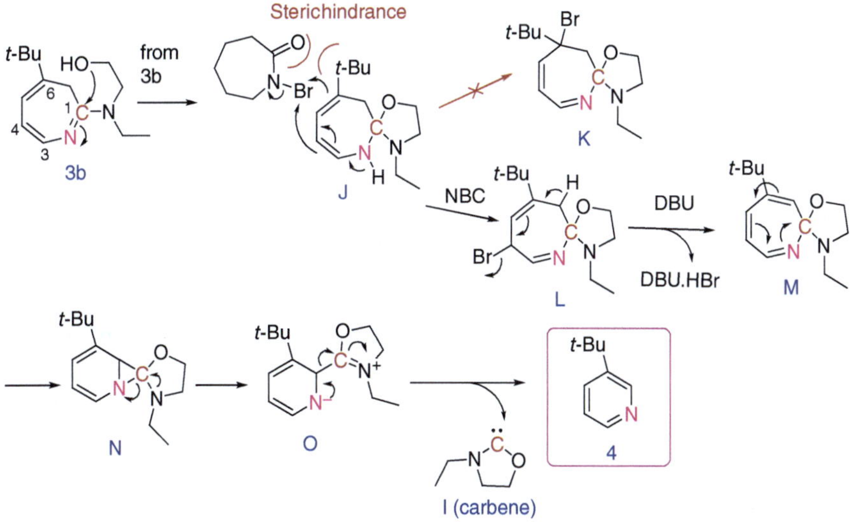

The fate of carbene **I** is eventually oxidized to the *N*-ethyl oxazolidinone **R**. This oxidation can occur either from adventitious dioxygen to dioxirane **P**, which is an oxidant and oxidizing another carbene **I** to give two molecules of oxazolidinone **R**. The carbine **I** could be also oxidized by Br^+ (from NBC) and forming the oxazolidinone **R** through hydrolysis upon aqueous workup.

Due to the tension of the enamine (azepine **C**) in seven-membered ring and its high reactivity, in the presence of ethylaminoethanol (**2**), there would be no chance to tautomerize to azepine **C′** via a 1,3-H shift. Consequently, the corresponding regiomeric product **S** was not observed.

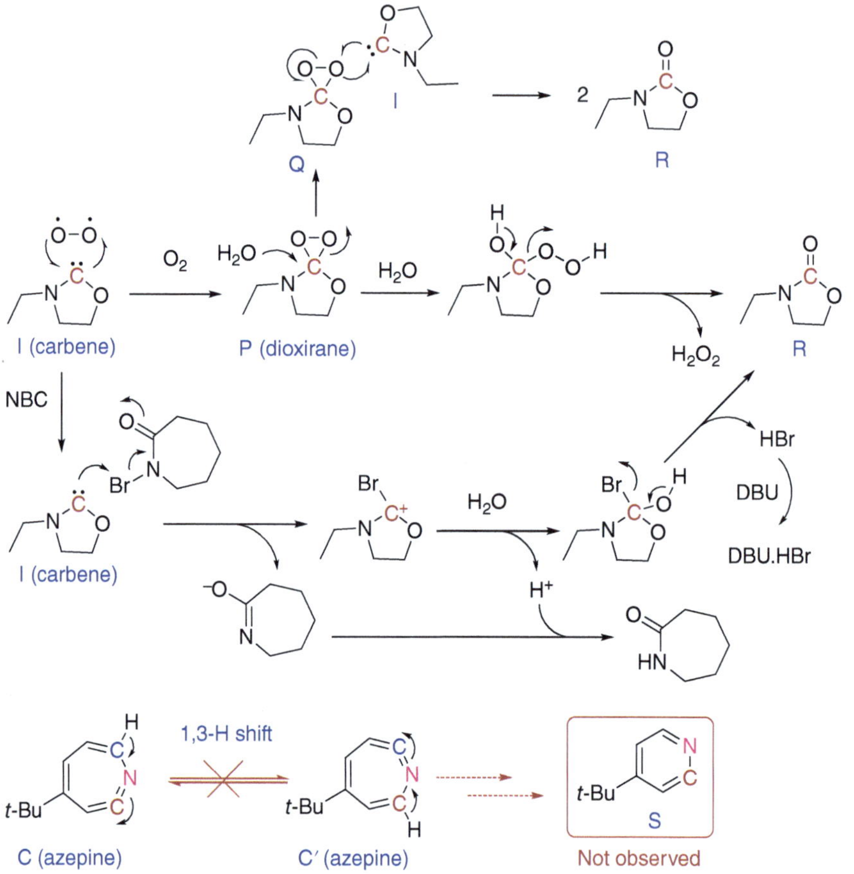

It is worth to note that, before finding NBC as a suitable oxidant, Levin's group tried NBS as an oxidant [187]. However, treatment of **8** with DBU and NBS led to the exclusive formation of the succinimide trapped oxidation product **9**, which further decomposed on heating.

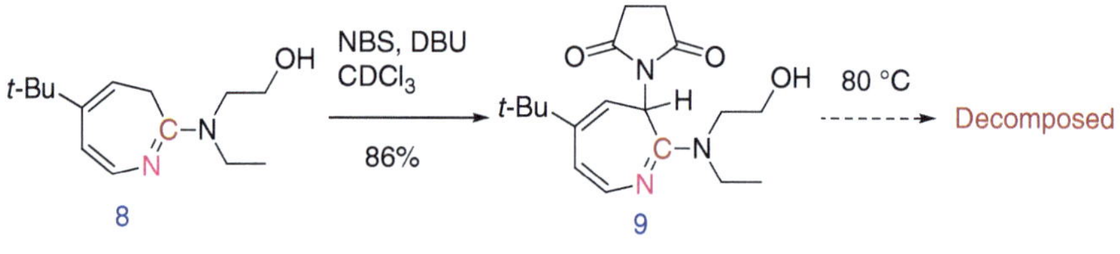

The phenomenon can be explained as follows. NBS is a stronger oxidizing Br^+ equivalent than NBC, it will brominate compound **8** to the bromide **T**. Subsequently, the succinimide anion displaces the Br to lead the product **9**.

Since caprolactam ($pK_a \sim 22$) is a much weaker acid than succinimide ($pK_a \sim 9.5$), it is not deprotonated by DBU (pK_a of $DBUH^+ \sim 13.5$) and thereby a strong nucleophile caprolactam anion is not generated. The amide substituted analog **U** is not formed. Therefore, replacing NBS with NBC directs the reaction along the desired pathway.

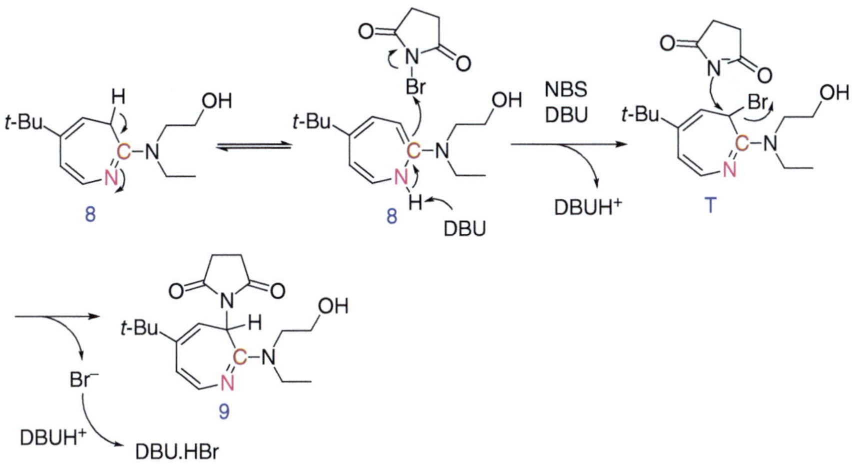

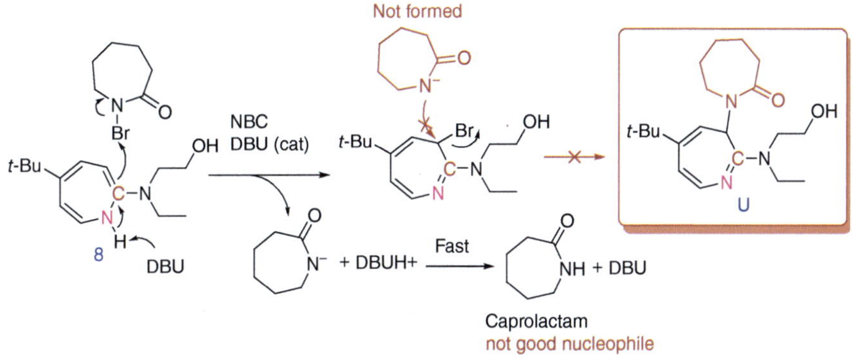

Problem 69: Synthesis of Multisubstituted Pyridines via HFIP Catalyzed Cyclization

Methoxatin (**4**), a natural product, has demonstrated the ability to mitigate damage caused by ischemia-reperfusion injury, which occurs during heart attacks or strokes. Although it is not currently classified as a vitamin, it is marketed as a dietary supplement due to its potential benefits [190]. Glinkerman and Boger developed a synthetic method for preparing di- and tetrahydroquinolines by reacting triazine (**1**) with an enamine, such as compound **2**, in hexafluoroisopropanol (HFIP) [191]. The resulting product (**3**) was subsequently converted into methoxatin (**4**) with a high overall yield.

Please suggest a mechanism for the formation of **3** from **1** and **2**.

Experimental Procedure (from *J. Am. Chem. Soc.*, **2016**, *138*, 12408)

Cycloaddition: A vigorously stirred solution of freshly prepared enamine **2** (0.965 mmol) in HFIP (9.65 mL) at 23 °C was treated with 1,2,3-triazine **1** (435 mg, 1.93 mmol). The vessel was sealed and heated at 60 C for 24 h. After 24 h, the reaction mixture was concentrated on a rotary evaporator and the residue was purified by preparative TLC (SiO_2, 50% EtOAc/hexanes) to provide **3** as a light yellow solid (356 mg, 95%): mp 97–99 °C; ^1H NMR ($CDCl_3$, 600 MHz) δ 11.72 (s, 1H),

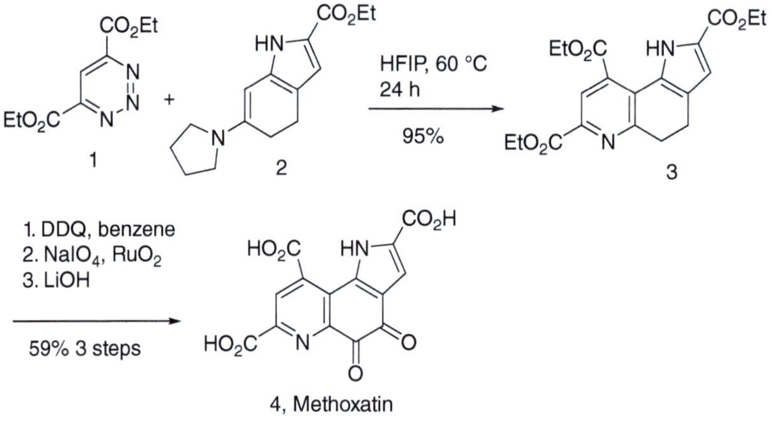

8.42 (s, 1H), 6.77 (d, $J = 2.4$ Hz, 1H), 4.52–4.48 (m, 4H), 4.36 (q, $J = 7.1$ Hz, 2H), 3.32–3.30 (m, 2H), 2.88–2.85 (m, 2H), 1.47–1.44 (m, 6H), 1.38 (t, $J = 7.1$ Hz, 3H); ^{13}CNMR ($CDCl_3$, 150 MHz) δ 168.3, 165.1, 161.0, 160.6, 143.8, 130.6, 128.1, 128.0, 127.0, 125.6, 125.2, 113.2, 63.4, 62.4, 61.0, 35.0, 21.1, 14.8, 14.7, 14.5; IR (film) ν_{max} 3328, 2981, 1709, 1368, 1251, 1227, 1186, 1098, 1022, 768 cm^{-1}; HRMS (ESI-TOF) m/z 387.1522 ($C_{20}H_{22}N_2O_6 + H^+$ requires 387.1551).

The Mechanism

The first stage is a Diels-Alder cycloaddition, followed by the extrusion of nitrogen gas. The final stage involves the elimination of pyrrolidine. Due to its strong hydrogen-bond-donating and Lewis acidic properties, HFIP forms hydrogen bonds with the 1,2,3-triazine, thereby activating it for the cycloaddition. In the final step, the acidity of HFIP may protonate the pyrrolidine moiety, thus promoting its elimination.

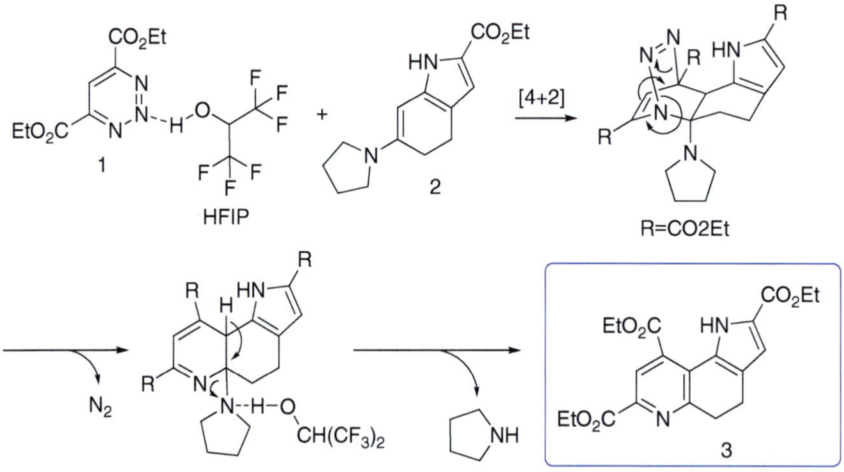

Problem 70: Preparation of Multisubstituted Pyridine

Recently, Chen et al. reported selective synthesis of poly-substituted fluorine-containing pyridines and dihydropyrimidines via cascade C—F bond cleavage protocol [192, 193]. As shown in the following representative example, benzylamine (**2**) reacted with 1,1,1-trifluoro-*N*,4-diphenylbut-3-yn-2-imine (**1**) in the presence of Cs_2CO_3 in refluxing THF to give the 2,4,5,6-tetrasubstituted pyridines (**3**) in high yield. However, when the reaction was carried out at 80 °C without a base, the uncyclized vinylogous amidine (**4**) was obtained. Upon addition of a base (Cs_2CO_3 or *t*-BuOK), continued heating of the reaction mixture at 80 °C also yielded pyridine (3). In contrast, lowering the temperature to −40 °C resulted exclusively in the formation of dihydropyrimidine (5).

The above methodology had been applied to the synthesis of Rinskor by Johnson [194]. Rinskor is a new 6-arylpicolinate herbicide that is used for postemergent control of grass, broadleaf, and sedge weeds in rice and other crops. In Johnson's paper, the core multisubstitued pyridine **8** was prepared from **6** and **7**, which was further transferred to Rinskor in four steps.

Please provide a mechanism for the formation of **3** and **5**.

Experimental Procedure (from *Org. Biomol. Chem.*, **2011**, *9*, 5682)

Representative Procedure for Pyridine 3 Synthesis. Alkynylimine **1** (0.4 mmol) was added to a solution of amines **2** (3 equiv.), Cs_2CO_3 (2.5 equiv.) in THF (2.0 mL). The solution was then stirred at 80 °C. After completion of reaction as indicated by TLC, the reaction crude was filtered and the filtrate evaporated. The residue was purified by flash chromatography on silica gel to provide the desired product **3**.

3-Fluoro-*N*,2,6-triphenylpyridin-4-amine (3a). White solid, mp 127–129 °C; ^{1}HNMR (300 MHz, $CDCl_3$) δ 8.02 (d, J = 7.1 Hz, 2H), 7.95 (d, J = 7.6 Hz, 2H), 7.56–7.14 (m, 12H), 6.34 (d, J = 3.3 Hz, 1H); ^{19}F NMR (282 MHz, $CDCl_3$) δ–152.37; ^{13}C NMR (100 MHz, $CDCl_3$) δ 153.3 (d, J = 5.9 Hz), 146.9 (d, J = 253.1 Hz), 144.0 (d, J = 8.1 Hz), 140.6 (d, J = 10.2 Hz), 139.5, 139.2, 136.0 (d, J = 5.1 Hz), 129.8, 129.0, 129.0 (d, J = 5.9 Hz), 128.7, 128.6, 128.4, 127.0, 124.7, 122.2, 104.3; LRMS (EI) m/z (relative intensity) 340 (99) [M^+], 39 (100); Anal. Calcd for $C_{23}H_{17}FN_2$: C, 81.16; H, 5.03; N, 8.23. Found: C, 80.89; H, 5.14; N, 8.15. IR (KBr): 3409, 3060, 1610, 1595, 1579, 1514, 1498, 1466, 1409, 1242, 694 cm^{-1}.

Procedure for 4a Synthesis. 1a (0.4 mmol) and benzylamine (2 equiv.) were added to 2.0 mL THF, the mixture was stirred at 80 °C for two hours. Then the solvent was evaporated, and the residue was purified by flash chromatography on silica, providing **4a** as a yellow solid in quantitative yield. Mp 76–77 °C; ^{1}HNMR (300 MHz, $CDCl_3$) δ 11.28 (br, 1H), 7.46–6.79 (m, 15H), 5.24 (s, 1H), 4.40 (s, 2H); ^{19}F NMR (282 MHz, $CDCl_3$) δ – 62.63; ^{13}C NMR (100 MHz, $CDCl_3$) δ 163.0, 152.6 (q, J = 26.4 Hz), 149.1, 139.0, 135.9, 129.5, 128.7, 128.6, 128.4, 128.0, 127.3, 126.6, 123.1, 120.3, 119.0 (q, J = 290.5 Hz), 89.9 (q, J = 3.7 Hz), 48.4; HRMS-ESI (m/z): $C_{23}H_{20}F_3N_2$ $[M + H]^+$ calcd 381.15731; found 381.15759; IR (KBr): 3029, 2927, 1616, 1554, 1483, 1281, 1231, 1182, 1137, 1089, 695 cm^{-1}.

Representative Procedure for Dihydropyrimidine 5 Synthesis. 1 (0.4 mmol) and amine (2 equiv.) were added to 2.0 mL THF, the mixture was stirred at 80 °C for two hours, and then cooled to −40 °C (room temperature for **5q**, **5w**), and *t*-BuOK (2.0 equiv.) was added slowly, and quenched with ice water after 5 min. The mixture was partitioned between ethyl acetate and water, the organic layer was washed with brine, dried over $MgSO_4$, and concentrated *in vacuo*. The residue was purified by column chromatography on silica gel to provide **5**.

6-(Difluoromethyl)-1,2,4-triphenyl-1,2-dihydropyrimidine (5a) Yellow solid, mp 106–107 °C; ^{1}HNMR (300 MHz, $CDCl_3$) δ 7.96–7.85 (m, 2H), 7.64 (d, J = 7.5 Hz, 2H), 7.50–7.11 (m, 11H), 6.64 (s, 1H), 6.57 (d, J = 2.5 Hz, 1H), 6.28 (t, J = 54.4 Hz, 1H); ^{19}F NMR (282 MHz, $CDCl_3$) δ–112.22 (dd, $J_{F\text{-}F}$ = 301.3 Hz, $J_{H\text{-}F}$ = 55.5 Hz, 1F), −121.41 (dd, $J_{F\text{-}F}$ = 301.2 Hz, $J_{H\text{-}F}$ = 53.5 Hz, 1F); ^{13}C NMR (100 MHz, $CDCl_3$) δ 160.1, 143.7, 142.9 (dd, J = 27.9 Hz, J = 22.0 Hz), 140.6, 137.2, 130.5, 129.6, 128.6, 128.3, 128.1, 127.2, 126.5, 126.4, 124.7, 110.2 (t, J = 242.1 Hz), 103.1 (t, J = 5.1 Hz), 79.8; HRMS-ESI (m/z): $C_{23}H_{19}F_2N_2$ $[M + H]^+$ calcd 361.15108; found 361.14966; IR (KBr): 3060, 3032, 2925, 1629, 1580, 1492, 1446, 1376, 1352, 1264, 1125, 1047, 765, 695 cm^{-1}.

The Mechanism

The first step involves the hydroamination of alkynylimine **1** with amine **2**, forming the intermediate vinylogous amidine **4**. This is followed by deprotonation and dehydrofluorination, generating a molecule (**A**) that contains both an anion and an imine. When the reaction is carried out at a low temperature with a soluble base (*t*-BuOK) (**path a**), the *in situ* generated amide (**B**) nucleophilic attacks the imine immediately without isomerization to form dihydropyrimidine (**E**) through a kinetically controlled pathway. After proton migration, the dihydropyrimidine (**5**) is produced. However, at elevated temperatures (**path b**), carbon nucleophilic addition becomes favorable, leading to the formation of a 1,2-dihydropyridine ring (**C**) under thermodynamic control. Subsequent proton migration, *β*-F elimination (to **D**) and final aromatization to a pyridine ring provide additional driving force for the formation of **3**. Meanwhile, the use of an insoluble base (Cs_2CO_3) effectively suppresses the kinetic pathway.

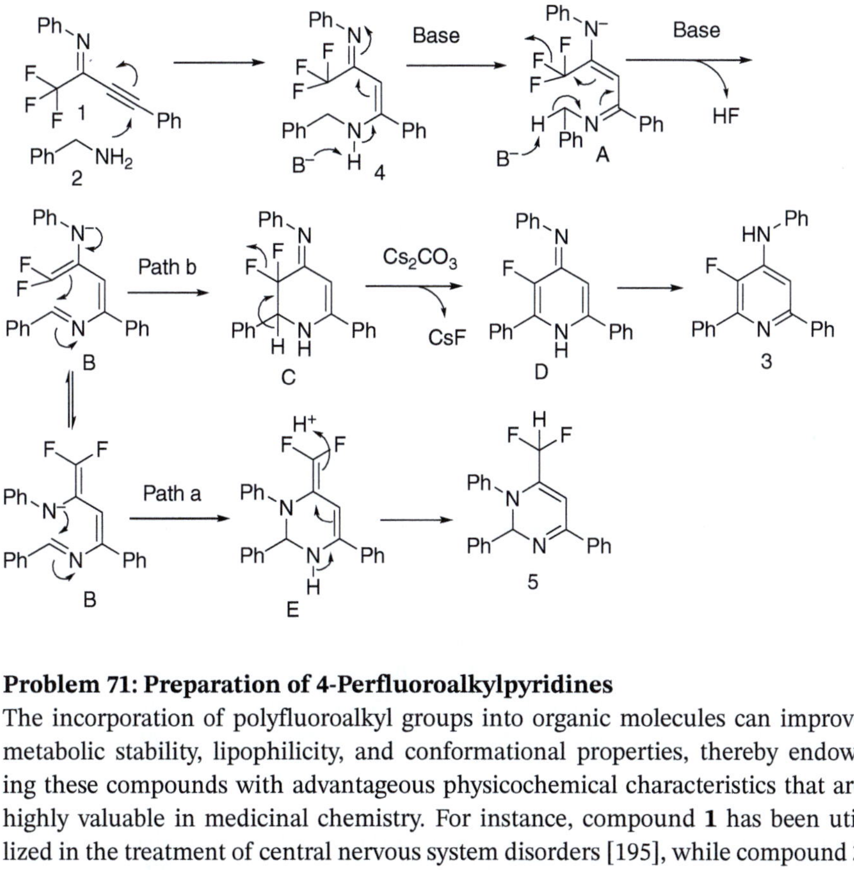

Problem 71: Preparation of 4-Perfluoroalkylpyridines

The incorporation of polyfluoroalkyl groups into organic molecules can improve metabolic stability, lipophilicity, and conformational properties, thereby endowing these compounds with advantageous physicochemical characteristics that are highly valuable in medicinal chemistry. For instance, compound **1** has been utilized in the treatment of central nervous system disorders [195], while compound **2** has been employed as an antifungal agent [196]. Consequently, the development of efficient and straightforward synthetic methods for polyfluoroalkyl pyridine derivatives is of significant importance in both organic synthesis and medicinal chemistry.

Recently Wang and Weng's group described the perfluorocarboxylic anhydrides (**3**) and enamide (**4**) triggered cyclization to 2,6-diaryl-4-perfluoroalkylpyridines (**5**) in one step [197].

Please provide a mechanism for the reaction.

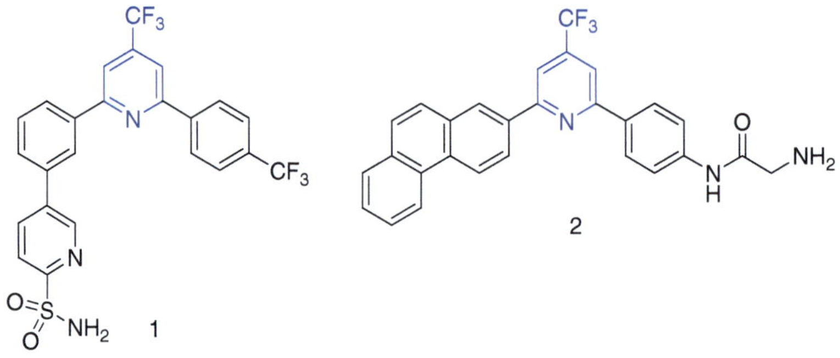

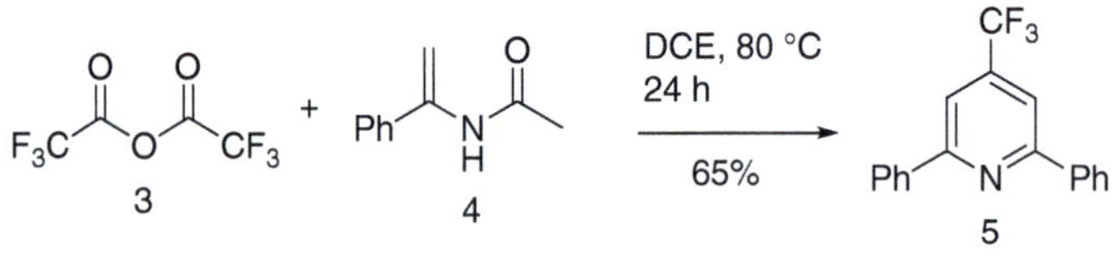

Experimental Procedure (from *J. Org. Chem.*, **2019**, *84*, 14926)

General Procedure for the Synthesis of 4-Polyfluoroalkylpyridines. The *N*-(1-phenylvinyl)acetamide derivatives (**4**) (1.0 mmol, 1.0 equiv), perfluorocarboxylic anhydride (**3**, 1.50 mmol, 1.5 equiv), and 1,2-dichloroethane (DCE) (4.0 mL) were added to a 5 mL reaction tube with a Teflon screw cap equipped with a stir bar. The mixture was stirred at the required temperature (60–120 °C) in an oil bath for 24 h. The reaction mixture was diluted with ethyl acetate (30 mL) and washed with saturated $NaHCO_3$ (30 mL) and saturated brine (30 mL). The organic phase was dried over $MgSO_4$. The solvent was removed by rotary evaporation and the resulting 4-polyfluoroalkylpyridine product was purified by column chromatography over a silica gel.

2,6-Diphenyl-4-(trifluoromethyl)pyridine (5a). Obtained as a white solid in 65% yield (97 mg). Mp: 81.7–82.5 °C. R_f (*n*-pentane/ethyl acetate = 40:1) = 0.40. ^{1}H NMR (400 MHz, $CDCl_3$): δ 8.23 (d, J = 7.4 Hz, 4H), 7.93 (s, 2H), 7.65–7.46 (m, 6H). ^{19}F NMR (376 MHz, $CDCl_3$): δ −64.6 (s, 3F). ^{13}C{^{1}H} NMR (101 MHz, $CDCl_3$): δ 158.2 (s), 140.0 (q, J = 33.4 Hz), 138.2 (s), 129.9 (s), 128.9 (s), 127.2 (s), 123.3 (q, J = 273.3 Hz), 114.0 (q, J = 3.5 Hz). IR (ATR): ν 3046, 2921, 2851, 1611, 1567, 1461, 1413, 1367, 1263, 1166, 1116, 1049, 873, 769, 687 cm^{-1}. GC-MS m/z: [M]$^+$ 299.

The Mechanism

Initially, the nucleophilic attack on the carbonyl of TFAA by the double bond of the enamide would form intermediate **A**, which smoothly undergoes β-H elimination to furnish intermediate **B**. Subsequently, the second nucleophilic attack on the carbonyl of **B**, which is much more electrophilic than the carbonyl of the acetamide, generates **C**, which is then poised to undergo a 6-*endo* cyclization, generating the intermediate **D**. Finally, the elimination of $AcNH_2$ and H_2O, and deacylation under acidic condition (the TFA is released from the first stage) from **D** furnishes the desired perfluoroalkylated product **3**.

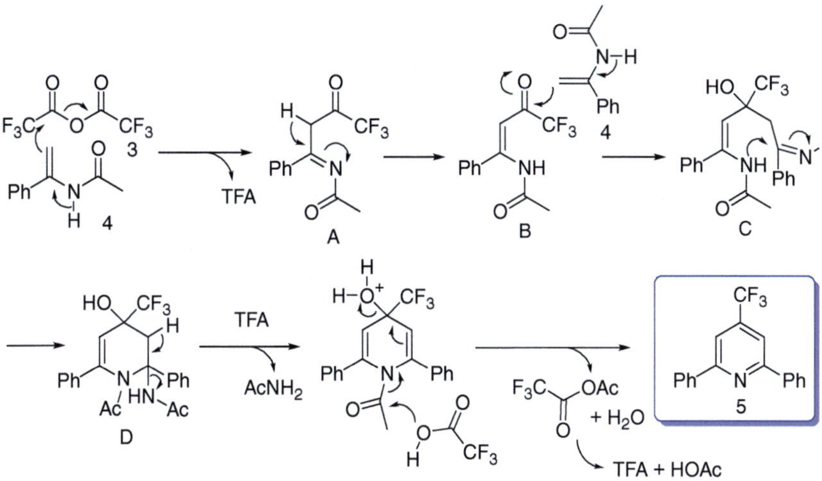

It could also be possible that the second enamide **4** attacks intermediate **B** in a Michael addition manner. While, this pathway would lead to the isomer, 2,4-diphenyl-6-(trifluoromethyl)pyridine **I**, that was not reported in Wang's paper [197].

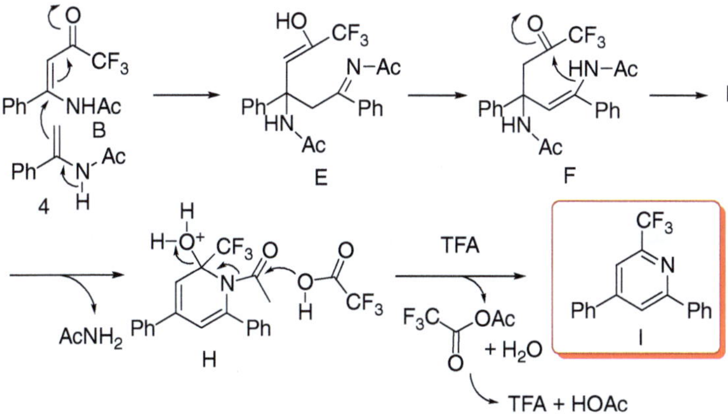

Problem 72: C-2 Amination of Pyridine

The 2-aminopyridine motif is a key structural feature found in numerous pharmaceuticals [198], including lemborexant, used to treat insomnia, and lasmiditan, a medication for migraines. Recently, Fier and colleagues developed a regioselective C2 amination of pyridines using *tert*-butyl ((3-chloro-5,6-dicyanopyrazin-2-yl)oxy) carbamate (**2**) as an amino source [199]. For instance, pyridine was successfully converted to *tert*-butyl pyridin-2-ylcarbamate (**3**) with a high yield of 93%. This approach falls under the category of late-stage C—H functionalization, a strategy increasingly applied to optimize pharmaceutical compounds [200]. As an example, 2-*N*-Boc amino roflumilast was synthesized from Roflumilast in 42% yield [201]. These innovative methodologies provide medicinal chemists with efficient and resource-effective tools to access novel analogs of bioactive compounds, significantly advancing drug discovery efforts.

Please provide a mechanism for the formation of **3** from **1** and **2**.

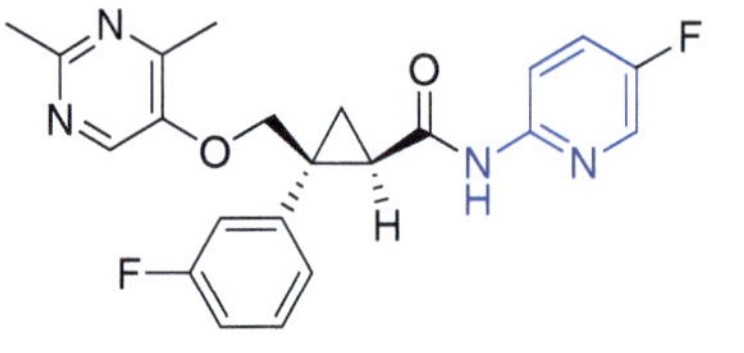

Lemborexant

Lasmiditan

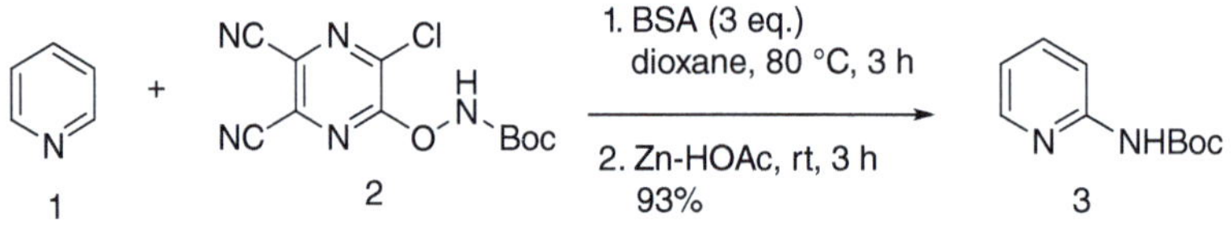

1. BSA (3 eq.) dioxane, 80 °C, 3 h
2. Zn-HOAc, rt, 3 h 93%

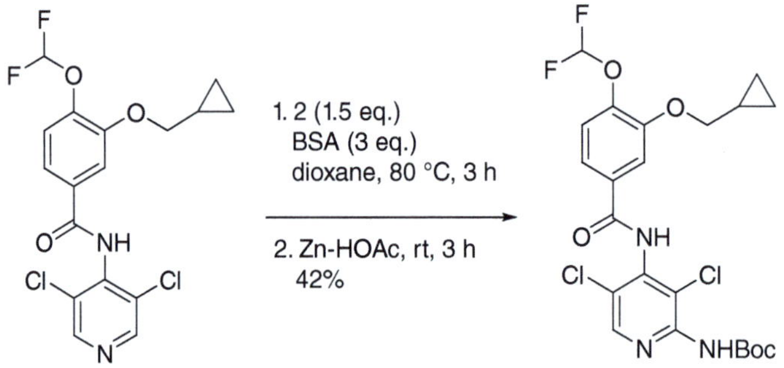

1 + 2 → 3

1. 2 (1.5 eq.) BSA (3 eq.) dioxane, 80 °C, 3 h
2. Zn-HOAc, rt, 3 h 42%

Roflumilast → 2-Boc-aminoroflumilast

Experimental Procedure (from *J. Am. Chem. Soc.*, **2020**, *142*, 8614, Supporting Information)

General Procedure for the C-2 Functionalization of Pyridines. To a screw cap vial with a magnetic stirbar was added pyridine substrate (**1**, 1.0 equiv), *N,O*-bis(trimethylsilyl)acetamide (3.0 equiv), and anhydrous dioxane (4 mL per mmol of **1**; 0.25 M). To the resulting mixture was added solid **2** (1.5 equiv), then the vial was sealed and heated at 80 °C for 3 h. The reaction mixture was cooled to ambient temperature, diluted with glacial AcOH (4 mL per mmol of **1**, equal volume as dioxane), and Zn powder (5.0 equiv) was added at once (note: a mild exotherm was observed on 5 mmol scale). The resulting slurry was stirred vigorously for 3 h at ambient temperature. The mixture was concentrated to dryness, and the product was purified on silica gel. For cases where assay yields of the products were determined, the entire reaction mixture was diluted in a volumetric flask with 3:1 MeCN: 0.1% aqueous H_3PO_4 and the resulting solution was analyzed on an HPLC instrument calibrated to the product using the method shown below.

***tert*-Butyl pyridin-2-ylcarbamate (3).** The compound was prepared using the standard procedure on 0.5 mmol scale. The assay yield was determined by HPLC to be 93% based on calibration with an authentic standard. NMR data for in situ characterization at the end of reaction in 1:1 1,4-dioxane-d_8 / CD_3COOD is provided below. ^{1}H NMR (600 MHz, 1:1 1,4-dioxane-d_8 / CD_3COOD) δ 8.28 (ddd, $J = 5.6, 1.9,$

and 0.8 Hz, 1H), 7.94 (ddd, $J = 9.0$, 7.4, and 1.9 Hz, 1H), 7.67 (d, $J = 8.6$ Hz, 1H), 7.19 (ddd, $J = 7.0$, 5.6, and 1.1 Hz, 1H), 1.45 (s, 9H). ^{13}C NMR (151 MHz, 1:1 1,4-dioxane-d_8 / CD_3COOD) δ 154.6, 152.2, 145.3, 143.5, 120.2, 115.8, 84.3, 28.4.

The Mechanism

The pyrazine chloride **2** is attacked by pyridine to give the pyridinium chloride **A**. Then cyclization occurs to provide the tricyclic intermediate **B** and HCl. *N,O*-bis-(trimethylsilyl)acetamide (BSA) promoted the desired reaction by scavenging HCl through the generation of TMSCl. Ring opening of **B** yields the hydroxyl pyrazine **C** which is protected by forming TMS ether **D**. Rearrangement of **D** via zwitterionic intermediate **E** affords *N*-pyrazinyl *N*-Boc-2-aminopyridine **F**. Finally, reductive cleavage of the pyrazinyl-N bond by zinc followed by acidic aqueous workup furnishes the *N*-Boc-2-aminopyridine **3**.

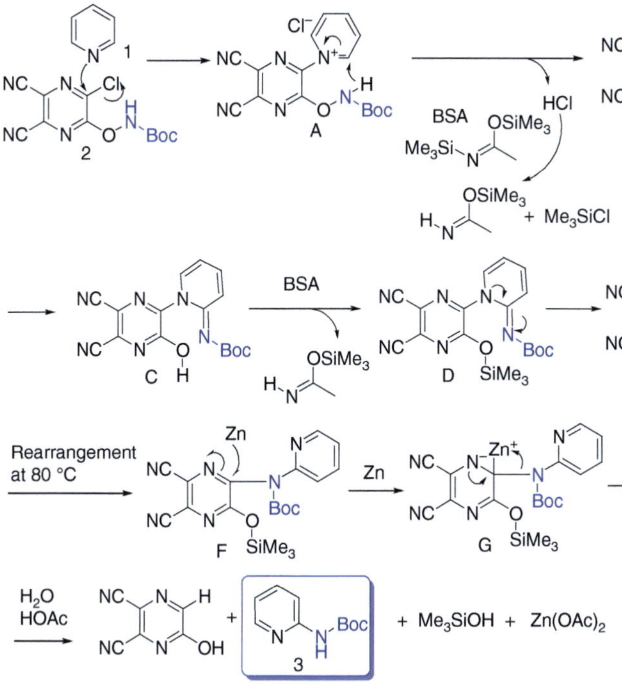

Problem 73: C-2 Amination of Pyridine *N*-oxide Derivatives

Lumacaftor (**3**) is a cystic fibrosis transmembrane (CFTR) corrector utilized in the treatment of cystic fibrosis. The combination therapy of lumacaftor and ivacaftor, marketed under the brand name ORKAMBI^{TM}, is specifically designed to treat cystic fibrosis patients who are homozygous for the F508del mutation in the CFTR conductance regulator gene. This mutation results in the production of a defective protein that is responsible for the disease. Developed by Vertex Pharmaceuticals, the lumacaftor/ivacaftor combination received FDA approval in 2015. The synthesis of lumacaftor has been detailed in a series of patents and patent applications by Vertex [202]. A key intermediate in the synthesis of lumacaftor is 2-aminopyridine (**2**).

The C-2 amination of pyridine *N*-oxide (**1**) was accomplished using acetonitrile as the amino source [202]. In a published study, 2-aminopyridine (2) was used as a crucial building block for constructing analogs of lumacaftor by the research group of S. G. Aller and M. Turlington [203]. Additionally, Hughes proposed a reaction mechanism in a review article [204].

Would you be able to propose your own mechanism without referring to the literature?

1. Ms_2O, Py, CH_3CN, 70–75 °C
2. $HOCH_2CH_2NH_2$

53%

Steps → Lumacaftor (3)

Experimental Procedure (from *Chem. Eur. J.*, **2019**, *25*, 3662, Supporting Information)

tert-Butyl 3-(6-amino-3-methylpyridin-2-yl)benzoate (**2**). 2-(3-(*tert*-Butoxycarbonyl)-phenyl)-3-methylpyridine 1-oxide **1** (2.62 g, 9.2 mmol, 1 equiv) was dissolved in MeCN (92 mL, 0.1 M). Pyridine (2.98 mL, 36.8 mmol, 4 equiv) was added, and the reaction was then heated to 70 °C under nitrogen atmosphere. Ms_2O (2.40 g, 13.8 mmol, 1.5 equiv) was dissolved in MeCN (23 mL) in a flame-dried pear flask, and was then transferred to a flame-dried slow addition funnel and was added dropwise. After addition, the reaction was stirred at 70 °C for 1.5 h. The reaction was then cooled to room temperature and ethanolamine (5.55 mL, 92 mmol, 10 equiv) was added dropwise via slow addition funnel. After stirring at room temperature for 2 h, the reaction was quenched with H_2O (100 mL) and stirred for 30 min. The reaction mixture was extracted with EtOAc (150 mL × 3), and the combined organic layers were dried with dried with Na_2SO_4, filtered, and concentrated via rotary evaporation. Flash silica gel column chromatography (100% hexanes to 50:50 hexanes/EtOAc) afforded the product as a tan solid in 53% yield (1.43 g). R_f = 0.62 (2:1 hexanes/EtOAc). IR (neat, cm^{-1}) 3438, 3177, 2971, 1698, 1628, 1487, 1295, 1139. ^{1}HNMR (400 MHz, $CDCl_3$) δ 8.11 (t, 1H, J = 1.4 Hz), 8.00 (dt, 1H, J = 7.8, 1.4 Hz), 7.65 (dt, 1H, J = 8.2 and 1.4 Hz), 7.47 (t, 1H, J = 7.4 Hz), 7.34 (d, 1H, J = 8.4 Hz), 6.45 (d, 1H, J = 8.2 Hz), 4.47 (bs, 2H), 2.19 (s, 3H), 1.60 (s, 9H). ^{13}CNMR (100 MHz, $CDCl_3$) δ 165.8, 156.4, 155.5, 141.1, 140.8, 133.0, 132.0, 130.0, 128.7, 128.1, 120.4, 107.9, 81.1, 28.3, 18.9. HRMS (TOF, ESI+, M) for $C_{17}H_{20}N_2O_2$: calc. mass 284.1525, found 284.1529.

The Mechanism

First, the *N*-oxide **1** is activated by methanesulfonic anhydride to form intermediate **A**. Subsequently, the nitrogen atom in acetonitrile attacks the α-carbon of **A**, leading to the formation of intermediate **B**. Deprotonation and elimination of methanesulfonate then restore the aromatic ring, generating intermediate **C**. The addition of methanesulfonate to **C** produces imidate **D**. Upon the introduction of ethanolamine, the methanesulfonate (MsO) group is displaced, resulting in the formation of amidine **E**. Further cyclization of **E** followed by hydrolysis yields 2-aminopyridine **2** and acetamide **F**. Notably, when a bromo substituent is present at the C-2 position of the pyridine *N*-oxide, C-4 amination is also observed [205].

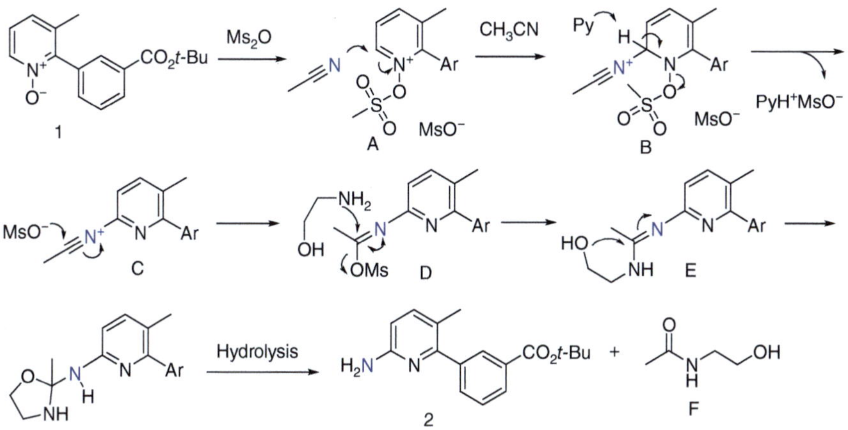

When 2-bromo-3-methylpyridine-1-oxide (**4**) was employed as a substrate, C-4 amination was also observed [205].

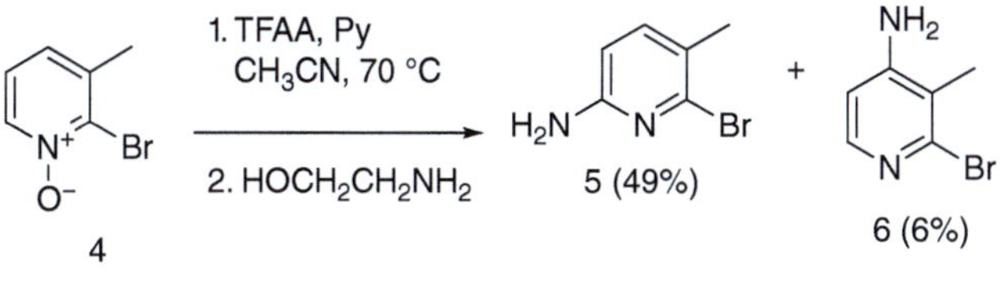

Problem 74: Nucleophilic Cyanation of Pyridines

In 2017, P. S. Fier designed (*Z*)-*N*-((methylsulfonyl)oxy)acetimidoyl chloride (**1**) assisted functionalization of pyridines [206]. As shown in the following reaction, 4-phenylpicolinonitrile (**3**) was prepared in 84%. This method was applied for the synthesis of a mixed progesterone agonist/antagonist **5**.

Could you please propose a mechanism for this reaction?

Experimental Procedure (from *J. Am. Chem. Soc.*, **2017**, *139*, 9499, Supporting Information)

General Procedure for the Conversion of Pyridines to 2-cyanopyridines. To a screw cap vial was added pyridine (**2**, 1.0 equiv), **1** (1.1 equiv), sodium trifluoromethanesulfonate (NaOTf, 1.2 equiv), and MeCN (1.0 mL per mmol of **1**). The vial was sealed and heated at 80 °C with rapid stirring for 1–8 h. The reaction mixture was cooled to room temperature and diluted with DMF (4 mL per mmol of **2**),

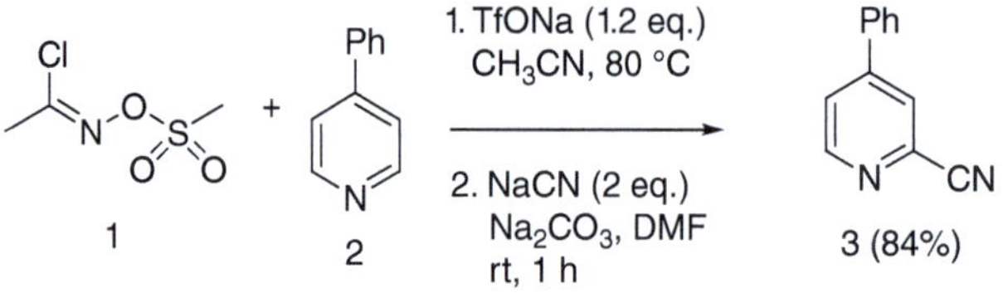

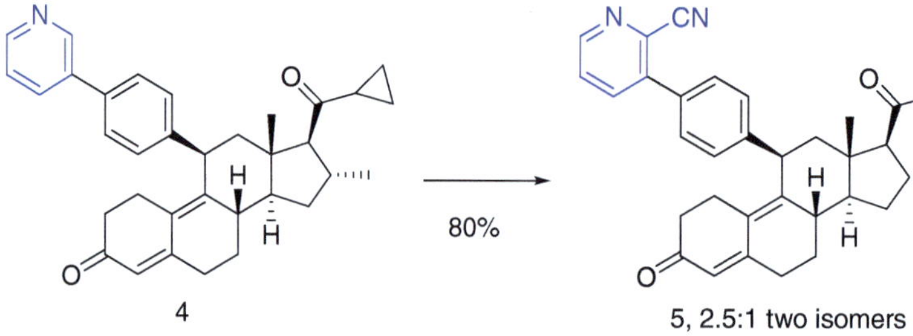

1. TfONa (1.2 eq.) CH_3CN, 80 °C
2. NaCN (2 eq.) Na_2CO_3, DMF rt, 1 h

1 + 2 → 3 (84%)

4 → 5, 2.5:1 two isomers (80%)

followed by the addition of Na_2CO_3 (2.0 equiv) and NaCN (2.0 equiv). The vial was sealed and aged at room temperature with rapid stirring for 1 h. For cases where assay yields were determined, the entire reaction mixture was diluted in a volumetric flask with 3:1 MeCN: 0.1% aqueous H_3PO_4 and the resulting solution was analyzed on an HPLC instrument calibrated to the product standard. The concentration of the 2-cyanopyridine product was determined and used to calculate the yield of the product. For cases where the product was isolated, the reaction mixture was diluted with isopropyl acetate, and washed twice with water and once with brine. The organic solution was dried over $MgSO_4$, and the product was purified by silica gel chromatography using a TeleDyne Isco CombiFlash purification system.

2-cyano-4-phenylpyridine (3). The compound was prepared using the standard procedure on 0.5 mmol scale. The product was purified on silica gel with a gradient of 0–50% EtOAc in hexanes, and obtained as a white solid in 84% yield. ^{1}H NMR (500 MHz, Acetonitrile-d_3) δ 8.72 (dd, J = 5.2 and 0.6 Hz, 1H), 8.11 (dd, J = 1.8 and 0.7 Hz, 1H), 7.86 (dd, J = 5.2 and 1.9 Hz, 1H), 7.79 – 7.72 (m, 2H), 7.59 – 7.48 (m, 3H). ^{13}C NMR (126 MHz, Acetonitrile-d_3) δ 152.55, 150.31, 136.85, 135.08, 131.12, 130.38, 128.10, 127.63, 125.83.

The Mechanism

Compound **1** functions as a bifunctional reagent, playing a dual role in the reaction mechanism. Initially, it acts as an electrophilic activator, promoting nucleophilic addition to the pyridine ring. Subsequently, in the presence of a base, compound **1** serves as a two-electron oxidant, facilitating the re-aromatization of the dihydropyridine intermediate (**C**). However, it was observed that the formation of the intermediate pyridinium chloride salt (**A**) was thermodynamically unfavorable, with the reaction equilibrium heavily favoring the starting materials. To address this, a stoichiometric amount of NaOTf was introduced, which enhanced the conversion to the pyridinium salt (**B**) by driving the precipitation of NaCl, thereby shifting the equilibrium toward the desired product.

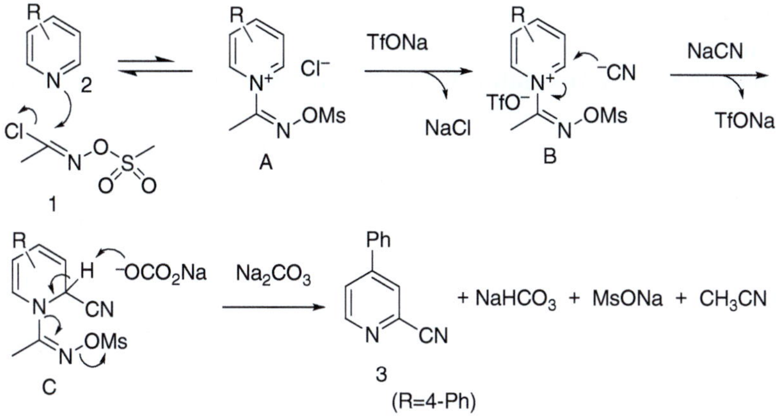

Pyridine substrates bearing 3-halogen and oxygen substituents predominantly yielded 2-cyano-3-substituted pyridines, with a selectivity ratio of 2-cyanation to 6-cyanation exceeding 95:5. In the case of isoquinolines, cyanation exclusively produced 1-cyano products. These findings align with the mechanism of nucleophilic addition occurring at the most electrophilic carbon of the pyridinium intermediate [207]. Substrates with bulkier or less polar groups at the C-3 position resulted in mixtures of two isomers. Notably, 4-cyanopyridine products were not detected under any conditions.

Problem 75: Synthesis of 2-Arylpyridines from Acetophenones

Vismodegib is a medication for treatment of basal-cell carcinoma. In a manufacturing production of vismodegib, 2-(2-chloro-5-nitrophenyl)pyridine (**3**) is a key building block. One of the large-scale (e.g., 128 kg) synthesis of **3** is the reaction of 1-(2-chloro-5-nitrophenyl)ethan-1-one (**1**) with 1,3-dimethyl-2-oxo-2,3-dihydropyrimidin-1-ium chloride (**2a**) and NH_4OAc in acetic acid [208]. 1,3-Dimethyl-2-oxo-2,3-dihydropyrimidin-1-ium hydrogen sulfate (**2b**) was also applied for preparation of **3** [209]. Neither articles mentioned the mechanism of this pyridine build-up reaction.

Please provide a mechanism for this reaction.

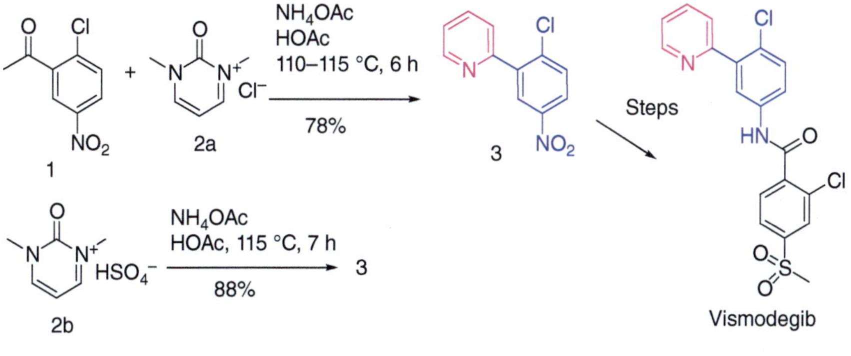

Experimental Procedure (from *Org. Process Res. Dev.*, **2016**, *20*, 1509, Supporting Information)

2-(2-chloro-5-nitrophenyl)pyridine 3: 1-(2-Chloro-5-nitrophenyl)ethan-1-one **1** (140.0 kg, 1.0 molar equiv), 1,3-dimethyl-2-oxo-2,3-dihydropyrimidin-1-ium chloride (**2**, 140.0 kg, 1.2 molar equiv) and ammonium acetate (360 kg) were added to a 3000 L reactor containing acetic acid (1000 kg), and the mixture was heated to 110–115 °C for 6 hours (IPC, **1** < 5 A% by HPLC). Then water (700 L) was added with cooling to 0–10 °C and stirred for at least 1 hour. The suspension was centrifuged, washed with a mixture of methanol (150 L) and water (150 L), and finally with water (480 L). The crude wet product was dissolved in warmed acetone (2000 L) at 40–50 °C, and the hot solution was polish filtered. After rinsing the used-line with acetone (50 L), water (660 L) was added dropwise into the filtrate. The mixture was then cooled to 0–10 °C and stirred for at least 1 hour. The solids were isolated by centrifugation and washed with a ~5 °C cooled mixture of acetone (75 L) and water (75 L). The cake was dried under vacuum at 50 °C for about 24 hours to give **3** as a pale yellow powder (128.4 kg, 78.0% yield, 99.9A% purity by HPLC). MP: 151.3 °C. ^{1}HNMR* (400 MHz, $CDCl_3$) δ 8.77 (1H, ddd, J = 4.8, 1.6, and 1.2 Hz), 8.52 (1H, d, J = 2.8 Hz), 8.20 (1H, dd, J = 8.8 and 2.8 Hz), 7.83 (1H, td, J = 8.0 and 2.0 Hz), 7.70 (1H, dt, J = 7.6 and 1.2 Hz), 7.66 (1H, d, J = 8.4 Hz), 7.38 (1H, ddd, J = 7.6, 4.8, and 1.2 Hz).

** The ^{1}HNMR data is interpreted by the author of this book according to the spectrum provided in the Supporting Information.*

The Mechanism

The first stage is an addition (condensation) of the enol form of **1** to **2** to give intermediate **A**, followed by ring opening to generate **B**. Nucleophilic addition of NH_3, from NH_4OAc, to the acid-activated **B** followed by elimination of 1,3-dimethylurea yields the keto-enamine **C**. Final cyclization and dehydration of **C** leads to the pyridine product **3**.

Problem 76: Preparation of *penta*-substituted Pyridone

Chemists form Takeda and Millennium developed a one-pot process for *penta*-substituted pyridone (**2**) from dimethyl-α-flouromalonate (**1**) [210]. Pyridone **2** is a key intermediate for synthesis of **TAK-733**, a new MEK inhibitor. In Fochon Pharma, we also employed this chemistry to build-up our MEK inhibitors, such as **FCN-159**, which was recently approved in China [211].

Please provide stepwise mechanisms for the process.

F
MeO2C CO2Me
1
1. DBU, malononitrile
2, aq. MeNH2
3. 10 N NaOH, MeOH
80%
OH
NC F
H2N N O
2 (3.00 kg)
Steps
TAK-733
Steps
FCN-159

Experimental Procedure (from *Org. Process Res. Dev.*, **2012**, *16*, 1652)

2-Amino-5-fluoro-4-hydroxy-1-methyl-6-oxo-1,6-dihydropyridine-3-carbonitrile (2). Dimethyl α-fluoro malonate (**1**, 3.00 kg, 20.0 mol) and malononitrile (1.32 kg, 20.0 mol) were dissolved in THF (15 L) and cooled to −25 ± 5 °C. DBU (6.09 L, 39.9 mol) was added over 5 h at <10 °C. Then the reaction mixture was allowed to slowly warm to room temperature over 2 h and stirred for another 16 h. The reaction progress was monitored with HPLC (Method 4.2.6.). Aqueous methylamine (40%, 14.1 L, 133 mol) was added dropwise, and the reaction mixture was stirred for another 2 h at room temperature. Then 10 N NaOH (3.0 L, 30 mol) was added, and the reaction was stirred at room temperature for 5 h. The methylamine and THF were removed by rotavap with a 35 °C bath under 600–650 mmHg vacuum. The resulting mixture was cooled to 0 ± 5 °C followed by pH adjustment to ~1–2 using conc. HCl (8.50 L). The resulting solid was collected by filtration, washed with water (2 × 6.0 L), and dried in vacuum oven at 60 °C until the moisture content of the product was below 10% by KF. Product **2** was obtained as a light-brown solid (3.00 kg, 80% yield, 99.6% purity, moisture content 9.28 wt %). ^{1}H NMR (400 MHz, DMSO-d_6) δ 11.71 (s, 1H), 7.29 (s, 2H), 3.27 (s, 3H); ^{13}C NMR (100 MHz, DMSD-d_6)

δ 154.7 (d, $J = 21.9$ Hz), 153.1, 151.4 (d, $J = 13.2$ Hz), 129.5 (d, $J = 211$ Hz), 115.4 (d, $J = 3.7$ Hz), 63.2 (d, $J = 2.9$ Hz), 28.7; ^{19}F NMR (376 MHz, DMSD-d_6) $\delta - 178.9$; MS $(M + H)^+$ m/z calcd 184.0, found 184.0.

The Mechanism

This one-pot transformation included three telescope reactions. The first step is a condensation between the dimethyl-α-flouromalonate and malononitrile catalyzed by DBU to form intermediate **A**.

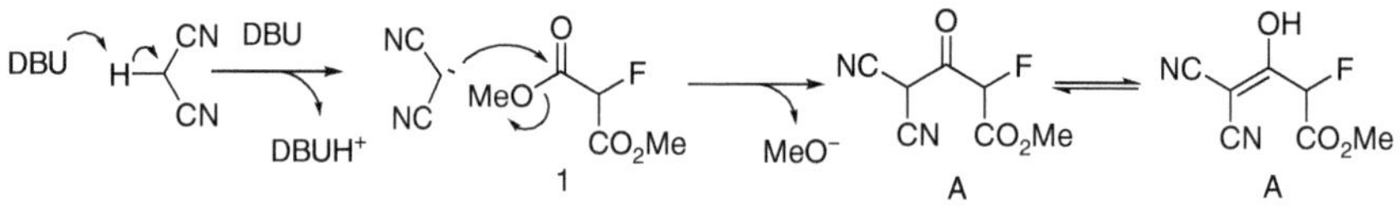

The next step is an amidation of the methyl ester with methylamine to provide intermediate **B**.

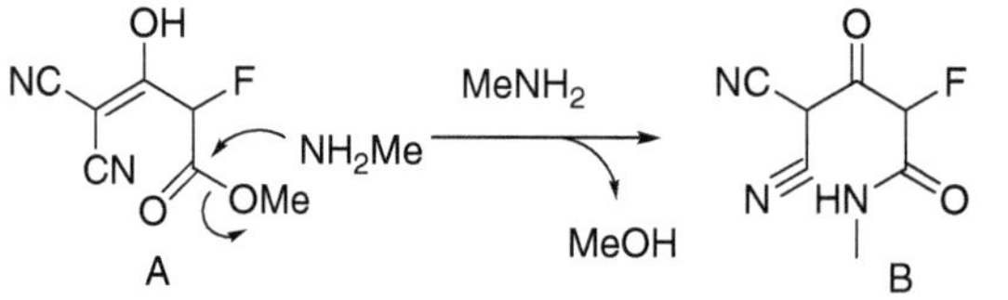

The final step is the intramolecular cyclization under basic condition to give product **2**.

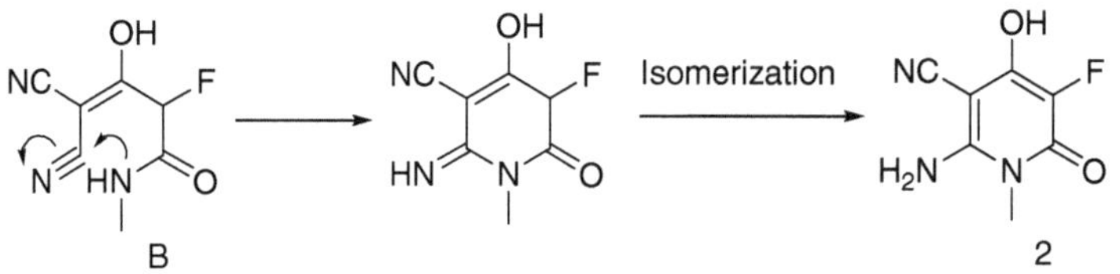

4.6.6 Imidazole

Problem 77: Formation of 2,4-disubstituted Imidazole

Imidazole subunit exits in many pharmaceuticals and bioactive compounds. For example, daclatasvir is an anti-HCV medication [212a]. Although the synthesis of these 2,4-disubtituted imidazoles by heating α-hydroxy ketone ester with ammonium acetate was described in many medicinal chemistry articles and patents [212, 213], the mechanism for this reaction has not been discussed in the literature. A representative preparation is shown in the following reaction [212b]. Heating 2-oxo-2-phenylethyl pent-4-ynoate (**3**) in refluxing xylene with excess ammonium acetate provided the imidazole product **4** in 98% yield.

Please provide a mechanism for this reaction.

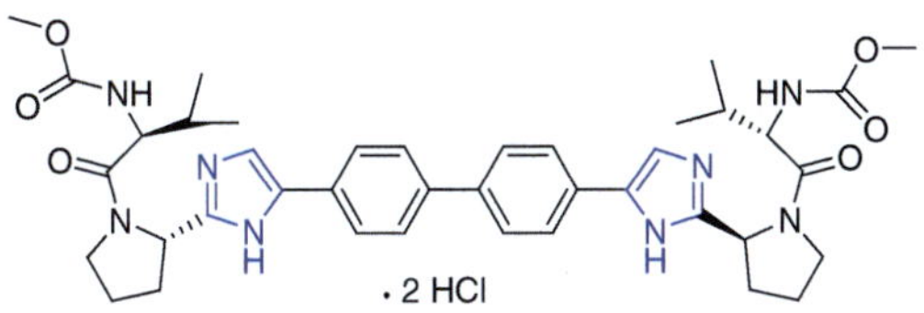

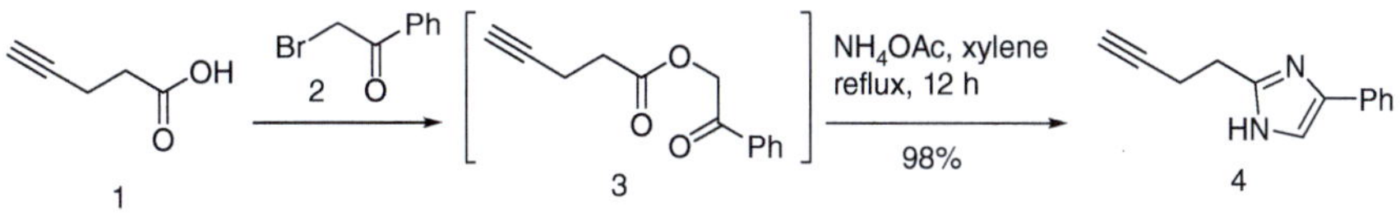

Experimental Procedure (from *Eur. J. Med. Chem.*, **2014**, *84*, 181)
Synthesis of 2-(but-3-yn-1-yl)-4-phenyl-1*H*-imidazole (**4**).

Step a. Cs_2CO_3 (0.5 eq, 5.10 mmol) was added to a solution of 4-pentynoic acid (**1**) (1 eq, 10.20 mmol) in EtOH/H_2O 1:1 (20 mL), and the whole stirred at room temperature for 1 h. The solvent was evaporated and the crude residue solubilized in dry DMF (12.5 mL); a solution of 2-bromoacetophenone (**2**, 1 eq, 10.20 mmol) in dry DMF (6.5 mL) was dropwise added and the whole stirred at room temperature for 15 min. The solvent was evaporated, the residue taken up with AcOEt and the inorganic salt was filtered off. The filtrate was dried over Na_2SO_4, and the solvent evaporated to afford a yellow oily residue.

Step b. The oily residue was solubilized in xylene (50 mL). NH_4OAc (15 eq, 153.00 mmol) was added and the resulting solution refluxed for 12 h using a Dean-Stark trap. The solution was cooled to room temperature, washed with H_2O and the organic phase evaporated to give a residue which was taken up with AcOEt. The organic phase was washed with saturated aqueous $NaHCO_3$ solution, brine, dried with Na_2SO_4 and evaporated to give 2-(but-3-yn-1-yl)-4-phenyl-1H-imidazole (**4**) as crude residue purified by FC (petroleum ether/AcOEt 6:4). Orange solid (1.960 g, 98%); R_f 0.10 (petroleum ether/AcOEt 6:4); mp 53–55 °C; ^1H NMR ($CDCl_3$) 7.67 (d, J = 7.4 Hz, 2H), 7.40–7.23 (m, 4H), 3.02 (t, J = 6.8 Hz, 2H), 2.63 (dt, J = 2.4, 6.8 Hz, 2H), 2.12 (t, J = 2.4 Hz, 1H).

The Mechanism

Ammonium acetate plays dual roles in this reaction, a catalyst and a nucleophile (NH_3). Ketone is activated by NH_4OAc, then NH_3 condenses with the carbonyl to form imine **A**, which is isomerized to enamine **B**. Then the NH_2 intramolecularly attacks the ester carbonyl to give the amide/enol **C**, which is isomerized to aldehyde **D**. The second condensation between NH_3 and the aldehyde group provides enamine **E**. Final cyclization and dehydration leads to product **4** (next page).

Problem 78: Synthesis of Substituted 2-Aryl-*N*-benzylbenzimidazoles

Benzimidazole and benzoxazole derivatives are pharmaceutical interesting compounds [214], such as veliparib [215], a PARP inhibitor, and prinaberel (ERB-041) [216], an estrogen receptor. Punniyamurthy's group reported a synthesis of substituted 2-aryl-*N*-benzylbenzimidazoles (**2**) from *N*-benzyl bisarylhydrazones (**1**) in the presence of $Cu(OTf)_2$ [217]. Similarly, 2-arylbenzoxazoles (**4**) were prepared from bisaryloxime ethers (**3**).

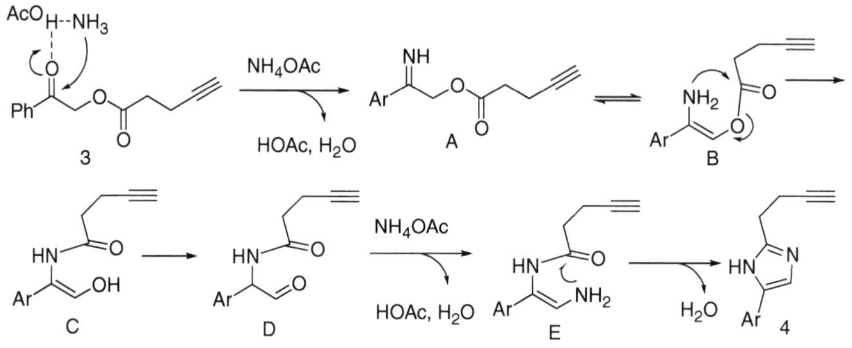

Please propose a mechanism for these reactions.

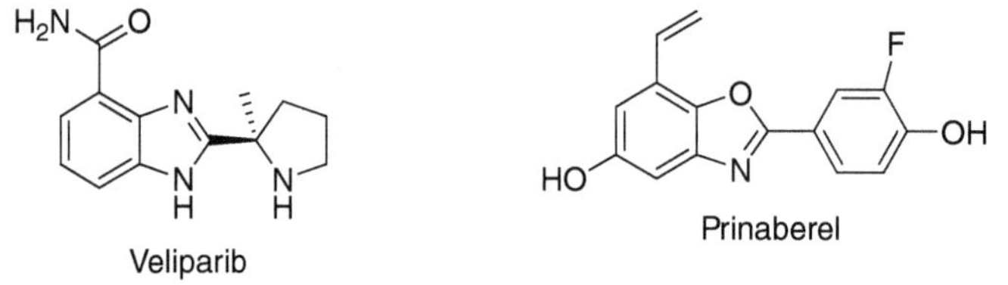

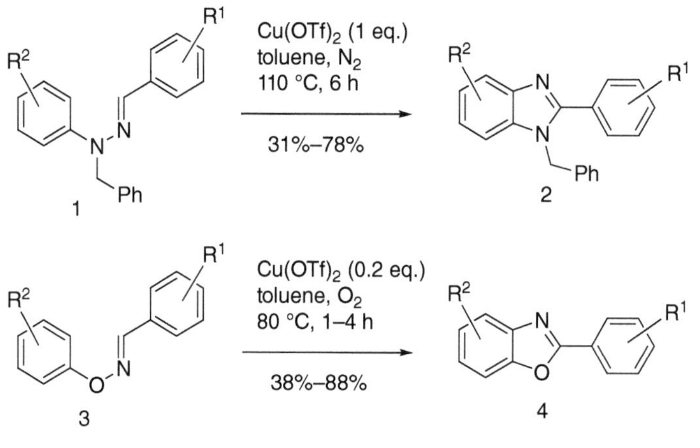

Veliparib

Prinaberel

Experimental Procedure (from *J. Org. Chem.*, **2011**, *76*, 5295)

General Procedure for the Cyclization of N-Benzyl Bisarylhydrazones 1a-y. Substrates **1a-y** (0.5 mmol) and $Cu(OTf)_2$ (0.5 mmol) were stirred at 110 °C in toluene (2 mL) under a nitrogen balloon (Table 2). Progress of the reaction was monitored by TLC using ethyl acetate and hexane as eluent. The reaction mixture was then cooled to room temperature and poured into 30% aqueous NH_4OH (5 mL). The solution was extracted with EtOAc (3 × 15 mL). Drying (Na_2SO_4) and evaporation of the solvent gave a residue that was purified on silica gel column chromatography using ethyl acetate and hexane as eluent to afford the desired *N*-benzyl 2-arylbenzimidazoles **2a-x** in analytically pure form.

1-Benzyl-2-phenyl-1H-benzo[d]imidazole 2a ($R^1 = R^2 = H$). Analytical TLC on silica gel, 1:9 ethyl acetate/hexane $R_f = 0.20$; colorless solid; yield 64%; mp 123–124 °C (lit.[30c] mp 132–133 °C); ^{1}H NMR (400 MHz,$CDCl_3$) δ 7.91 (d, $J = 8.0$ Hz, 1H), 7.70 (d, $J = 7.6$ Hz, 2H), 7.47 (d, $J = 7.2$ Hz, 3H), 7.36–7.30 (m, 4H), 7.26–7.21 (m, 2H), 7.12 (d, $J = 6.8$ Hz, 2H), 5.47 (s, 2H); ^{13}C NMR (100 MHz, $CDCl_3$) δ 154.3, 143.0, 136.5, 136.1, 132.7, 130.2, 129.5, 129.2, 128.9, 128.0, 126.1, 123.3, 123.0, 120.1, 110.7, 48.5; FTIR (KBr) 3060, 3029, 2962, 2928, 1706, 1600, 1508, 1492, 1469, 1449, 1441, 1392, 1361, 1330, 1314, 1261, 1177, 1163, 1097, 1028, 1002 cm^{-1}; (ESI-MS) m/z 285.14 $[M + H]^+$. Anal. Calcd for $C_{20}H_{16}N_2$: C, 84.48; H, 5.67; N, 9.85. Found: C, 84.60; H, 5.65; N, 9.75.

General Procedure for the Cyclization of Bisaryloxime Ethers 3a-z. Substrates **3a-z** (0.5 mmol) and $Cu(OTf)_2$ (20 mol %) were stirred at 80 °C in toluene (2 mL) under an oxygen balloon (Table 3). Progress of the reaction was monitored by TLC using ethyl acetate and hexane as eluent. The reaction mixture was then cooled to room temperature and passed through a short pad of silica gel using hexane followed by a mixture of ethyl acetate and hexane as eluent to afford the titled compounds **4a-z** in analytically pure form.

2-(2-Bromophenyl)benzo[d]oxazole 4b (R^1 = 2-Br, R^2 = H). Analytical TLC on silica gel, 1:19 ethyl acetate/hexane $R_f = 0.35$; colorless solid; yield 60%; mp 54–55 °C; ^{1}H NMR (400 MHz, $CDCl_3$) δ 8.07 (d, $J = 7.6$ Hz, 1H), 7.85–7.83 (m, 1H), 7.77 (d, $J = 8.0$ Hz, 1H), 7.62–7.60 (m, 1H), 7.47–7.33 (m, 4H); ^{13}C NMR (100 MHz, $CDCl_3$) δ 161.6, 150.7, 141.7, 134.8, 132.3, 132.1, 127.7, 127.6, 125.7, 124.8, 122.0, 120.6, 110.9; FTIR (KBr) 2925, 1606, 1565, 1530, 1469, 1450, 1427, 1307, 1262, 1251, 1235, 1081, 1016 cm^{-1}; (ESI-MS)m/z 273.99, 275.99 $[M + H]^+$. Anal. Calcd for $C_{13}H_8BrNO$: C, 56.96; H, 2.94; N, 5.11. Found: C, 56.91; H, 2.96; N, 5.14.

The Mechanism

The imine/oxime nitrogen in the substrate **1/3** chelates with the Lewis acid catalyst $Cu(OTf)_2$ and then the aniline/cyclohexa-2,4-dien-1-one attacks the catalyst ($Cu(OTf)_2$) to form the organocopper coordination complex **A**. Cyclization of **A** provides the six-membered organocopper compound **B**. After reductive elimination of copper metal, the product **2/4** is generated. The copper(0) species may undergo reaction with trifluoromethanesulfonic acid to regenerate the catalyst. In the case of the reaction **3** to **4**, the oxygen may accelerate the oxidation of the Cu(0) to $Cu(OTf)_2$.

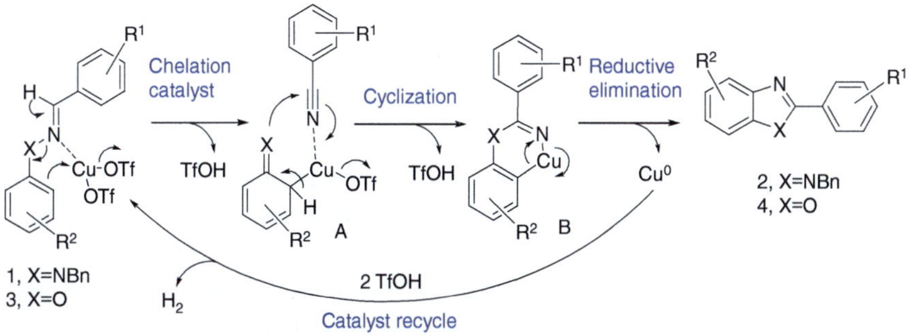

Problem 79: Preparation of 1-Isopropenyl-1,3-dihydro-2H-benzo[d]imidazol-2-one

The reaction between benzene-1,2-diamine (**1**) and ethyl acetoacetate (**2**) was initially documented by Sexton in 1942 [218]. However, due to the limitations of analytical techniques available at the time, the structure of the resulting product was mistakenly identified as 1-(1*H*-benzo[d]imidazol-2-yl)propan-2-one (**4**). It was not until 1960 that Davoll correctly determined the structure of the product to be 1-(prop-1-en-2-yl)-1,3-dihydro-2H-benzo[d]imidazol-2-one (**3**). This conclusion was reached by comparing the hydrogenation product (**5**) with an authentic sample synthesized from *N*-isopropylbenzene-1,2-diamine (**6**) and phosgene [219]. In 1968, Israel, Jones, and Modest further explored the mechanistic details of this reaction [220]. Over time, this "classic" reaction has found renewed relevance in medicinal chemistry [221, 222], particularly in the synthesis of flibanserin [223, 224], a drug used to treat hypoactive sexual desire disorder in premenopausal women. The one-pot synthesis of compound **3** is illustrated as follows [224].

Please provide your own mechanism for the formation of **3** from **1** and **2**.

Experimental Procedure (from *Org. Proc. Res. Dev.*, **2016**, *20*, 1576)

1-(Prop-1-en-2-yl)-1H-benzo[d]imidazol-2(3H)-one (3). KOH solution in ethanol (1.8 g, 9.2 mL) and subsequently ethyl acetoacetate (**2**, 132.3 g, 1.017 mol) in xylene (50 mL) was added dropwise within 2.5 h to a refluxing solution of **1** (100 g, 0.924 mol) in xylene (400 mL) under nitrogen, while the water produced during the reflux was removed using a Dean-Stark trap. It was refluxed for another 5 h and then cooled to 40 °C. An aqueous solution of KOH (71.8 g, 250 mL) was added, and the mixture was stirred at 40 °C for 0.5 h. The organic layer was separated, and acetic acid (111 mL) was added slowly. The resulting slurry was filtered, washed with *n*-heptane, and dried under vacuum to give compound **3** (89.7 g) as a pale brown solid with yield of 56%. ^{1}H NMR (400 MHz, $CDCl_3$) δ 10.57 (s, 1H), 7.18–7.12 (m, 1H), 7.12–7.04 (m, 3H), 5.42 (d, $J = 1.5$ Hz, 1H), 5.26 (s, 1H), 2.25 (s, 3H). ESI-MS (m/z): 175.4 $[M + H]^+$, 173.3 $[M - H]^-$; HPLC: retention time of 12.5 min, 99.5% purity.

The Mechanism

The first stage of the reaction is the normal condensation between the amino and the carbonyl followed by intramolecular amidation to give 4-methyl-1,3-dihydro-2*H*-benzo[b][1,4]diazepin-2-one (**3a**). This was converted to the product **3** via a [1,3]-sigma tropic rearrangement at high temperature.

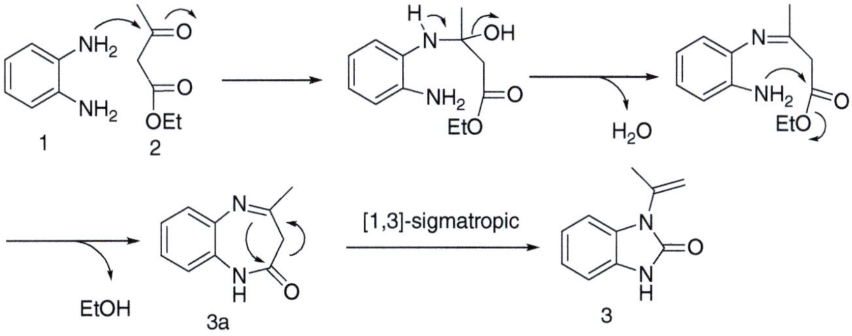

4.6.7 Isoxazole

Problem 80: Formation of 5-Chloro-3-phenylbenzo[c]isoxazole

In 1960, Davis and Pizzini reported the preparation of 3-phenylbenzo[c]isoxazoles from nitrobenzenes and phenylacetonitriles in the presence of KOH [225]. The product **3** was a key intermediate for synthesizing the drug Diazepam [226]. Benzo[c]isoxazole is the pharmacophore of a series of IP6K inhibitors for the treatment of obesity and obesity-induced metabolic dysfunctions [227]. Some SF_5 or CF_3 group containing analogs were also prepared by this reaction [228].

Please propose a mechanism for this reaction.

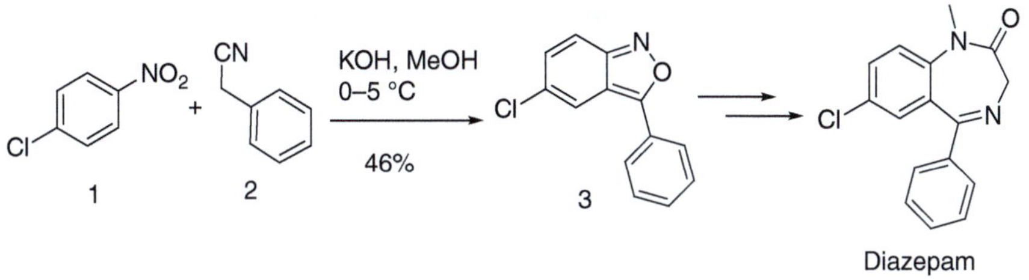

Experimental Procedure (from *Beil. J. Org. Chem.*, **2013**, 9, 411, Supporting Information)

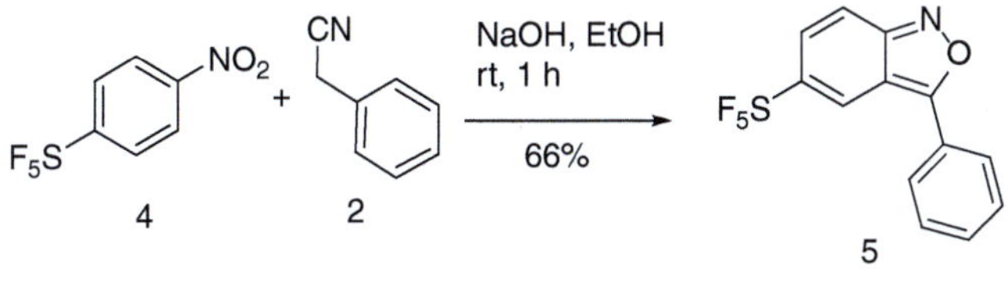

General Procedure for the Synthesis of Benzisoxazoles. Powdered NaOH (0.8 g, 20 mmol, 10 equiv) was stirred with ethanol (10 mL) at room temperature for 20 min. A mixture of **4** (500 mg, 2 mmol) and arylacetonitrile **2** (3–4 mmol, 1.5–2 equiv) was added, and the mixture was stirred in a closed reaction flask at room temperature for the given time (1–2 h) and then poured into water (70 mL). The crude product was extracted into EtOAc (3 × 30 mL). The combined organic phase was washed with saturated NH_4Cl (20 mL), dried and the solvent was removed under reduced pressure. Flash chromatography using silica gel (EtOAc–PE) provided pure products.

3-Phenyl-5-(pentafluorosulfanyl)benzo[c]isoxazole (5). A white solid (66% yield); mp 129–130 °C; R_f 0.40 (EtOAc–PE, 4:96); IR (film) ν_{max} 3128, 3090, 3050, 1631, 1547, 1521, 1495, 1466, 1447, 1368, 1144, 1064, 935, 839; ^{1}H NMR (500 MHz, $CDCl_3$) δ_H 7.56–7.64 (m, 3H), 7.65 (dd, 1H, $^{4}J_{HH} = 9.8$ Hz, $^{4}J_{HH} = 1.8$ Hz), 7.67 (br d, 1H, $^{3}J_{HH} = 9.8$ Hz), 7.98 (m, 2H), 8.31 (br dd, 1H, $^{4}J_{HH} = 1.8$ Hz, $^{4}J_{HH} = 0.8$ Hz); ^{13}C NMR (125.7 MHz, $CDCl_3$) δ_C 112.4, 116.3, 121.3 (quint., $^{3}J_{CF} = 5.3$ Hz), 127.0, 127.2, 127.5 (quint., $^{3}J_{CF} = 4.1$ Hz), 129.6, 131.5, 150.4 (quint., $^{2}J_{CF} = 18.1$ Hz), 156.4, 168.6; ^{19}F NMR (470.3 MHz, $CDCl_3$) δ_F 62.8 (d, 4F, $^{2}J_{FF} =$ 150.4 Hz), 83.6 (quint., 1F, $^{2}J_{FF} =$ 150.4 Hz); MS (EI) m/z (rel. int.) 322 (32), 321 (100) $[M]^+$, 213 (55), 194 (24), 185 (35), 184 (21), 166 (55), 140 (21), 139 (32), 105 (18), 77 (70), 51 (19); HRMS (EI) m/z calcd for $C_{13}H_8F_5NOS$ $[M]^+$ 321.0247, found 321.0244.

The Mechanism

The reaction necessitates one equivalent of KOH, leading to the elimination of one equivalent of potassium cyanide (KCN) and two equivalents of H_2O. The mechanism of this transformation was elucidated by Davis and Pizzini [225]. According to their interpretation, the process initiates with the condensation of nitrobenzene (**1**) and phenylacetonitrile (**2**), wherein the phenylacetonitrile anion attacks the ortho position relative to the nitro group, generating intermediate **I**. Following this, water is eliminated under basic conditions, producing the nitroso intermediate **II**. Deprotonation of **II** yields the potassium salt **III**, which can subsequently rearrange to form **IV**. Finally, ring closure, potentially via salt **V**, results in the formation of the product **3**.

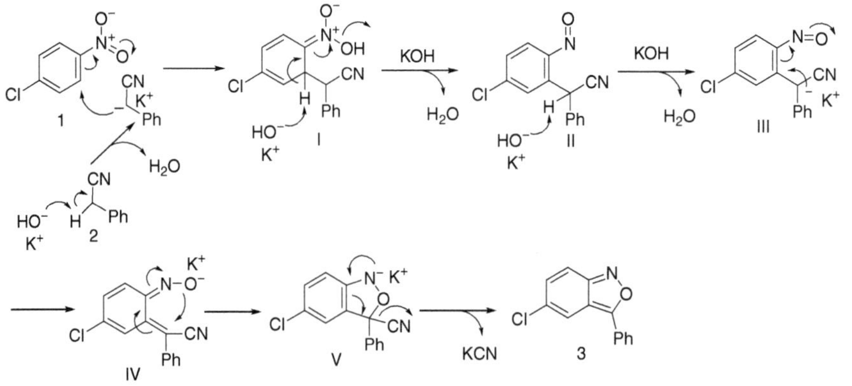

The reaction can be facilitated by potassium methoxide, although this is not the critical factor. I believe the process might follow a different sequence. Specifically, I propose that after the deprotonated phenylacetonitrile reacts with nitrobenzene, the intermediate **I** is likely the potassium salt **A**, as illustrated below. Under strong basic conditions, the salt **A** has little opportunity to be protonated to form intermediate **I**. The sodium salt of phenylnitromethane was previously synthesized in ethanol through the reaction of phenylnitromethane with sodium ethoxide [229]. However, the anionic oxygen serves as a potent nucleophile, displacing the CN group (with the elimination of KCN) to generate the cyclized intermediate **B**. Following aromatization and a hydrogen shift, the neutral species **C** is formed. Finally, the base-catalyzed elimination of one molecule of H_2O yields the product **3**.

$$MeOH + KOH \rightleftharpoons MeOK + H_2O$$

Problem 81: Preparation of Isoxazoles from Nitro Compounds and Oxetanes
Isoxazoles has been used as a functional moiety in the synthesis of polo-like kinase 1 (Plk1) binding inhibitors [230]. Carreira's group reported a facile synthesis of isoxazoles by base-mediated rearrangement of substituted oxetanes [231].

Please propose a mechanism for this reaction.

1 + O_2N–R (2) → 1. Et_3N (0.2 eq.) 2. MsCl (1.1 eq.) Et_3N (2.0 eq.) 3. DIPEA (1.0 eq.) → 3, 58–85%

Experimental Procedure (from *Angew. Chem. Int. Ed.*, **2011**, *50*, 5379, Supporting Information)
General Procedure for the One-Pot Reaction Sequence. In a 10 mL oven-dried cone-shaped flask were mixed under Ar the nitro compound (0.75 mmol, 1.0 equiv), and oxetan-3-one (0.98 mmol, 1.3 equiv), and the mixture was cooled to 0 °C. Et_3N (0.15 mmol, 0.2 equiv) was added, and the reaction mixture was stirred at 0 °C for 10 min, warmed to room temperature and stirred for 90 min.

In most cases, formation of a solid was observed. The reaction mixture was diluted with THF (7.5 mL). Et_3N (1.5 mmol, 2.0 equiv) was added and the solution was cooled to −78 °C. MsCl (0.83 mmol, 1.1 equiv) was added dropwise and the mixture was stirred at 78 °C for 30 min, then allowed to slowly warm to room temperature over 60 min. iPr_2NEt (0.75 mmol, 1.0 equiv) was added and the mixture was stirred at room temperature for 12 h. It was diluted with CH_2Cl_2 (15 mL), quenched with H_2O (5 mL), and the aqueous phase was extracted with CH_2Cl_2 (3 × 5 mL). The combined organic phases were washed with brine (10 mL), dried (Na_2SO_4), filtered, and concentrated *in vacuo*. The pure product was obtained after purification by flash column chromatography on silica gel.

3-Benzylisoxazole-4-carbaldehyde (3a). Following the general procedure using (2-nitroethyl)benzene (113 mg, 0.75 mmol, 1.0 equiv), oxetan-3-one (70 mg, 0.98 mmol, 1.3 equiv), Et_3N (21 μL, 0.15 mmol, 0.2 equiv), then Et_3N (0.21 mL, 1.5 mmol, 2.0 equiv), MsCl (64 μL, 0.83 mmol, 1.1 equiv), and iPr_2NEt (130 μL, 0.75 mmol, 1.0 equiv). The crude product was purified by FC (SiO_2; hexanes: EtOAc 8:1) to afford the pure title compound. Yield: 112 mg (0.60 mmol, 80%). Colorless oil. TLC: $R_f = 0.34$ (hexanes: EtOAc 4:1; UV, DNP); ^{1}H-NMR (400 MHz, $CDCl_3$): $\delta = 9.92$ (s, 1H), 8.91 (s, 1H), 7.52–6.91 (m, 5H), 4.29 (s, 2H); ^{13}C-NMR (101 MHz, $CDCl_3$): $\delta = 182.7$, 165.5, 160.5, 136.2, 129.4, 129.0, 127.4, 121.4, 31.4; IR (neat): 3099, 3039, 1689, 1575, 1455, 1421, 1392, 1129, 874, 744, 720, 694, 671 cm^{-1}; HRMS (EI): exact mass calculated for $C_{11}H_9NO_2$ (M^+), 187.0628; found 187.0629.

The Mechanism

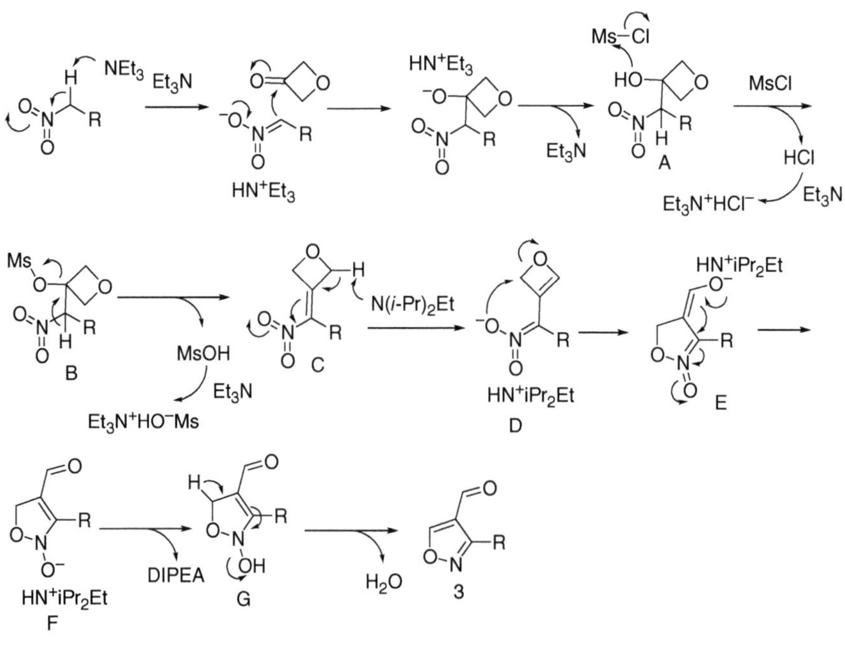

The first step is a Henry reaction between nitroalkane (**2**) and the oxetan-3-one (**1**) to give the β-nitroalcohol **A**. The next elimination is promoted by adding MsCl, via formation of the mesylate **B**, to afford the α,β-unsaturated nitro compound **C**. In the presence of base, DIPEA, **C** is isomerized to **D**, which is readily cyclized to isoxazole **E**/**F**. Finally, one molecule of H_2O is eliminated to yield the product **3**.

4.6.8 Thiazole

Problem 82: Preparation of Thiazole by Thiourea

Compound **4** is a TRPV1 inhibitor developed by Johnson & Johnson. The key process to build-up the thiazolo[5,4-d]pyrimidine core is shown by the following reaction [232].

Please suggest the mechanism for this reaction.

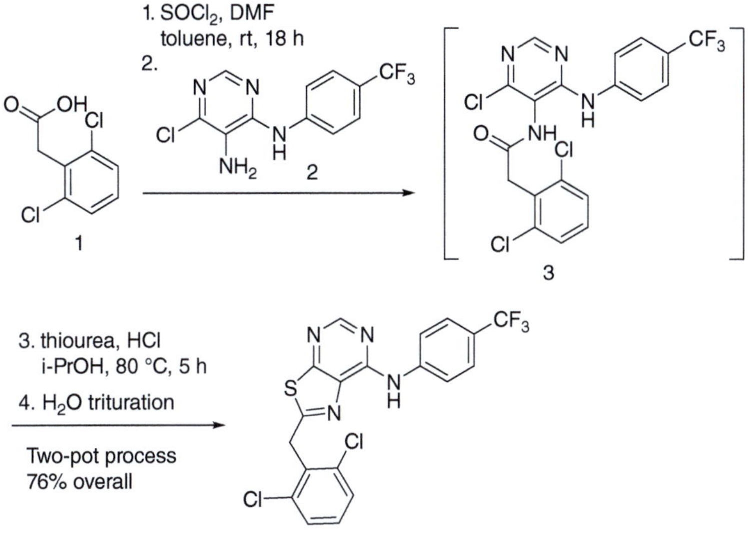

Experimental Procedure (from *Org. Process Res. Dev.*, **2011**, *15*, 382)

2-(2,6-Dichlorobenzyl)-*N*-(4-(trifluoromethyl)phenyl)thiazolo[5,4-d]pyrimidin-7-amine (4). 2,6-Dichlorophenylaceticacid (**1**, 62.5 g, 0.305 mol, 1.17 equiv) and DMF (1.60 mL) were dissolved in toluene (160 mL). Thionyl chloride (26.6 mL, 0.36 mol, 1.38 equiv) was added slowly over 10 min. The reaction mixture was stirred at 20 °C for 18 h, then added over 1.5 h to a solution of 6-chloro-N^4-(4-trifluoromethyl-phenyl)pyrimidine-4,5-diamine hemihydrochloride (**2**, 80.0 g, 0.261 mol, 1.00 equiv) in DMA (80 mL). The reaction was stirred at 20 °C for 2 h. Thiourea (31.6 g, 0.416 mol, 1.60 equiv) was added in one batch, followed by HCl in 2-propanol (5–6 M, 504 mL, ~2.77 mol, ~10.6 equiv) over 1 h via an addition funnel. The resulting solution was heated to 80 °C for 4 h. Water (340 mL) was added over 20 min and the resulting suspension was cooled to room temperature. The precipitate was collected by filtration and washed with ethanol (500 mL) to afford crude 4 (90 g, 76%) as a light-pink solid.

Final Purification of 4. A suspension of crude 2-(2,6-dichlorobenzyl)-*N*-(4-(trifluoromethyl)phenyl)thiazolo[5,4-d]pyrimidin-7-amine (**4**, 98.0 g, 0.215 mol) in acetone (1.55 L) was heated to reflux and filtered. The filtrate was heated to reflux again. Water (620 mL) was added over 5 min, and the suspension was cooled to 20 °C over 4 h. The solid was collected by filtration and dried in a vacuum oven at 50 °C overnight to give **4** as a white crystalline solid (91.5 g, 93% yield). ^{1}H NMR (400 MHz, d_6-DMSO, δ): 10.43 (s, 1H), 8.58 (s, 1H), 8.19 (d, J = 8.4 Hz, 2H), 7.71 (d, J = 8.5 Hz, 2H), 7.63 (d, J = 8.1 Hz, 2H), 7.48 (t, J = 8.1 Hz, 1H), 4.83 (s, 2H); ^{13}C NMR (126 MHz, d_6-DMSO, δ): 165.7, 162.8, 153.5, 152.4, 142.7, 135.4, 132.4, 130.8, 130.8, 128.9, 125.6 (q, J = 3.7 Hz), 124.4 (q, J = 271.2 Hz), 123.0 (q, J = 37.0 Hz), 121.1, 36.0; MS (ESI$^+$): calculated for $C_{19}H_{12}Cl_2F_3N_4S$ $[M + H]^+$, 455.0; m/z found, 455.1; HPLC retention time: 6.076 min; Anal. Calculated for $C_{19}H_{11}Cl_2F_3N_4S$: C, 50.12; H, 2.44; N, 12.31; found: C, 49.92; H, 2.27; N, 12.23.

The Mechanism

At first, the pyrimidine in **3** is activated by HCl and then attacked by thiourea to give the adduct **A**. Elimination of HCl from **A** provides the SN_{Ar} intermediate **B**. Subsequently, isopropanol as a nucleophile attacks the carbamimidothioate in **C** to release the thiopyridine **C** and isopropyl carbamimidate, which should be the HCl salt. Cyclization of the thiol with the amide moiety affords **D**. Finally, water is eliminated to yield the thiazole product **4**.

Addition thiourea; 3 → A; − HCl → B (i-PrOH); → C + i-Pr–O–C(=NH)NH$_2$; → D; − H_2O → 4

Another example for this kind of thiazole formation is described in a patent application [233].

Thiourea, HCO_2H, EtOH, 85 °C, 15 h; 78%

Problem 83: Preparation of Thiazole from Sulfoxide During Ceftibuten Synthesis

In 1970, Cooper and Jose reported a novel rearrangement of penicillin V sulfoxide (**1**) to a thiazole product **2** [234]. Treatment of **1** with trimethyl phosphite in refluxing benzene for 30 hours provided **2** in high yield (>80%). The mechanism of this reaction was also briefly discussed. This reaction has been applied to the synthesis of ceftibuten [235], seco-isopenicillin N [236], intermediates of cefaclor [237] and cefovecin [238].

Please suggest a mechanism for this reaction.

Experimental Procedure (from CN111763221 A, **2020***)

4-Nitrobenzyl (*R*)-2-((1*R*,5*R*)-3-benzyl-7-oxo-4-thia-2,6-diazabicyclo[3.2.0]hept-2-en-6-yl)-3-methylbut-3-enoate (6b). To the reaction vessel was added anhydrous toluene (450 kg) followed by **5b** (100 kg). The addition port was rinsed with anhydrous toluene (20 kg). The mixture was heated to 90–100 °C and stirred to form a clear solution. Trimethyl phosphite (51.1 kg) was added. The reaction temperature was controlled between 90 and 100 °C till the reaction was completed. The mixture was cooled to 30 °C under the atmosphere of nitrogen. Then triethylamine (27 kg) was added while the reaction temperature was controlled below 30 °C. The mixture was stirred by avoiding light till reaction completed. Aqueous HCl (5%, 200 kg) was added and the mixture was stirred for 5 minutes. The mixture was left to stand until layers formed. The layers were separated. The aqueous layer was extracted with DCM (250 kg). Aqueous was discarded. The combined organic solutions were washed with aqueous NaCl (12.5%, 150 kg). The solvents were evaporated under reduced pressure to give the crude product. This was crystallized from methanol, after being dried, to yield the product, 74.4 kg, yield 80%, theoretical yield 92.99 kg.

** Translated from Chinese by the author of this book.*

The Mechanism

The ring of sulfoxide **1** is opened to the sulfenic acid **A** through a reversible rearrangement. Then sulfenic acid **A** is reduced by trimethyl phosphite to the thiol **D** via **B** and **C**. The final stage is a condensation of the thiol with the amido side chain, eliminating one molecule of water, to give the thiazoline ring **2**.

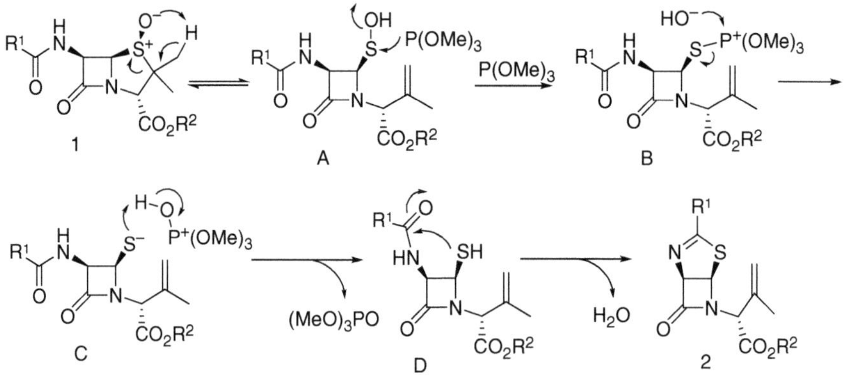

4.6.9 Pyrazole

Problem 84: Synthesis of Trifluoromethylpyrazoles

A regioselective synthesis of trifluoromethylpyrazoles was developed by Zhu and Jiang's group [239]. As shown below, 5-phenyl-3-(trifluoromethyl)-1H-pyrazole (**3**) was prepared from benzaldehyde (**1**), 4-methylbenzenesulfonohydrazide, and 2-bromo-3,3,3-trifluoroprop-1-ene (BTP) (**2**) in 88% yield. Notably, the reactions were successfully carried out on a 100 mmol scale, efficiently yielding the key intermediates necessary for the synthesis of celecoxib [240], SC-560 [241], and AS-136A [241].

Please suggest a mechanism for this reaction.

Ph–CHO + TsNHNH$_2$ + 2 (Br, CF$_3$) → DBU, toluene, 60 °C, 6 h, 88% → 3

Celecoxib
COX-2 inhibitor

SC-560
COX-2 inhibitor

AS-136A
measles virus inhibitor

Experimental Procedure (from *Org. Lett.*, **2020**, *22*, 809, Supporting Information)
General Procedure for the Synthesis of 3-Trifluoromethylpyrazole. To a 25 mL round bottom flask equipped with a magnetic stirring bar and a refluxing condenser was added aldehyde **1** (1 mmol), 4-methylbenzenesulfonohydrazide (1.2 mmol), toluene (6 mL), 1,8-diazabicyclo[5.4.0]undec-7-ene (DBU) (3 mmol), and 2-bromo-3,3,3-trifluoropropene (**2**, 2 mmol). The resulting mixture was vigorously stirred at 60 °C for 6 h. Then water was added to the mixture (15 mL) and extracted with EtOAc (15 mL × 3). The combined organic phases were dried over anhydrous Na_2SO_4, filtered and concentrated *in vacuo*. The obtained crude product was then purified by flash column chromatography on silica gel (eluting with petroleum ether/ethyl acetate = 6/1) to provide the product **3**.

5-Phenyl-3-(Trifluoromethyl)-1*H*-Pyrazole (3a). 186.6 mg, 88% yield; yellow solid, mp: 112–113 °C; ^{1}H NMR (400 MHz, $CDCl_3$) δ 13.25 (brs, 1H), 7.50–7.52 (m, 2H), 7.33–7.39 (m, 3H), 6.60 (s, 1H); ^{13}C NMR (100 MHz, $CDCl_3$) δ 145.3, 143.4 (q, $^2J_{F-C}$ = 38.0 Hz), 129.4, 129.2, 127.8, 125.5, 121.1 (q, $^1J_{F-C}$ = 267.2 Hz), 100.8; ^{19}F NMR (376 MHz, $CDCl_3$) δ – 62.0 (s, 3F); IR (KBr): 3127, 2881, 1592, 1498, 1437, 1271, 1158 cm^{-1}; HRMS (ESI, m/z): $[M + H]^+$ Calcd. for $C_{10}H_7F_3N_2$ + H, 213.0634; found, 213.0635.

The Mechanism

The condensation of benzaldehyde **1** with *p*-toluenesulfonyl hydrazide generates *N*-tosylhydrazone **A**, which then decomposes to diazo compound **B** in the presence of bases. The regioselective [3 + 2] cycloaddition of diazo compound **B** with BTP

H$_2$NNHTs, 1, H_2O, A, DBU, DBUH$^+$, Tol, Tol–SO$_2^-$ DBUH$^+$, B, 2, [3+2], C, HBr, D, H-shift, 3

(**2**) affords intermediate **C**, which is further converted to trifluoromethylated diazo intermediate **D** via the elimination of HBr in the presence of base. Finally, **D** rearranges to the conjugated pyrazole product **3**.

Problem 85: Preparation of Pyrazole by Sulfur Extrusion

In 1977, Bulka's group reported preparation of 5-aminopyrazoles (**2**) via sulfur extrusion from 2,3-dihydro-6H-1,3,4-thiazine (**1**) in high yield [242]. That method was applied to the synthesis of a CB_1 antagonist CE-178,253 [243], and to the synthesis of p38 mitogen-activated protein kinase inhibitors [244, 245].

Please suggest a mechanism for this reaction from compounds **3** and **4** to **5**.

EtONa
EtOH, reflux
10 examples
yield 81–96%

1. EtOH, 0 °C
2. 48% HBr, 80 °C
70%

CE-178,253–26

KH_2PO_4, EtOH-H_2O
40 °C, 22 h

8 (30.6 kg)

9.HBr (16.7 kg)
p38 MAPK inhibitor

Experimental Procedure (from *Tetrahedron*, **2009**, *65*, 3292)

3-(2-Chlorophenyl)-4-(4-chlorophenyl)-1H-pyrazol-5-amine (5). Bromoketone **3** (3.41 g, 9.95 mmol) and 4-benzhydrylthiosemicarbazide (**4**) (2.58 g, 10.0 mmol) were added to EtOH (120 mL) at 0 °C. The suspension was allowed to warm to room temperature overnight. HBr (2 mL of a 48% aqueous solution) was added and the reaction mixture was heated to 80 °C for 4 h. HCl (10 mL of a 6 N solution) was added and heating was continued for 16 h. The mixture was cooled to room temperature and concentrated. The residue was partitioned between EtOAc (100 mL) and 2 N NaOH (50 mL). The mixture was filtered and the layers were separated. The organic layer was concentrated and the crude product purified by chromatography on silica gel (4:1 EtOAc/hexanes) to afford 3-(2-chlorophenyl)-4-(4-chlorophenyl)-1Hpyrazol-5-amine (**5**) as a slightly tan solid (2.10 g, 70%), mp 156–157 °C. IR (thin film) cm^{-1} 3425, 1603, 1563, 1347, 1175, 1125, 865, 780, 630. ^{1}H NMR (400 MHz, $CDCl_3$): δ 7.40 (d, J = 8.3 Hz, 2H), 7.22 (d, J = 8.7 Hz, 2H), 7.16–7.30 (m, 5H), 7.09 (d, J = 9 Hz, 2H). ^{13}C NMR (100 MHz, $CDCl_3$): δ 133.8,

133.1, 132.7, 132.6, 132.5, 130.7, 130.5, 130.2, 130.0, 129.9, 129.7, 129.2, 128.6, 128.3, 127.4. Anal. Calcd for $C_{15}H_{11}Cl_2N_3$: C, 59.23; H, 3.65; N, 13.81. Found: C, 58.98; H, 3.57; N, 13.61.

The Mechanism

The reaction sequence likely begins with the substitution of the bromine atom by a sulfur atom, yielding intermediate **A** and releasing one equivalent of HBr. This intermediate can be isolated as the HBr salt. The charge distribution in **A** makes the terminal NH_2 group of the hydrazine the most nucleophilic site, facilitating cyclization with the ketone to form aminal **B**. Subsequent dehydration of **B** generates the 5,6-diaryl-6*H*-1,3,4-thiadiazin-2-amine **C**, which may also exist in its tautomeric form **D**. The benzhydryl protecting group plays a crucial role by introducing steric effects that enhance regioselectivity. Upon further addition of HBr, sulfur extrusion from **C** and/or **D** via intermediate **E** leads to the formation of protected aminopyrazole **F**. Finally, the benzhydryl group is removed in the presence of aqueous HCl to yield the desired product **5**.

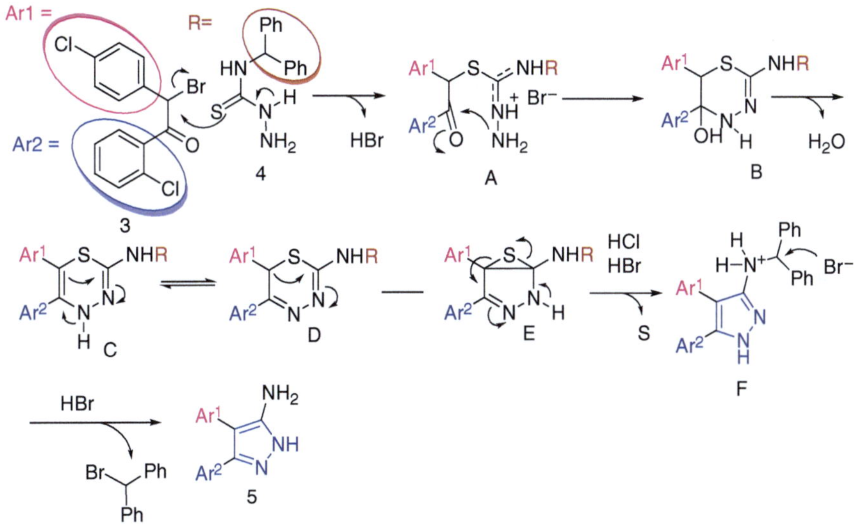

Other sulfur extrusion examples were also reported [246, 247].

Problem 86: Synthesis of 4-(3-Aminopropyl)-5-amino-1-methylpyrazole

FR259647 (**1**) shows superiority to current carbapenems in the antibacterial activity against *P. aeruginosa* and also strong resistance to β-lactamase. 4-(3-Aminopropyl)-5-amino-1-methylpyrazole (**2a**), the side chain of **1**, was synthesized from 3-cyanopyridine (**3**) via an intermediate, 1,4,5,6-tetrahydro-pyridine-3-carbonitrile (**4**), and *N*-methylhydrazine [248]. The ratio of the amino pyrazole products **2a**/**2b** influenced by the equivalents of HCl used. For example, 1 equivalent of HCl provided **2a**/**2b** in a 90:9.5 ratio, while 2 equivalents of HCl gave **2a**/**2b** in around 1:1.

Please provide a mechanism for the transformation from **4** to **2a**/**2b**.

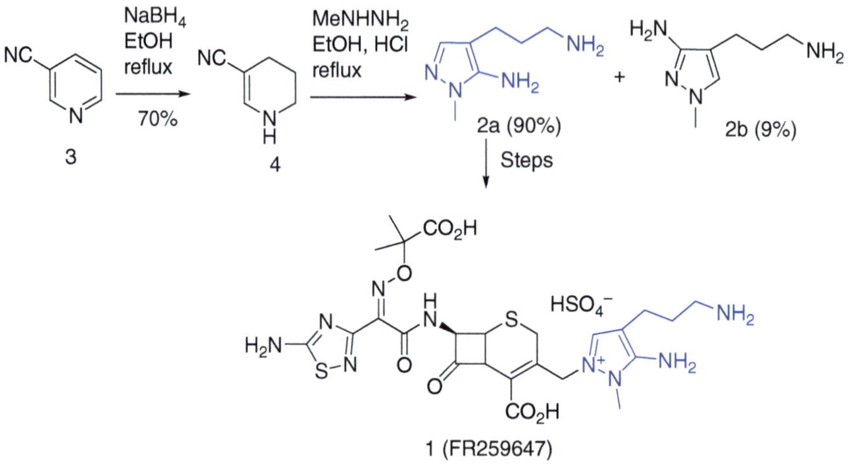

Experimental Procedure (from *Org. Process Res. Dev.*, **2006**, *10*, 159)

4-(3-Aminopropyl)-5-amino-1-methylpyrazole bihydrochloride (2a). To a mixture of compound **4** (5.0 g, 46.2 mmol) and ethyl alcohol (50 mL) were added *N*-methylhydrazine (2.34 g, 50.8 mmol) and 35% (w/w) hydrochloric acid (4.8 g, 46.2 mmol) at between 0 and 10 °C. The reaction mixture was refluxed for 69 h, then cooled below 10 °C, followed by addition of 35% (w/w) hydrochloric acid (7.22 g, 69.3 mmol) and evaporation at room temperature to dryness. The residue was treated with isopropyl alcohol (IPA) (30 mL) and subsequently concentrated to dryness under reduced pressure. The residue was suspended with isopropyl ether (IPE) (100 mL). The crystals were filtrated, washed with IPE (10 mL), and subsequently dried under reduced pressure to give compound **2a** (11.22 g, 106.8%). ^{1}H NMR (D_2O): δ 1.88–2.00 (2H, m), 2.49 (2H, t, $J = 7.4$ Hz), 3.02 (2H, t, $J = 7.5$ Hz), 3.73 (3H, s), 7.65 (1H, s). Side product derivative **2b** was estimated at 9 mol % by integration of proton at 7.55 ppm (1H, s). Mass (*e*/*z*): 155 (M + H$^+$). IR (cm^{-1}): 2880 (NH_2). Chloride ion: calcd 32.0%, found 29.35% (bis-hydrochloride).

4-(3-Aminopropyl)-3-amino-1-methylpyrazole bihydrochloride (2b). To a mixture of compound **4** (5.0 g, 46.2 mmol) and ethyl alcohol (50 mL) were added *N*-methylhydrazine (2.34 g, 50.8 mmol) and 35% (w/w) hydrochloric acid (9.6 g, 92.4 mmol) at between 0 and 10 °C. The reaction mixture was refluxed for 3 h, and solids appeared after 1 h. The precipitate was filtered at between 0 and 10 °C, washed with IPA (10 mL), and then dried under reduced pressure overnight to give compound **2b** (4.20 g, 40% yield). ^{1}H NMR (D_2O): δ 1.83–1.99 (2H, m), 2.51 (2H, t, $J = 7.5$ Hz), 3.02 (2H, t, $J = 7.6$ Hz), 3.77 (3H, s), 7.52 (1H, s). Mass (*e*/*z*): 155 (M + H$^+$). IR (cm^{-1}): 2800 (NH_2). Chloride ion: calcd 32.0%, found 31.19% (bis-hydrochloride).

The Mechanism

The primary amino moiety of *N*-methylhydrazine attacks the most electrophilic carbon of cyclic cyanoenamine **4**, and then the secondary nitrogen moiety spontaneously attacks the nitrile group followed by aromatization to preferably provide the thermodynamically stable pyrazole **2a**. While the secondary amino group of *N*-methylhydrazine attacks the carbon of cyclic cyanoenamine **4** will lead the minor isomer **2b**. Increasing the equivalents of HCl prompted the increased formation of regioisomer **2b**. It was assumed that excess acid accumulated around the primary amine of hydrazine, reducing its nucleophilicity and reaction selectivity.

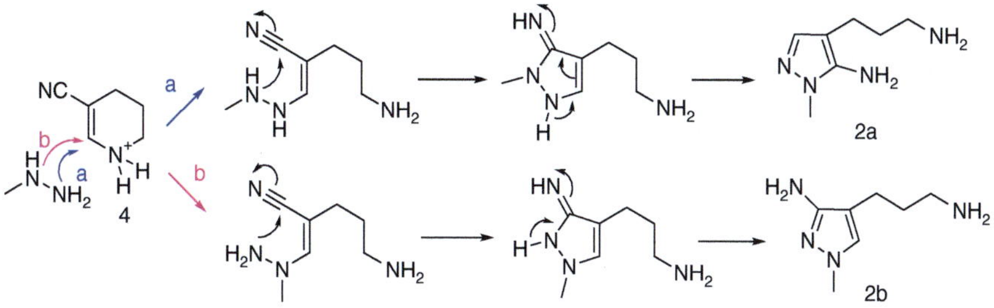

4.6.10 Pyrimidine

Problem 87: Formation of 5-Hydroxypyrimidin-4(3*H*)-one by 3,3-sigmatropic Rearrangement

In 1971, Heindel and Chun documented a thermal rearrangement of the adduct (**2**) formed between benzamide oxime (**1**) and DMAD. However, due to the limitations of analytical techniques available at the time, they incorrectly identified the product as the wrong isomer (**3**) [249]. This error was later rectified in 1979 by Culbertson, who correctly determined the structure to be methyl 5,6-dihydroxy-2-phenylpyrimidine-4-carboxylate (**4**) by comparing it with a synthetic standard of its decarboxylation derivative [250].

The 5-hydroxypyrimidin-4(3H)-one moiety is a significant structural motif in various medicinal chemistry projects [251–256]. A notable example is raltegravir potassium, marketed as Isentress (**7**), which became the first FDA-approved HIV integrase inhibitor in 2007. The highly functionalized hydroxypyrimidinone core (**6**) in Isentress (**7**) was synthesized through a thermal rearrangement of amidoxime-DMAD adducts (**5**) [251].

Please suggest a mechanism for the formation of **6** from **5**.

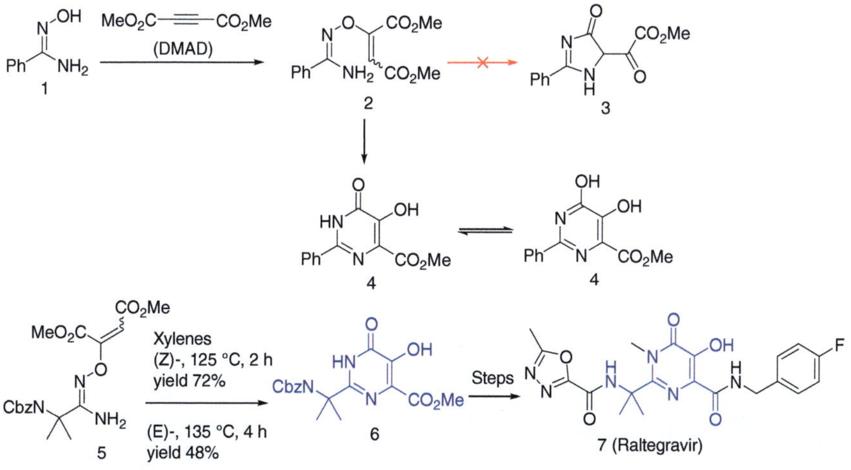

Experimental Procedure (from *Org. Process Res. Dev.*, **2011**, *15*, 73)

2-(1-Benzyloxycarbonylamino-1-methyl-ethyl)-5,6-dihydroxy-pyrimidine-4-carboxylic acid methyl ester (6). A slurry of amidoxime (63.1 kg, 251 mol) and methanol (180 kg) was cooled to 15 °C. Dimethyl acetylenedicarboxylate (38.8 kg, 273 mol) was added over 30 min, maintaining the batch temperature between 15 and 25 °C. The resultant solution was aged at 25 °C for 2 h to obtain >99% conversion and gave Michael adducts **5Z/5E** (ratio ~65:35 by ^{1}H NMR). For characterization purposes, a small sample was purified by flash chromatography (silica gel, EtOAc/hexanes) to give ***5Z*** and ***5E***. ***5Z*** as colorless needles, mp 76.5–77.0 °C. ^{1}H NMR ($CDCl_3$, 400 MHz) *δ*: 7.38–7.26 (m, 5H), 5.75 (s, 1H), 5.66 (br s, 3H), 5.08 (s, 2H), 3.82 (s, 3H), 3.71 (s, 3H), 1.54 (s, 6H). ^{13}C NMR ($CDCl_3$, 100 MHz) *δ*: 165.5, 163.2, 161.2, 155.6, 154.9, 136.0, 128.6 (2 C), 128.3, 128.1 (2 C), 102.3, 66.9, 54.3, 52.6, 51.5, 25.9 (2 C). HRMS (ESI) calculated for $C_{18}H_{23}N_3O_7$ $(M + Na)^+$ 416.1434, found 416.1438. ***5E*** as colorless needles, mp 65.0–66.1 °C. ^{1}H NMR ($CDCl_3$, 400 MHz) *δ*: 7.38–7.26 (m, 5H), 5.83 (s, 1 H), 5.40 (br s, 2 H), 5.08 (s, 1 H), 5.07 (s, 2 H), 3.88 (s, 3 H), 3.69 (s, 3 H), 1.56 (s, 6 H). ^{13}C NMR (CDC13, 100 MHz) *δ*: 166.9, 163.4, 161.3, 160.6, 155.6, 135.8, 128.6 (2 C), 128.3 (2 C), 128.1, 94.9, 67.0, 54.5, 52.8, 51.4, 26.0 (2 C). HRMS (ESI) calculated for $C_{18}H_{23}N_3O_7$ $(M + Na)^+$ 416.1434, found 416.1430.

The solution was concentrated to about 160 L under reduced pressure and solvent switched, by feed and bleed at constant volume, to xylenes. The final batch temperature was maintained below 70 °C during the solvent switch. The mixture was heated to 125 °C, aged for 2 h, warmed to 135 °C for 5 h to complete the reaction and then cooled to 60 °C. Methanol (45 kg) was added and the resultant slurry aged at 35 °C for 1 h. MTBE (145 kg) was added over 1 h at 20–25 °C. The slurry was cooled to −10 °C over 2 h and aged overnight. The slurry was filtered, the cake washed with 9:1 of MTBE: methanol (140 kg) and dried at 40 °C under reduced pressure to afford **7** (49.6 kg, >99.5 wt %, 98.4 LCAP%) in 54% overall yield, mp 186.3–187.0 °C. ^{1}H NMR ($CDCl_3$, 400 MHz) *δ*: 12.23 (br s, 1H), 10.75 (br s, 1H), 7.24 (m, 5H), 6.10 (br s, 1H), 4.98 (s, 2H), 3.98 (s, 3H) 1.69 (s, 6H). ^{13}C NMR (CDC13, 100 MHz) *δ*: 169.9, 159.4, 155.1, 153.5, 149.4, 136.2, 128.3 (2 C), 128.0, 127.9 (2 C), 126.1, 66.7, 55.6, 53.2, 26.4 (2 C). HRMS (ESI) calculated for $C_{17}H_{19}N_3O_6$ $(M + H)^+$ 362.1352, found 362.1352.

The Mechanism

This a thermal 3,3-sigmatropic rearrangement (Claisen rearrangement) followed by cyclization to form **6**.

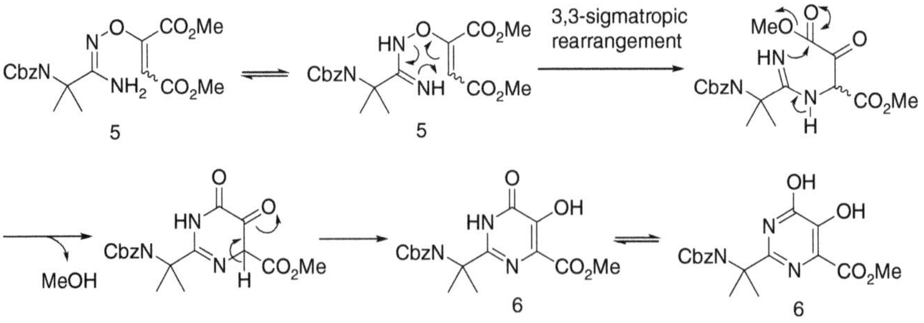

Subsequent regioselective *N*-methylation of **6** was achieved by addition of $Mg(OMe)_2$, which was supposed to chelate with the amide oxygen to form a five-membered ring. Chelation with the ester oxygen would form a six-membered ring and is, therefore, less favorable.

Mg(OMe)2
MeOH
Favorable
MeI
Aq. workup
8 (78%)
6
Mg(OMe)2
MeOH
Less favorable
MeI
Aq. workup
9 (22%)

Problem 88: Regioselective Chlorination of 1-Methylpyrimidine-2,4,6(1*H*,3*H*,5*H*)-trione

6-Chloro-3-methyluracil (**2**) is an important building block for constructing some pharmaceutical molecules [257]. For examples, alogliptin (**4**) [258] and trelagliptin (**5**) [259] are marketed medications for treatment of type 2 diabetes. Compound **2** was prepared by chlorination of 3-methylbarbituric acid (**1**) with $POCl_3$-H_2O. There was no report of regiomer, 4-chloro-3-methyluracil (**3**), isolated [258a, 260].

Why does this reaction show such a regioselectivity?

$POCl_3$-H_2O, 80 °C, 4 h, 80%
1
2 (80% yield)
3 (not reported)
4, R=H, alogliptin
5, R=F, trelagliptin

Experimental Procedure (from *J. Heterocycl. Chem.*, **2014**, *51*, 594)

6-Chloro-3-methyl-2,4(1H,3H)-pyrimidinedione (2). To a stirred mixture of the pyrimidinetrione **1** (213 g, 1.5 mol) in phosphorus oxychloride (600 mL, 6.4 mol) at 0 °C, water (10 mL, 0.55 mol) was added slowly, and the mixture was then stirred at 80 °C for 4 h. The excess phosphorus oxychloride was removed under reduced pressure, and the residue was poured into ice water. The resulting precipitate was collected by suction filtration and washed with 100 mL water. Gave **2** (195 g, 81%) as a white solid. ^1H NMR (DMSO-d_6, 300 MHz): δ 3.07 (s, 3H), 5.87 (s, 1H), 12.35 (s, 1H); ms: m/z 161.03 (M + 1).

The Mechanism

Under the reaction condition, 3-methylbarbituric acid (**1**) would enolyze to give **A** or **B**. Next, the enol **A** reacts with $POCl_3$ to provide the intermediate **C**, which could form a 6-member ring via the hydrogen bond. That makes the intermediate **C** favorable to be formed. While enol **B** reacts with $POCl_3$ to form intermediate **D**, the absence of a hydrogen bond in intermediate **D** makes it less favorable. Consequently, region isomer **2** was obtained as the major product.

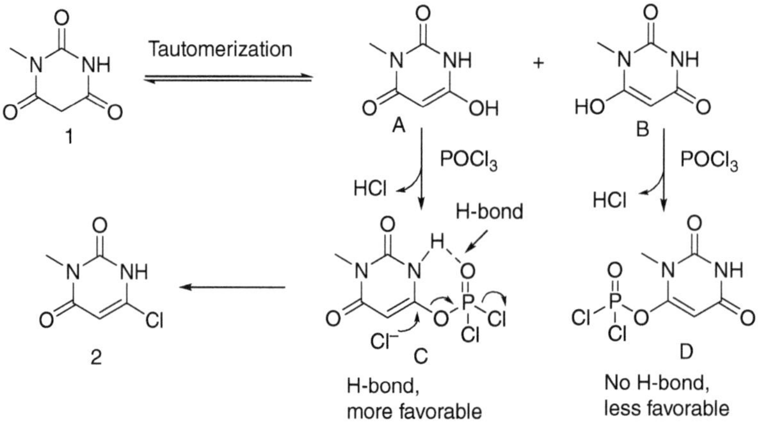

Tautomerization

A + B

$POCl_3$, HCl

H-bond

C: H-bond, more favorable

D: No H-bond, less favorable

2

Problem 89: An Industrial Synthesis of Orotic Acid

Orotic acid (OA) molecule is itself required for the regulation of genes that are important in the development of cells, tissues, and organisms [261]. In a patent application, Shi described a one-pot preparation of OA (**3**) from hydantoin **1** and glyoxylic acid **2** [262].

Please suggest a mechanism for the reaction.

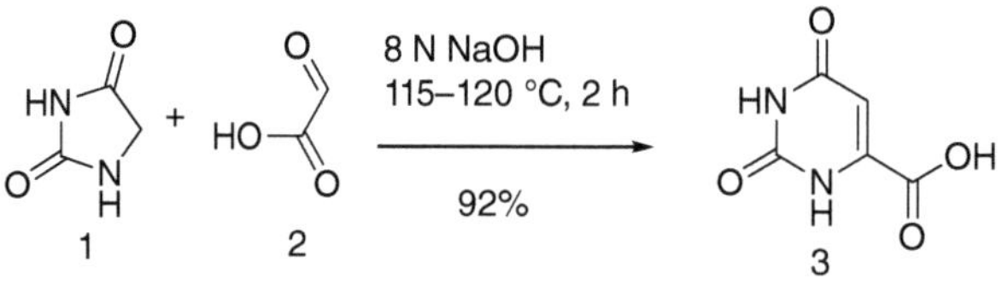

1 + 2 → 3 (8 N NaOH, 115–120 °C, 2 h; 92%)

Experimental Procedure (from CN109761916, **2019**, A)

An Improved Synthesis Method of OA: Maintain the temperature in the reaction channel at 115–120 °C, adjust the flow rate of the dual pump. Simultaneously pump in 1.5 kg of hydantoin, 10 kg of water, 2.5 kg of a 50% mixed solution of glyoxylic acid, and 7.5 liters of 8 N sodium hydroxide solution. After heating and circulating the mixture for 2 hours, stop heating. Once the temperature drops to 70–80 °C, transfer the reaction mixture to a 50-liter purification reactor. Maintain the internal temperature at 60–65 °C, add concentrated hydrochloric acid dropwise to adjust the pH to 9–9.5, and centrifuge the precipitated solid. A crude white sodium orotate is obtained, and the crude product is added to 6 liters of water. Incubate at 60–65 °C, add concentrated hydrochloric acid to adjust the pH to 1–2, stir for half an hour, centrifuge the precipitated solid to obtain 3 kg of crude OA; S2. Add the above crude sodium orotate to 90 liters of water, raise the temperature to 90–95 °C, add concentrated ammonia water to adjust the pH to 4–5. To the solid solution, add 50 g of acidic activated carbon, decolorize for 15 minutes, then filter while hot. The filtrate was added with concentrated hydrochloric acid at 80 °C to adjust the pH to 1, and after slowly stirring for 2 minutes, the temperature was naturally lowered. The crystal is allowed to stand, and when the temperature is lowered to 50–55 °C, centrifugation is started, and the filter cake is washed with water. The air was dried at 100 °C to obtain 2.15 kg of anhydrous OA, and the yield was 91.9%.

The spectra data of OA were reported in another article [263]. IR (cm^{-1}): ν (NH) 3152 br, 3110s; δ (N^1H) 1515s, δ (N^3H) 1406s; ν_s (OCO)1481s, ν_{as} (OCO)1675 sh, D ν 194; $\nu(C^4{=}O)$ 1652s, $\nu(C^2{=}O)$1733s; ^{1}H NMR (DMSO-d_6): δ 11.60 (s,1H, N^1-H); 11.2 (s,1H, N^3-H); 5.86 (s,1H, C5-H); 12.94 (s,1H, (COOH)); ^{13}CNMR δ (ppm) (DMSO-d_6): 162.00 (OCO); 157.00(C2); 167.08(C4); 107.00 (C5); 146.08(C6).

The Mechanism

The first step should be the aldol condensation between hydantoin **1** and glyoxylic acid **2** under the strong base condition. Then the hydantoin ring in intermediate **A** is opened by elimination of the acidic α-hydrogen of the carboxylic group to form the conjugated intermediate **B**. The more basic enol salt **B** is hydrolyzed to the free enol **C** and then isomerized to the ketone form. Recyclization between the NH_2 and the carbonyl provides the intermediate **D**. Finally, dehydration and acidic aqueous workup leads to OA **3**.

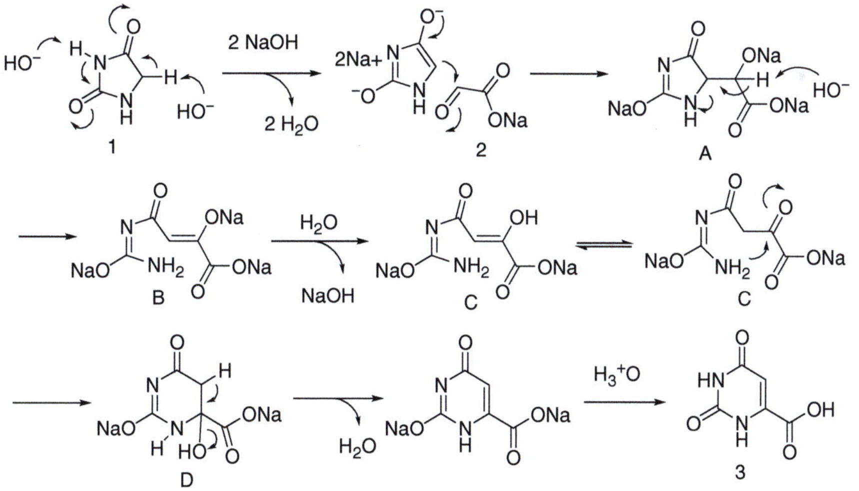

It should be noted that the following mechanisms are not reasonable.

a.

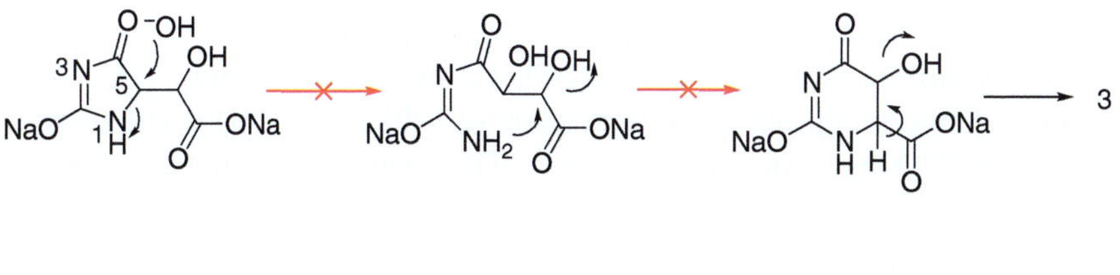

At the first step, C-5 is not electrophilic under strong basic conditions. For the second step, hydroxyl is not a leaving group at basic condition.

b.

NaO, OH, ONa, HN, NH, H_2O, H_3^+O, 3

Carboxylic acid sodium salt would not cyclize with urea under strong basic conditions.

Problem 90: 2-Thioxo-2,3-dihydropyrimidin-4(1H)-one to Pyrimidine-2,4 (1H,3H)-dione

In 1976, Basnak and Farkas reported the transformation of thiouracil to uracil by heating with chloroacetic acid in water [264]. The condition has been applied on the large-scale (>100 kg) synthesis of compound **2**, an intermediate for the synthesis of ATR inhibitor ceralasertib [265].

Please provide a mechanism for the reaction from **1** to **2**.

H_2O, 95 °C, 7 h

83%

1

2

Ceralasertib

Experimental Procedure (from *Org. Process Res. Dev.*, **2021**, *25*, 43)

6-[1-(Methylsulfanyl)cyclopropyl]-2,4(1H,3H)-pyrimidinedione (2). 6-[1-(Methylsulfanyl)cyclopropyl]-2-sulfanylidene-2,3-dihydro-4(1H)-pyrimidinone (**1**, 147.5 kg, 679.8 mol) was charged to water (1066 kg) with stirring at 10–30 °C.

Chloroacetic acid (316.8 kg, 3352 mol, 4.9 equiv) was charged at 10–30 °C. The contents of the vessel were stirred at approximately 95 °C for 6 h and then cooled to approximately 5 °C. The resulting solid was collected by filtration and then slurried in aqueous hydrochloric acid solution (2 M, 463 kg). The resulting solid was collected by filtration and dried at approximately 40 °C to yield 6-[1-(methylsulfanyl)-cyclopropyl]-2,4(1H,3H)-pyrimidinedione (**2**) as a white solid (113.9 kg, 97.7% w/w, 561.4 mol, 83% yield). ^{1}H NMR (500 MHz, DMSO, 27 °C): 1.03–1.07 (m, 2H), 1.22–1.26 (m, 2H), 2.06 (s, 3H), 5.40 (s, 1H), 10.93 (s, 2H); ^{13}C NMR (126 MHz, DMSO, 27 °C): 14.53, 15.59, 26.69, 98.99, 151.69, 155.34, 164.05; HRMS (ESI): calcd for $C_8H_{11}N_2O_2S$ $[M + H]^+$, 199.0536; found, 199.0540.

The Mechanism

The initial step involves the alkylation of thiouracil **1** with chloroacetic acid **2** to produce compound **A**. This is followed by a cyclization process, leading to the formation of a five-membered ring intermediate **B**. The final step is hydrolysis, which releases the desired product along with thioglycolic acid. The reaction is driven by two key factors:

(a) Thermodynamic considerations: By calculating the energy changes using bond energies, it is evident that the reaction is exothermic. The energy input includes 530 kJ/mol (C=S), 330 kJ/mol (C—Cl), and 467 kJ/mol (H—O), totaling 1327 kJ/mol. The energy output comprises 800 kJ/mol (C=O), 260 kJ/mol (C—S), 430 kJ/mol (H—Cl), and 347 kJ/mol (S—H), summing up to 1837 kJ/mol. The overall energy change is 1837 kJ/mol – 1327 kJ/mol = 510 kJ/mol, indicating an exothermic process.

(b) Acidity differences: Thioglycolic acid ($pK_a = 3.83$) is a weaker acid compared to chloroacetic acid ($pK_a = 2.87$), which further drives the reaction forward.

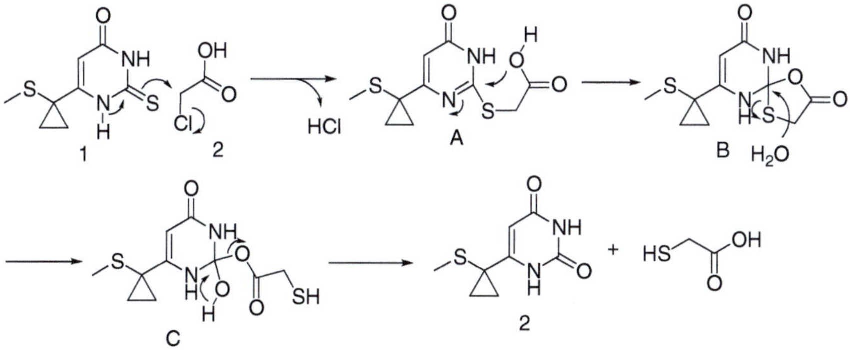

Problem 91: Preparation of *N,N′*-Dicyclohexylbarbituric Acid from DCC and Malonic Acid

Developed by GSK, daprodustat is a medication for treatment of anemia of chronic kidney disease in adults on dialysis and approved by the US FDA in 2023 [266]. The core unit is *N,N′*-dicyclohexylbarbituric acid (**3**). The synthesis of **3** was firstly reported by Bose and Garrat from dicyclohexylcarbodiimide (DCC) and malonic acid in 1962 [267]. O'Brien and Ricci developed a flow process and briefly discussed the mechanism [268].

Please propose your own mechanism for the reaction.

1, malonic acid + 2, DCC → (THF, 0 °C - rt, 2 h, 72%) 3, N,N′-dicyclohexylbarbituric acid → → Daprodustat

Experimental Procedure (from *Org. Proc. Res. Dev.*, **2018**, *22*, 399)
Representative Procedure for the Preparation of 3 in Continuous Flow. The experiment was performed in a Syrris Asia flow reactor fitted with a PFA reactor coil (1.6 mm OD, 1.0 mm ID) and an un-heated Vapourtec acid-resistant back-pressure regulator cartridge. Feeds were switched using Swagelock 2-way valves. Dry (<100 ppm H_2O) THF was used for all feeds. The following feeds were prepared and connected to the inputs of two reciprocating syringe pumps as described in Figure 4: (a) a 0.675 M solution of DCC (139 g, 675 mmol) in THF (758 g, 1.0 L) prepared with stirring under N_2; (b) a 0.675 M solution of both malonic acid (35.1 g, 337.5 mmol) and phosphorus oxychloride (51.7 g, 31.5 mL, 337.5 mmol) in THF (500 mL); (c) a 0.800 M solution of phosphorus oxychloride (28.2 g, 18.9 mL, 202 mmol) in THF (205 g, 230 mL); (d) THF for flushing. The phosphorus oxychloride–THF and THF flushing feeds were pumped at 1.72 mL/min and 3.44 mL/min respectively, mixing via a T-union, into an 18.5 mL tubing reactor at 80 °C under 4 bar back-pressure (residence time 3.6 min, with flow rates confirmed by measuring the collected mass of the output over time). After flushing for 7 min, the feeds were switched to malonic acid–phosphorus oxychloride–THF and DCC–THF respectively at the same flow rates, and **3** was collected as a solution over a typical run time of >4 h. The assay yield was determined by HPLC analysis of the concentration of **3** in the product stream. The product **3** could be isolated by addition of the product stream (1 volume) to *i*PrOH (4 volumes) with stirring. After 30 s, a white solid formed, which could be collected by filtration, washed with *i*PrOH (1 volume), and dried to give *N,N′*-dicyclohexylurea **3** as a white solid: IR ν_{max} (THF solution) 1704, 1689, 1417, 1372, 1193 cm^{-1}; ^{1}H NMR (400 MHz, $CDCl_3$) δ 4.60 (tt, J = 12.5, 3.5 Hz, 2H), 3.60 (s, 2H), 2.26 (qd, J = 12.5, 3.5 Hz, 4 H), 1.85 (m, 4H), 1.65 (m, 6H), 1.36 (m, 6H); ^{13}C NMR (100 MHz, $CDCl_3$) δ 165.0, 151.2, 55.1, 40.9, 29.0, 26.2, 25.0.

The Mechanism

The diimine carbon in DCC is electrophilic, thus the malonic acid should attack it first to form **A**, which is easily rearranged to acyl urea **B**. Then **B** attacks the second DCC to form **C**. Finally, intramolecular cyclization occurs to provide the product *N,N′*-dicyclohexylbarbituric acid and release the by-product, *N,N′*-dicyclohexyl urea (DCHU).

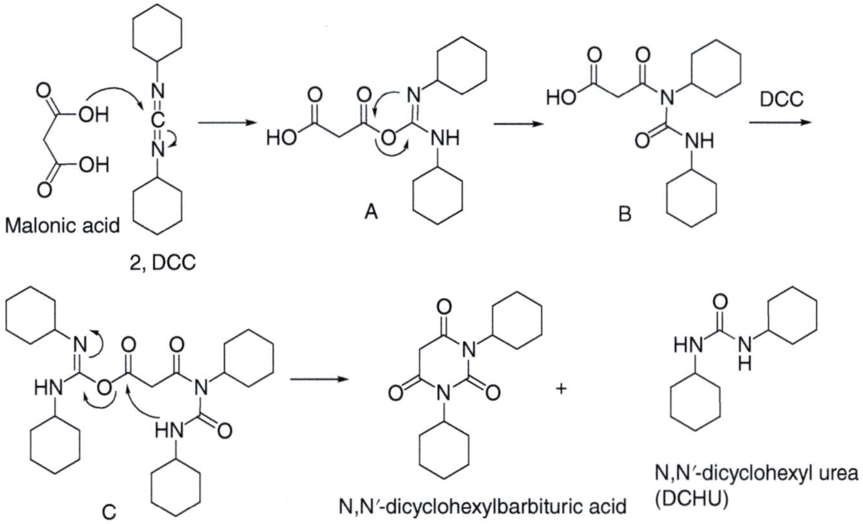

Malonic acid

2, DCC

A

B

C

N,N′-dicyclohexylbarbituric acid

N,N′-dicyclohexyl urea (DCHU)

DCC would be regenerated from DCHU by $POCl_3$.

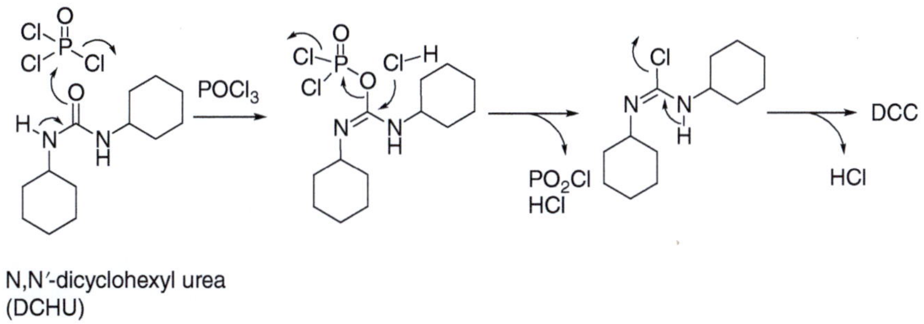

N,N′-dicyclohexyl urea (DCHU)

4.6.11 Pyrazine

Problem 92: Synthesis of Ethyl 3-(trifluoromethyl)pyrazine-2-carboxylate
Ethyl 3-(trifluoromethyl)pyrazine-2-carboxylate (**1**) is a key intermediate for synthesizing the new fungicide pyraziflumid (**2**) [269]. One of the reported synthesis of **1** is shown below [270].

Please suggest a mechanism for the formation of **1** from the oxime **4**.

Experimental Procedure (from *Org. Process Res. Dev.*, **2017**, *21*, 448)
One-Pot Procedure. Ethyl 3-(Trifluoromethyl)pyrazine-2-carboxylate (1). To ethyl 4,4,4-trifluoro-3-oxobutanoate (**3**, 3.70 g, 20.1 mmol, 1.0 equiv) at 0 °C were added butyl nitrite (2.81 mL, 24.0 mmol, 1.2 equiv) and benzoic acid (366 mg, 3.00 mmol, 0.15 equiv), and the mixture was stirred at 0 °C for 2 h and then at room temperature for 24 h. Vacuum (10 mbar, 40 °C) was applied for 0.5 h to remove the excess butyl nitrite. In a separate flask, a mixture of benzoic acid (8.18 g, 67.0 mmol, 3.3 equiv), 3-picoline (38 mL, 0.39 mol, 19 equiv), and ethylenediamine (1.74 mL, 26.1 mmol, 1.3 equiv) was prepared. Upon addition of the amine, a precipitate formed

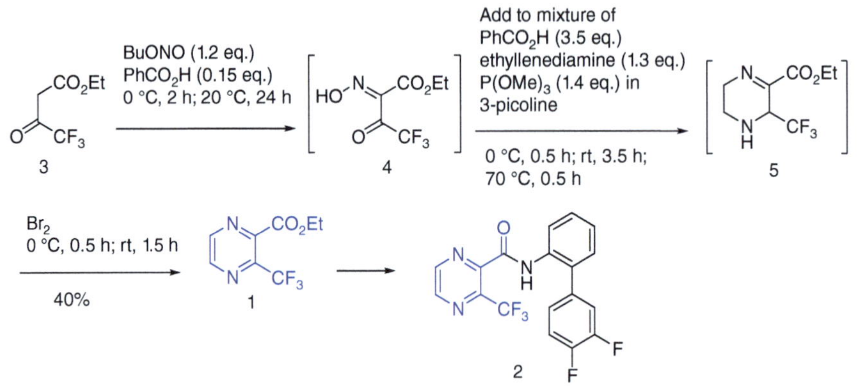

(probably the salt of ethylenediamine and benzoic acid). This mixture could still be thoroughly stirred with a magnetic stirrer. After cooling to 0 °C, trimethyl phosphite (3.31 mL, 28.0 mmol, 1.4 equiv) was added, followed by the dropwise addition of the crude oxime. The oxime container was rinsed with 3-picoline (3.0 mL, 31 mmol, 1.5 equiv), and this solution was also added to the reaction mixture. After stirring at 0 °C for 0.5 h and at room temperature for 3.5 h, the mixture was heated to 70 °C for 0.5 h and then cooled to 0 °C. Bromine (2.56 mL, 50.0 mmol, 2.5 equiv) was then added dropwise within 5 min. The mixture was stirred at 0 °C for 0.5 h and then at room temperature for 1.5 h. The mixture was cooled to 0 °C and added to an ice-cold mixture of concentrated aqueous hydrochloric acid (67 mL, 0.70 mol, 35 equiv) and water (200 mL). After adding sodium bisulfite (3.12 g, 30.0 mmol, 1.5 equiv), the product was extracted with butyl acetate (3 × 50 mL), and the combined extracts were washed with brine (100 mL), with a solution of potassium carbonate (15 g, 0.11 mol) in water (100 mL), with brine (100 mL), dried ($MgSO_4$), and concentrated under reduced pressure. The residue was purified by chromatography over silica gel (30 g; gradient elution with heptane/ethyl acetate). Pyrazine **1** was obtained as an oil (1.81 g, 40% yield, 99% pure by ^{1}H NMR content determination with triisobutyl phosphate as internal standard). ^{1}H NMR ($CDCl_3$, 400 MHz) δ 8.85 (m, 1H), 8.81 (m, 1H), 4.52 (q, J = 7 Hz, 2H), 1.44 (t, J = 7 Hz, 3H); ^{13}C NMR ($CDCl_3$, 100 MHz) δ 163.7, 146.3, 145.3, 144.7, 141.6 (q, J = 37 Hz), 120.6 (q, J = 274 Hz), 63.2, 13.9; ^{19}F NMR ($CDCl_3$, 377 MHz) δ − 66.2; DSC (4 K/min) dec onset 173 °C, dec energy −11 J/g.

The Mechanism

The ketooxime **4** isomerizes to the enol form **A**. Then trimethyl phosphite attacks the N=O group to give the adduct **B**. Cyclization of **B** provides the highly electrophilic oxazaphosphole **C**, which is in turn attacked by ethylenediamine (mono-benzoate form) to generate the ring open intermediate **D** with elimination of trimethyl phosphate and benzoic acid. After **D** being isomerized to oxime **E**, recyclization occurs to yield intermediate **5**. Finally bromine-mediated dehydrogenation leads to pyrazine **1**.

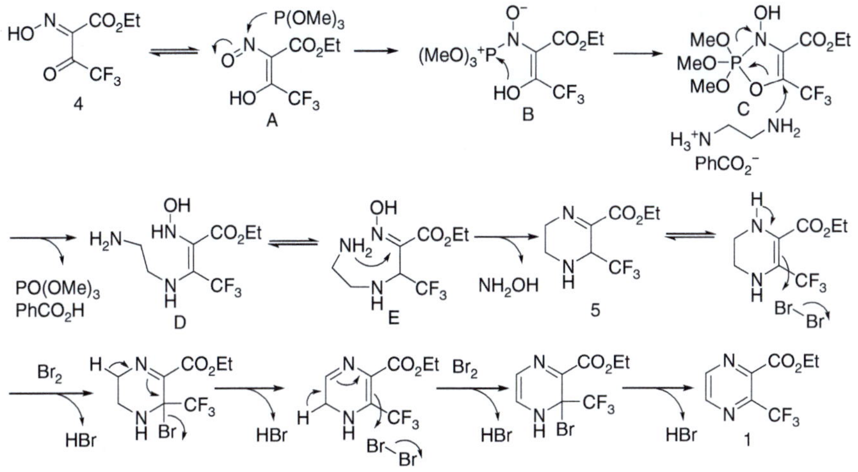

4.6.12 Triazole

Problem 93: Preparation of [1,2,4]Triazolo[1,5-a]pyridin-2-amine

In 1983, Tisler's group reported the preparation of a series of heterocyclic compounds [271], including [1,2,4]triazolo[1,5-a]pyridin-2-amine (**2**) from the thiourea **1**. Recently, Tisler triazolopyrimidine cyclization had been applied to the synthesis of JAK inhibitors CEP-33779 [272, 273], filgotinib [274], VEGFR-2 kinase inhibitor (**3**) [275], as well as crop protection agent (**4**) [276].

Please suggest a mechanism for the reaction from **1** to **2**.

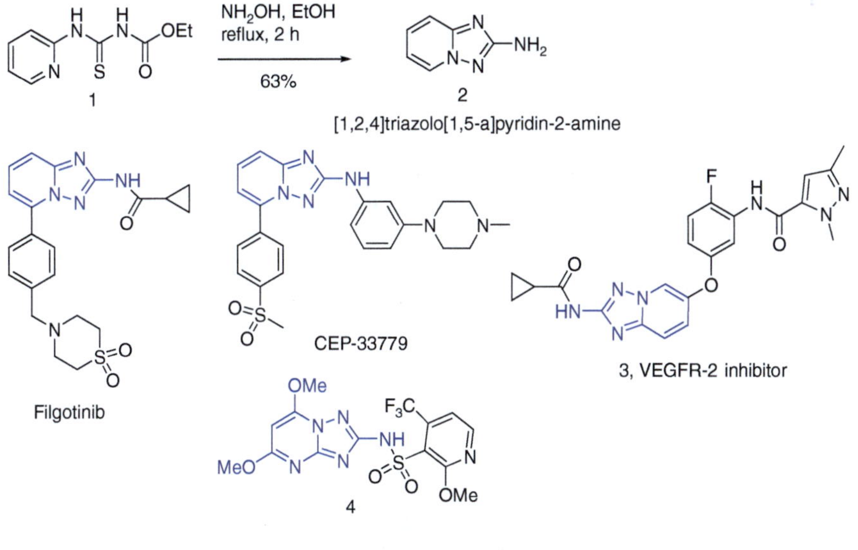

NH_2OH, EtOH reflux, 2 h; 63%

[1,2,4]triazolo[1,5-a]pyridin-2-amine

Filgotinib

CEP-33779

3, VEGFR-2 inhibitor

4

Experimental Procedure (from *Monatsh. Chem.*, **1983**, *114*, 789)

[1,2,4]Triazolo[1,5-a]pyridin-2-amine (2). 2.25 g of **1** was suspended in 100 mL of ethanolic solution of free hydroxylamine (prepared from 0.7 g sodium, ethanol and 3.5 g of hydroxylamine hydrochloride) and the mixture was heated under reflux for 2 h. During the reaction, hydrogen sulfide was evolved. The solvent was then evaporated *in vacuo* and the residue was extracted with $CHCl_3$. The oily residue was crystallized from benzene (0.84 g, 63%), m. p. 110–112 °C (Lit [274]. m. p. 108–109 °C). MS (m/e): 134 (M^+). NMR ($CDC1_3$): 7.88 (m, H5), 6.50 (m, H6), 7.00 (m, H7 and H8), 4.60 (broad, NH_2).

The Mechanism

The first step is the thio group replaced by hydroxylamine to give the corresponding hydroxyimino derivative **A**, which is subsequently cyclized with the participation of the ethoxycarbonyl group to form **B**. Upon heating, with the participation of the nitrogen in pyridine, elimination of carbon dioxide from **B** provides the product **2**.

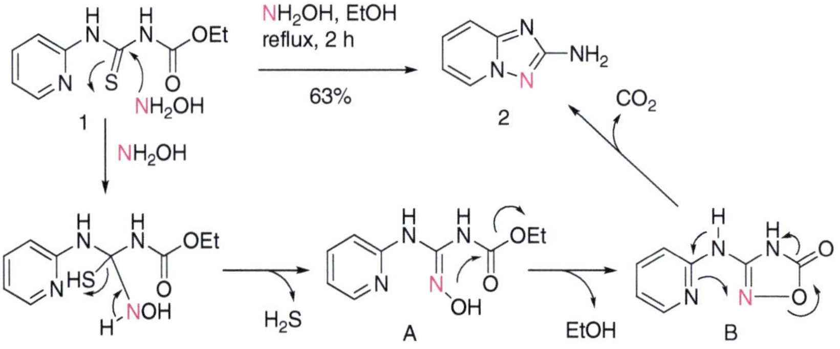

As the by-product H_2S was easily oxidized by air during the reaction, some products contained a certain level of sulfur. The amount of residual sulfur decreased when the reaction was conducted under N_2 atmosphere [276].

4.6.13 Oxadiazole

Problem 94: Formation of 1,3,4-Oxadiazole

1,3,4-Oxadiazole **3** is an important building block for constructing raltegravir. It was synthesized from acetonitrile in a three-step sequence [277]. Treatment of acetonitrile with NaN_3 in the presence of $AlCl_3$ provided the tetrazole **1** [278]. Acylation of tetrazole **1** with ethyl oxalyl chloride afforded intermediate **2**, which was heated at 70 °C to form the oxadiazole ester. Direct KOH-mediated hydrolysis afforded the corresponding potassium salt **3** as a crystalline solid, in 91% overall yield.

Please propose mechanisms for the synthesis of **3** from acetonitrile.

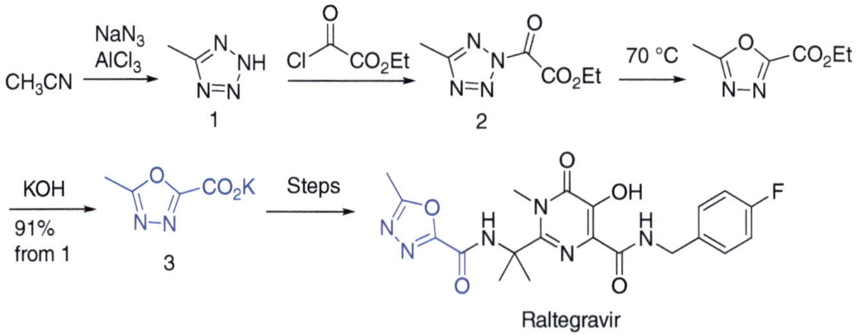

Experimental Procedure (from *Org. Process Res. Dev.*, **2011**, *15*, 73)

Potassium 5-Methyl-[1,3,4]oxadiazole-2-carboxylate (3). Ethyl oxalyl chloride (4.01 kg, 29.4 mol) was slowly added to a mixture of methyl tetrazole **1** (2.50 kg, 29.7 mol; while 5-methyl-tetrazole is not rated an explosive material (see: http://www.chemicalbook.com/ChemicalProductProperty_EN_CB4290179.htm), it is an irritant and highly flammable. It should be handled only after all hazard data has been reviewed and proper PPE obtained) and triethylamine (3.03 kg, 30.0 mol) in toluene (32 L) at 0 °C at such a rate that the temperature stayed below 5 °C at all times. (Caution: **2** is a potentially explosive intermediate that will quickly generate nitrogen gas at $<$50 °C. Conduct this chemistry only after training for handling explosive compounds, and conduct the heating only behind a shield.) The slurry was stirred for 1 h at 0–5 °C. The triethylamine hydrogen chloride salt was filtered off. The solid was washed with 27 L of cold toluene (5 °C). The combined filtrates were kept at 0 °C at all times and were slowly added to a hot solution of toluene (50 °C, 15 L) over 40–50 min. (Caution: N_2 gas evolves. Use the same precautions during heating as above.) Most of the intermediate **2** had been converted to the corresponding non-explosive oxadiazole ester at this moment. The solution was continually stirred at 70 °C for 1 h. After cooling to 20 °C, the toluene solution was washed with 5 L of 10% brine, and solvent was switched to ethanol. EtOH was added to adjust the final volume to approximately 40 L. The solution was cooled to 10 °C and aqueous KOH (8.0 L) was added over 30 min. The resulting thick slurry was then stirred for 40 min at room temperature while the oxadiazole K salt crystallized. The solid was collected by filtration and washed with EtOH (11 L) and MTBE (15 L). The crystalline solid was dried overnight under vacuum at 20 °C with nitrogen sweep to give oxadiazole K salt **3** (4.48 kg) in 91% yield, mp 258.3 °C dec. ^{1}H NMR (400 MHz, CD_3OD/D_2O) 2:1) δ: 2.62 (S, 3H); ^{13}C NMR (100 MHz, CD_3OD/D_2O) 2:1) δ: 165.8, 161.8, 158.3, 9.8. MS (FAB) m/z 129 (corresponding to free acid, M + H, 100%).

The Mechanism

The first stage is a [3+2] cycloaddition. The acylated intermediate **2** was heated to release one molecule of N_2, followed by recyclization to afford the oxadiazole ester.

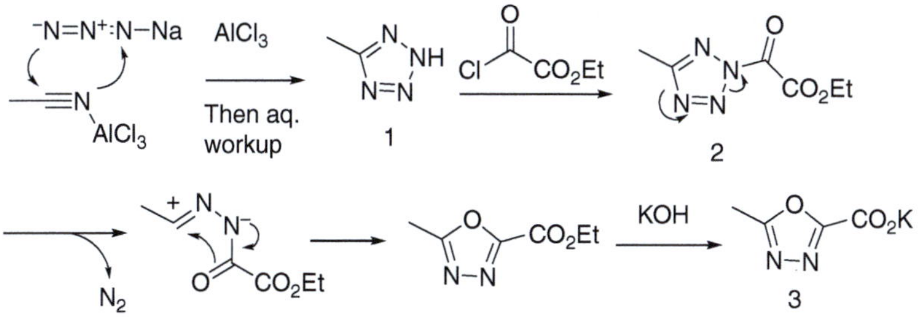

Problem 95: Preparation of 1,2,4-Oxadiazole

1,2,4-Oxadiazole is an important subunit in some drugs such as opicapone, a medication for Parkinson's disease. In a patent application [279], 5-(2-nitrophenyl)-1,2,4-oxadiazole (**2**) was prepared from 2-nitrobenzamide (**1**) by the following reactions.

Please provide mechanisms for the transformation from **1** to **2**.

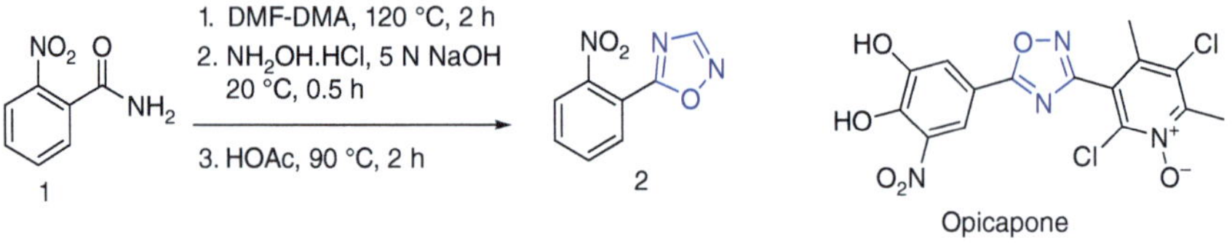

1. DMF-DMA, 120 °C, 2 h
2. $NH_2OH.HCl$, 5 N NaOH 20 °C, 0.5 h
3. HOAc, 90 °C, 2 h

Opicapone

Experimental Procedure (from WO2018/144620, **2018**, A1)

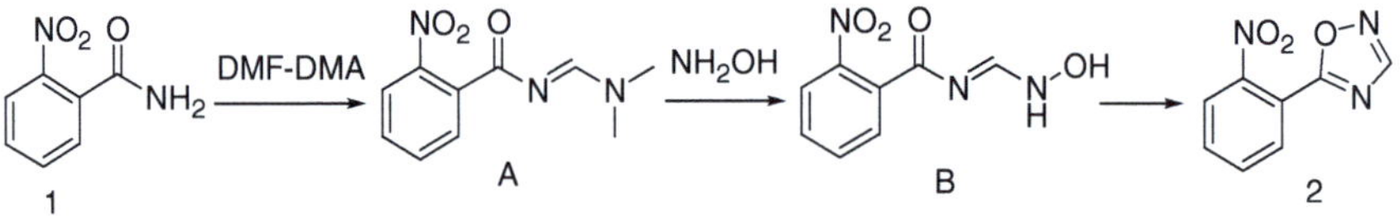

***N*-[(*E*)-(dimethylamino)methylidene]-2-nitrobenzamide (A).** 2-Nitrobenzamide (**1**, 1.00 g, 6.02 mmol) and 1,1-dimethoxy-*N*,*N*-dimethylmethanamine (4.0 mL, 30 mmol) were stirred at 100 °C for 1 h. The mixture was filtered, washed with hexane, and dried under reduced pressure at 50 °C to give 1.03 g (**A**, 100% purity, 78% yield) of the target compound, which was used without further purification. LCMS (Method 2): $R_t = 0.76$ min; MS (ESIpos): m/z = 222 $[M + H]^+$. ^1H-NMR (400 MHz, DMSO-d_6): δ [ppm] = 3.03 (d, 3H), 3.20 (s, 3H), 7.63–7.73 (m, 2H), 7.78–7.82 (m, 1H), 7.92 (dd, 1H), 8.58 (t, 1 H).

5-(2-Nitrophenyl)-l,2,4-oxadiazole (2). To a solution of hydroxylamine hydrochloride (0.98 g, 14.1 mmol, 1.3 equiv.) in sodium hydroxide (5 M, 2.8 mL, 1.3 equiv.) was added (*E*)-*N*-(dimethylaminomethylene)-2-nitrobenzamide (**A**, 2.4 g, 10.8 mmol, 1.0 equiv.) portionwise over 5 minutes at 20 °C and the mixture stirred at 20 °C for 0.5 hours. The mixture was diluted with water (10 mL) and extracted with DCM (10 mL × 3). The organic layer was dried and concentrated under reduced pressure to give (*E*)-*N*-((hydroxyamino)methylene)-2-nitrobenzamide (**B**, 1.20 g) which was used directly without further purification. A mixture of (*E*)-*N*-((hydroxyamino)methylene)-2-nitrobenzamide (**B**, 0.9 g, 4.30 mmol, 1.0 equiv.) in acetic acid (7 mL) and dioxane (7 mL) was stirred at 90 °C for 2 hours. The mixture was concentrated under reduced pressure and the residue was purified by silica gel chromatography (petroleum ether-ethyl acetate 10: 1 to 5: 1) to give the desired product as a light yellow solid (0.4 g, 49%); ^1H NMR (DMSO-d_6, 400 MHz) δ 9.25 (s, 1H), 8.23–8.21 (m, 1H), 8.13–7.11 (m, 1H), 7.99–7.95 (m, 2H).

The Mechanism

Step 1. Formation of (*E*)-*N*-((dimethylamino)methylene)-2-nitrobenzamide (**A**)
Step 2. Hydroxylamine replaced dimethylamine, formation of (*E*)-*N*-((hydroxyamino)methylene)-2-nitrobenzamide (**B**)
Step 3. Cyclization and dehydration to form 5-(2-nitrophenyl)-1,2,4-oxadiazole (**2**)

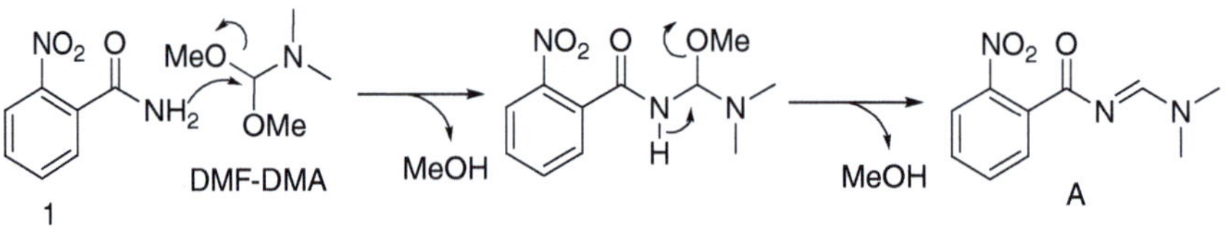

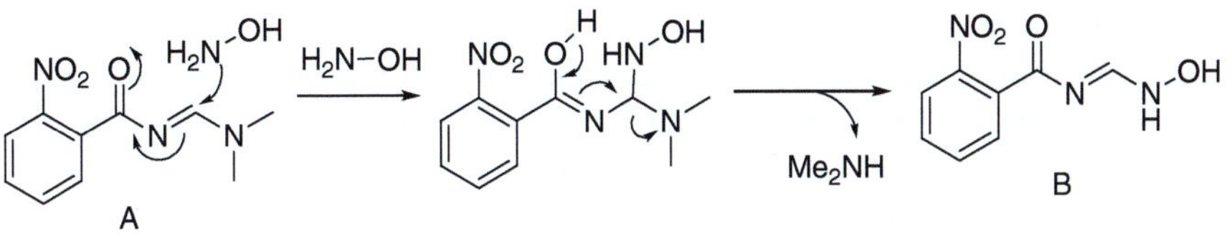

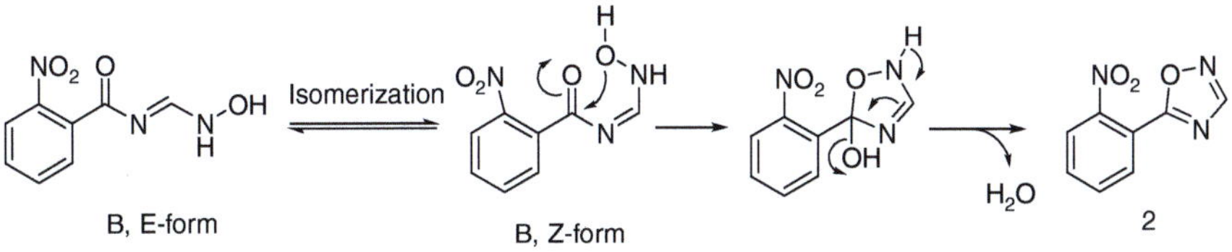

4.6.14 Thiadiazole

Problem 96: Formation of 1,2,4-Thiadiazole

Z-Isomer of 2-(5-amino-1,2,4-thiadiazol-3-yl)-2-(methoxyimino)acetic acid (**1**) is a side chain of cephem antibiotics such as SCE2787 [280]. Tatsuta's group reported the synthesis of the key intermediate (**2**) from malononitrile (**3**) by the following reaction sequence [281].

Please provide reaction mechanisms from **3** to **2**.

Experimental Procedure (from *Bull. Chem. Soc. Jpn.*, **1994**, *76*, 1701)

3,3-Dimethoxyacrylonitrile (4). To a stirred solution of malononitrile (**3**, 12.0 g, 182 mmol) and methanol (7.35 mL, 182 mmol) in dry ether (240 mL) was bubbled hydrogen chloride gas (8.5 g, 230 mmol) during 1 h at −5 °C. After stirring at the same temperature for 1 h and at room temperature for 6 h, the formed crystals were filtered off, washed with ether, and dried *in vacuo* to give methyl cyanoacetimidate hydrochloride (23.8 g, 98%).

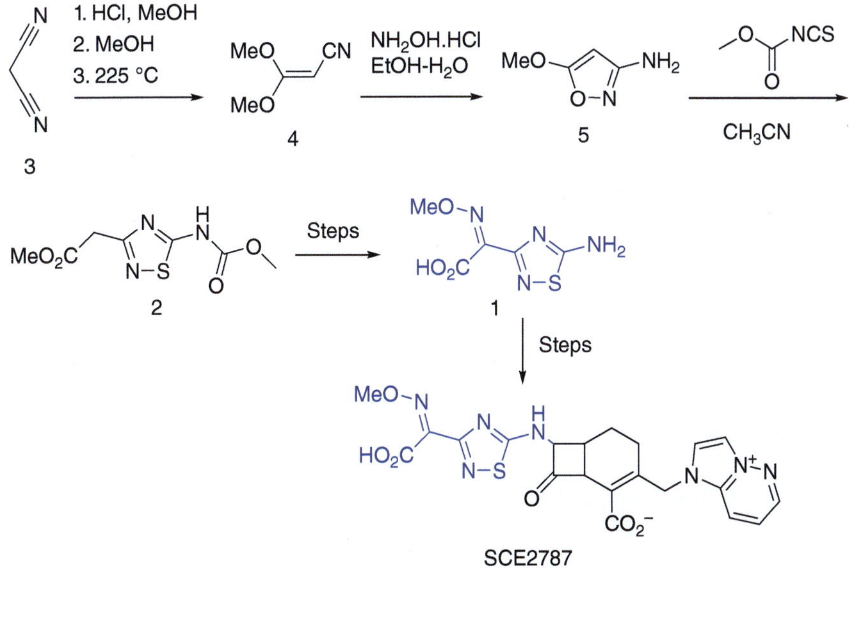

A suspension of methyl cyanoacetimidate hydrochloride (23.8 g, 177 mmol) in methanol (177 mL) was stirred at room temperature for 12 h and the reaction mixture was concentrated to a residue, which was dissolved in EtOAc. The solution was washed with saturated aqueous Na_2CO_3 and saturated NaCl, dried, and concentrated. The crude syrup obtained was purified by distillation under reduced pressure to give 3,3,3-trimethoxypropionitrile (22.1 g, 86%) as a syrup: Bp 110–112 °C/2533 Pa; ^{1}H NMR $\delta = 2.87$ (2H, s, CH_2) and 3.36 (9H, s, 3 Me).

A syrup of 3,3,3-trimethoxypropionitrile (22.1 g, 153 mmol) was stirred at 222–225 °C for 7 min. The reactant was distilled under reduced pressure to give **4** as a syrup (14.2 g, 70% from **3**): Bp 112–115 °C/2533 Pa; ^{1}H NMR $\delta = 3.50$ (1H, s, H-2) 3.74 (3H, s, OMe), and 3.80 (3H, s, OMe). Found: C, 52.91; H, 6.10; N, 12.04%. Calcd for $C_5H_7NO_2$: C, 53.09; H, 6.24; N, 12.38%.

3-Amino-5-methoxyisoxazole (5). To a stirred solution of hydroxylamine hydrochloride (1.55 g, 22.3 mmol) in water (3.41 mL) was added 8 M NaOH solution (3.35 mL, 26.8 mmol) at room temperature. After the solution warmed to 45 °C, a solution of **4** (2.27 g, 20.1 mmol) in methanol (5.50 mL) was added dropwise during 30 min, and the resulting mixture was stirred at the same temperature for 1 h. After the consumption of 4 was completed, 8 M NaOH solution (1.25 mL, 10.1 mmol) was added at 45 °C, and the reaction mixture was stirred at 60 °C for 6 h and then concentrated until precipitates of inorganic compound separated out. The mixture was extracted with EtOAc and then the extracts were dried and evaporated to a residue. Recrystallization from hexane-EtOAc gave **5** as colorless crystals (1.71 g, 75%): Mp 82–83 °C; IR ($CHCl_3$) 3840, 1628, and 1487 cm^{-1}; ^{1}H NMR ($CDCl_3 + D_2O$) $\delta = 3.93$ (3H, s, OMe), and 4.83 (1H, s, H-4). Found: C, 41.74; H, 5.06; N, 24.35%. Calcd for $C_4H_6N_2O_2$: C, 42.11; H, 5.30; N, 24.55%.

Methyl 2-(5-methoxycarbonylamino-1,2,4-thiadiazol-3-yl)acetate (2). A suspension of methyl chloroformate (0.927 mL, 12 mmol) and potassium thiocyanate (KSCN) (1.26 g, 13 mmol) in MeCN (10 mL) was stirred at 70 °C for 30 min and, after ice-cooling, **5** (1.14 g, 10 mmol) was added under stirring. The reaction mixture was further stirred at 5 °C for 10 min and at room temperature for 15 min, and poured into ice water (18 mL). The formed precipitates were filtered off, washed successively with water and ether, and dried *in vacuo*. Recrystallization from MeOH gave **2** as colorless crystals (1.98 g, 86%): Mp 167–169 °C; IR ($CHCl_3$) 3406, 1736, and 1549 cm^{-1}; ^{1}H NMR $\delta = 3.73$ (3H, s, COOMe), 3.95 (NCOOMe), 3.97 (2H, s, CH_2), and 10.50 (1H, br s, NH). Found C, 36.40; H, 3.94; N, 18.11%. Calcd for $C_7H_9N_3O_4S$: C, 36.36; H, 3.92; N, 18.17%.

The Mechanism

The process begins with the addition of methanol to the acid-activated nitrile group in malononitrile (**3**), forming acetimidate **A**. A second methanol addition to **A** yields 3-amino-3,3-dimethoxypropanenitrile (**B**), which is then converted to 3,3,3-trimethoxypropionitrile (**C**) by replacing the amino group with a methoxy group. At elevated temperatures, the elimination of one molecule of methanol from **C** produces 3,3-dimethoxyacrylonitrile (**4**).

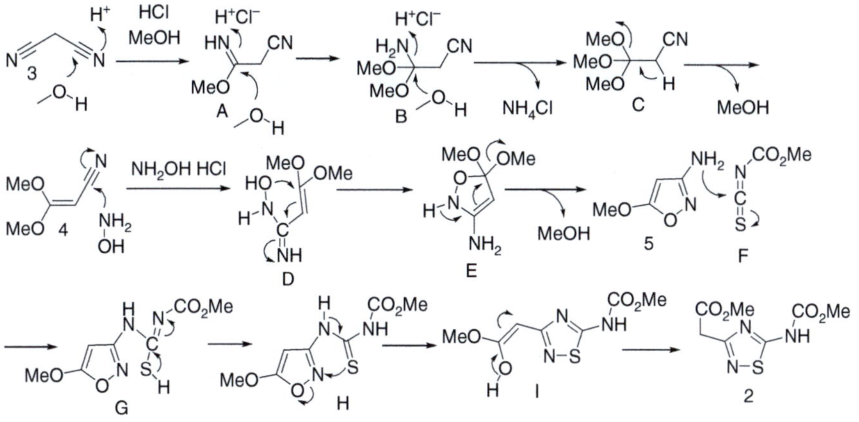

Next, hydroxylamine reacts with the nitrile group in **4** to form imidamide **D**. An intramolecular Michael addition of the hydroxyl group to the C=C double bond results in the formation of dihydroisoxazole **E**. Following the elimination of another methanol molecule, 5-methoxyisoxazol-3-amine (**5**) is generated.

The final step involves the reaction of the amino group in **5** with *O*-methyl carbonisothiocyanatidate (**F**), leading to the formation of thiourea **H** via intermediate **G**. Subsequently, the thio group attacks the isoxazole ring, forming 1,2,4-thiadiazole **I**, which readily tautomerizes to yield the final product **2**.

Direct elimination of NH_3 in the presence of HCl from intermediate **C** will also generate intermediate **4**.

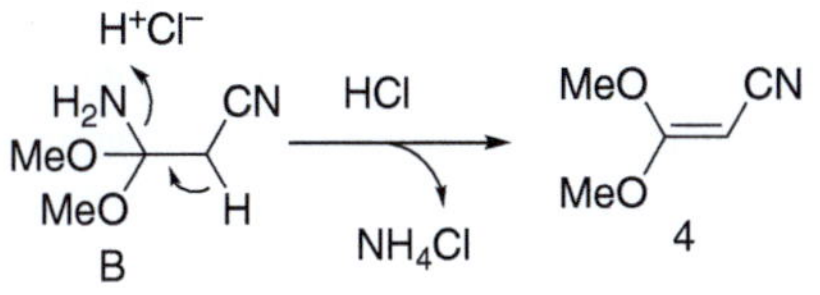

Intermediate **F** is generated in situ from methyl chloroformate and KSCN.

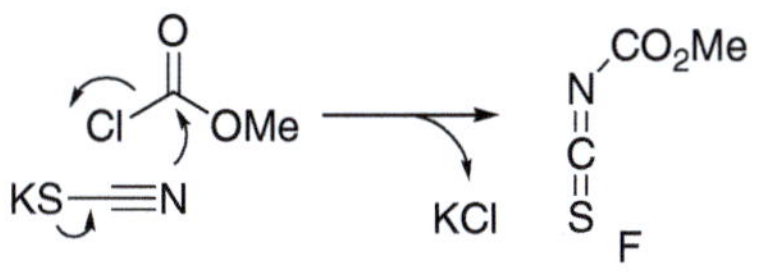

4.6.15 Triazine

Problem 97: Synthesis of 1,2,3-Triazines

Methoxatin is a natural product and shows the ability to reduce damage in ischemia-reperfusion injury of a heart attack or stroke. 1,2,3-Triazines (**6**) is applied in the synthesis of methoxatin. Boger's group reported the preparation of a series of 1,2,3-triazines (**6**) by the following reactions [282, 283]. One of the reported preparations of compound **3** in 95% yield is achieved just by heating both starting materials (**1** and **2**) neat at 80 °C for 30 minutes [284].

Please provide mechanisms stepwise for the synthesis from starting materials (**1** and **2**) to **3** and **6**.

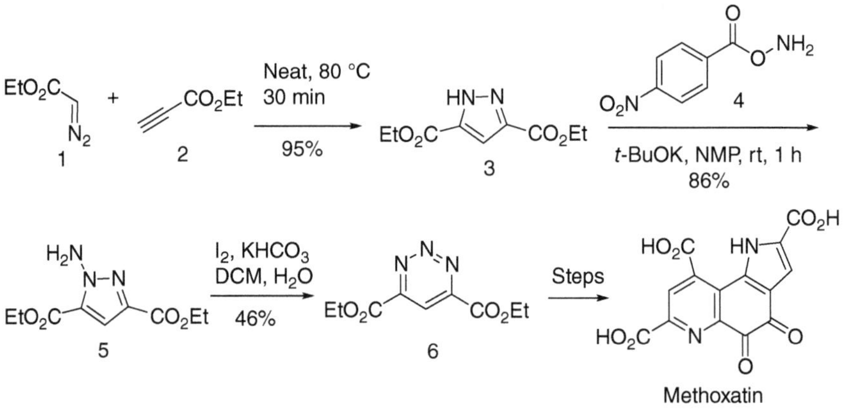

Experimental Procedure (from *Org. Lett.*, **2014**, *16*, 5084, Supporting Information)
Diethyl 1-amino-1*H*-pyrazole-3,5-dicarboxylate (5). A solution of *t*-BuOK (775 mg, 6.91 mmol) in N-methyl-2-pyrrolidone (NMP) (6.9 mL) was added to a stirred solution of 3,5-dicarboethoxypyrazole (**3**, 1.33 g, 6.28 mmol) in NMP (8.1 mL). The reaction mixture was stirred for 20 min at room temperature and then a solution of *O*-4-nitrobenzoylhydroxylamine (**4**, 1.14 g, 6.28 mmol) in NMP (4.3 mL) was added dropwise while keeping the internal reaction temperature below 25 °C. The reaction mixture was stirred at room temperature for 1 h. After 1 h, 10% aqueous NaCl (25 mL) and EtOAc (25 mL) were added to the reaction mixture. The organic layer was separated and the aqueous layer was extracted with EtOAc (25 mL). The combined organic extracts were washed with 5% aqueous $NaHCO_3$ (15 mL) and H_2O (3 × 15 mL), dried over Na_2SO_4, and concentrated on a rotary evaporator. Flash chromatography (SiO_2 30% EtOAc/hexanes) provided the title compound (1.22 g, 86%) as an off-white solid: mp 63–64 °C; ^1H NMR ($CDCl_3$, 600 MHz) δ 7.18 (s, 1H), 6.84 (bs, 2H), 4.33 (q, $J = 7.2$ Hz, 2H), 4.31 (q, $J = 7.2$ Hz, 2H), 1.33 (t, $J = 7.2$ Hz, 3H), 1.31 (t, $J = 7.2$ Hz, 3H); ^{13}C NMR ($CDCl_3$, 100 MHz) δ 161.2, 159.8, 137.4, 127.54, 127.49, 111.89, 111.84, 61.8, 61.2, 14.3, 14.1; IR (film) ν_{max} 1714, 1532, 1270, 1218, 1200, 1160, 1109, 1078, 998, 960, 926, 853, 757, 718 cm^{-1}; HRESI-TOF *m/z* 228.0980 ($C_9H_{13}N_3O_4 + H^+$ requires 228.0979).

Diethyl 1,2,3-Triazine-4,6-dicarboxylate (6). Diethyl 1-amino-1*H*-pyrazole-3,5-dicarboxylate (**5**, 340 mg, 1.50 mmol) was dissolved in CH_2Cl_2–20% aqueous $KHCO_3$ (15 mL, 1:1), cooled in an ice bath, and stirred while I_2 (570 mg, 2.25 mmol) in CH_2Cl_2 (15 mL) was added by an addition funnel. The reaction mixture was stirred at room temperature for 1 h. After 1 h, saturated aqueous Na_2SO_3 was added to quench the reaction mixture and the organic layer was separated. The aqueous layer was extracted with CH_2Cl_2 (2 × 25 mL). The combined organic layers were washed with H_2O (25 mL), dried over Na_2SO_4, and concentrated on a rotary evaporator. Flash chromatography (SiO_2, 3% acetone/toluene) provided **6** (142 mg, 42%; 40–46%) as a yellow solid: mp 85 °C (hexanes); ^{1}H NMR ($CDCl_3$, 400 MHz) δ 8.55 (s, 1H), 4.61 (q, $J = 7.2$ Hz, 4H), 1.51 (t, $J = 7.2$ Hz, 6H); ^{13}C NMR ($CDCl_3$, 100 MHz) δ 162.6, 151.4, 117.7, 64.3, 14.6; IR (film) ν_{max} 1721, 1384, 1368, 1351, 1263, 1092, 1014, 957, 857, 803, 761, 718 cm^{-1}; HRESI-TOF *m/z* 226.0822 ($C_9H_{11}N_3O_4 + H^+$ requires 226.0822).

The Mechanism

The formation of **3** is a [3+2] cycloaddition.

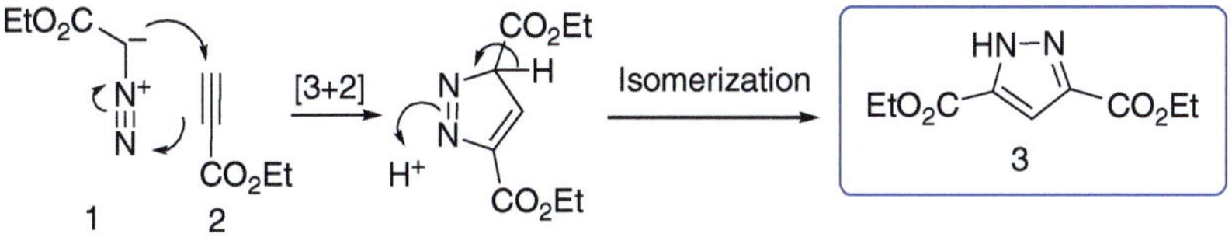

The formation of **5** is an amination of the pyrazole **3** by the reagent **4** under strong base.

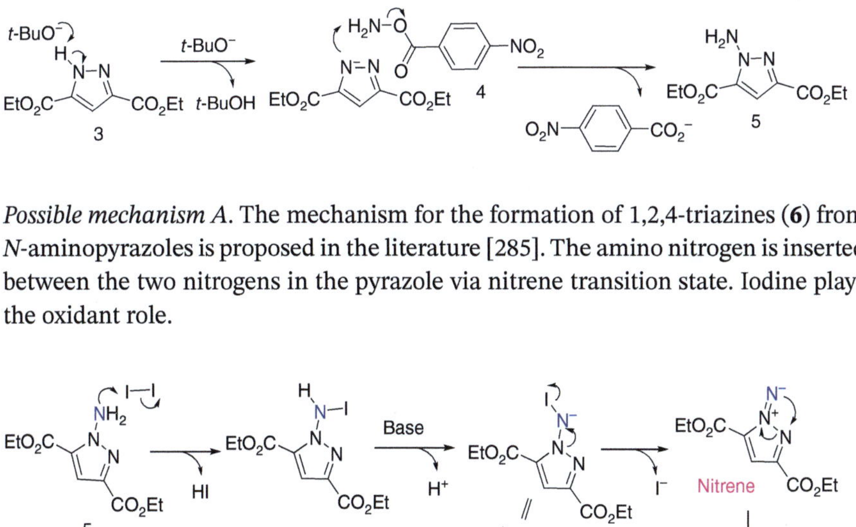

Possible mechanism A. The mechanism for the formation of 1,2,4-triazines (**6**) from *N*-aminopyrazoles is proposed in the literature [285]. The amino nitrogen is inserted between the two nitrogens in the pyrazole via nitrene transition state. Iodine plays the oxidant role.

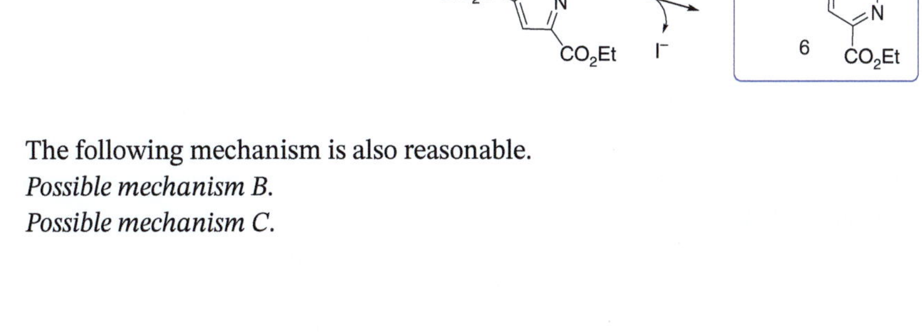

The following mechanism is also reasonable.

Possible mechanism B.

Possible mechanism C.

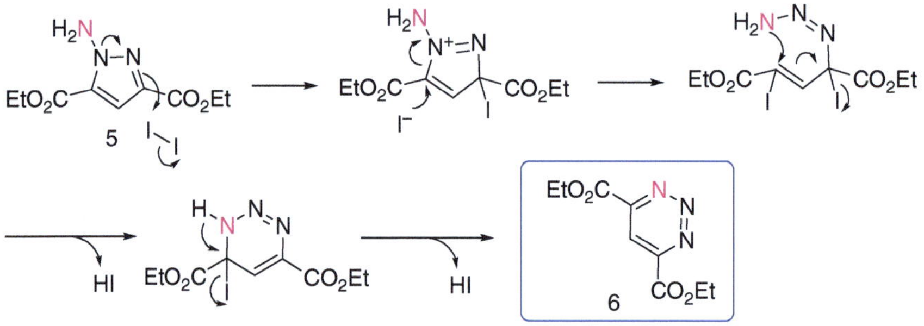

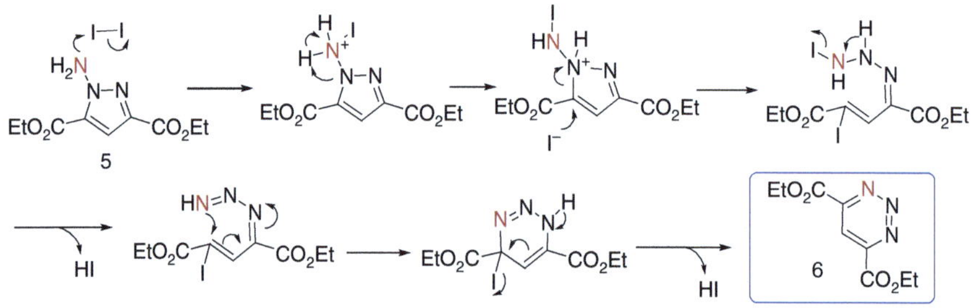

If ^{15}N labeled reagent **4** is employed for preparation of amino pyrazole **5**, then a ^{15}N NMR will address the mechanism uncertain.

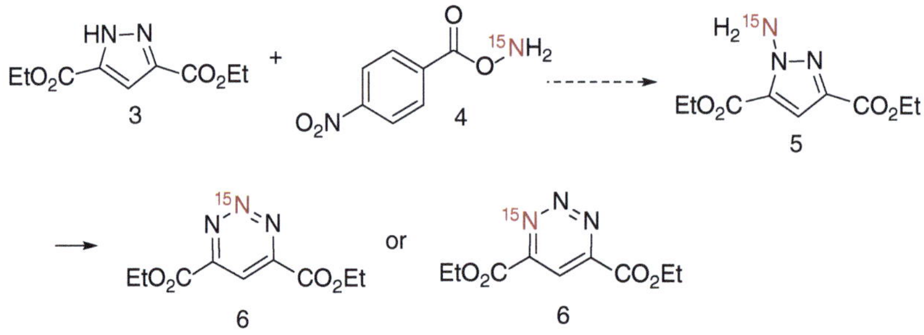

4.7 Formation of Bicyclic and Polycyclic Aromatic Heterocycles

4.7.1 Indole

Problem 98: Formation of 2,3-Disubstituted Indole

3-(2-Carboxyethyl)-1*H*-indole-2-carboxylic acid derivatives are antagonists of the strychnine-insensitive glycine [286, 287] and GPR17 [288]. The synthesis of those 2,3-disubstituted indoles is based on Japp-Klingemann condensation followed by Fischer indole synthesis [289]. Recently, a series of MCL-1 inhibitors were prepared from 2,3-disubstituted indoles, such as compound **3**. In a patent application [290], compound **3** was prepared from aniline **1** and ethyl 2-oxocyclopentane-1-carboxylate **2** as shown below.

Please suggest a mechanism for the transformation from **1** to **3**.

1. $NaNO_2$, HCl, H_2O
0 °C
2. NaOAc
0 °C - rt, 2 h
3. EtOH, H_2SO_4, 95 °C, 144 h
Steps
1
2
3, 42%
MCL-1 inhibitor

Experimental Procedure (from WO**2019**/096922, A1)
Ethyl-7-bromo-3-(3-ethoxy-3-oxopropyl)-6-fluoro-1 H-indole-2-carboxylate (3). To a stirred solution of 2-bromo-3-fluoroaniline **1** (CAS 111721-75-6, 9.50 g, 50.0 mmol, 1.00 eq.) in an aqueous hydrochloric acid solution (12.5 mL conc. HCI in 80.0 mL of water, 150 mmol, 3.00 eq.) was added dropwise a 2.5 M solution of sodium nitrite (20.0 mL, 50.0 mmol, 1.00 eq.) in water at a temperature of 0 °C. After complete addition, a 4.5 M solution of sodium acetate in water (62.4 mL, 281 mmol, 5.62 eq.) was added via dropping funnel, followed by dropwise addition of ethyl-2-oxocyclopentanecarboxylate **2** (CAS 611-10-9, 7.40 mL, 50.0 mmol, 1.00 eq.). The resulting yellow suspension was maintained at 0 °C for 15 minutes and then warmed to room temperature and stirred for 2 hours. The reaction mixture was extracted thrice with DCM (100 mL each) and the combined organic extracts were dried over magnesium sulfate, filtered and concentrated under reduced pressure to give the crude hydrazone as a red oil (18.1 g). The residue was dissolved in ethanol (50.0 mL, 1.00 M), after which sulfuric acid (6.63 mL, 125 mmol, 2.50 eq.) was added dropwise. The dark orange solution was heated at 95 °C for 6 days and then cooled to room temperature. The dark brown solution was poured onto ice/water (200 mL) and extracted thrice with DCM (200 mL each). The combined organic extracts were washed with saturated aqueous bicarbonate solution (200 mL), dried over magnesium sulfate, filtered and concentrated under reduced pressure to give a brown solid. The residue was purified by flash column chromatography (0–30% ethyl acetate/hexane gradient) and then recrystallized from hot ethyl acetate/hexane (9 : 1) to give the title compound as a light yellow solid (8.35 g, 42%). $R_f = 0.22$ (15% ethyl acetate/hexane, UV).

The Mechanism
The first stage is a Sandmeyer diazotization of the aniline **1** followed by the addition of the deprotonated form of **2** to the N ≡ N triple bond of the diazonium salt **A** to give intermediate **B**. Under the basic condition and the participation of sodium acetate, **B** undergoes retro-Dieckmann reaction to generate the ester acid **C**. **C** is isomerized to **D** and **E**, then a Fischer indole synthesis, and meanwhile esterification of the terminal acid, leads to the product **3**.

Sandmeyer reaction
NaNO$_2$, HCl
NaCl
EtOH
H_2SO_4
AcONa
AcOH
retro-Dieckmann condensation
Hydrolysis of anhydride
Fischer indole synthesis
EtOH
H_2SO_4
NH_3
H_2SO_4
NH_4HSO_4

Problem 99: A Novel Synthesis of Indoles by Vinyl Grignard Reagent with Nitroarenes

4-Bromo-7-chloro-1*H*-pyrrolo[2,3-c]pyridine (**3e**) is used as a core structure of fostemsavir [291], a recently marketed anti-HIV medication. Compound **3e** was prepared from the pyridine **1** and vinyl magnesium bromide **2** [292].

Please provide mechanism for the transformation.

THF
−78–20 °C
35%
1e
2
3e
Fostemsavir

Experimental Procedure (from *J. Med. Chem.*, **2018**, *61*, 6308, Supporting Information)

4-Bromo-7-chloro-1*H*-pyrrolo[2,3-c]pyridine (**3e**): At −78 °C, to a solution of 5-bromo-2-chloro-3-nitropyridine (**5**, 250 g, 0.97 mol) in THF (5 L), a solution of freshly prepared vinyl magnesium bromide in THF (3.92 L, 3.92 mol) was added via a cannula slowly, maintaining the temperature lower than −40 °C. The reaction was carried out at −78 °C for 1.5 hours, followed by adding 1.5 L of saturated aqueous NH_4Cl solution. The resulting mixture was stirred for 16 hours and the layers were separated. The aqueous phase was extracted with EtOAc (3 × 1 L). The organic extracts were combined, washed with brine, and concentrated under vacuum to give a residue which was dried under vacuum. 1 L of EtOAc was added to the residue to form a suspension, from which ~800 mL of EtOAc was removed under vacuum. The residual suspension was left in a refrigerator for 1 hour. The solid was collected via filtration, washed with EtOAc (2 × 50 mL) and EtOAc/hexanes (1:1, 50 mL and 100 mL), dried under high vacuum to yield **3e** (52.3 g, 23%) as a brown solid. ^{1}H NMR (500 MHz, DMSO-d_6) δ ppm 12.5 (br, 1H), 8.05 (s, 1H), 7.79 (s, 1H), 6.58 (s, 1H). ^{13}C NMR (125 MHz, DMSO-d_6) δ ppm 137.2, 134.4, 132.9, 131.6, 129.3, 110.9, 102.2. MS *m/z*: $(M + H)^+$ calcd. for $C_7H_4BrClN_2$: 230.93; found 231.15.

(from *J. Org. Chem.*, **2002**, *67*, 2345)

General Procedure for the Preparation of Azaindoles as Exemplified by the Preparation of 7-chloro-6-azaindole (3c). 2-Chloro-3-nitropyridine **1c** (5.0 g, 31.5 mmol) was dissolved in dry THF (200 mL) under N_2, and the solution was cooled to −78 °C. Excess vinyl magnesium bromide (1.0 M in THF, 100 mL, 100 mmol) was added and the reaction mixture stirred at −20 °C for 8 h before the reaction was quenched with 20% NH_4Cl (150 mL). The aqueous phase was extracted with EtOAc (3 × 150 mL) and the combined organic layer dried over $MgSO_4$, filtered, and concentrated. The crude product was purified by silica gel column chromatography to afford 7-chloro-6-azaindole **3c** (1.5 g, 31%).

4-Bromo-7-chloro-6-azaindole 3e (35%). ^{1}H NMR (500 MHz, DMSO-d_6) δ 8.05 (s, 1H), 7.79 (d, 1H, J = 3.00 Hz), 6.58 (d, 1H, J = 3.00 Hz), 3.40 (b, 1H); ^{13}C NMR (125 MHz, DMSO-d_6) δ 137.2, 134.3, 132.9, 131.6, 129.3, 110.9, 102.2; HRMS m/z: $(M-H)^-$ calcd for $C_7H_3BrClN_2$ 228.9168, found 228.9173.

The Mechanism

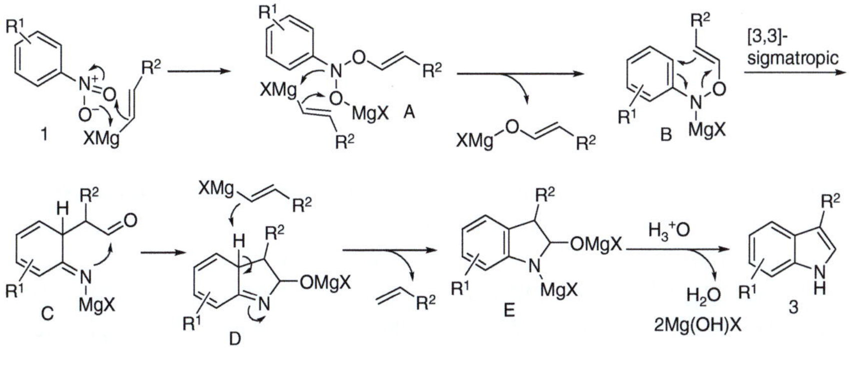

The reaction in question was originally developed by an Italian research group in 1989 [293]. The same group also investigated the mechanism of this reaction, which can be outlined as follows [294]: In the initial step, the Grignard reagent targets the oxygen atoms of the nitroarene, forming intermediate **A**. A second equivalent of the Grignard reagent then reduces intermediate **A** to produce the *N*-aryl-*O*-vinylhydroxylamino magnesium salt **B**. This intermediate undergoes a [3,3]-sigmatropic rearrangement, resulting in the formation of compound **C**. Subsequently, a rapid ring closure occurs, generating the bicyclic salt **D**. A third equivalent of the Grignard reagent acts as a base on this bicyclic intermediate, facilitating the re-aromatization of the six-membered ring. Finally, an acidic aqueous workup, followed by the elimination of water, yields the indole product **3**.

Problem 100: Deprotection of 2-Hydroxylethyl from Indole Derivative

PNU-142731A (**3**) is a potential lead compound as a treatment for asthma in Pharmacia. In a process chemistry, the protection group 2-hydroxylethyl in compound **1** was removed by the reactions as shown below [295].

Please suggest a mechanism for the deprotection.

1. MsCl, DCM TEA
2. NaCN, DMSO H_2O, 115 °C
93%

1. BuLi
2. Br–CH2–C(O)–N(pyrrolidine)
3. HCl
83%

1 2 3 (PNU-142731A)

Experimental Procedure (from *Org. Process Res. Dev.*, **2001**, *5*, 144)

2,4-Di-1-pyrrolidinyl-5H-pyrimido[4,5-b]indole (2). A stirred, −10 °C solution of hydroxyethyl intermediate **1** (125 g, 0.36 mol) and triethylamine (36.4 g, 0.36 mol) in 2000 mL of CH_2Cl_2 was treated with a solution of methanesulfonyl chloride (40.8 g, 0.36 mol) in 200 mL of CH_2Cl_2, added over 30 min. After an additional 30 min, TLC analysis showed a small amount of **1** remaining, and therefore, an additional 5 mL of Et_3N and 1 mL of MsCl were added, and stirring at −10° was continued for 15 min. The reaction mixture was then poured into 1 L of 10% aqueous sodium bicarbonate and extracted thoroughly with CH_2Cl_2. The extracts were washed with brine, dried over anhydrous sodium sulfate, and concentrated, thereby affording the crude solid mesylate. This material (containing small amounts of Et_3N) was used directly in the cyanide displacement/elimination which follows. (For other purposes, residual triethylamine and trace by-products could easily be removed from the mesylate by trituration with 60/40 EtOAc/hexane.)

A solution of the above mesylate and sodium cyanide (186 g, 3.8 mol) in DMSO (3.3 L) and water (300 mL) was heated at 115 °C for 12 h. Approximately two-thirds of the DMSO was then removed *in vacuo* (bath temperature 100 °C), and the residue was cooled to 25 °C and partitioned between CH_2Cl_2 (2 L) and 10% aqueous $NaHCO_3$ (800 mL). The aqueous layer was extracted with additional CH_2Cl_2 (3 × 300 mL), and the combined extracts were washed with brine (800 mL) and dried with Na_2SO_4. The residue resulting from removal of the solvents was triturated with acetone (400 mL), filtered, washed with cold acetone, and dried (0.05 mm, 40 °C, 48 h) to afford **9** (101.5 g, 93%) as a white solid, completely clean by TLC and NMR: mp 209–211 °C; ^{1}H NMR δ 9.70 (s, 1H), 7.87 (m, 1H), 7.22 (m, 1H), 7.15–7.06 (m, 1H), 3.95 (m, 4H), 3.67 (m, 4H), 2.02–1.95 (m, 8H); ^{13}C NMR δ 159.8, 158.2, 158.0, 135.9, 122.1, 121.8, 120.6, 119.8, 110.4, 90.4, 49.6, 46.8, 25.7; MS (FAB) 308 (M^+ + H), 307. Anal. Calcd for $C_{18}H_{21}N_5$: C, 70.33; H, 6.89; N, 22.78. Found: C, 70.28; H, 7.20; N, 22.67.

The Mechanism

The primary alcohol **1** is converted to mesylate **A**. Then the OMs is substituted by CN to form **B**. Finally, acrylonitrile is eliminated via a retro-Michael reaction at basic condition to give product **2**.

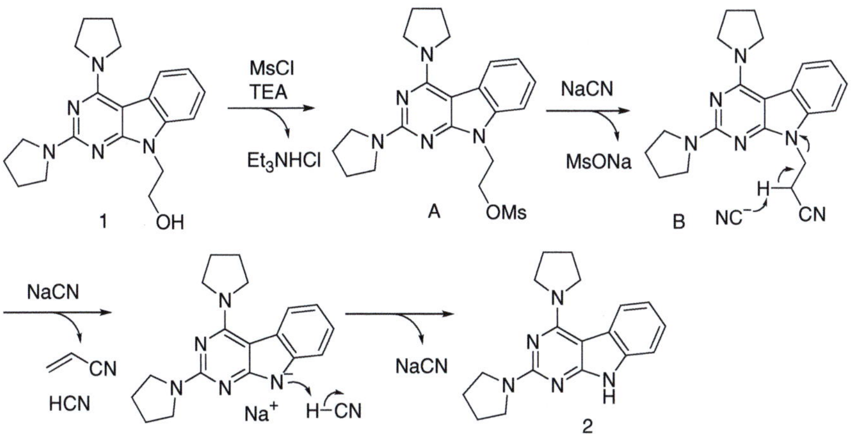

Couty's group reported a similar removal of hydroxyethyl group from tertiary amines [296]. Due to the phenyl attached to the hydroxyethyl moiety, the retro-Michael reaction was carried at room temperature. Without the phenyl attachment, the reaction did not work at the conditions.

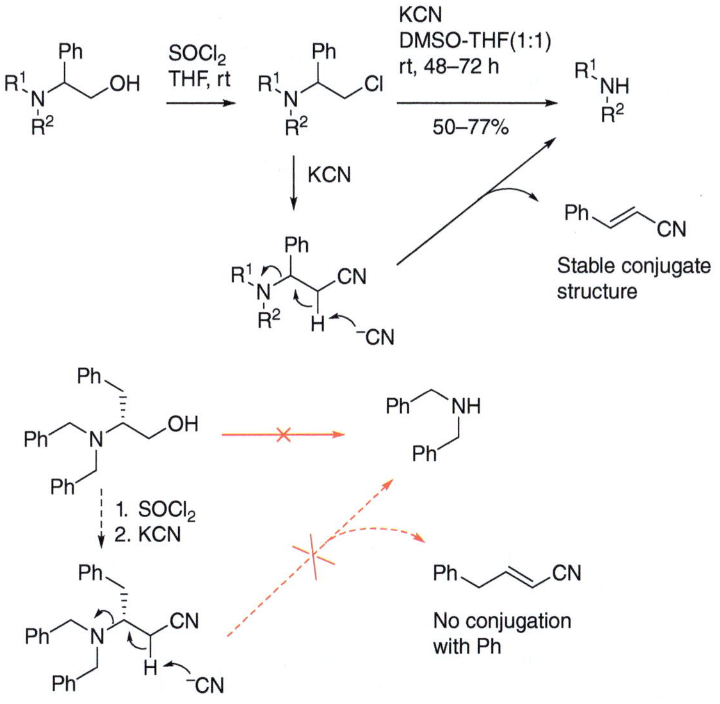

4.7.2 Indolizine

Problem 101: Access to 2-Iodoindolizines via Iodine-Mediated Cyclization of 2-pyridylallenes

Indolizines are key motifs in many biologically active compounds [297, 298]. Martinez et al. reported a straightforward preparation of 2-iodoindolizines via iodine-mediated cyclization of 2-pyridylallenes [299]. An example is shown below.

Please propose a mechanism for this reaction.

Ph C t-Bu Ph N 1 — I_2, K_2CO_3, MeCN, 70 °C, 19 h, 92% → Ph I Ph N 2

Experimental Procedure (from *Org. Process Res. Dev.*, **2020**, *24*, 817, Supporting Information)

2-Iodo-1,3-diphenylindolizine (2). In anhydrous glassware, a solution of allene **1** (1.0 g, 3.07 mmol, 1 equiv), iodine (1.56 g, 6.15 mmol, 2 equiv.), and potassium carbonate (850 mg, 6.15 mmol, 2 equiv) in acetonitrile (61 mL) was heated at 70 °C over 20 h. The mixture was quenched with a 5 w% aqueous $NaHSO_3$ solution (50 mL). The solid was filtered, washed with a 5 w% aqueous $NaHSO_3$ solution (2 × 25 mL), and dried under high vacuum to afford compound **2** as a green solid (1.11 g, 92% yield). m.p. = 128–129 °C. IR (neat): (cm^{-1}) = 3063, 3018, 1949, 1597, 1530, 1520, 1441, 1339, 1303, 1243, 1176, 1140, 1071, 1028, 753, 736, 72, 698, 671. ^{1}H NMR (400 MHz, $CDCl_3$) δ 7.92 (dt, J = 7.2, 0.8 Hz, 1H, H_1), 7.64 – 7.51 (m, 6H, H_{Ar}), 7.53 – 7.43 (m, 4H, H_{Ar} and $_4$), 7.38 (tt, J = 6.9, 1.2 Hz, 1H, H_{Ar}), 6.68 (ddd, J = 9.1, 6.5, 1.0 Hz, 1H, H_3), 6.47 – 6.35 (m, 1H, H_2). ^{13}C NMR (101 MHz, $CDCl_3$) δ 135.4 (C, C_{Ar}), 131.6 (C, C_{Ar}), 131.2 (2CH, C_{Ar}), 130.9 (C, C_{Ar}), 130.5 (2CH, C_{Ar}), 129.2 (2CH, C_{Ar}), 128.8 (CH, C_{Ar}), 128.5 (2CH, C_{Ar}), 126.9 (CH, C_{Ar}), 126.6 (C, C_{Ar}), 122.8 (CH, C_1), 118.5 (CH, C_3), 118.4 (C, C_{Ar}), 117.9 (CH, C_4), 111.4 (CH, C_2), 77.0 (C, C_7). HRMS (ESI) calcd. for $C_{20}H_{14}IN$ ($[M + H]^+$) 396.0199 found 396.0230.

The Mechanism

Addition of iodine to 1,2-diene in **1** gives iodonium transition state **A**. Then cyclization of **A** provides indolizinium intermediate **B**. Under basic condition, here is K_2CO_3, elimination of 2-methylprop-1-ene leads to the product **2** (next page).

4.7.3 Quinolone

Problem 102: Oxidative Rearrangement of Carboline to Quinolone

R301249 (RWJ387273) and R290629 (RWJ444772) are two PDE-V inhibitors active in the treatment of erectile dysfunction. In a synthesis of those compounds, oxidation of β-carboline **1** with dimethyldioxirane, generated in situ from Oxone and acetone, followed by further oxidation with *m*CPBA gave the desired keto-lactam **2** and an impurity hydroxy ketone **3** [300]. Intramolecular condensation of **2** provided quinolone **4**.

Ph t-Bu Ph N 1 I_2 A B KO O K+ H
K_2CO_3 KI, $KHCO_3$ 2-methylprop-1-ene 2

Please propose mechanisms for the formation of **2** and **3**.

R301249 (RWJ387273) R290629 (RWJ444772)

1. Oxone, acetone 2. mCPBA N-CBZ 1 2 3 NaOH EtOH 4 R301249 (RWJ387273) R290629 (RWJ444772)

Experimental procedure (from *Org. Process Res. Dev.* **2006**, *10*, 1275)

Benzyl (3*R*)-3-(2,3-Dihydro-1-benzofuran-6-yl)-2,7-dioxo-1,2,3,5,6,7-hexahydro-4*H*-1,4-benzodiazonine-4-carboxylate (2). A 250 mL vessel was charged successively with 1-(*R*)-1-(2,3-dihydrobenzofuran-5-yl)-2,3,4,9-tetrahydro-1*H*-*β*-carboline (5.22 g, 0.018 mol), ethyl acetate (150 mL), and triethylamine (2.1 g, 0.0208 mol). The mixture was heated to 40–45 °C. Then the heating was stopped, and benzylchloroformate (3.55 g, 0.0208 mol) was added dropwise. The addition was slightly exothermic, and the temperature rose from 42 to 51 °C. The mixture was stirred for another 2 h at 51 to 42 °C. Water (18 mL) was added, and the layers were separated while keeping the temperature at 35–40 °C. To the organic layer, acetone (107 mL) was added, and the mixture was cooled to 20–25 °C. Subsequently, water (36 mL) and $NaHCO_3$ (4.82 g, 0.057 mol) were added, and the mixture was further cooled to 0–5 °C. A solution of oxone (15.5 g, 0.025 mol) in water (60 mL) was added by means of a dosing pump over a 2 h period at 0–5 °C. After the addition was complete, the mixture was stirred for another 2 h at 0–5 °C. Subsequently, a solution of *m*-chloroperbenzoic acid 71% (5.34 g, 0.217 mol) in DCM (62 mL) was added in 15 min at 0–5 °C. The mixture was stirred for another 1.5 h at 0–5 °C and then washed successively with water (76 mL) and with a solution of $NaHCO_3$ (6 g) in water (76 mL). After checking for peroxides, the organic layer was evaporated to nearly dryness, and the resulting oily residue was crystallized from ethyl acetate (40 mL). After stirring for 18 h at 10–25 °C the precipitate was filtered, washed with ethyl acetate (2 mL), and dried *in vacuo* at 50 °C for 20 h yielding 3 g (36.5%) of the title compound. ^{1}H NMR (400 MHz, DMSO-d_6) showed a mixture of rotamers: δ ppm [2.75–2.90] (m, 2H); [3.08–0.019] (m, 3H); [4.45–4.53] (o, 3H); [5.01–5.24] (4 × d (o), 2H); 5.93, 5.96 (2 × s, 1H); [6.67–6.72] (2 × d (o), 1H); [6.84–6.90] (2 × d (o)n 1H); 7.01, 7.03 (2 × s, 1H); [7.19–7.67] (o.m. 9H); 10.84, 10.86 (2 × s, 1H).

Probably other rotamers visible: [3.38–4.28] (m); [4.60–4.77] (2 × d); 5.54, 5.61 (2 × s); 9.68, 9.73 (2 × s). o: overlapping.

The Mechanism

The first step is an epoxidation of the 2,3-double bond of indole in **1** by dimethyldioxirane, generated in situ from acetone and Oxone. The epoxide **A** in turn rearranges to hydroxy-imine **B**. Further oxidation of the imine with *m*CPBA gives the peracid ester **C**. Subsequently, cleavage of the 2,3-C—C bond provides the keto-lactam **2**.

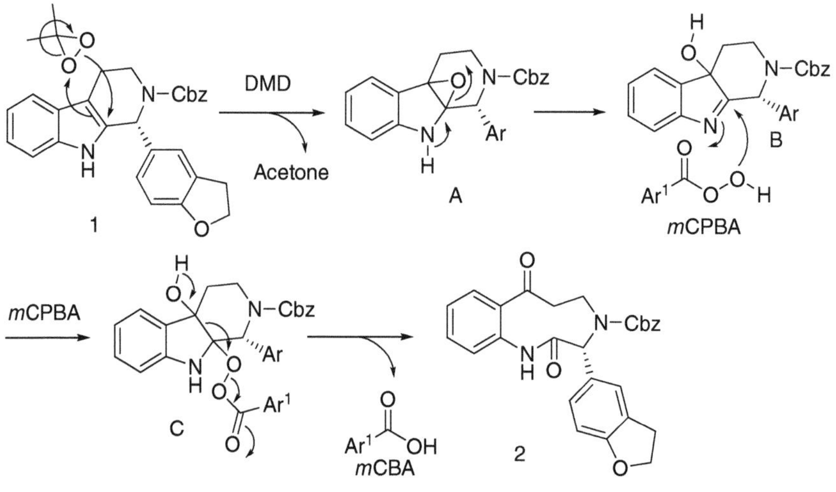

Once the epoxide **A** is formed, the elimination of a proton could also take place from the piperidine side to generate the enamine **D**. Then the double bond is further oxidized by *m*CPBA to give epoxide **E**. The epoxide **E** is opened in the assistance of the α-amino group to afford the iminodiol **F**. Finally, **F** is rearranged to hydroxy ketone **3**.

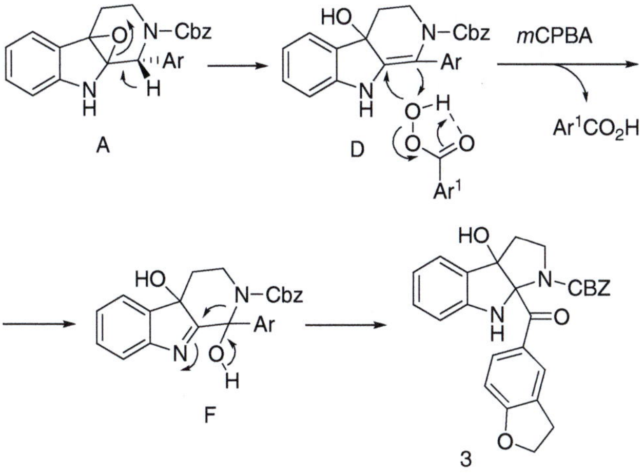

4.7.4 Indazole

Problem 103: A Special Cadogan Reductive Cyclization-Preparation of Indazole

2*H*-indazoles as pharmacophores in drug discovery has been exemplified in several cases such as MK-4827 and pazopanib. Genung's group reported a convenient Cadogan reductive cyclization by using tri-*n*-butylphosphine as the reducing reagent [301]. For example, (*E*)-1-(2-nitrophenyl)-*N*-phenylmethanimine **1** is transformed into 2-phenyl-2*H*-indazole **2** in 87% yield.

Please propose a mechanism for this reaction.

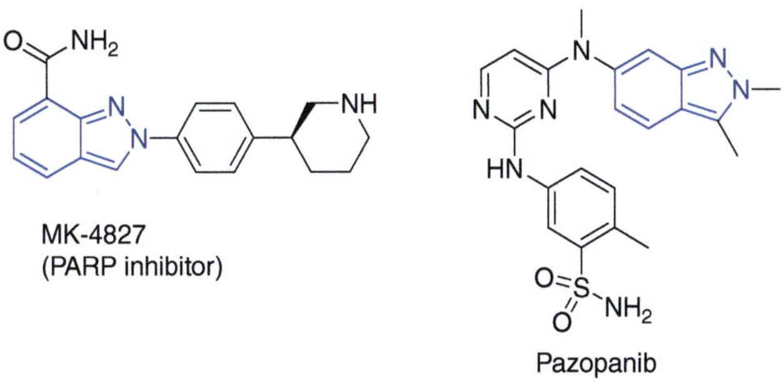

MK-4827
(PARP inhibitor)

Pazopanib

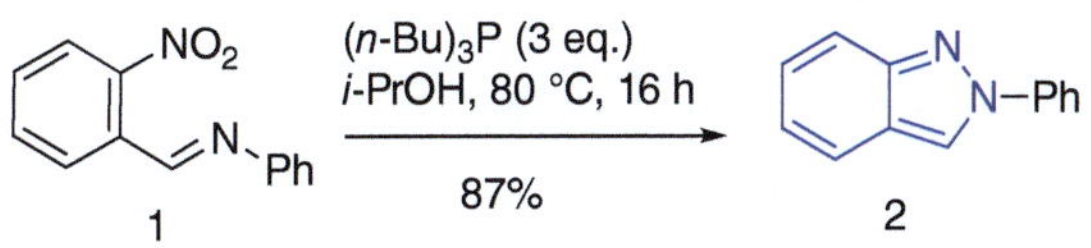

$(n\text{-Bu})_3P$ (3 eq.)
i-PrOH, 80 °C, 16 h
87%

Experimental Procedure (from *Org. Lett.*, **2014**, *16*, 3114, Supporting Information)
Representative Procedure for the One-Pot Synthesis of 2*H*-indazoles: 2-Phenyl-2H-indazole (2): To a 2-dram vial equipped with a stir bar was added 2-nitrobenzldehyde (200 mg, 1.30 mmol) and *i*PrOH (3.0 mL). Aniline (133 mg, 0.130 mL, 1.43 mmol) was added in one portion and the resulting solution was heated to 80 °C in a sealed 2-dram vial with stirring for 4 h. The mixture was cooled to room temperature and tri-*n*-butylphosphine (790 mg, 0.980 mL, 3.91 mmol) was added in one portion followed by stirring at 80 °C in a sealed 2-dram vial for 16 h. The mixture was cooled to room temperature and diluted with EtOAc (25 mL). The organics were washed with ammonium chloride (15 mL), brine (15 mL), dried over $MgSO_4$, filtered, and concentrated *in vacuo*. The residue was purified by silica gel chromatography (0–25% EtOAc/heptane) to afford the desired product **2** (213 mg, 85%) as a white solid. The spectroscopic data obtained for the product matched previously reported data: ^{1}H NMR ($CDCl_3$, 600 MHz) $\delta = 8.43$ (s, 1H), 7.92 (d, $J =$ 8.8 Hz, 2H), 7.81 (d, $J = 8.8$ Hz, 1H), 7.73 (d, $J = 8.2$ Hz, 1H), 7.55 (t, $J = 8.2$ Hz, 2H), 7.46–7.38 (m, 1H), 7.34 (dd, $J = 7.3, 8.5$ Hz, 1H), 7.13 (dd, $J = 7.0, 8.2$ Hz, 1H).

The Mechanism

The tributylphosphine reduces the nitro group sequentially through addition to the NO bond in both starting material **1** and the assumed intermediate **A** to give a nitrene intermediate **B**. Then cyclization occurs to provide product **2**.

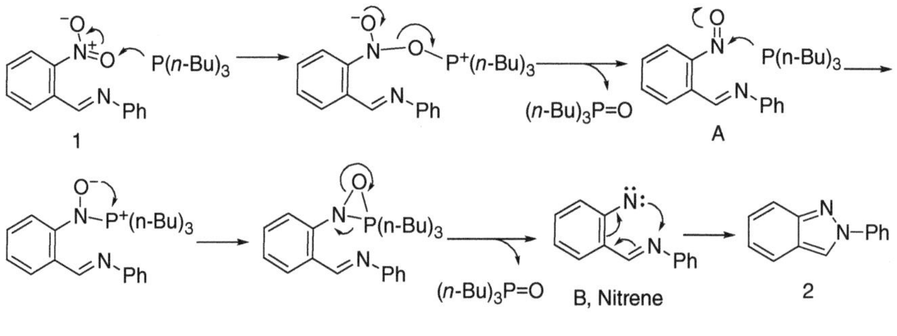

Another possible mechanism may be via the *N*-oxide intermediate **C**.

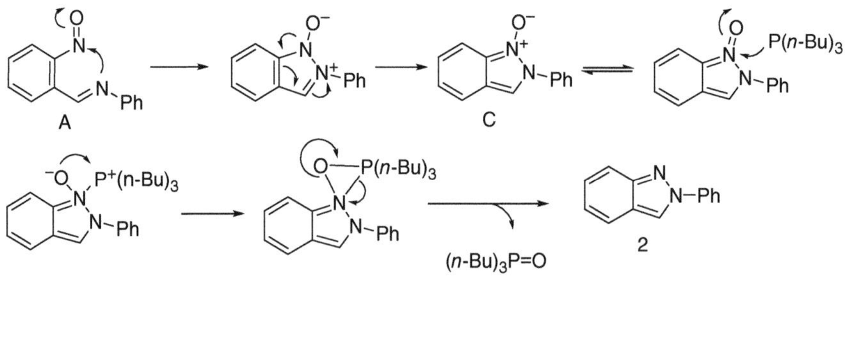

The product **2** could also be generated via 1,4-addition of tributylphosphane to the *N*-oxide **C**.

C P(n-Bu)3 → P+(n-Bu)3 → n-Bu n-Bu P–n-Bu → (n-Bu)3P=O 2

Problem 104: Preparation of Indazoles from 2-Methylanilines

Compound **1** is a very active anti-HIV agent [302]. To develop an efficient synthesis of **1**, the key indazole core was prepared from 2-alkylanilines. The reaction from **2** to **3** was reported by Sun's group [303].

Could you please provide the mechanism for the reaction?

1, anti-HIV active

Ac_2O, KOAc, amylnitrite, 18-C-6, $CHCl_3$, 58%

Experimental Procedure (from *J. Org. Chem.*, **1997**, *62*, 5627)

1-Acetyl-5-(acetoxymethyl)indazole (3). A mixture of **2** (16.58 g, 0.12 mol), acetic anhydride (34.0 mL, 0.36 mol), and potassium acetate (23.71 g, 0.24 mol) in 240 mL of $CHCl_3$ was stirred at room temperature for 3 h, refluxed for 2 h, and stirred at room temperature overnight. Then *n*-amyl nitrite (32 g, 0.27 mol) and 18-crown-6 (1.59 g, 6.0 mmol) were added and the mixture heated at reflux for 28 h. After being cooled to room temperature, the reaction mixture was added to acetic anhydride (10 mL) and stirred at room temperature overnight. The reaction mixture was diluted with CH_2Cl_2 (400 mL), washed with saturated $NaHCO_3$ (200 mL), water, and brine, and dried (Na_2SO_4) and the solvent evaporated to give a dark brown solid. Chromatography (silica gel, 15% EtOAc/hexane) gave **3** as a yellow solid (16.98 g, 58%). **3**: mp 73–74 °C; ^{1}H NMR ($CDCl_3$) δ 8.44 (d, J = 8.8 Hz, 1 H), 8.13 (d, J = 0.8 Hz, 1 H), 7.75 (d, J = 0.7 Hz, 1H), 7.56 (dd, J = 8.8, 1.5 Hz, 1H), 5.23 (s, 2H), 2.79, (s, 3H), 2.12 (s, 3H); CIMS (NH_3) *m/z* 267 (M + 2 X NH_3 + H^+, 100), 250 (M + NH_4^+). Anal. Calcd for $C_{12}H_{12}N_2O_3$: C, 62.06; H, 5.22; N, 12.06. Found: C, 62.07; H, 5.07; N, 11.91.

The Mechanism

This reaction is actually an old one, initially reported by Jacobson and Huber in 1908 [304]. According to the literature [305, 306], the nitroso intermediate **A** undergoes isomerization to form the diazoacetate **B**. This intermediate can then cyclize either through the formation of the diazonium salt **C** or via a synergistic reaction to yield intermediate **D**. Following isomerization and acetylation of the indazole, the final product **3** is obtained.

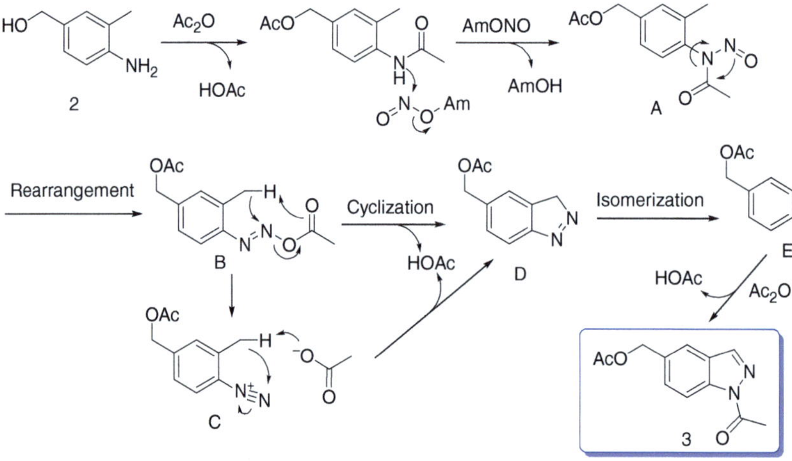

Problem 105: Synthesis of 7-aza-Indazoles by Diels-Alder/Retro-Diels-Alder Cascade

A method for synthesizing 7-aza-indazoles by Diels-Alder/retro-Diels-Alder cascade was developed [307, 308]. For example, ethyl 5-fluoro-1H-pyrrolo[2,3-b]pyridine-3-carboxylate (**3**) was prepared in quantitative yield from commercially available starting materials **1** and **2**. Compound **3** was further converted to compound **4**, which is the key intermediate for synthesizing the marketed drug vericiguat [309].

Please suggest a mechanism from **1**, **2** to **3**.

Experimental Procedure (from *J. Am. Chem. Soc.*, **2019**, *141*, 15901, Supporting Information)

Synthesis of Vericiguat Intermediate 4

Procedure. To a solution of 5-fluoro-2-hydrazinylpyrimidine (**1**, 500 mg, 3.90 mmol, 1.10 equiv.) in THF (0.55 M) were added ethyl 2-oxo-4-(trimethylsilyl) but-3-ynoate (**2**, 704 mg, 3.55 mmol, 1.00 equiv.) and a catalytic amount of TFA (0.05 mL, 0.71 mmol, 0.20 equiv.) under nitrogen. The reaction mixture was warmed to 60 °C for 20 min at which point TLC and ^{1}H NMR analysis of an aliquot demonstrated the total consumption of the starting materials. 3-Pentanone (1.12 mL, 10.64 mmol, 3.00 equiv.) and TFAA (1.48 mL, 10.64 mmol, 3.00 equiv.) were then added dropwise and stirring was continued for 1 h. TLC and ^{1}H NMR demonstrated the total consumption of the trifluoroacetylated hydrazone previously formed.

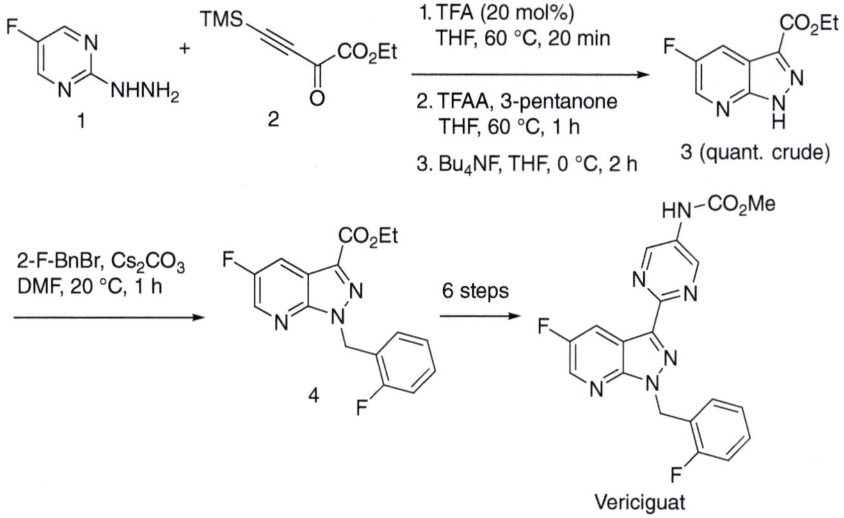

The reaction mixture was cooled to room temperature, diluted with EtOAc and washed twice with an aqueous saturated solution of Na_2CO_3. The organic layer was dried over $MgSO_4$, filtered, and the solvent evaporated under vacuum. The crude residue was directly engaged in the desilylation step. To a 0 °C solution of the crude residue in THF (0.25 M) under nitrogen was added dropwise TBAF (3.55 mL, 1 M in THF, 3.55 mmol, 1.10 equiv.). After 2 hours, TLC showed total conversion of the intermediate. The mixture was diluted with EtOAc and washed with water. The organic layer was dried over $MgSO_4$, filtered, and the solvent evaporated under vacuum. The crude residue was directly engaged in the protection step. To a solution of the residue in DMF (0.22 M) at room temperature was added cesium carbonate (873 mg, 4.05 mmol, 1.10 equiv.) followed by 1-(bromomethyl)-2-fluorobenzene (0.49 mL, 4.05 mmol, 1.10 equiv.). After 1 h, the solution was dissolved with EtOAc and washed with water. The organic layer was dried over $MgSO_4$, filtered, and the solvent was evaporated under vacuum. The crude product was analyzed by ^{1}H NMR and consisted of a clean mixture of two isomers resulting from the *N*-benzylation reaction (N1/N2 = 60 : 40). After purification on silica gel chromatography, ethyl 5-fluoro-1-(2-fluorobenzyl)-1*H*-pyrazolo[3,4-*b*]pyridine-3-carboxylate (494.2 mg) and ethyl 5-fluoro-2-(2-fluorobenzyl)-2*H*-indazole-3-carboxylate (330.5 mg) were isolated as colorless oils (67% combined yield).

Characterization. Ethyl 5-fluoro-1-(2-fluorobenzyl)-1*H*-pyrazolo[3,4-*b*]pyridine-3-carboxylate (4). ^{1}H NMR ($CDCl_3$, 500 MHz, ppm): δ 8.42 (q, $J = 1.0$ Hz, 1H), 8.06 (dd, $J = 8.0$ Hz, $J = 3.0$ Hz, 1H), 7.19–7.14 (m, 1H), 7.09–7.06 (m, 1H), 7.00–6.93 (m, 2H), 5.81 (s, 2H), 4.46 (q, $J = 7.0$ Hz, 2H), 1.41 (t, $J = 7.0$ Hz, 3H); These data are in agreement with the reported ^{1}H NMR for this compound. ^{13}C NMR ($CDCl_3$, 125 MHz, ppm): δ 161.6, 160.3 (d, $J = 248.2$ Hz, 1C), 156.5 (d, $J = 250.1$ Hz, 1C), 147.9, 139.8 (d, $J = 38.3$ Hz, 1C), 134.5 (d, $J = 6.5$ Hz, 1C), 129.8 (d, $J = 10.1$ Hz, 1C), 129.8 (d, $J = 4.3$ Hz, 1C), 124.2 (d, $J = 4.5$ Hz, 1C), 122.9 (d, $J = 18.1$ Hz, 1C), 115.6 (d, $J = 26.8$ Hz, 1C), 115.5, 115.4 (d, $J = 26.3$ Hz, 1C), 61.4, 45.3 (d, $J = 6.0$ Hz, 1C), 14.4; ^{19}F NMR ($CDCl_3$, 471 MHz, ppm): δ – 117.8, –133.6.

The Mechanism

The first reaction is a condensation of the hydrazine **1** with the ketone **2** to form intermediate **A**. An intramolecular Diels-Alder reaction then occurs between the triple bond and the pyrimidine to give the bridged intermediate **B**. Subsequently, a retro Diels-Alder reaction occurs with the elimination of HCN to yield intermediate **C**. The HCN could be trapped by 3-pentanone. Finally, desilylation of **C** by TBAF leads to the product **3**.

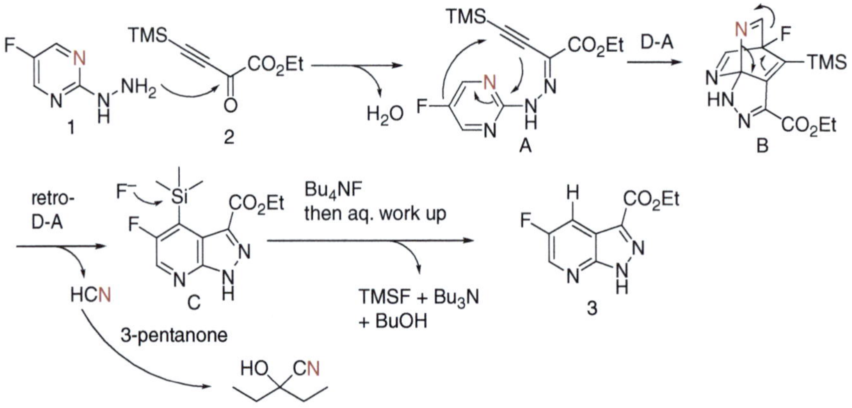

Problem 106: Synthesis of Indazole by Davis-Beirut Reaction

BI-0282 (**1**) is a very potent inhibitor of the MDM2-p53 interaction [310]. The final step of the synthesis of **1** incorporates a Davis-Beirut reaction [311] as shown below.

Please suggest a mechanism for this transformation.

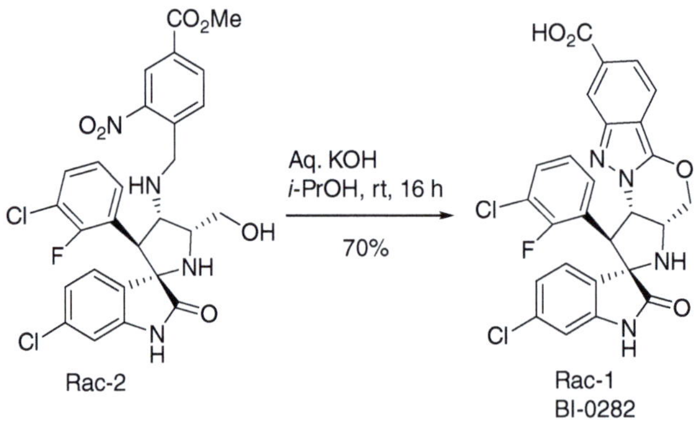

Experimental Procedure (from *Org. Process Res. Dev.*, **2022**, *26*, 2526)

Compound 1. Intermediate **rac-2** (28.8 g, 44.8 mmol) is dissolved in isopropanol (300 mL) and a solution of KOH (39.0 g, 694.9 mmol) in water (95 mL) is slowly added. After stirring for 16 h at ambient temperature, the solvents are partially removed under reduced pressure. The residue is diluted with ethyl acetate and treated with a diluted aqueous solution of citric acid. After extraction of the aqueous layer with ethyl acetate, the organic layers are combined, dried with sodium sulfate, and the solvent is removed under reduced pressure. Purification by normal phase column chromatography using DCM and methanol as solvents yields **rac-1** (25.8 g, 43.5 mmol) in 70% yield as an amorphous white solid.

^{1}H NMR (500 MHz, DMSO-d_6): δ 12.64 (br s, 1H), 10.29 (s, 1H), 7.67 (s, 1H), 7.47 (d, J = 8.83 Hz, 2H), 7.29–7.36 (m, 1H), 7.26 (d, J = 7.88 Hz, 1H), 7.21 (dd, J = 1.26, 8.83 Hz, 1H), 7.12 (t, J = 8.04 Hz, 1H), 6.92 (dd, J = 1.89, 7.88 Hz, 1H), 6.48 (d, J = 1.89 Hz, 1H), 5.86 (t, J = 9.14 Hz, 1H), 4.59–4.68 (m, 1H), 4.52 (dd, J = 7.88, 11.35 Hz, 1H), 4.23–4.32 (m, 1H), 4.20 (d, J = 10.09 Hz, 1H), 2.27 (dd, J = 7.57, 13.08 Hz, 1H), 2.13 (dd, J = 5.83, 13.08 Hz, 1H), 0.47–0.62 (m, 1H), 0.26–0.37 (m, 1H), 0.11–0.20 (m, 1H), −0.04 to 0.04 (m, 1H), −0.25 (s, 1H). ^{13}C{^{1}H} NMR (125 MHz, DMSO-d_6): δ 177.5, 168.1, 156.1 (d, $^1J_{C,F}$ = 248.7 Hz), 146.3, 145.3, 144.0, 134.1, 130.3, 129.7, 129.5, 126.8, 126.7, 125.4 (d, $^3J_{C,F}$ = 4.4 Hz), 123.5 (d, $^2J_{C,F}$ = 13.2 Hz), 122.5, 120.0, 119.9, 119.7 (d, $^2J_{C,F}$ = 18.3 Hz), 118.7, 110.0, 107.3, 76.4, 69.2, 57.5, 56.8, 54.2, 51.2, 11.6, 5.5, 4.1. HRMS (ESI) m/z: $[M + H]^+$ calcd for $C_{30}H_{24}Cl_2FN_4O_4$, 593.1153; found, 593.1165.

The Mechanism

The mechanism of Davis-Beirut reaction was first published in 2005 by Kurth, Olmstead, and Haddadin (see below) [312]. The reaction begins with the base deprotonation of benzylic hydrogen adjacent to the secondary amine group, creating a carbanion. The carbanion then extracts an oxygen from the nitro group.

That mechanism is widely cited in some online sources such as Wikipedia, YouTube, and Google. However, I do not agree with that mechanism since the negative oxygen on nitro group is almost impossible as an electrophile attacked by a carbanion nucleophile.

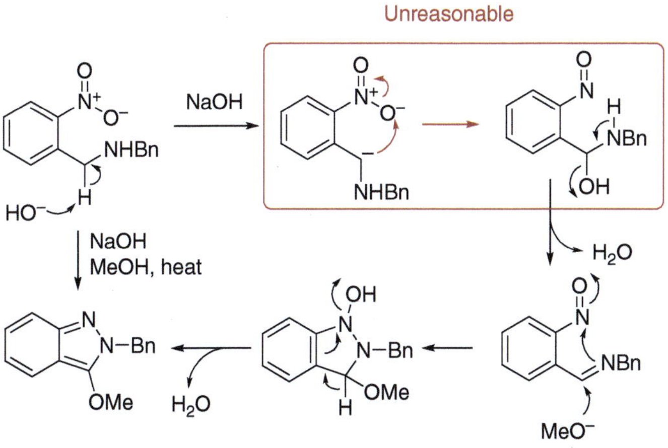

In this context, I propose an alternative mechanism. As expected, the methyl ester undergoes facile hydrolysis to yield the corresponding carboxylic acid. In the presence of KOH, this acid is converted into its potassium salt. The Davis-Beirut reaction initiates with the amine acting as a nucleophile, attacking the nitrogen atom within the nitro group to form intermediate **A**. Under strongly basic conditions, a molecule of water is eliminated, resulting in intermediate **B**. Subsequently, the hydroxyl group of the primary alcohol adds to the indazole *N*-oxide, leading to the formation of the *N*-hydroxy dihydroindazole **C**. Under the same strongly basic conditions, another molecule of water is eliminated, ultimately producing product **1**.

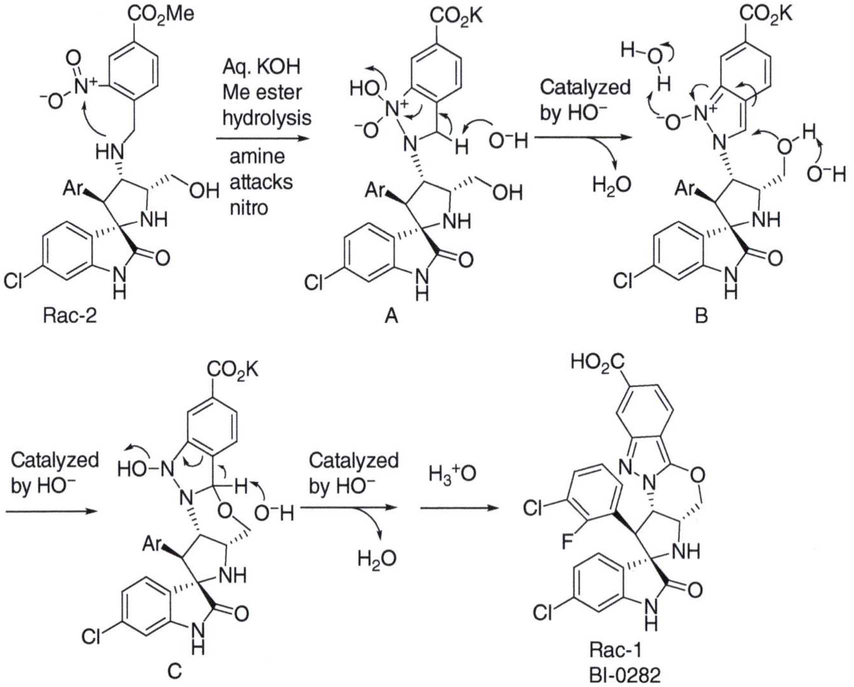

Upon revisiting the mechanism proposed by Kurth, Olmstead, and Haddadin [312], a slight modification leads to a plausible pathway. The carbanion in intermediate **I** attacks the nitro group from the oxygen side of the N=O double bond, resulting in the formation of the benzoisoxazole intermediate **II**. Subsequent ring opening of the oxazole generates the nitroso intermediate **IIIa/IIIb**. Elimination of water from **IIIb** then produces the nitroso imine **IV**. The remaining steps align with the mechanism originally proposed by Kurth et al.

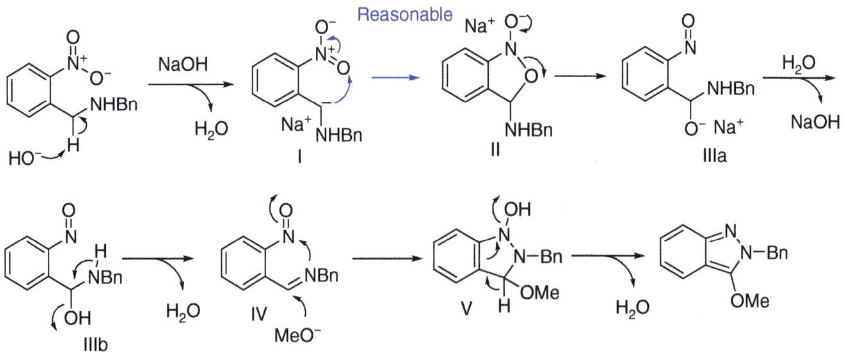

The following mechanism is also reasonable.

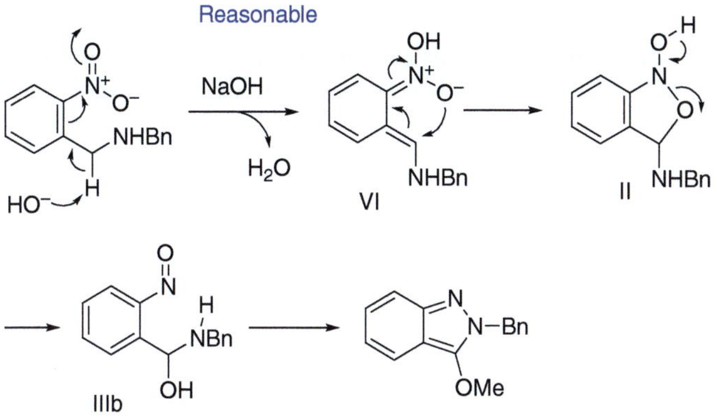

4.7.5 Quinazoline

Problem 107: Synthesis of Quinazolines

Quinazoline is an important structural motif occurring in several marketed drugs such as anticancer medication vandetanib. Yu's group reported a preparation of quinazolines (**2o**) by the following reaction [312].

Please provide a mechanism for the reaction.

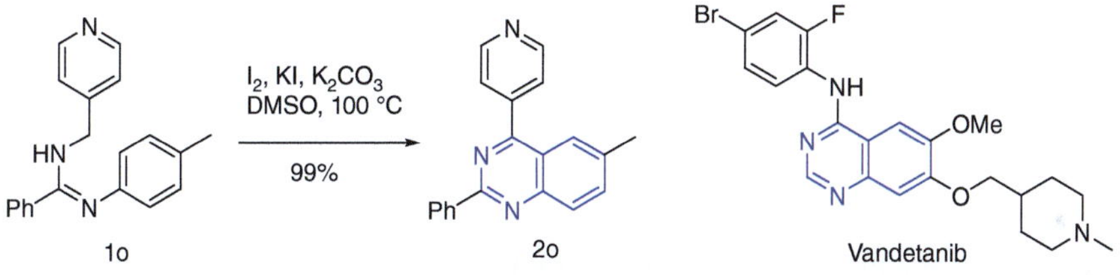

Experimental Procedure (from *J. Org. Chem.*, **2016**, *81*, 9924)

General Procedure for the Synthesis of Products (2). A mixture of iodine (244 mg, 0.96 mmol) and KI (160 mg, 0.96 mmol) in DMSO (2.5 mL) was stirred at room temperature for 10 min and then treated with substrate **1** (0.40 mmol), followed by addition of K_2CO_3 (139 mg, 1 mmol) and a further 2.5 mL of DMSO. The reaction was heated to 100 °C until TLC indicated the conversion was complete. After cooling to room temperature, it was quenched with 5% $Na_2S_2O_3$ (15 mL) and then extracted with EtOAc (15 mL × 3). The combined organic layer was dried over anhydrous Na_2SO_4, concentrated, and then purified through silica gel column chromatography to afford the desired product **2**.

6-Methyl-2-phenyl-4-(pyridin-4-yl)quinazoline (2o). 3.5 h; eluent: EtOAc/PE 80:20; yield: 117 mg, 99%; brown solid, mp 134–136 °C; ^{1}H NMR (400 MHz, $CDCl_3$) δ 8.88 (d, J = 5.6 Hz, 2H), 8.66–8.63 (m, 2H), 8.09 (d, J = 9.2 Hz, 1H), 7.78–7.57 (m, 4H), 7.55–7.50 (m, 3H), 2.54 (s, 3H); ^{13}C NMR (100 MHz, $CDCl_3$) δ 164.9, 159.7, 150.7, 150.0, 145.5, 138.0, 137.9, 136.4, 130.6, 129.2, 128.6, 128.5, 124.6, 124.4, 121.2, 22.0; HRMS (m/z) $[M + H]^+$ calcd for $C_{20}H_{16}N_3$, 298.1339, found 298.1337.

The Mechanism

The initial step involves the oxidation (dehydrogenation) of phenylbenzimidamide (**1**) using I_2 in the presence of a base. This is followed by an oxidative cyclization facilitated by I_2 and the base.

Base, Py, H, N, Ph, N, 1, I—I---IK, K_2CO_3, I_2, $KHCO_3$, KI, Py, H, Base, N, Ph, N, I, K_2CO_3, $KHCO_3$, KI, IK, I, Py, N, Ph, N, K_2CO_3, I_2

Py, H, Base, I, N, Ph, N ⊕, I ⊖, K_2CO_3, $KHCO_3$, KI, Py, H, Base, I, N, Ph, N, K_2CO_3, $KHCO_3$, KI, N, N, Ph, N, 2

Problem 108: Carbon Dioxide Mediated Synthesis of Quinazoline-2,4(1H,3H)-dione

Quinazoline-2,4(1*H*,3*H*)-diones, such as compound **2**, is an important pharmacophore in several drugs such as alfuzosin, prazosin, and doxazosin. Recently, Rasal and Yadav developed a process for preparation of quinazoline-2,4(1*H*,3*H*)-diones from 2-aminobenzonitriles mediated by CO_2 with DMF in water. When an electron withdrawing group (F, Cl, NO_2) attached on the benzene ring, some side products were also obtained [313].

Please suggest a mechanism for the reaction.

NH_2, N, N, N, O, O, N, H, O, O — Alfuzosin; NH_2, N, N, N, N, O, O, O, O — Prazosin; NH_2, N, N, N, N, O, O, O, O, O — Doxazosin

R, CN, NH_2, 1 — DMF (5 eq.), CO_2 (5 MPa), H_2O, 150 °C, 5 h → R, O, NH, N, H, O — 2 (85–99%); R, N, N, H, O — 3 (3–10%, R = F, Cl, NO_2)

Experimental Procedure (from *Org. Process Res. Dev.*, **2016**, *20*, 2067)

Reaction Procedure. All reactions were carried out in a 100 mL autoclave as discussed in our recent report. In a typical experiment, 10 mmol of 2-aminobenzonitrile (ABN) (or another substrate) and 5 equiv of DMF in 30 mL of water were taken in the reactor. The reactor was flushed two to three times with CO_2 to remove unwanted air. The temperature was increased to the desired value (150 °C), and 5 MPa of CO_2 was then introduced in the reactor and maintained for 5 h. After completion of the reaction, the reactor was cooled naturally to room temperature and then CO_2 was released. The product was separated from the reaction mass by simple filtration. Then, recrystallization of the product was done in water, and the product was weighed to obtain the yield of the reaction. The product was analyzed by HPLC (area normalization method) and characterized by mass and NMR.

6,7-Dimethoxy Quinazoline-2,4(1*H*,3*H*)-dione (2). ^{1}H NMR (500 MHz, DMSO-d_6) δ 11.09 (s, 1H), 10.94 (s, 1H), 7.21 (d, $J = 6.1$ Hz, 1H), 6.66 (s, 1H), 3.81 (s, 3H), 3.76 (s, 3H); ^{13}C NMR (126 MHz, DMSO-d_6) δ 162.86, 155.31, 150.88, 145.44, 136.95, 107.52, 106.6, 98.17, 56.17.

The Mechanism

A plausible mechanism was initially proposed by the authors in the literature [314]. However, several inaccuracies were identified in the published mechanism, which does not meet the criteria for a well-defined mechanism. Below is my interpretation of the authors' proposed pathway. Under high-pressure conditions (5 MPa) in water, CO_2 is partially converted into carbonic acid, which exhibits greater electrophilicity compared to CO_2. Subsequently, carbonic acid is attacked by DMF, forming an activated species **A**, analogous to the Vilsmeier–Haack reagent. This activated species is then attacked by the -NH_2 group in ABN, **1**, followed by the elimination of Me_2NH, yielding the formylated intermediate **B**. Intermediate **B** undergoes hydrolysis in a mild acidic medium (H_2CO_3) to produce intermediate **C**. The hydrolyzed product then cyclizes to form the hydroxyl intermediate **D**. Further oxidation of hydroxyl intermediate **D** leads to the formation of quinazoline-2,4(1H,3H)-dione (**2**). Dehydration of **D** could result in the side product quinazolin-4(3H)-one (**3**). In this mechanism for the formation of **2**, DMF serves as the carbon source, while CO_2 primarily acts as an activator.

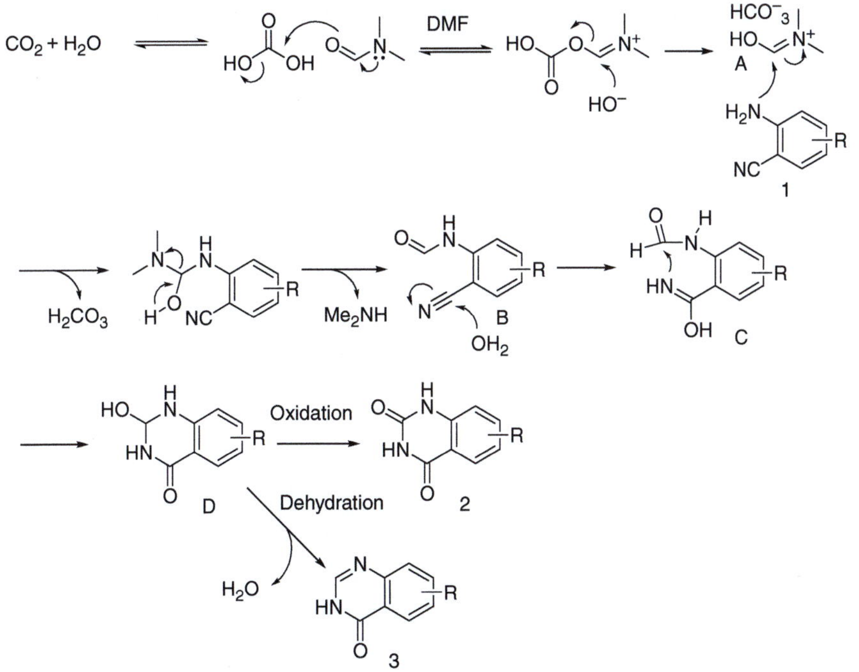

The proposed mechanism appears plausible for the formation of the side product quinazolin-4(3H)-one (**3**). However, I find it less convincing in explaining the formation of quinazoline-2,4(1H,3H)-dione (**2**), as several critical issues remain unaddressed.

First, the authors did not provide evidence confirming dimethylamine as a by-product. Dimethylamine could have been easily trapped by HCl, yet this was not verified. Second, if a ^{13}C -labeled DMF ($Me_2N^{13}CHO$) had been used, ^{13}C NMR analysis could have clarified whether the labeled ^{13}C was incorporated into the product. Third, the oxidation state of the carbonyl carbon in DMF is +2, whereas in product **2**, it is +4. If DMF were the carbon source, an oxidation step must have occurred. The authors did not elaborate on how the hydroxyl intermediate **D** was oxidized to form product **2**. What was the oxidant involved, and where did it originate? These unresolved questions cast doubt on the proposed mechanism for the formation of compound **2**.

I propose an alternative mechanism for the formation of compound **2**, as outlined below. Initially, carbonic acid is activated by DMF to generate intermediate **E**. Subsequently, the nitrile group is activated by **E**, and the carbon in the nitrile is attacked by the hydroxyl group (HO-) from **E**, resulting in the formation of benzimidic acid **F**. This intermediate then undergoes isomerization to yield benzamide **G**. Up to this stage, DMF and CO_2 function as catalysts. The amino group in **F** then attacks the carbonyl group in **E**, releasing DMF and forming carbamic acid **H**. Finally, cyclization occurs to produce the target compound **2**. In this mechanism, CO_2 serves as the carbon source, while DMF acts as both a catalyst and an activator. Notably, no dimethylamine is generated during the formation of **2**; dimethylamine is instead a by-product in the formation of the side product **3**. In comparison to another study where DMF was not added, the products were still obtained in high yields (80–93%) but under more severe conditions (14 MPa CO_2, 160 °C, 18–48 h) [314].

CO_2 + H_2O ⇌ HO–C(=O)–OH + DMF ⇌ E (1) → F (–CO_2, DMF) ⇌ G → (E; –DMF, H_2O) H → Cyclization (–H_2O) → 2

Problem 109: Synthesis of 3-Alkylquinazolin-4-ones Promoted by Mukaiyama Reagent

Quinazolinones possess a wide range of biological properties. For example, febrifugine (**1**), isolated from the Chinese herb *Dichroa febrifuga*, is used as an antimalarial drug [315]. Wacharasindhu's group reported a synthesis of 3-alkylquinazolin-4-ones (**4**) from quinazolin-4-ones (**2**) and amines (**3**) promoted by Mukaiyama reagent [316].

Please suggest a mechanism for this reaction.

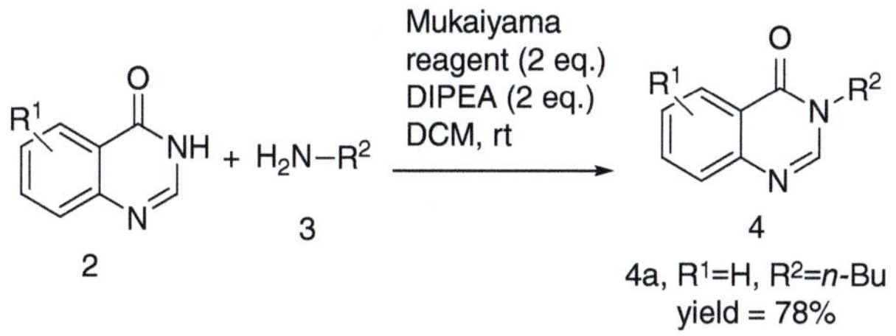

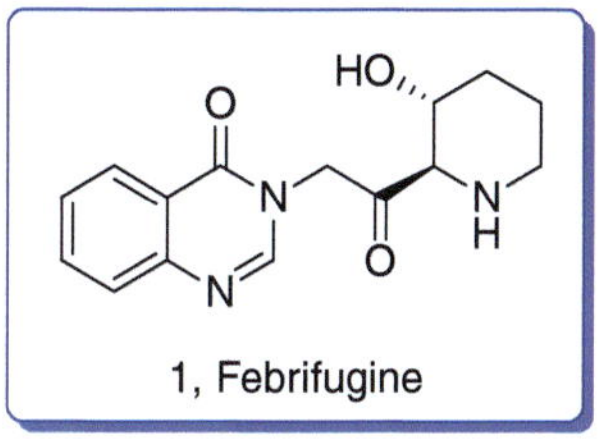

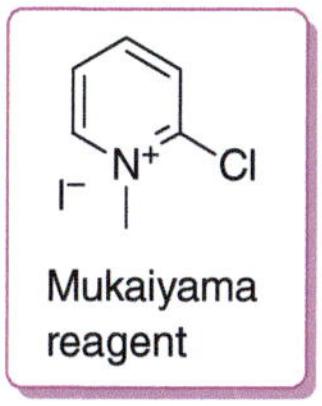

Experimental Procedure (from *Tetrahedron Lett.*, **2010**, *51*, 1713)

General Procedure: To a solution of quinazolin-4-one (**2**, 73.1 mg, 0.5 mmol) in CH_2Cl_2 (5 mL) at room temperature was added 2-chloro-1-methylpyridinium iodide (255.49 mg, 1.0 mmol) followed by diisopropylethylamine (435.48 μL, 2.5 mmol). The solution was stirred at room temperature for 1 h after which was added *n*-butylamine (248 μL, 2.5 mmol). The mixture was stirred at room temperature overnight and then concentrated and purified by column chromatography on silica gel (eluent: EtOAc/hexanes, 50%) to give 3-butylquinazolin-4(3H)-one. ^{1}H NMR (400 MHz, $CDCl_3$) δ ppm 8.32 (1H, d, J = 7.96 Hz), 8.03 (1H, s), 7.81–7.65 (2H, m), 7.51 (1H, t, J = 7.42 Hz), 4.01 (2H, t, J = 7.31 Hz), 1.87–1.70 (2H, m), 1.51–1.32 (2H, m), 0.97 (3H, t, J = 7.34 Hz). ^{13}C NMR (100 MHz, $CDCl_3$) δ ppm 160.4, 147.6, 146.2, 133.5, 126.8, 126.6, 126.1, 121.6, 46.2, 30.9, 19.4, 13.2. HRMS [M + H^+]: calcd for $C_{12}H_{13}N_2O$ 203.1184, found 203.1007.

The Mechanism

The 3-NH in quinazolinone as a nucleophile attacks the Mukaiyama reagent to form a pyridinium-quinazolinone intermediate **A**, in which the carbonyl is activated. The primary amine attacks the carbonyl on the amide moiety resulting in the ring-opening product, enamine **B**. Cyclization followed by elimination of 1-methylpyridin-2(1*H*)-imine **D** gives the product **4**.

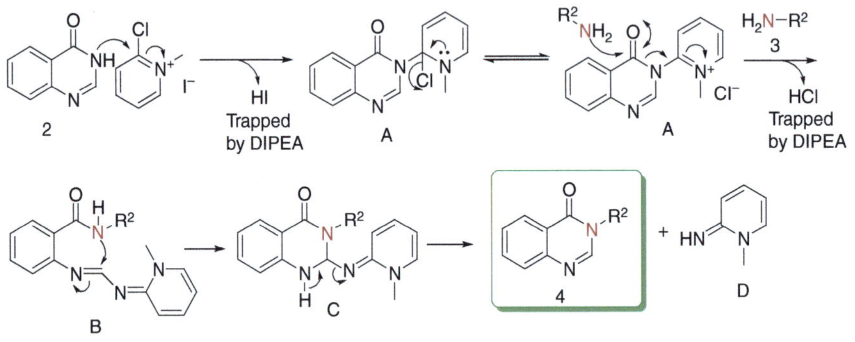

Problem 110: Dimroth Rearrangement in the Preparation of Gefitinib and Erlotinib

The 4-anilinoquinazolines gefitinib (**1**, Iressa) and erlotinib (**2**, Tarceva) have been launched for the treatment of non-small cell lung cancer. The final step in the synthesis of **1** involves the reaction of precursor **3** with 3-chloro-4-fluoroaniline (**4**) in hot acetic acid [317].

Please suggest a mechanism for this reaction.

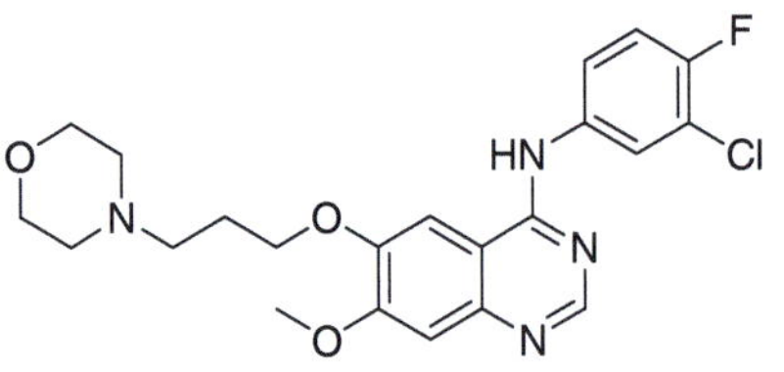

1 (gefitinib)

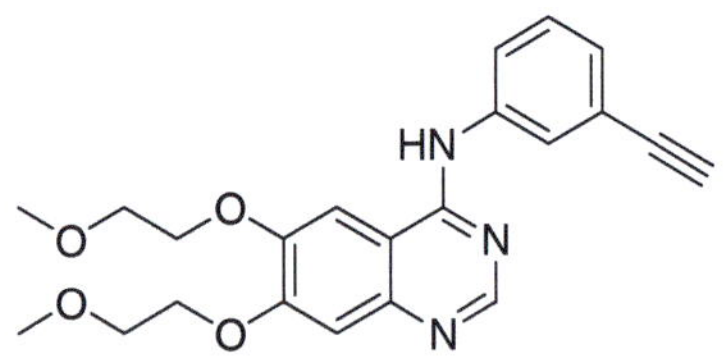

2 (erlotinib)

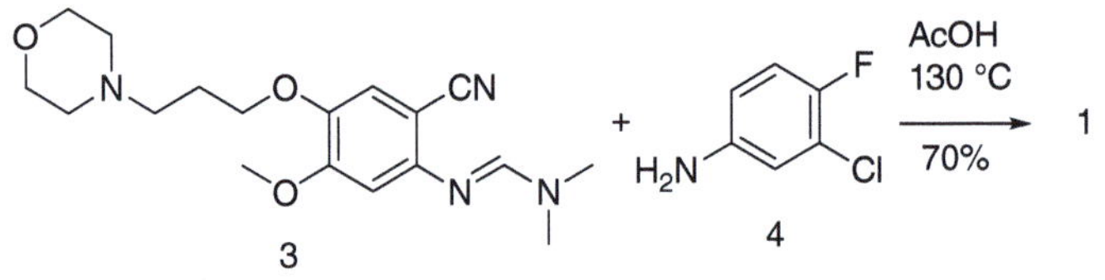

AcOH 130 °C 70%

Experimental Procedure (from *Org. Process Res. Dev.*, **2007**, *11*, 813)

4-(3′-Chloro-4′-fluoroanilino)-7-methoxy-6-(3-morpholinopropoxy)quinazoline (1). To a reaction flask fitted with a condenser and a Dean-Stark apparatus were added 2-amino-4-methoxy-5-(3-morpholinopropoxy)benzonitrile (283 g, 0.97 mol), toluene (2.5 L), acetic acid (3 mL), and DMF-DMA (280 mL, 2.10 mol). The reaction mixture was heated to 105 °C and stirred for 3 h. While stirring, methanol was collected using the Dean-Stark apparatus. Toluene was completely stripped off under vacuum to obtain *N*′-[2-cyano-5-methoxy-4-{3-(4-morpholinyl) propoxy}phenyl]-*N*,*N*-dimethylformamidine (**3**) as a brown liquid. IR (thin film): 864, 1010, 1118, 1384, 2214, 2812, 2950 cm^{-1}. ^{1}H NMR ($CDCl_3$): δ 2.0 (m, 2H), 2.47 (m, 6H), 3.06 (s, 6H),

3.72 (t, J = 4.5 Hz, 4H), 3.87 (s, 3H), 4.03 (t, J = 6.5 Hz, 2H), 6.45 (s, 1H), 6.98 (s, 1H), 7.57 (s, 1H). HRMS: calcd for $C_{18}H_{26}N_4O_3$ (M + H) 347.2083, found 347.2081.

To the residue (**3**) were added acetic acid (2.5 L) and 3-chloro-4-fluoroaniline (**4**, 175 g, 1.20 mol). The reaction mixture was heated to 125–130 °C and stirred for 3 h. The reaction mixture was then cooled to 25 °C, quenched in ice water (4 L), and adjusted pH ~9 with ammonia solution. To this reaction mass was added ethyl acetate (1 L), and the mixture was stirred for 1 h. The solid precipitate was filtered to obtain crude product **1**. Crude material was suspended in MeOH (4 L), and cooled to 20 °C. To this reaction mass was added concentrated HCl (190 mL) slowly with efficient stirring. The precipitate was filtered and washed with chilled methanol (200 mL) to give gefitinib hydrochloride. The solid was suspended in H_2O (5 L), stirred for 1 h at room temperature, cooled to 5 °C, filtered, and washed with chilled water (100 mL) to obtain an off-white solid of gefitinib hydrochloride. The solid was then suspended in water (2 L), and the suspension was adjusted to pH ~8 using ammonia solution, filtered, and dried at 50 °C to give **1** as off-white solid (304 g, 70% yield). Mp 193–195 °C. HPLC purity >99%. ^{1}H NMR ($CDCl_3$, 200 MHz): δ 2.11 (m, 2H), 2.46–2.59 (m, 6H), 3.74 (dd, J = 4.5 and 4.4 Hz, 4H), 3.98 (s, 3H), 4.17 (t, J = 6.5 Hz, 2H), 7.09 (s, 1H), 7.16 (t, J = 8.8 Hz, 1H), 7.26 (s, 1H), 7.34 (brs, 1H, exchangeable with D_2O), 7.50–7.58 (m, 1H), 7.84–7.88 (m, 1H), 8.66 (s, 1H). MS (m/z): 446(M^+), 128, 100. Elemental Anal. Calcd for $C_{22}H_{24}ClFN_4O_3$: C, 59.19; H, 5.38; N, 12.55. Found: C, 59.17; H, 5.21; N, 12.33.

The Mechanism

The formation of **1** from compound **3** and 3-chloro-4-fluoroaniline (**4**) is believed to take place via Dimroth rearrangement. Firstly, the aniline substitutes the dimethylamino and cyclizes to give the imine **A**. Then the pyrimidine ring is opened by acetic acid at the C—N_3 bond with subsequent rotation of 180° around the C_{10}—C_4 bond. Recyclization of the molecule and elimination of acetic acid provides the product **1**.

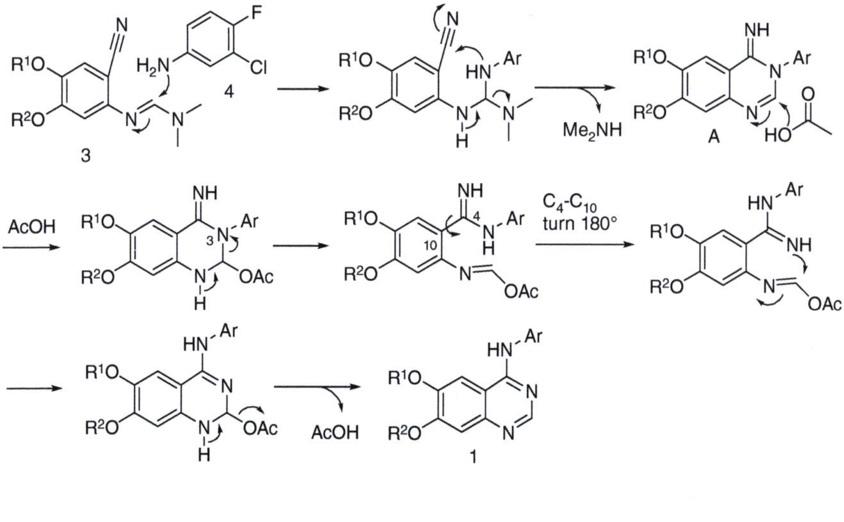

4.7.6 Thienoquinolizine

Problem 111: A [3+2] Cycloaddition Followed by Sulfur Extrusion

In order to explore novel benzodiazepine receptors, Widmer's group synthesized a series of dihydrothienoquinolizinone, such as compound **5**. The first two steps included the reaction from epoxide **1** and 5,6-dihydrothieno[2,3-c]pyridine-7(4*H*)-thione **2** to intermediates **3** and **4** as shown below [318].

Please suggest mechanisms for the formation of **3** and **4**.

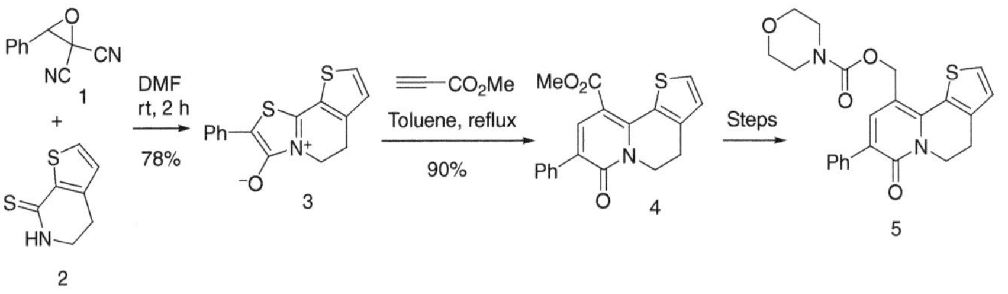

Experimental Procedure (from *Helv. Chim. Acta*, **1990**, *73*, 763)

5,6-Dihydro-2-phenylthiazolo[3.2-a] thieno[2,3-c]pyridin-4-ium 3-oxide (3). To a solution of 4,5-dihydrothieno[2,3-c]pyridine-7(6*H*)-thione (**2**; 101.4 g, 0.6 mol) in DMF (600 mL) under Ar, 3-phenyloxirane-2,2-dicarbonitrile (**1**; 122.5 g, 0.72 mol) was added with stirring. After ca. 15 min, the product began to crystallize from the dark red solution. After 2 h, the dark red crystals of **3** were filtered off, washed with AcOEt, and dried at 75°/0.1 Torr. Evaporation of the mother liquor to about half of its volume and trituration with AcOEt gave another crop of **3** (total yield: 133.2 g, 78%) which was sufficiently pure for the subsequent cycloaddition step. Recrystallization from dioxane/MeCN afforded anal. pure **3.** M.p. 225° (dec.). IR: 1615s, 1590s, 1500s, 1140m, 758m. ^{1}H NMR (60 MHz): 3.13 (t, J = 8.0, 2 H-C(6)); 4.33 (t, J = 8.0, 2 H-C(5)); 6.98 (d, J = 5.0, H-C(7)); 7.42 (d, J = 5.0, H-C(8)); 6.90–7.50 (m, 3 arom. H); 7.75–8.05 (m, 2 arom. H). MS: 285 (100, M^+), 121 (48).

1,3-Dipolar Cycloaddition of Mesoionic Thiuzole Derivatives with Methyl prop-2-ynoate. Under Ar, **3** (20 mmol) in toluene (200 mL) in the presence of methyl prop-2-ynoate (2.0 mL, 24 mmol) was heated to reflux temperature for 8 h. The solvent was evaporated and **4** was separated by chromatography.

Methyl 4,5-dihydro-7-oxo-8-phenyl-7H-thieno[2.3-a]quinolizine-10-carboxylate (4). Yield 90%. M.p. 117–118° (AcOEt/(*i*-Pr)$_2$O). IR: 1720s, 1645s, 15353, 1510s, 1270s, 1250s, 1200s, 1132s, 794m, 748m,700m. ^{1}H-NMR (80 MHz): 3.00 (t, J = 7.0, 2 H-C(4)); 3.95 (s, CH_3O); 4.47 (t, J = 7.0, 2 H-C(5)); 7.02 (d, J = 5.5, H-C(3)); 7.34–7.50 (m, 3 arom. H); 7.58 (d, J = 5.5, H-C(2)); 7.75–7.88 (m, 2 arom. H); 7.90

The Mechanism

The thioamide group in compound **2** acts as a nucleophile, attacking the epoxide **1** to form the 2-hydroxymalonitrile intermediate **A**. This is followed by an intramolecular displacement of one of the cyano (CN) groups by the nitrogen atom in the dihydropyridine ring, resulting in the cyclized species **B**. Upon the elimination of two molecules of hydrogen cyanide (HCN), the 1,3-dipolar compound **3** is generated. Subsequently, a [3+2] cycloaddition reaction between **3** and methyl propiolate produces the bridged compound **C**. Finally, the extrusion of sulfur from **C** yields the product **4**.

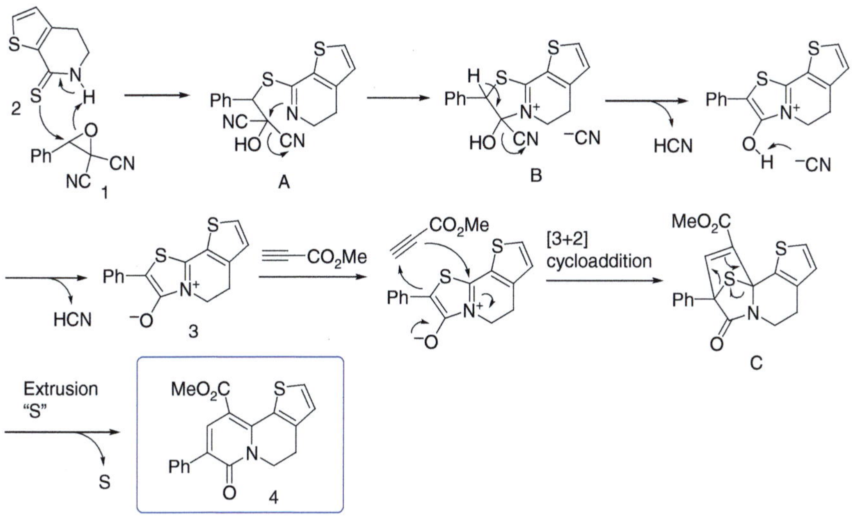

4.7.7 Pyridopyrimidine

Problem 112: An Extended Dimroth Rearrangement Catalyzed by MeONa
Trametinib, developed by GSK, is a MEK1 and MEK2 inhibitor with anticancer activity. In May 2013, it was approved by the FDA as a single-agent for the treatment of patients with V600E mutated metastatic melanoma. In a patent application [319], the final step was a rearrangement of precursor **1** to trametinib catalyzed by MeONa.

Please propose a mechanism for this reaction.

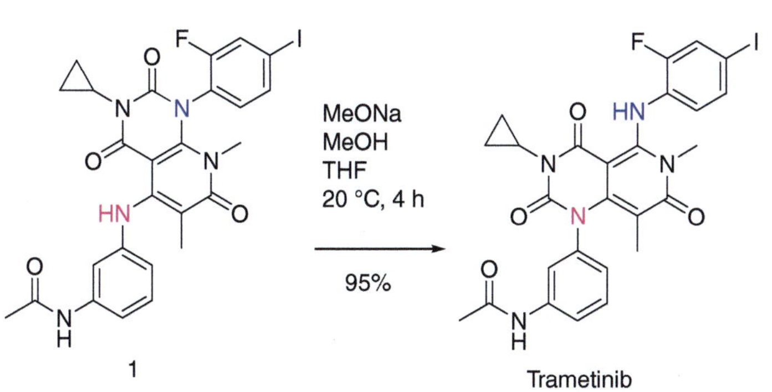

Experimental Procedure (from WO **2005**, 121142 A1)
Under a nitrogen atmosphere, to a solution (1.57 g) of 28% sodium methoxide in methanol was added THF (40 mL), *N*-{3-[3-cyclopropyl-1-(2-fluoro-4-iodophenyl)-6,8-dimethyl-2,4,7-trioxo-1,2,3,4,7,8-hexahydro-pyrido[2,3-d]pyrimidin-5- ylamino]

phenyl}acetamide **1** (5.00 g) obtained in Step 7, and the mixture was stirred at room temperature for 4 h. Acetic acid (0.56 mL) was added, and the mixture was stirred at room temperature for 30 min. Water (40 mL) was added and the mixture was further stirred for 1 h. The crystals were collected by filtration and dried to give *N*-{3-[3-cyclopropyl-5-(2-fluoro-4-iodophenylamino)-6,8-dimethyl-2,4,7-trioxo-3,4,6,7-tetrahydro-2*H*-pyrido[4,3-d]pyrimidin-l-yl]phenyl}acetamide **2** (4.75 g, yield 95%) as colorless crystals. MS ESI m/e: 616 (M + H), 614 (M-H). ^{1}H-NMR (DMSO-d$_6$, 400 MHz) δ 0.63–0.70 (m, 0.91–1.00 (m, 2H)), 1.25(s, 3H), 2.04(s, 3H), 2.58–2.66 (m, 1H), 3.07(s, 3H), 6.92 (t, J = 8.8 Hz, 1H), 7.00–7.05 (m, 1H), 7.36 (t, J = 8.2 Hz, 1H), 7.52–7.63(m, 3H), 7.79 (dd, 10.4 Hz, 1H), 10.10(s, 1H), 11.08 (s, 1H).

The Mechanism

This reaction can be regarded as a Dimroth rearrangement via ring-opening, ring-closing catalyzed by MeONa. The driving force of the reaction may be due to the salt **C** which is a weaker base than salt **A**. This means **NH** in trametinib was more acidic than the **NH** in **1**.

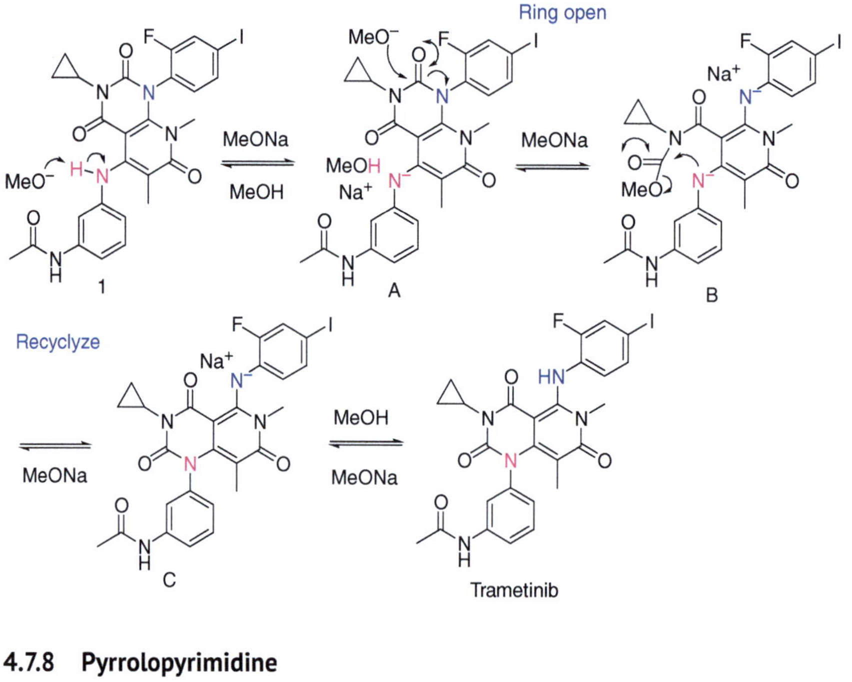

4.7.8 Pyrrolopyrimidine

Problem 113: One-Pot Synthesis of 2-(7H-Pyrrolo[2,3-d]pyrimidin-4-yl) acetates

4-(1*H*-Pyrazol-4-yl)-7-((2-(trimethylsilyl)ethoxy)methyl)-7*H*-pyrrolo[2,3-d]pyrimidine (**5**) is an important intermediate for synthesizing of ruxolitinib [320] and baricitinib [321]. In a patent application, Qi et al. reported a new process for **5**, in which a one-pot procedure for preparation of 2-(7*H*-pyrrolo[2,3-d]pyrimidin-4-yl)acetates (**4**) from cyanoacetyl acetate **1**, 2-bromo-1,1-diethoxyethane **2** and formimidamide hydrochloride **3** was described [322].

Please suggest a mechanism for the formation of **4**.

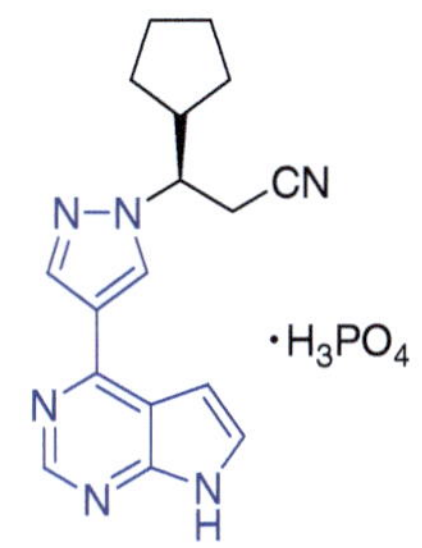

Ruxolitinib phosphate

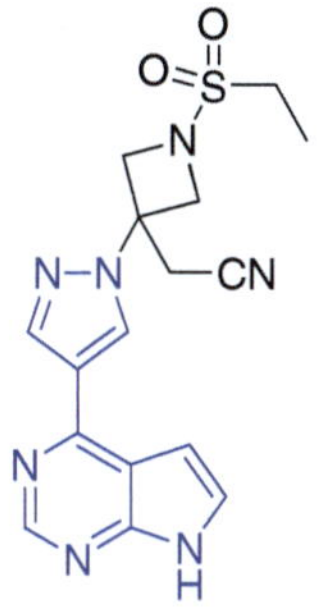

Barcitinib

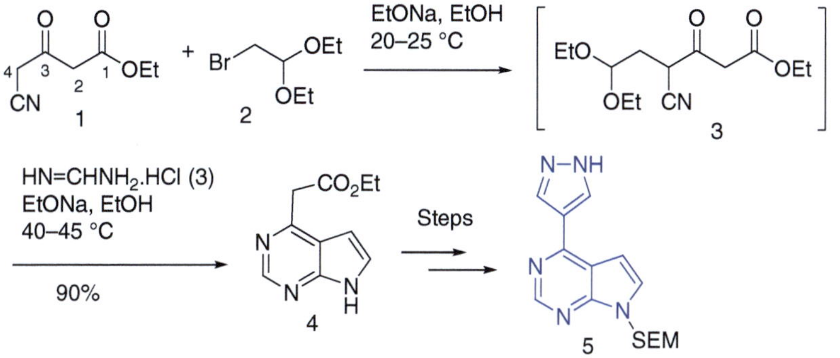

Experimental Procedure* (from CN **2019**/109651424, A)
To a 500 mL four-necked flask, charged with 150 g of *N,N*-dimethylformamide, were added ethyl cyanoacetate (**1**, 15.5 g, 0.1 mol) and 30% sodium ethoxide ethanol solution (27.5 g, 0.12 mol). Controlling the reaction temperature between 20 and 25 °C, bromoacetaldehyde diethyl acetal (**2**, 19.5 g, 0.1 mol) was added dropwise, and the addition was completed in about 30 minutes. Then the reaction was stirred at 20–25 °C for 6 hours. Gas chromatography confirmed the complete consumption of bromine acetaldehyde diethyl acetal and the formation of ethyl 1,1-diethoxy-3-cyano-4-oxo-*n*-hexanoate (**3**). Formamidine hydrochloride (10.0 g, 0.12 mol) and a 30% ethanolic sodium solution (27.5 g, 0.12 mol) were added. The reaction mixture was stirred at 40–45 °C for 4 hours. Saturated ammonium chloride aqueous solution (20 g) was added to the mixture and the pH value was adjusted to 3–4. The reaction was stirred at 30–35 °C for 3 hours. Five hundred grams of ice water was added to the reaction. The precipitates were collected by filtration to give a filter cake. This was recrystallized from isopropanol (80 g) to obtain 18.6 g of 2-(7-hydropyrrole[2,3-d]pyrimidin-4-yl)acetate, the HPLC purity was 99.5%, and the yield was 90.7%.

* *Translated from Chinese by the author of this book.*

The Mechanism

The C4-H (pK_a ~ 10) in **1** is more acidic than C2-H (pK_a ~ 11); thus, the first step alkylation regioselectively occurs at the C4 position to give **A**. At the formimidamide attack stage, the carbonyl is more reactive than the nitrile group, and thus provides **B**. Water is eliminated from **B** to afford **C** and then the formimidamide cyclizes toward the nitrile to generate **D**. The produced amino group further cyclizes with the acetal group to yield the dihydropyrrole **E**. Final elimination of EtOH leads to product **4**.

EtO⁻ H 4 3 2 1 OEt CN 1 — EtONa, EtOH → Br EtO OEt 4 3 2 1 OEt CN 2 — NaBr → EtO OEt N A CO₂Et H₂N HN 3

EtO OEt HO N H ⁻OEt N HN B CO₂Et — H₂O → EtO OEt N HN C CO₂Et → EtO EtO H₂N N N D CO₂Et

— EtOH → EtO⁻ H EtO N H N N E CO₂Et — EtOH → N H N N 4 CO₂Et

4.7.9 Pyrazolopyrimidine

Problem 114: Cyclization of Diazo Compound with Aldehyde

In a synthesis of SHP2 inhibitors, the core bicyclic heterocycles were prepared from a diazo compound **1** and aldehyde **2** [323]. For example, the following compound **3** was prepared in 39% yield [324].

Please suggest a mechanism for this reaction.

PMB N N H N O N O 1 — Cl Cl CHO N 2; Piperidine, DMF, *i*-PrOH, 90 °C; 39% → PMBN N H N O N O Cl Cl N 3 — Steps → H N N N N N O H₂N O Cl N H N SHP389

Experimental Procedure (from *J. Med. Chem.*, **2019**, *62*, 1781)

To a suspension of 6-((4-methoxybenzyl)diazenyl)-3-methylpyrimidine-2,4(1*H*,3*H*)-dione (**1**, 0.68 g, 2.479 mmol) in DMF (8 mL) and *i*PrOH (4 mL) was added 2,3-dichloroisonicotinaldehyde (**2**, 0.436 g, 2.479 mmol) and piperidine (0.241 mL,

2.430 mmol). The resultant mixture was stirred at 85 °C for 1 h. At completion, the reaction mixture was partitioned between water (50 mL) and ethyl acetate (50 mL). The layers were separated, and the aqueous layer was washed with ethyl acetate (50 mL). The combined organics were washed with brine (40 mL), dried with Na_2SO_4, filtered, and concentrated. The crude was purified by flash chromatography over silica gel (0–10% methanol/DCM eluent) to afford 3-(2,3-dichloropyridin-4-yl)-2-(4-methoxybenzyl)-5-methyl-2H-pyrazolo[3,4-d]pyrimidine-4,6-(5H,7H)-dione as a yellow solid (420 mg, 39% yield, 90% purity). LCMS: m/z 432 $(M + 1)^+$, $R_t = 1.29$ min.

The Mechanism

The aldehyde **2** is activated by piperidine to form a 1-arylidenepiperidin-1-ium hydroxide **A**. Then aldol reaction occurs between **1** and **A** to give intermediate **B**. There may be three different forms of **B: B-1, B-2,** and **B-3**. Cyclization of **B-3** followed by elimination of piperidine provides the product **3**.

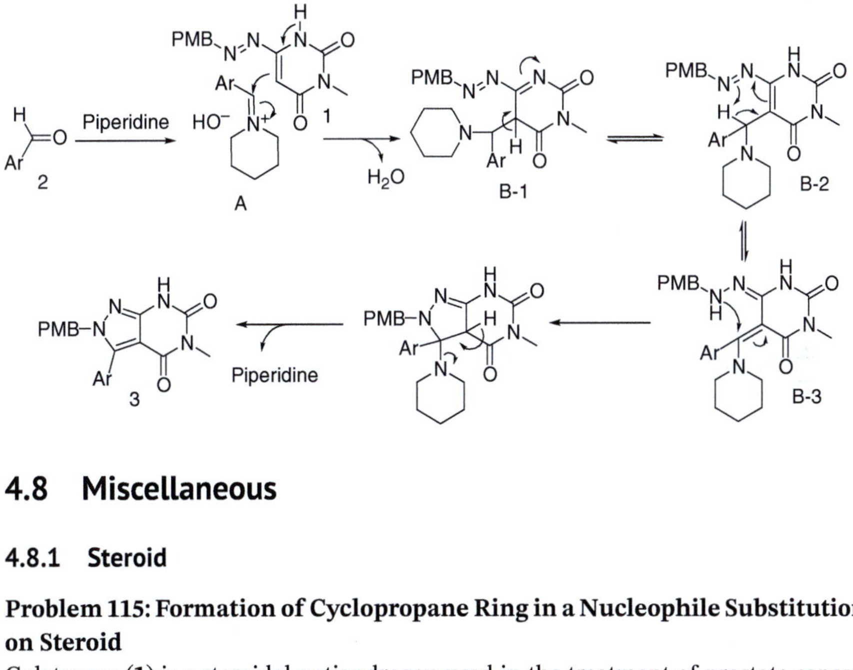

4.8 Miscellaneous

4.8.1 Steroid

Problem 115: Formation of Cyclopropane Ring in a Nucleophile Substitution on Steroid

Galeterone (**1**) is a steroidal antiandrogen used in the treatment of prostate cancer. In an effort to explore novel analogs, Njar's group synthesized 3β-(1H-imidazole-1-yl)-17-(1H-benzimidazole-1-yl)-androsta-5,16-diene (galeterone 3β-imidazole (**2**) in a four-step process, achieving an overall yield of 63% starting from compound **1** [325]. However, the initial results were less than optimal. When 3β-mesyl galeterone (**3**) was treated with imidazole in refluxing toluene, the desired compound **2** was obtained in only 11% yield, along with the 3α-isomer (**4**, 3%), the 6β-isomer of 3α, 5α -cycloandrostane (**5**, 35%), and several uncharacterized elimination products [326].

Please suggest a mechanism for the formation of **2**, **4**, and **5**.

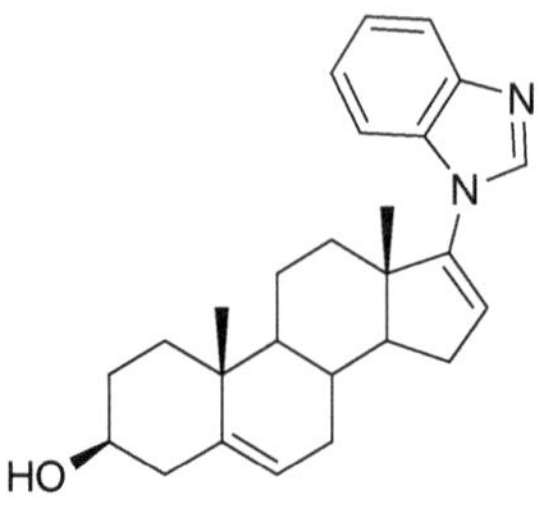

1, Galeterone

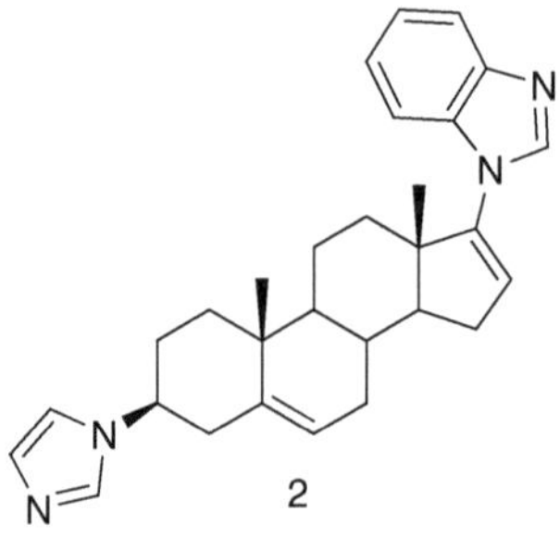

2

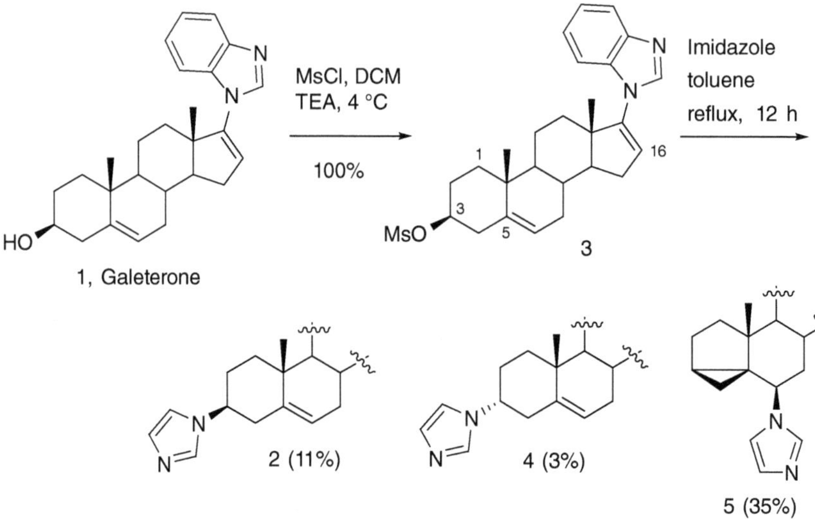

1, Galeterone

MsCl, DCM
TEA, 4 °C
100%

3

Imidazole
toluene
reflux, 12 h

2 (11%)

4 (3%)

5 (35%)

Experimental Procedure (from WO2014, 165815, A2)
6*β*-Imidazol-lyl-3,5-cycloandrostan-17-benzimidazole-l-yl-16-ene (5), 3α-(l*H*-imidazol-l-yl)-17-(l*H*-benzimidazol-l-yl)androsta-5,16-diene (4) and 3*β*-(l*H*-imidazol-l-yl)-17-(l*H*-benzimidazol-l-yl)androsta-5,16-diene (2). A round bottom flask equipped with Dean-Stark apparatus and condenser is charged with **3** (0.6 g, 1.3 mmol), imidazole (0.3 g, 4.4 mmol) and toluene. Reaction mixture is refluxed for 12 h. Reaction mixture is cooled, solvent is evaporated under vacuum, and the residue is washed with water to get a crude product. Purification by FCC [1% ethanol in DCM with traces of TEA] to get product of three components **2**, **4**, and **5** (0.23 g, 61%): This three-component mixture was subjected to preparative HPLC for the isolation of individual compounds. HPLC separation: Preparative HPLC separation was performed on Waters Prep Nova-Pak (7.8 × 300 mm, 60 A, 6 μm) HR C-18 reversed-phase HPLC column (Waters, Milford, MA) coupled with Waters 2489 UV/visible detector operated at 254 nm. Elution was performed using Waters model 2535 Quartenary Gradient Module pump to deliver a constant flow rate of 6 mL/min. The solvent system consisted of water/MeOH/CH_3CN (200:500:300, v/v/v + 500 μ of TEA per 1000 mL of mobile phase) and maintained isocratically. Sample stock

solution was prepared by dissolving 200 mg (three-component mixture obtained by FCC) in 20 mL of mobile phase. A total of 20 injections, each with a volume of 1 mL and a run time of 13 minutes, was performed. Compound **5** was collected between 5.12 and 6.73 min, **4** collected between 7.77 and 8.93 min and **2** was collected between 10.07 and 12.09 min. Figure shows an example of HPLC chromatogram of a three-component mixture. The retention times for **5**, **4** and **2** were 5.14, 7.87 and 10.07 min, respectively. See Figure 11. 33063

6*β*-Imidazol-lyl-3,5-cycloandrostan-17-benzimidazole-l-yl-16-ene (5). Retention time 5.14 min; 0.13 g, 34.6%: mp 104–106 °C; ^{1}H NMR (400 MHz, DMSO-d$_6$) δ ppm 0.67 (s, 3H), 0.91 (s, 3H), 3.73 (br. s., 1H), 6.09 (s, 1H), 6.93 (br. s., 1H), 7.21–7.33 (m, 2H), 7.36 (s, 1H), 7.57 (d, J = 7.82 Hz, 1H), 7.70 (d, J = 7.58 Hz, 1 H), 7.85 (br. s., 1 H), 8.25 (s, 1 H); ^{13}C NMR (400 MHz, CDC1$_3$) 5147.35, 143.24, 141.49, 136.41, 134.47, 128.62, 123.52, 123.48, 122.54, 120.21, 1 18.13, 1 1 1.07, 58.61, 55.37, 48.47, 47.42, 43.24, 35.18, 34.90,, 34.34, 33.34, 30.14, 29.19, 24.78, 24.24, 22.01, 19.72, 16.25, 14.31; HRMS calcd 439.2856 ($C_{29}H_{34}N_4O$)H$^+$, found 439.2858.

3α-l*H*-imidazol-l-yl)-17-(lH-benzimidazol-l-yl)androsta-5,16-diene (4). Retention time, 8.87 min. 0.012 g, 3.2%. m.p. 199–200 °C. ^{1}H NMR (400 MHz, DMSO-d$_6$) δ 0.97 (s, 3H, 18-CH$_3$), 1.10 (s, 3H, 19-CH$_3$), 4.42 (br. s., 1H, 3β-H), 5.53 (br. s., 1H, 6-H), 6.06 (br. s., 1H, 16-H), 6.86 (s, 1H), 7.19–7.33 (m, 3H), 7.56 (d, J = 7.83 Hz, 1H), 7.69 (d, J = 7.34 Hz, 2H), 8.24 (s, 1H); ^{13}C NMR (400 MHz, CDC1$_3$) δ 147.11, 143.29, 141.64, 139.13, 134.58, 128.56, 124.18, 123.41, 123.36, 122.48, 120.22, 118.70, 1 11.13, 55.74, 53.00, 50.19, 47.23, 37.27, 35.97, 34.74, 32.29, 31.12, 30.23, 30.20, 29.71, 28.39, 20.27, 19.32, 16.01; HRMS calcd 439.2856 found 439.2857.

3*β*-(l*H*-imidazol-l-yl)-17-(l*H*-benzimidazol-l-yl)androsta-5,16-diene (2). Retention time 10.074 min; 0.04 g, 10.6%: mp 198–200 °C; ^{1}H NMR (400 MHz, DMSO-d$_6$) δ ppm 0.99 (s, 3H), 1.12 (s, 3H), 3.93–4.03 (m, 1H), 5.47 (d, J = 4.65 Hz, 1H), 6.07 (s, 1H), 6.88 (s, 1H), 7.23–7.32 (m, 2H), 7.33 (s, 1H), 7.57 (d, J = 7.58 Hz, 1H), 7.71 (d, J = 7.58 Hz, 1H), 7.74 (s, 1H), 8.27 (s, 1H). ^{13}C NMR (400 MHz, CDC1$_3$) δ 147.25, 143.40, 141.72, 140.20, 134.68, 129.27, 124.21, 123.55, 122.63, 122.39, 120.34, 116.97, 111.26, 57.54, 55.92, 50.62, 47.35, 40.67, 38.00, 37.03, 34.94, 31.17, 30.41, 30.02, 29.83, 20.73, 19.47, 16.15, 0.14; HRMS calcd 439.2856 (C_{29}H34N_4O)H$^+$, found 439.2858.

The Mechanism

Direct substitution of the mesylate by imidazole (SN2 mechanism) provides **4**. Due to the participation of C5=C6 double bond, ionization of mesylate produces homoallylic hybrid carbonium ion intermediate **A**. The nucleophile (imidazole) attacks the C3 position to generate products **2** and **4**, while attack at the C6 position leads to product **5**. However, since the reaction is highly stereoselective, favoring the formation of the 6β-isomer, a synergistic process is also reasonable.

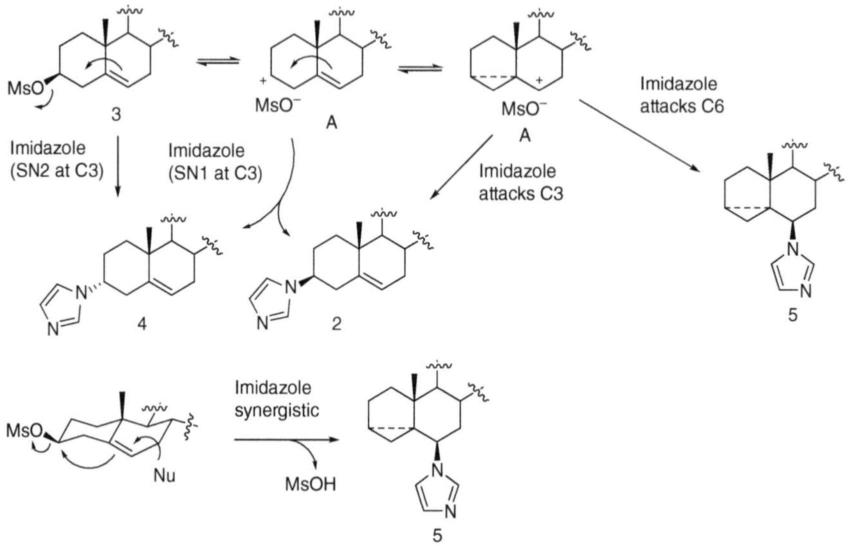

Nucleophilic substitution of this type of homoallylic hybrid carbonium ion has been previously reported [327–329].

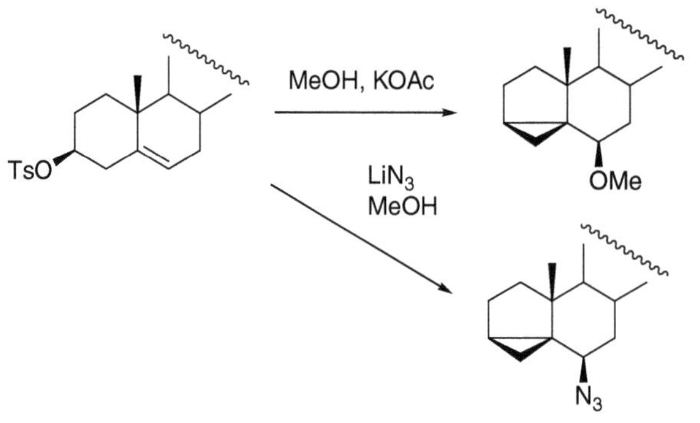

Problem 116: Preparation of 21-Hydroxypregna-4,9(11), 16-triene-3,20-dione-21-acetate

Dexamethasone (**3**) is a steroid drug first approved in 1958 and has shown mortality reduction in severe COVID-19 patients [330]. In a synthesis of compound **3**, 21-hydroxypregna-4,9(11),16-triene-3,20-dione-21-acetate (**2**) was a key intermediate [331]. This intermediate (**2**) was also prepared from **1** in 3 steps via precursor **4** [332].

Please propose a mechanism for the reaction from **4** to **2**.

Experimental Procedure (from US 4,216,159, August 5, 1980)

21-Hydroxypregna-4,9(11),16-triene-3,20-dione-21-acetate (2). Anhydrous sodium acetate (5.8 g.) and DMF are stirred and heated at 120 °C under nitrogen. Crystalline 20-chloropregna-4,9(11),17(20)-triene-3-one-21-al (**4**, 12 g) was added in portions of 2 gm every 20 minutes. The mixture was stirred for 90 minutes at 120 °C, then cooled, and toluene (100 mL) was added. The mixture was extracted with sodium chloride (5%, 2 × 100 mL) and backwashed with toluene (2 × 20 mL). The toluene phase was dried over magnesium sulfate and concentrated under reduced pressure to give the title compound, m.p. 120–124 °C, PMR ($CDCl_3$) 0.89, 1.35, 2.18, 4.95, 5.55, 5.72 and 6.74 δ.

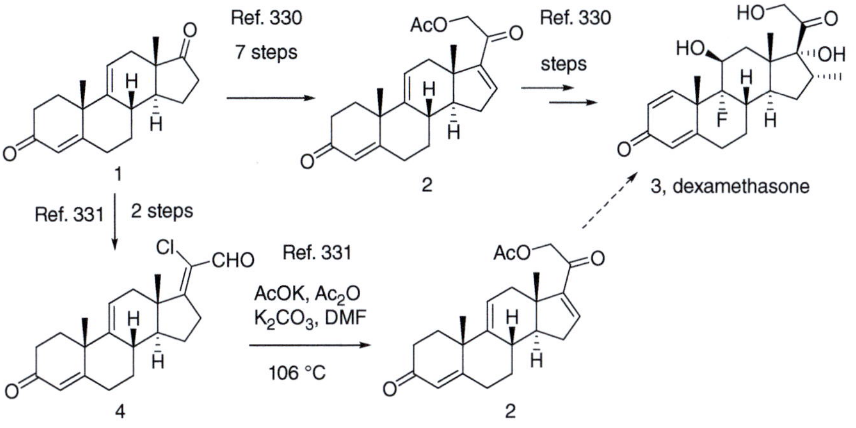

The Mechanism

The first reaction should be a Michael addition of KOAc to the α,β-unsaturated aldehyde **4** to give the enol salt **A**. Then Ac migrates to the enol via intramolecular transesterification to provide potassium *tert*-alkoxide **B**. Intramolecular addition of the oxygen anion to the double bond in **B**, followed by elimination of KCl via intermediate **C**, generates methylene oxirane **D**. Finally, rearrangement of the oxirane, possibly catalyzed by K_2CO_3, leads to the product **2**.

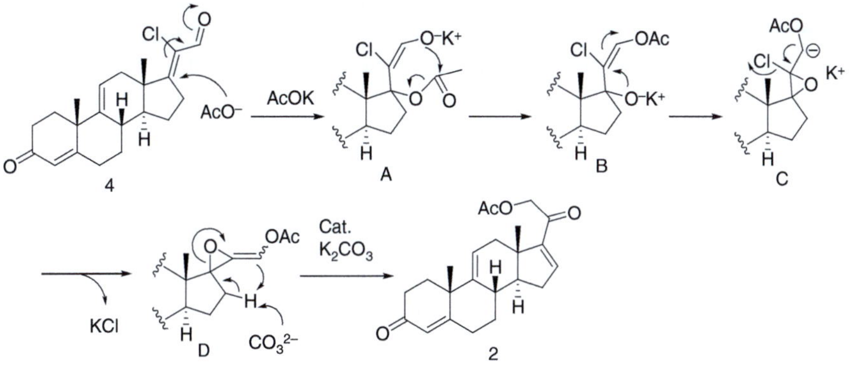

4.8.2 Macrolide

Problem 117: An Ascomycin Rearrangement

Ascomycin **1**, also known as immunomycin, is an ethyl analog of tacrolimus that exhibits potent immunosuppressive activity. It has been investigated for its potential in treating autoimmune disorders, skin conditions, and in preventing organ transplant rejection. When compound **1** was reacted with triethylamine (Et_3N) and a catalytic amount of KOH in refluxing acetonitrile, a rearranged product, compound **2**, was obtained with a yield of 65% [333].

Please suggest a mechanism for this rearrangement.

Et_3N (6 eq.)
KOH (0.016 eq.)
CH_3CN
reflux, 16 h

65%

1 (Ascomycin)

2

Experimental Procedure (from *Org. Process Res. Dev.*, **2001**, 5, 211)

Procedure for the Ascomycin Rearrangement Step. Crystallized ascomycin (**1**, 31.8 kg (98.1%) and 19.3 kg (97.0%)) was dissolved in 529 L toluene at 70 °C jacket temperature in a 630 L stainless steel reaction vessel and the solvent was distilled at 60 °C jacket temperature to effect azeotropic water removal. The residue was redissolved in 427 L of acetonitrile, and 37.96 kg of triethylamine and 56.5 g of KOH (90%) were added. The suspension was refluxed at 74 °C internal temperature (corresponding to a jacket temperature of about 100 °C) for 7 h. Afterward, the suspension was cooled to an internal temperature of 50 °C, and a second portion of 56.5 g of KOH (90%) was added. The suspension was again refluxed at 74 °C internal temperature for 5 h, and finally the solvents were removed by distillation at a jacket temperature of 60 °C/process vacuum. The residue was dissolved in 136 L of ethyl methyl ketone at 60 °C internal temperature, and 80.0 L heptane was added. On cooling to room temperature (20–25 °C), the product precipitated within 2.5 h. The suspension was filtered via a centrifuge (80 cm i.d.) equipped with a Meraklon filter plate. The product was washed with 34 L of ethyl methyl ketone/heptane) 1/4 (V/V) and dried at 60 °C/<20 mbar. Yield: 33.6 kg (assay: 78.9%, purity: 82.5%, **21-epimer**: 11.8%).

The Mechanism

At the basic conditions, the semi-acetal or tetrahydropyran is cleaved to release the carbonyl compound **A**, which is epimerized to **C** via enol **B**. At this stage, there are two possible pathways to go further. The path **A** is the C-14 hydroxyl cyclized with C-9 carbonyl to form a seven-membered semi-acetal **D** followed by C-8 acyl shifts to C-10 to give lactone **E**. Intramolecular ester condensation between C-2 and C-9 leads to product **2**.

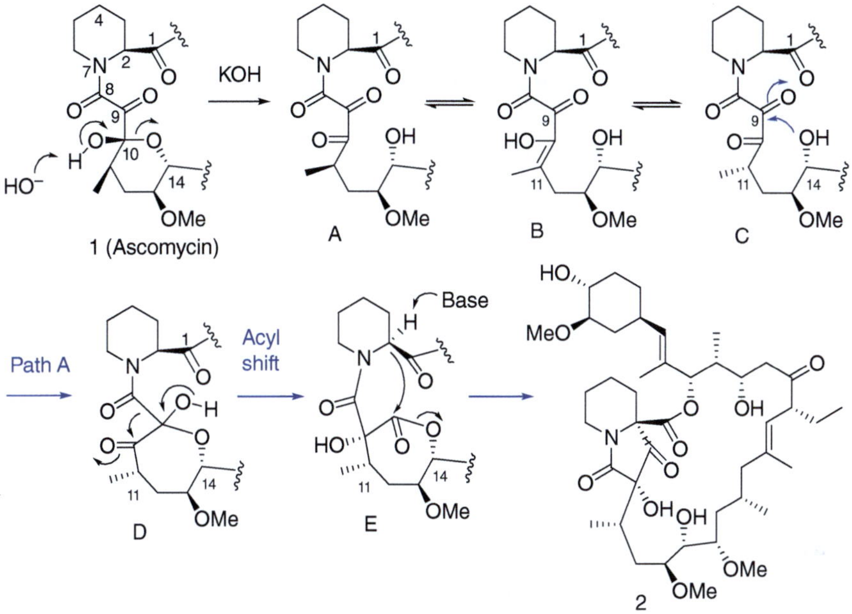

From intermediate **C**, another possible pathway (path B) is the aldol condensation between C-2 and C-10. This would generate intermediate **F**. Then alkyl shifts from C-10 to C-9 provides product **2**.

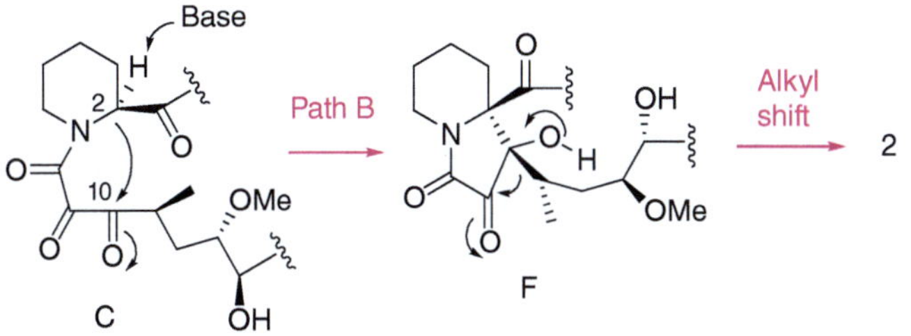

4.8.3 Gibberellin

Problem 118: An Acyloin Rearrangement in Gibberellin

Gibberellins (GAs) are plant hormones that regulate various developmental processes, including stem elongation, germination, dormancy, flowering, flower development, and leaf and fruit senescence. In a synthesis of gibberellin natural product, compound **2** was obtained in high yield from compound **1** by isomerization in the presence of MeONa [334].

Please propose a mechanism for this transformation.

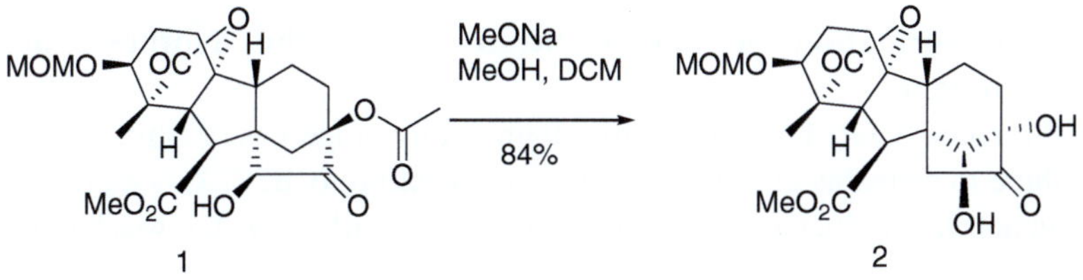

Experimental Procedure (from *Tetrahedron*, **1998**, *54*, 11637)

***ent*-10β,13,14α-Trihydroxy-3α-methoxymethoxy-16-oxo-17,20-dinorgibberellane-7,19-dioic acid 7-methylester 19,10-lactone (2).** 5 mg of NaH (60% in mineral oil) was added to a solution of ketol **1** (10 mg) in CH_2Cl_2 (1 mL) under a N_2 atmosphere at 25 °C followed by one drop of MeOH. The reaction was quenched with saturated KH_2PO_4 and the organic layer washed with brine, dried, reduced to dryness, and chromatographed on silica gel (EtOAc/hexane, 1 : 1) to afford **2** as a white solid (8.4 mg, 84% yield), m.p. 235–236 °C. Calcd. For $C_{21}H_{28}O_9$ C, 59.43; H, 6.65; found C, 59.14; H, 6.36%. ^1H NMR (300 MHz, $CDCl_3$) δ 1.22 (3H, s, 4-Me), 2.23 (1H, d, J = 18.7 Hz, H15), 2.41 (1H, d, J = 18.7 Hz, H15), 3.15 (1H, d, J = 9.4 Hz, H6), 3.27 (1H, d, J = 9.4 Hz, H5), 3.38 (3H, s, OMe), 3.68 (1H, br s, H3), 3.73 (3H, s, CO_2**Me**), 4.28 (1H, s, H14), 4.63 (1H, d, J = 7.1 Hz, OCH_2O), 4.74 (1H, d, J = 7.1 Hz, OCH_2O). ^{13}C NMR (75 MHz, $CDCl_3$) δ 14.4 (C18), 15.8 (C11), 24.6 (C1), 26.8 (C2), 32.5(C12), 44.3 (C15), 49.8 (C6), 51.2 (C8), 52.4 (CO_2**Me**), 54.0, 55.6 (C9, C5), 54.3 (C4), 55.8 (OMe), 75.4, 76.7 (C14, C3), 81.5 (C13), 93.0 (C10), 95.8 (OCH_2O), 173.4 (C7), 177.4 (C19), 216.8 (C16). MS (EI) m/z 424 (M+, 14%), 406 (6), 293 (35), 365 (97), 332 (100), 316 (30), 301 (20), 157 (40), 143 (27), 129 (26), 117 (18), 105 (30), 91 (34), 79 (21). HRMS found 424.1734, $C_{21}H_{28}O_9$ requires 424.1733.

The Mechanism

Treatment of acetate **1** with MeONa leads to the hydrolysis of the 13-acetate, followed by an acyloin rearrangement to yield product **2**.

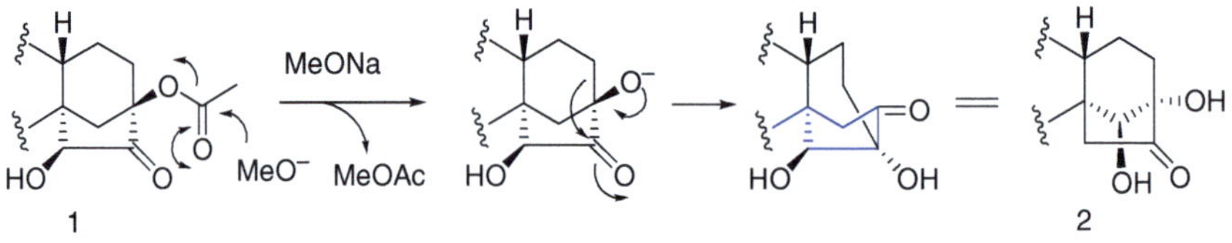

Another possible pathway, as shown below, is less reasonable since the four-membered intermediate is less stable than the five-membered intermediate.

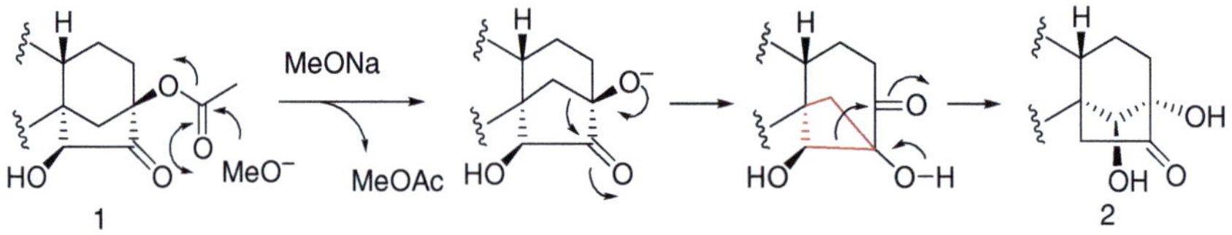

4.8.4 Cannabicitran

Problem 119: Formation of Cannabicitran During the Synthesis of Cannanbichromene

Cannanbichromene (CBC, **1**) exhibits anti-inflammatory properties in vitro, which may contribute to cannabis analgesic effects. Cannabicitran (CBTC, **2**) is a phytocannabinoid natural product as a trace component of *Cannabis sativa*. It was found to reduce intraocular pressure in tests on rabbits. In a biomimetic-like synthesis of cannanbichromene (**1**), the mixture of citral (**3**) and olivetol (**4**) in water was heated at reflux to give cannanbichromene (**1**, yield 45%) and cannabicitran (**2**, yield 5%) [335]. When 0.2 equivalent of NH_4Cl was added, the reaction time was reduced to 24 hours and the yield of **1** was improved to 75%.

Please propose mechanisms for the formation of **1** and **2**.

Citral (3)

+

OH

HO n-C_5H_{12}

Olivetol (4)

H_2O (0.045 M)
reflux, 51 h

OH

O n-C_5H_{12}

Cannanbichromene
(1, 45%)

+

H

O O

n-C_5H_{12}

Cannabicitran
(2, yield 5%)

Experimental Procedure (from *J. Org. Chem.*, **2021**, *86*, 3344)

Reaction of Citral with Phenolic Compounds in Water: General Procedure. To a round bottom flask containing citral were added the phenolic compound and water. The mixture was then heated at refluxing temperature (oil bath) until the disappearance of the starting material. Then, brine was added, and the aqueous phase was extracted with EtOAc (×3). The organic layer was dried with Na_2SO_4 and concentrated under a vacuum. The residue was purified by column chromatography.

Using Olivetol as a Phenolic Compound. According to the general procedure, the reaction of citral (150 mg 0.8 mmol) with olivetol (300 mg, 2.4 mmol) in refluxing water (22.5 mL) for 52 h provided after flash chromatography (H/MTBE, 5:1) 45% of **1** as a colorless oil (113 mg, 0.43 mmol) and 5% of compound **2** as a colorless oil (13 mg, 0. 04 mmol).

Cannabicitran (2): ^{1}H NMR (600 MHz, $CDCl_3$) δ 6.32 (s, 1H,), 6.27 (s, 1H,), 2.85 (bs, 1H), 2.50 (t, $J = 7.7$ Hz, 2H), 2.21 (dt, $J = 13.1$, 4.1 Hz, 1H), 2.01 (ddd, $J = 11.3$, 5.0, 2.7 Hz, 1H), 1.82 (d, $J = 13.1$ Hz, 1H), 1.76 (bd, $J = 15.3$ Hz, 1H), 1.59–1.53 (m, 2H), 1.51 (s, 3H,), 1.41 (dt, $J = 14.5$, 7.0 Hz, 1H), 1.37 (s, 3H), 1.34–1.20 (m, 5H), 1.01 (s, 3H), 0.87 (t, $J = 7.0$ Hz, 3H), 0.65–0–56 (m, 1H); $^{13}C\{^1H\}$ NMR (151 MHz, CDCl3) δ 156.9, 156.5, 142.5, 114.0, 109.7, 108.8, 83.5, 74.4, 46.8, 37.3, 36.1, 35.3, 31.4, 30.9, 29.7, 29.0, 28.1, 23.7, 22.5, 22.1, 14.0; HRMS (ESI-TOF) m/z $[M + H]^+$ calcd for $C_{21}H_{30}O_2$ 315.2324, found 315.2324.

The spectra data of cannanbichromene (CBC, **1**) reported in a recent article [336] (Supporting Information) are listed here. R_f – 0.20 (20:1 Hexane:EtOAc). ^{1}H NMR (500 MHz, $CDCl_3$) δ 6.61 (d, $J = 10.0$ Hz, 1H), 6.25 (s, 1H), 6.12 (s, 1H), 5.49 (d, $J = 10.1$ Hz, 1H), 5.10 (t, $J = 7.2$ Hz, 1H), 4.73 (s, 1H), 2.44 (t, $J = 7.8$ Hz, 2H), 2.14 – 2.07 (m, 2H), 1.75 – 1.68 (m, 1H), 1.66 (s, 3H), 1.64–1.61(m, 1H), 1.58 (s, 3H), 1.57–1.50 (m, 2H), 1.38 (s, 3H), 1.31 (td, $J = 8.5$, 7.3, 4.9 Hz, 4H), 0.88 (t, $J = 6.8$ Hz, 3H). ^{13}C NMR (126 MHz, $CDCl_3$) δ 154.19, 151.11, 144.90, 131.77, 127.38, 124.33, 116.91, 109.28, 107.80, 107.11, 78.31, 41.19, 36.04, 31.61, 30.77, 26.39, 25.81, 22.85, 22.68,

17.76, 14.16. HRMS – (ES+, m/z) $[M + H]^+$ calcd. for $C_{21}H_{31}O_2$ 315.2324; found, 315.2314. IR – (ATR, neat, cm^{-1}): 3404 (br), 2962 (s), 2928 (m), 2858 (s), 1624 (m), 1576 (m), 1430 (m), 1083 (m), 774 (m).

The Mechanism

Due to its 1,3-dihydroxylbenzene function, olivetol (**4**) acts as a good nucleophile, undergoing aldol condensation with citral (**3**) to form intermediate **A**. After dehydration of **A**, intermediate **B** is produced, which undergoes an oxo-6π electrocyclization to afford cannanbichromeme (CBC, **1**). The weak acidic ammonium chloride plays dual functions as an activator and a catalyst for both the aldol reaction and the dehydration, thus accelerating the formation of cannanbichromene (**1**).

The phenol from CBC could be an equilibrium with its quino form of CBC (CBC-quinone). An intramolecular Diels-Alder reaction of the CBC-quinone will lead to the tetracyclic compound **2**.

4.8.5 Serratezomine

Problem 120: Transformation of Serratinine into Serratezomine A
Lycopodium alkaloids exhibiting marked anticholinesterase activity [337]. In a related study, Morita and Kobayashi transformed serratinine (**1**) into serratezomine A (**2**) and a minor product **3** by the following reaction [338]. Serratezomine A (**2**) was converted to compound **3** by treatment with *p*-TsOH. They also proposed the mechanism for the transformation.

Could you please draw your own mechanism for the formation of **2** and **3**?

1, *m*CPBA, DCM, 0 °C, 1 h
2, TFAA, –20 °C, 1 h
3, $NaBH_3CN$, MeOH, 0 °C, 15 min

1

2 (48%)

3 (27%)

p-TsOH, $CHCl_3$
MeOH, 50 °C

Experimental Procedure (from *J. Org. Chem.*, **2002**, 67, 5378)
A Modified Polonovski Reaction for Serratinine (1). *m*-Chloroperbenzoic acid (8 mg, 0.047 mmol) in 0.1 mL of CH_2Cl_2 was added at 0 °C to a stirred solution of serratinine (**1**, 10 mg, 0.036 mmol) in 0.1 mL of CH_2Cl_2. After 1 h, the reaction mixture was concentrated and then trifluoroacetic anhydride (0.02 mL, 0.14 mmol) was added to a stirred solution of the residue in dry CH_2Cl_2 (0.2 mL) under N_2 at −20 °C. After 1 h, excess solvent and TFAA were immediately distilled off *in vacuo* at 0 °C. The residue was dissolved in MeOH (0.1 mL), and excess $NaBH_3CN$ was added at 0 °C. After 15 min, the reaction mixture was poured into H_2O (10 mL) and extracted with $CHCl_3$. C18 HPLC (15% CH_3CN/0.1% TFA) of the residue afforded compounds **2** (4.8 mg, 48%) and **3** (2.7 mg, 27%). **2**: $[\alpha]_D$ +12° (c 0.6, MeOH), spectral data and $[\alpha]_D$ value were identical with those of serratezomine A (**2**), $[\alpha]_D$ +13° (*c* 0.5, MeOH) [339]. **3**: colorless solid; $[\alpha]_D$ +15° (*c* 0.5, MeOH); FABMS *m/z* 280 (M + H)$^+$; HRFABMS *m/z* 280.1926 (M + H; calcd for $C_{16}H_{26}NO_3$, 280.1939); IR (neat) ν_{max} 3374, 3148, 2928, 1780, 1680, and 1195 cm^{-1}; ^{1}H NMR (CD_3OD) δ 3.16 (1H, m, H-1a), 3.41 (1H, ddd, 10.1, 10.1, 10.1, H-1b), 2.15 (2H, m, H-2), 2.22 (1H, m, H-3a), 2.38 (1H, m, H-3b), 4.02 (1H, dd, 6.8, 12.3, H-4), 2.53 (1H, m, H-6a), 2.61 (1H, dd, 14.1, 16.5, H-6b), 2.35 (1H, m, H-7), 4.44 (1H, dd, 5.4, 12.4, H-8), 2.96 (1H, dt, 3.6, 13.3, H-9a), 3.14 (1H, m, H-9b), 1.74 (1H, brd, 14.5, H-10a), 2.52 (1H, dd, 7.4, 16.5, H-10b), 1.65 (1H, brd, 13.7, H-11a), 1.88 (1H, dd, 5.0, 14.1, H-11b), 3.95 (1H, dd, 4.8, 11.5, H-13), 1.82 (1H, ddd, 1.9, 4.8, 14.6, H-14a), 2.03 (1H, ddd, 5.3, 11.6, 16.7, H-14b), 2.46 (1H, m, H-15), 1.04 (3H, d, 7.2, H-16); ^{13}C NMR (CD_3OD) δ 54.0 (C-1), 19.7 (C-2), 24.0 (C-3), 63.9, (C-4), 177.5 (C-5), 36.8 (C-6), 43.6 (C-7), 80.8 (C-8), 47.8 (C-9), 21.4 (C-10), 30.4 (C-11), 42.2 (C-12), 76.0 (C-13), 32.3 (C-14), 31.3 (C-15), 11.6 (C-16).

Chemical Conversion of Serratezomine A (2) into Compound 3. A solution of **2** (1 mg) and *p*-toluenesulfonic acid (0.1 mg) in 1:1 $CHCl_3$/MeOH (0.1 mL) was stirred at 50 °C for 3 h and then concentrated under reduced pressure. The residue was dissolved in $CHCl_3$ and washed with saturated aqueous NaCl and 1% NH_4OH and then dried over anhydrous Na_2SO_4. Removal of the solvent afforded the residue, which was subjected to a silica gel column (9:1 $CHCl_3$/MeOH) to give compound **3** (0.4 mg).

The Mechanism

This describes a modified Polonovski reaction [340]. Initially, the tertiary amine **1** is oxidized to the *N*-oxide **A** using *m*CPBA. Subsequent trifluoroacetylation of **A** is followed by an intramolecular attack of the hydroxy group at C13 on the ketone at C5, leading to the formation of hemiketal **C**. This step is accompanied by the cleavage of the C4—C5 bond, resulting in the formation of a lactone ring between C5 and C13. Another lactone ring, **E**, may form through intramolecular transesterification. The rigid fused tetracyclic ring system of serratinine (**1**) is anticipated to facilitate fragmentation via the Polonovski reaction. Reduction of the unstable iminium intermediate **D** or **E** with $NaBH_3CN$ proceeds with hydride addition from the R-face of the molecule due to steric hindrance, ultimately generating an R-configuration at C4.

4.8.6. Vincamine

Problem 121: Preparation of Vincamine from Tabersonine

(+)-Vincamine (**4**) exhibits valuable cerebrovascular and cerebroprotective activities. One of the preparation in a patent application is the transformation of relatively abundant natural product tabersonine (**1**) to **4** by the following process [341].

Please provide a mechanism for the transformation from **3** to **4**.

1 (Tabersonine) → H_2, Pd-C, THF, rt, 18 h, 95% → 2 (Vincadifformine) → mCPBA, THF, <35 °C → 3 → Ph_3P, HOAc, 35–45 °C, 16 h → 4 (vincamine)

Experimental Procedure (from patent CN **2020**, 110950859 A*)
Example Name 1.1.1.7. Once the final reaction step was completed, the temperature was maintained under 30 °C. Glacial acetic acid (4.1 kg) was slowly added followed by triphenylphosphine (3.7 kg) to the reaction solution. The reaction was continued at 35–45 °C for 16–20 hours, the temperature was checked every 2 hours. After the reaction was completed, the solvents were evaporated under vacuum, bath temperature was controlled at 45–55 °C. DCM (30.0 kg) was added to the residue, the pH was adjusted to 8–10 with 15% sodium hydroxide solution at a temperature below 35 °C. Layers were separated, the organic phase was concentrated under vacuum at 45-55 °C. Anhydrous methanol (6.0 kg) was added to the residue, and the mixture was stirred at room temperature for 0.5 hours to crystallize. After centrifugal filtration, the filter cake was rinsed with anhydrous methanol (2.0 kg) to give crude vincamine **4**. The crude vincamine was added to the spin flask and anhydrous methanol (4.0 kg) was added. The crystal was spin washed at room temperature for 0.5 hours. The product was collected by centrifugal filtration, rinsed with 2.0 kg of anhydrous methanol to afford pure vincamine **4**.

The reported characterization data of (+)-vincamine (**4**) in another article [342] are listed as follows. White crystals: mp 230–232 °C; $[\alpha]^{22}_{D}$ +44.0 (c = 1.4, pyridine); IR (KBr) 3400–2400, 1748 cm^{-1}; ^1H NMR (200 MHz) δ 0.91 (t, J = 7.5 Hz, 3H), 1.30–1.80 (m, 5H), 2.10 (d, J = 14.3, 1H), 2.24 (d, J = 14.2, 1H), 2.27 (m, 1H), 2.42–2.70 (m, 3H), 2.90–3.10 (m, 1H), 3.20–3.40 (m, 2H), 3.82 (s, 3H), 3.92 (s, 1H), 4.64 (s, 1H), 7.08–7.18 (m, 3H), 7.45–7.52 (m, 1H); ^{13}C NMR (50 MHz) δ 7.6 (CH_3), 16.8 (CH_2), 20.8 (CH_2), 25.1 (CH_2), 28.9 (CH_2), 35.1 (C), 44.4 (CH_2), 44.6 (CH_2), 50.9 (CH_2), 54.2 (CH_3), 59.1 (CH), 81.9 (C), 105.9 (C), 110.3 (CH), 118.4 (CH), 120.2 (CH), 121.6 (CH), 129.0 (C), 131.4 (C), 134.4 (C), 174.4 (C). Anal. Calcd for $C_{21}H_{26}N_2O_3$: C, 71.16; H, 7.33; N, 7.90. Found: C, 71.05; H, 7.40; N, 7.80.

* *This was translated from Chinese by the author of this book.*

The Mechanism

The first stage should be the reduction of the *N*-oxide **3** to free amine **A** by Ph_3P. Acid-catalyzed rearrangement of **A** to **B** is followed by ring-opening to form **C**, and subsequent recyclization yields product **4**.

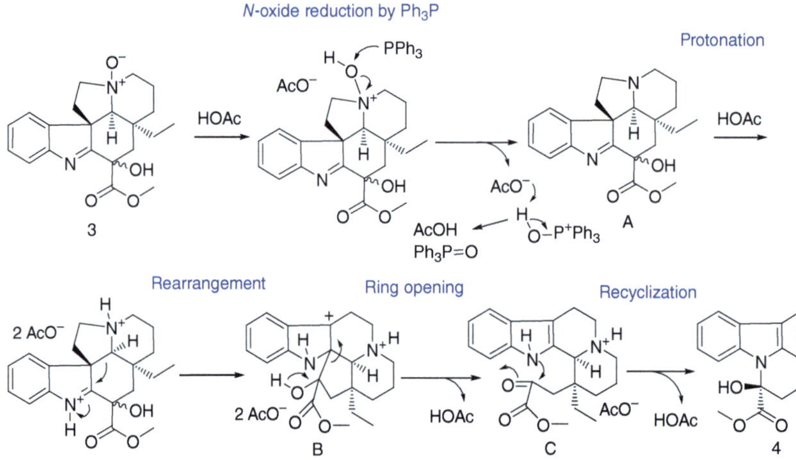

References

1 St. Jean, D. J. Jr.; Ashton, K. S.; Bartberger, M. D.; Chen, J.; Chmait, S.; Cupples, R.; Galbreath, E.; Helmering, J.; Hong, F.-T.; Jordan, S. R.; Liu, L.; Kunz, R. K.; Michelsen, K.; Nishimura, N.; Pennington, L. D.; Poon, S. F.; Reid, D.; Sivits, G.; Stec, M. M.; Tadesse, S.; Tamayo, N.; Van, G.; Yang, K.; Zang, J.; Norman, M. H.; Fotsch, C.; Lloyd, D. J.; Hale, C. *J. Med. Chem.*, **2014,** *57*, 325.

2 St. Jean, D. J. Jr.; Bourbeau, M. P.; Ashton, K. S.; Yan, J. *J. Org. Chem.*, **2014,** *79*, 3684.

3 a) Brown, C. A.; Yamashita, A. *J. Am. Chem. Soc.*, **1975,** *97*, 891. b) Abrams, S. R.; Shaw, A. C. *Org. Synth.* **1988,** *66*, 127; *Coll. Vol. 8*, p. 146.

4 Toutov, A. A.; Betz, K. N.; Schuman, D. P.; Liu, W.-B.; Fedorov, A.; Stoltz, B. M.; Grubbs, R. H. *J. Am. Chem. Soc.*, **2017,** *139*, 1668.

5 Rochat, R.; Yamamoto, K.; Lopez, M. J.; Nagae, H.; Tsurugi, H.; Mashima, K. *Chem. - Eur. J.*, **2015,** *21*, 8112.

6 a) Hodgson, D. M.; Miles, S. M. *Angew. Chem., Int. Ed.* **2006,** *45*, 935. b) Hodgson, D. M.; Humphreys, P. G.; Miles, S. M.; Brierley, C. A. J.; Ward, J. G. *J. Org. Chem.*, **2007,** *72*, 10009.

7 Polniaszek, R.; Pfeiffer, S.; Yu, R.; Cullen, A.; Dowdy, E.; Tran, D.; Kent, K.; Zhou, Z.; Cordeau, D.; Easton, L. WO2010115000A2, **2010**.

8 Feng, J.; Gwaltny, S. L.; Stafford, J. A.; Zhang, Z.; Elder, B. J.; Isbester, P. K.; Palmer, G. J.; Salsbury, J. S.; Ulysse, L. WO 2007035629 A2, 29 March **2007**.

9 Wang, W.; Li, T. US 8, 648, 073 B2, **2014**.

10 Follmann, M.; Ackerstaff, J.; Redlich, G.; Wunder, F.; Lang, D.; Kern, A.; Fey, P.; Griebenow, N.; Kroh, W.; Becker-Pelster, E.-M.; Kretschmer, A.; Geiss, V.; Li, V.;

Straub, A.; Mittendorf, J.; Jautelat, R.; Schirok, H.; Schlemmer, K.-H.; Lustig, K.; Gerisch, M.; Knorr, A.; Tinel, H.; Mondritzki, T.; Trübel, H.; Sandner, P.; Stasch, J.-P. *J. Med. Chem.*, **2017,** *60*, 5146.

11 Dorsch, D.; Schadt, O.; Stieber, F.; Meyring, M.; Grädler, U.; Bladt, F.; Friese-Hamim, M.; Knühl, C.; Pehl, U.; Blaukat, A. *Bioorg. Med. Che. Lett.*, **2015,** *25*, 1597.

12 Mohite, A. R.; Phatake, R. S.; Dubey, P.; Agbaria, M.; Shames, A. I.; Lemcoff, N. G.; Reany, O. *J. Org. Chem.*, **2020,** *85*, 12901.

13 Gledhill, A. P.; McCall, C. J.; Threadgill, M. D. *J. Org. Chem.*, **1986,** *51*, 3196.

14 Glase, S. A.; Akunne, H. C.; Georgic, L. M.; Heffner, T. G.; MacKenzie, R. G.; Manley, P. J.; Pugsley, T. A.; Wise, L. D. *J. Med. Chem.*, **1997,** *40*, 1771.

15 Wang X.; Porco, Jr. J. A. *J. Org. Chem.*, **2001,** *66*, 8215.

16 Yamaguchi, T.; Kouko, T.; Noguchi, S.; Yamane, Y.; Kondo, F.; Aoki, T.; Takeda, T.; Sakanishi, K.; Sato, H.; Ueda, T.; Matuura, S.; Kurahashi, K.; Kitagawa, Y.; Nakamura, T. WO 2019044947 A1, **2019.**

17 Brown, D. G.; and Boström, J.; *J. Med. Chem.*, **2016,** *59*, 4443.

18 Iranpoor, N.; Firouzabadi, H.; Nowrouzi, N.; Khalili, D. *Tetrahedron*, **2009,** *65*, 3893.

19 Shioiri, K.; Kawai, N. *J. Org. Chem.*, **1978,** *43*, 2936.

20 Dragovich, P. S.; Prins, T. J.; Zhou, R.; Webber, S. E.; Marakovits, J. T.; Fuhrman, S. A.; Patick, A. K.; Matthews, D. A.; Lee, C. A.; Ford, C. E.; Burke, B. J.; Rejto, P. A.; Hendrickson, T. F.; Tuntland, T.; Brown, E. L.; Meador, J. W., III; Ferre, R. A.; Harr, J. E. V.; Kosa, M. B.; Worland, S. T. *J. Med. Chem.*, **1999,** *42*, 1213.

21 Tao, J.; McGee, K. *Org. Process Res. Dev.*, **2002,** *6*, 520.

22 Billek, G. *Organic Syntheses*, Wiley & Sons: New York, **1973**; *Coll. Vol. 5*, 627.

23 Albericio, F.; Kruger, H. G. *Future Med. Chem.*, **2012,** *4*, 1527.

24 Yang, Z.; Maeng, J. H.; Manning, D.; Masih, L.; Cao, Y.; Pattamana, K.; Bois, F.; Molino, B. *Synthesis*, **2012,** *44*, 63.

25 Tona, V.; de la Torre, A.; Padmanaban, M.; Ruider, S.; Gonzalez, L.; Maulide, N.; *J. Am. Chem. Soc.*, **2016,** *138*, 8348.

26 de la Torre, A.; Kaiser, D.; Maulide, N. *J. Am. Chem. Soc.*, **2017,** *139*, 6758.

27 Wada, C. K.; Frey, R. R.; Ji, Z.; Curtin, M. L.; Garland, R. B.; Holms, J. H.; Li, J.; Pease, L. J.; Guo, J.; Glaser, K. B.; Marcotte, P. A.; Richardson, P. L.; Murphy, S. S.; Bouska, J. J.; Tapang, P.; Magoc, T. J.; Albert, D. H.; Davidsen, S. K.; Michaelides, M, R. *Bioorg. Med. Chem. Lett.*, **2003,** *13*, 3331.

28 Tsunoda, T.; Ozaki, F.; Ito, S. *Tetrahedron Lett.*, **1994,** *35*, 5081.

29 Mosallanejad, A.; Lorthioir, O. *Tetrahedron Lett.*, **2018,** *59*, 1708.

30 Velcicky, J.; Soicke, A.; Steiner, R.; and Schmalz, H.-G. *J. Am.Chem. Soc.*. **2011,** *133*, 6948.

31 Fleming, F. F.; Yao, L.; Ravikumar, P. C.; Funk, L.; Shook, B. C. *J. Med. Chem.*, **2010,** *53*, 7902.

32 Fleming, F. F.; Yao, L.; Ravikumar, P. C.; Funk, L.; Shook, B. C. *J. Med. Chem.*, **2010,** *53*, 7902.

33 Ishii, G.; Moriyama, K.; Togo, H. *Tetrahedron Lett.*, **2011,** *52*, 2404.

34 Sehon, C. A.; Wang, G. Z.; Viet, A. Q.; Goodman, K. B.; Dowdell, S. E.; Elkins, P. A.; Semus, S. F.; Evans, C.; Jolivette, L. J.; Kirkpatrick, R. B.; Dul, E.; Khandekar,

S. S.; Yi, T.; Wright, L. L.; Smith, G. K.; Behm, D. J.; Bently, R.; Doe, C. P.; Hu, E.; Lee, D. *J. Med. Chem.*, **2008,** *51*, 6631.

35 Huber, V. J.; Bartsch, R. A. *Tetrahedron*, **1998,** *54*, 9281.

36 Ledeboer, M.W.; Pierce, A.C.; Duffy, J.P.; Gao, H.; Messersmith, D.; Salituro, F.G.; Nanthakumar, S.; Come, J.; Zuccola, H.J.; Swenson, L.; Shlyakter, D.; Mahajan, S.; Hoock, T.; Fan, B.; Tsai, W.-J.; Kolaczkowski, E.; Carrier, S.; Hogan, J. K.; Zessis, R.; Pazhanisamy, S.; Bennani, Y.L. *Bioorg. Med. Chem. Lett.*, **2009,** *19*, 6529.

37 Meng, C.; Miculka, C.; Soll, M.; Paulini, R.; Pohlman, M.; Sorgel, S.; Bastiaanns, H. M. M.; Thompson, S.; Ebuenga Doyog, C.; Malveda Umali, A.; Suiza Cosare, R.; Palmer, C.; Hokama, T. WO **2015**/161224, A1.

38 Mills, L. R.; Rousseaux, S. A. L.; *Tetrahedron*, **2019,** *75*, 4298.

39 Cannata, V.; Tamerlani, G.; Marconi, P.; Zagnoni, G. US 5,097,058, **1992**.

40 Li, J.; Zhao, H. B.; Wang, Z. W. WO2019/047915, A1, **2019**.

41 Ceradini, D.; Cacivkins, P.; Ramos-Llorca, A.; Shubin, K. *Org. Process Res. Dev.*, **2022,** *26*, 2937.

42 a) Birum, G. H. *J. Org. Chem.*, **1974,** *39*, 209; b) Oleksyszyn, J.; Subotkowska, L.; Mastalerz, P. *Synthesis*, **1979,** *12*, 985.

43 Davies, T. Q.; Tilby, M. J.; Skolc, D.; Hall, A.; Willis, M. C. *Org. Lett.*, **2020,** *22*, 9495.

44 Zetterberg, F.; Peterson, K.; Nilsson, U. WO **2022**/144274 A1.

45 Vidal, J.; Drouin, J.; Collet, A. *J. Chem. Soc., Chem. Comm.*, **1991,** 435.

46 Vidal, J.; Damestoy, S.; Guy, L.; Hannachi, J.-C.; Aubry, A.; Collet, A. *Chem. Eur. J.*, **1997,** *3*, 1691.

47 Armstrong, A.; Jones, L. H.; Knight, J. D.; Kelsey, R. D. *Org. Lett.*, **2005,** *7*, 713.

48 Armstrong, A.; Cooke, R. S. *Chem. Comm.*, **2002,** 904.

49 Urban, F. J.; Anderson, B. G.; Orrill, S. L.; Daniels, P. J. *Org. Process Res. Dev.*, **2001,** *5*, 575.

50 Dow, R. L.; Kelly, R. C.; Schletter, I.; Wierenga, W. *Synth. Commun.*, **1981,** *11*, 43.

51 Hantzsoh, A.; Thomson, K. J., *Chem. Ber.*, **1905,** *38*, 2266.

52 Brecknell, D. J.; Carman, R. M.; Deeth, H. C.; Kibby, J. J. *Aust. J. Chem.*, **1969,** 22, 1915.

53 Laurent, A.; Rose, Y. WO2014/005217, A1, **2014**.

54 Zhao, X.; Zhang, W.; Chen, Z.; Chen, L.; Wang, X.; Li, Z.; Tan, R.; Yang, L.; Tan, H.; Liu, B.; Ran, K.; ZOU, Z.; Lin, M.; Sun, J.; Wang, W., **WO2017**/219955, A1, **2017**.

55 Pirrung, M. C.; Dunlap, S. E.; Trinks, U. P. *Helv. Chim. Acta*, **1989,** *72*, 1301.

56 Adam, J.-M.; Foricher, J.; Hanlon, S.; Lohri, B.; Moine, G.; Schmid, R.; Stahr, H.; Weber, M.; Wirz, B.; Zutter, U. *Org. Process Res. Dev.*, **2011,** *15*, 515.

57 Kitaori, K.; Mikami, M.; Furukawa, Y.; Yoshimoto, H.; Otera J. *Synlett*, **1998,** 499.

58 Kulinkovich, O. G.; Sviridov, S. V.; Vasilevskii, D. A.; Pritytskaya, T. S. *Zh. Org. Khim.*, **1989,** *25*, 2244.

59 Kulinkovich, O. G.; Kananovich, D. G., *Eur. J. Org. Chem.*, **2007,** 2121.

60 Chaplinski, V.; De Meijere, A., *Angew. Chem. Int. Ed.*, **1996,** *35*, 413.

61 Pearson-Long, M. S. M.; Beauseigneur, A.; Karoyan, P.; Szymoniak, J.; Bertus, P., *Synthesis*, **2010,** 3410.

62 Bertus, P.; Szymoniak, J., *J. Org. Chem.*, **2002,** *67*, 3965.

63 Laroche, C.; Harakat, D.; Bertus, P.; Szymoniak, J., *Org. Biomol. Chem.*, **2005,** *3*, 3482.

64 Laroche, C.; Bertus, P.; Szymoniak, J.; *Tetrahedron Lett.*, **2003,** *44*, 2485.

65 Zheng, X.; Zhang, Y.; Fu, C., CN 109748902 A, **2019**.

66 Beaulieu, P. L.; Gillard, J.; Bailey, M. D.; Boucher, D.; Duceppe, J.-S.; Simoneau, B.; Wang, X.-J.; Zhang, L.; Grozinger, K.; Houpis, I.; Farina, V.; Heimroth, H.; Krueger, T.; Schnaubelt, J. *J. Org. Chem.*, **2005,** *70*, 5869.

67 Overman, L. E.; Humphreys, P. G.; Welmaker, G. S., *Org. React.*, **2011,** *75*, 747.

68 Tremblay, M. R.; Nevalainen, M.; Nair, S. J.; Porter, J. R.; Castro, A. C.; Behnke, M. L.; Yu, L.-C.; Hagel, M.; White, K.; Faia, K.; Grenier, L.; Campbell, M. J.; Cushing, J.; Woodward, C. N.; Hoyt, J.; Foley, M. A.; Read, M. A.; Sydor, J. R.; Tong, J. K.; Palombella, V. J.; McGovern, K.; Adams, J. *J. Med. Chem.*, **2008,** *51*, 6646.

69 Austad, Brian C.; Hague, A, B.; White, P.; Peluso, S.; Nair, S. J.; Depew, K. M.; Grogan, M. J.; Charette, A. B.; Yu, L.-C.; Lory, C. D.; Grenier, L.; Lescarbeau, A.; Lane, B. S.; Lombardy, R.; Behnke, M. L.; Koney, N.; Porter, J. R.; Campbell, M. J.; Shaffer, J.; Xiong, J.; Helble, J. C.; Foley, M. A.; Adams, J.; Castro, A. C.; Tremblay, M. R. *Org. Process Res. Dev.*, **2016,** *20*, 786.

70 Charette, A. B.; Beauchemin, A., *Org. React.*, **2001,** *58*, 1.

71 Daiichi Sankyo Company, Limited; Nakamura, Y. US2014/0094623, A1, **2014**.

72 Snider, B. B.; Allentoff, A. J.; Kulkarni, Y. S., *J. Org. Chem.*, **1988,** *53*, 5320.

73 Ravn, M. M.; Wagaw, S. H.; Engstrom, K. M.; Mei, J. Z.; Kotecki, B.; Souers, A. J.; Kym, P. R.; Judd, A. S.; Zhao, G. *Org. Process Res. Dev.*, **2010,** *14*, 417.

74 Ayesa, S.; Belda, O.; Bjorklund, C.; Nilsson, M.; Russo, F.; Sahlberg, C.; Wiktelius, D., WO2013/095275 A1, **2013**.

75 Meshram, H. M.; Reddy, P. N.; Sadashiv, K.; Yadav, J. S. *Tetrahedron Lett.*, **2005,** *46*, 623.

76 Favorskii, A. E. *J. Russ. Phys. Chem. Soc.*, **1894,** *26*, 590.

77 Watson, T. J. N.; Curran, T. T.; Hay, D. A.; Shah, R. S.; Wenstrup, D. L.; Webster, M. E. *Org. Process Res. Dev.*, **1998,** *2*, 357.

78 Curran, T. T.; Hay, D. A.; Koegel C. P.; Evans, J. C. *Tetrahedron*, **1997,** *53*, 1983.

79 Purohit, V. C.; Allwein, S. P.; Bakale, R. P. *Org. Lett.*, **2013,** *15*, 1650.

80 Allwein, S. P.; Mowrey, D. R.; Petrillo, D. E.; Reif, J. J.; Purohit, V. C.; Milkiewicz, K. L.; Bakale, R. P.; Christie, M. A.; Olsen, M. A.; ‡ Neville, C. J.; Gilmartin, G. J. *Org. Process Res. Dev.*, **2017,** *21*, 740.

81 Taylor, E. C.; Conley, R. A.; Katz, A. H.; McKillop, A. J. *J. Org. Chem.*, **1984,** *49*, 3840.

82 Covarrubias-Zuniga, A.; Gonzalez-Lucas, A.; Dominguez, M. M. *Tetrahedron*, **2003,** *59*, 1989.

83 Golinski, J.; Makosza, M. *Tetrahedron Lett.*, **1978,** *19*, 3495.

84 Makosza, M.; Golinski, J.; Baran, J., *J. Org. Chem.*, **1984,** *49*, 1488.

85 Makosza, M.; Winiarski, J. *Acc. Chem. Res.*, **1987,** *20*, 282.

86 Kuang, R.; Wu, H.; Ting, P. C.; Aslanian, R. G.; Cao, J.; Kim, D. W.; Lee, J. F.; Schwerdt, J.; Zhou, G.; Herr, R. J.; Zych, A. J.; Yang, J.; Lam, S. Q.; Jenkins, D. M.; Sakwa, S. A.; Wainhaus, S.; Black, T. A.; Cacciapuoti, A.; McNicholas, P. M.; Xu, Y.; Walker, S. S. *Bioorg. Med. Chem. Lett.*, **2012,** *22*, 5268.

87 Welch, W. M.; Kraska, A. R.; Sarges, R.; Koe, B. K. *J. Med. Chem.*, **1984,** *27*, 1508.

88 Vukics, K.; Fodor, T.; Fischer, J.; Fellegvari, I.; Levai, S. *Org. Process Res. Dev.*, **2002,** *6*, 82.

89 Taber, G. P.; Pfisterer, D. M.; Colberg, J. C. *Org. Process Res. Dev.*, **2004,** *8*, 385.
90 Lee, S. H.; Kim, I. S.; Li, Q. R.; Dong, G. R.; Jeong, L. S.; Jung, Y. H. *J. Org. Chem.*, **2011,** *76*, 10011.
91 DeClerq, P. J. *Chem. Rev.*, **1997,** *97*, 1755.
92 Warm, A.; Naughton, A. B.; Saikali, E. A. *Org. Process Res. Dev.*, **2003,** *7*, 272.
93 Ikunaka, M.; Matsumoto, J.; Fujima, Y.; Hirayama, Y. *Org. Process Res. Dev.*, **2002,** *6*, 49.
94 Cohen, N.; Banner, B. L.; Laurenzano, A. J.; Carozza, L. *Org. Synth.*; Wiley & Sons: New York, **1985**; *63*, 127.
95 Dunigan, J.; Weigel, L. O. *J. Org. Chem.*, **1991,** *56*, 6225.
96 Batra, H.; Moriarty, R. M.; Penmasta, R.; Sharma, V.; Stanciuc, G.; Staszewski, J. P.; Tuladhar, S. M.; Walsh, D. A. *Org. Process Res. Dev.*, **2006,** *10*, 887.
97 Batra, H.; Moriarty, R. M.; Penmasta, R.; Sharma, V.; Stanciuc, G.; Staszewski, J. P.; Tuladhar, S. M.; Walsh, D. A. *Org. Process Res. Dev.*, **2006,** *10*, 484.
98 Baldwin, J. E. *J. Chem. Soc., Chem. Commun.*, **1976,** 734.
99 Dong, Y.; Wittlin, S.; Sriraghavan, K.; Chollet, J.; Charman, S. A.; Charman, W. N.; Scheurer, C.; Urwyler, H.; Tomas, J. S.; Snyder, C.; Creek, D. J.; Morizzi, J.; Koltun , M.; Matile, H.; Wang, X.; Padmanilayam, M.; Tang, Y.; Reto Brun, A. D.; Vennerstrom, J. L. *J. Med. Chem.*, **2010,** 481.
100 Li, T.; Yang, Y.-T.; Li, Y.-L. *J. Chem, Res.*, **1993,** 30.
101 Yang, J.; Li, T.; Li, Y.-L. *Chin. Chem. Lett.*, **1996,** *7*, 997.
102 Parkerr, W.; Raphael, A.; Wilkinson, D. I. *J. Chem. Soc.*, **1958,** 3871.
103 Shin, S. S.; Byun, Y.; Lim, K. M.; Choi, J. K.; Lee, K.-W.; Moh, J. H.; Kim, J. K.; Jeong, Y. S.; Kim, J. Y.; Choi, Y. H.; Koh, H.-J.; Park, Y.-H.; Oh, Y. I.; Noh, M.-S.; Chung, S. *J. Med. Chem.*, **2004,** *47*, 792.
104 Yu, R. H. C.; Brown, B. H.; Polniaszek, R. P.; Graetz, B. R.; Sujino, K.; Tran, D. D.-P.; Triman, A. S.; Kent, K. M.; Pfeiffer, S. US8,987,437, B2, **2015**.
105 Standley, E. A.; Bringley, D. A.; Calimsiz, S.; Ng, J. D.; Sarma, K.; Shen, J.; Siler, D. A.; Ambrosi, A.; Chang, W.-T. T.; Chiu, A.; Davy, J. A.; Doxsee, I. J.; Esanu, M. M.; Garber, J. A. O.; Kim, Y.; Kwong, B.; Lapina, O.; Leung, E.; Lin, L.; Martins, A.; Phoenix, J.; Phull, J.; Roberts, B. J.; Shi, B.; St-Jean, O.; Wang, X.; Wang, L.; Wright, N.; Yu, G. *Org. Process Res. Dev.*, **2021,** *25*, 1215.
106 Clinch, K.; Watt, D. K.; Dixon, R. A.; Baars, S. M.; Gainsford, G. J.; Tiwari, A.; Schwarz, G.; Saotome, Y.; Storek, M.; Belaidi, A. A.; Santamaria-Araujo, J. A. *J. Med. Chem.*, **2013,** *56*, 1730.
107 Huang, X. J.; O'Brien, E.; Thai, F.; Cooper, G. *Org. Process Res. Dev.*, **2010,** *14*, 592.
108 Pennell, A. M. K.; Aggen, J. B.; Wright, J. J. K.; Sen, S.; Chen, W.; Dairaghi, D. J.; Zhang, P. WO2005/084667 A1, **2005**.
109 Huy, P.; Neudorfl, J.-M.; Schmalz, H.-G. *Org. Lett.*, **2011,** *13*, 216.
110 Perni, R. B.; Farmer, L. J.; Cottrell, K. M.; Court, J. J.; Courtney, L. F.; Deininger, D. D.; Gates, C. A.; Harbeson, S. L.; Kim, J. L.; Lin, C.; Lin, K.; Luong, Y.-P.; Maxwell, J. P.; Murcko, M. A.; Pitlik, J.; Rao, B. G.; Schairer, W. C.; Tung, R. D.; Van Drie, J. H.; Wilson, K.; Thomsonet, J. A. *Bioorg. Med. Chem. Lett.*, **2004,** *14*, 1939.
111 DeSolms, S. J. *US* 5,872,135, **1999**.

112 Angell, P. T.; Lewandowski, B.; Littler, B. J.; Nugent, W. A.; Smith, D.; Studley, J. WO2019/028228 A1, **2019**.

113 Lai, J. T.; Westfalh, J. C. *J. Org. Chem.*, **1980,** *45*, 1513.

114 Lind, H.; Winkler, T. *Tetrahedron Lett.*, **1980,** *21*, 119.

115 Hughes, D. L. *Org. Process Res. Dev.*, **2019,** *23*, 2302.

116 Tanaka, M.; Ubukata, M.; Matsuo, T.; Yasue, K.; Matsumoto, K.; Kajimoto, Y.; Ogo, T.; Inaba, T. *Org. Lett.*, **2007,** *9*, 3331.

117 Tanaka, M.; Sagawa, S.; Hoshi, J.; Shimoma, F.; Yasue, K.; Ubukata, M.; Ikemoto, T.; Hase, Y.; Takahashi, M.; Sasase, T.; Ueda, N.; Matsushita, M.; Inaba, T. *Bioorg. Med. Chem.*, **2006,** *14*, 5781.

118 Challis, B. C.; Rzepa, H. S. *J. Chem. Soc., Perkin Trans. 2*, **1977,** 281.

119 Gollner, A.; Rudolph, D.; Arnhof, H.; Bauer, M.; Blake, S. M.; Boehmelt, G.; Cockroft, X.-L.; Dahmann, G.; Ettmayer, P.; Gerstberger, T.; Karolyi-Oezguer, J.; Kessler, D.; Kofink, C.; Ramharter, J.; Rinnenthal, J.; Savchenko, A.; Schnitzer, R.; Weinstabl, H.; Weyer-Czernilofsky, U.; Wunberg, T.; McConnell, D. B. *J. Med. Chem.*, **2016,** *59*, 10147.

120 a) Poornachandran, M.; Raghunathan, R. *Synth. Commun.* **2007,** *37*, 2507; b) Lakshmi, N. V.; Thirumurugan, P.; Perumal, P. T. *Tetrahedron Lett.*, **2010,** *51*, 1064; c) Alimohammadi, K.; Sarrafi, Y.; Tajbakhsh, M.; Yeganegi, S.; Hamzehloueian, M. *Tetrahedron*, **2011,** *67*, 1589.

121 Gillard, J.; Abraham, A.; Anderson, P. C.; Beaulieu, P. L.; Bogri, T.; Bousquet, Y.; Grenier, L.; Guse, I.; Lavallee, P. *J. Org. Chem.*, **1996,** *61*, 2226.

122 Abderson, P. C.; Soucy, F.; Yoakim, C.; Lavallee, P.; Beaulieu, P. L. Eur. Patent Appl., 560268, **1993.**

123 Hays, S. J.; Malone, T. C.; Johnson, G. *J. Org. Chem.*, **1991,** *56*, 4084.

124 Larsen, S. D.; Grieco, P. A.; Fobare, W. F. *J. Am. Chem. Soc.*, **1986,** *108*, 3512.

125 Fernando, D.; Lacasse, S. M.; Wiglesworth, K. WO2019/102311, A1, **2019**.

126 Rodriguez, J.; Dulcere, J.-P. *Synthesis*, **1993,** 1177.

127 Schwab, L. S. *J. Med. Chem.*, **1980,** *23*, 698.

128 Marton, J.; Simon, C.; Hosztafi, S.; Szabo, Z.; Marki, A.; Borsodi, A.; Makleit, S. *Bioorg. Med. Chem.*, **1997,** *5*, 369.

129 Smissman, E. E.; Makriyannis, A. *J. Org. Chem.*, **1973,** *38*, 1652.

130 Thavaneswaran, S.; McCamley, K.; Scammells P. J. *Nat. Prod. Comm.*, **2006,** *1*, 885.

131 Chung, J. Y. L.; Meng, D.; Shevlin, M.; Gudipati, V.; Chen, Q.; Liu, Y.; Lam, Y.-H.; Dumas, A.; Scott, J.; Tu, Q.; Xu, F. *J. Org. Chem.*, **2020,** *85*, 994.

132 Baldwin, J. E.; Adlington, R. M.; Godfrey, C. R. A.; Gollins, D. W.; Vaughan, J. G. *J. Chem. Soc., Chem. Commun.*, **1993,** 1434.

133 Wentland, M. P.; Lu, Q.; Lou, R.; Bu, Y.; Knapp, B. I.; Bidlack, J. M. *Bioorg. Med. Chem. Lett.*, **2005,** *15*, 2107.

134 Sun, D.; Blaine, P.; Jonathan, A. E. *Chem. Sci.*, **2019,** *10*, 535.

135 Grewe, R.; Friedrichsen, W. *Chem. Ber.*, **1967,** *100*, 1550.

136 Chen, C.; Kozikowski, A. P.; Wood, P. L.; Reynolds, I. J.; Ball, R. G.; Pang, Y.-P. *J. Med. Chem.*, **1992,** *35*, 1634.

137 Cohen, L. A.; Witkop, B. *J. Am. Chem. Soc.*, **1955,** *77*, 6595.

138 Sosnovsky, G.; Brown, J. H. *Chem. Rev.*, **1966,** *66*, 529.

139 Chang, H. S.; Edward, J. T. *Can. J. Chem.*, **1963,** *41*, 1233.

140 Ronn, M.; McCubbin, Q.; Winter, S.; Veige, M. K.; Grimster, N.; Alorati, T.; Plamondon, L. *Org. Process Res. Dev.*, **2007,** *11*, 241.

141 Miyano, S. *Synthesis*, **1978,** 701.

142 McIntosh, J. M. *Can. J. Chem.*, **1980,** *58*, 2604.

143 Alder, R. W.; Arrowsmith, R. J.; Boothby, C. St. J.; Heilbronner, E.; Yang, Z.-Z. *J. Chem. Soc., Chem. Commun.*, **1982,** 940.

144 Ronn, M.; McCubbin, Q.; Winter, S.; Veige, M. K.; Grimster, N.; Alorati, T.; Plamondon, L. *Org. Process Res. Dev.*, **2007,** *11*, 241.

145 Abdel-Aziz, A. A.-M.; El-Azab, A. S.; Abou-Zeid, L. A.; Eltahir, K. E. H.; Abdel-Aziz, N. I.; Ayyad, R. R.; Al-Obaid, A. M. *Eur. J. Med. Chem.*, **2016,** *115*, 121.

146 Mahmoodi, N. O.; Emadi, S. *Rus. J. Org. Chem.*, **2004,** *40*, 377.

147 Brickner, S. J.; Hutchinson, D. K.; Barbachyn, M. R.; Manninen, P. R.; Ulanowicz, D. A.; Garmon, S. A.; Grega, K. C.; Hendges, S. K.; Toops, D. S.; Ford, C. W.; Zurenko, G. E. *J. Med. Chem.*, **1996,** *39*, 673.

148 Clark, J. D.; Collins, J. T.; Kleine, H. P.; Weisenburger, G. A.; Anderson, D. K. *Org. Process Res. Dev.*, **2004,** *8*, 571.

149 Samanen, J. M.; Ali, F. E.; Barton, L. S.; Bondinell, W. E.; Burgess, J. L.; Callahan, J. F.; Calvo, R. R.; Chen, W.; Chen, L.; Erhard, K.; Feuerstein, G.; Heys, R.; Hwang, S.-M.; Jakas, D. R.; Keenan, R. M.; Ku, T. W.; Kwon, C.; Lee, C.-P.; Miller, W. H.; Newlander, K. A.; Nichols, A.; Parker, M.; Peishoff, C. E.; Rhodes, G.; Ross, S.; Shu, A.; Simpson, R.; Takata, D.; Yellin, T. O.; Uzsinskas, I.; Venslavsky, J. W.; Yuan, C.-K.; Huffman, W. F. *J. Med. Chem.*, **1996,** *39*, 4867.

150 Andrews, I. P.; Atkins, R. J.; Breen, G. F.; Carey, J. S.; Forth, M. A.; Morgan, D. O.; Shamji, A.; Share, A. C.; Smith, S. A. C. Walsgrove, T. C.; Wells, A. S. *Org. Process Res. Dev.*, **2003,** *7*, 655.

151 Jefford, C. W.; Jaggi, D.; Boukouvalas, J.; Kohmoto, S. *J. Am. Chem. Soc.*, **1983,** *105*, 6497.

152 Jefford, C. W.; Jaggi, D.; Boukouvalas, J.; Kohmoto, S. *J. Chem. Soc., Chem. Commun.*, **1984,** 523.

153 Tiwari, M. K.; Chaudhary, S. *Med. Res. Rev.*, **2020,** *40*, 1220.

154 Wu, L.; Abreu, B. L.; Blake, A. J.; Taylor, L. J.; Lewis, W.; Argent, S. P.; Poliakoff, M.; Boufroura, H.; George, M. W. *Org. Process Res. Dev.*, **2021,** *25*, 1873.

155 Turconi, J.; Griolet, F.; Guevel, R.; Oddon, G.; Villa, R.; Geatti, A.; Hvala, M; Rossen, K.; Goller, R.; Burgard, A. *Org. Process. Res. Dev.*, **2014,** *18*, 417.

156 a) Schenck, G. O.; K. Ziegler, *Naturwiss.* **1944,** *32*, 157; b) Schaffner, K. *Angew Chem. Int. Ed.*, **2003,** *42*, 2932.

157 Hock, H.; Lang, S. *Ber. Dtsch. Chem. Ges.*, **1944** (A and B Series), *77*, 257.

158 Marcuccio, S. M.; Epa, R.; White, J. M.; Deadman, J. J. *Org. Process Res. Dev.*, **2011,** *15*, 763.

159 Shainyan, B. A.; Kirpichenko, S. V.; Freeman, F. *J. Am. Chem. Soc.*, **2004,** *126*, 11456.

160 Sugen, Inc. US2004/266843, A1, **2004**.

161 Wang, H.; Chen, M.; Wang, L. *Chem. Pharm. Bull.*, **2007,** *55*, 1439.

162 Xu, K.; Li, W.; Sun, R.; Luo, L.; Chen, X.; Zhang, C.; Zheng, X.; Yuan, M.; Fu, H.; Li, R.; Chen, H. *Org. Lett.*, **2020,** *22*, 6107.

163 Carson, J. R.; Jetter, M. C.; Lee, J. S.; Youngman, M. A. US2004/0192720, A1, **2004**.
164 Van Der Plas, H. C. *Acc. Chem. Res.*, **1978,** *11*, 462.
165 Furet, P.; Guagnano, V.; Fairhurst, R. A.; Imbach-Weese, P.; Bruce, I.; Knapp, M.; Fritsch, C.; Blasco, F.; Blanz, J.; Aichholz, R.; Hamon, J.; Fabbro, D.; Caravatti, G. *Bioorg. Med. Chem. Lett.*, **2013,** *23*, 3741.
166 Kataja, A. **WO2018**/002415 A1.
167 Allegretti, M.; Anacardio, R.; Cesta, M. C.; Curti, R.; Mantovanini, M.; Nano, G.; Topai, A.; Zampella, G. *Org. Process Res. Dev.*, **2003,** *7*, 209.
168 Wuhan QR Pharmaceuticals, CN114644614A, **2022**.
169 Brandange, S.; Lindblom, L. *Acta Chem. Scand. Ser. B*, **1976,** *30*, 93.
170 Haslego, M. L.; Maryanoff, C. A.; Scott, L.; Sorgi, K. L. *Heterocycles*, **1993,** *35*, 643.
171 Sorgi, K. L.; Maryanoff, C. A.; McComsey, D. F.; Maryanoff B. E. *Org. Synth.*, **1998,** *75*, 215.
172 Watanabe, M.; Okachi, T.; Kawahara, M.; Nagasawa, H.; Sato, N.; Takita, T.; Hasegawa, G. US**2016**/0096803 A1, **2016.**
173 Foley, L. H. *Tetrahedron Lett.*, **1994,** *35*, 5989.
174 Murineddu, G.; Gignarella, G.; Chelucci, G.; Loriga, G.; Pinna, G. A. *Chem. Pharm. Bull.*, **2002,** *50*, 754.
175 Arikawa, Y.; Nishida, H.; Kurasawa, O.; Hasuoka, A.; Hirase, K.; Inatomi, N.; Hori, Y.; Matsukawa, J.; Imanishi, A.; Kondo, M.; Tarui, N.; Hamada, T.; Takagi, T.; Takeuchi, T.; Kajino, M. *J. Med. Chem.*, **2012,** *55*, 5989.
176 Ball, M.; Jones, M. F.; Kenley, F.; Pittam, J. D. *Org. Process Res. Dev.*, **2017,** *21*, 310.
177 Hartmann, H. *Synthesis*, **1984,** 275.
178 Andersen, H. S.; Olsen, O. H.; Iversen, L. F.; Sørensen, A. L. P.; Mortensen, S. B.; Christensen, M. S.; Branner, S.; Hansen, T. K.; Lau, J. F.; Jeppesen, L.; Moran, E. J.; Su, J.; Bakir, F.; Judge, L.; Shahbaz, M.; Collins, T.; Vo, T.; Newman, M. J.; Ripka, W. C.; Møller, N. P. H. *J. Med. Chem.*, **2002,** *45*, 4443.
179 Alvi, K. A.; Casey, A.; Nair, B. G. *J. Antibiotics*, **1998,** *51*, 515.
180 Mardjan, M. I. D.; Parrain, J.-L.; Commeiras, L. *Adv. Synth. Catal.*, **2016,** *358*, 543.
181 Mardjan, M. I. D.; Perie, S.; Parrain, J.-L.; Commeiras, L. *Org. Biomol. Chem.*, **2017,** *15*, 3304.
182 Mardjan, M. I. D.; Mardjan, D.; Mayooufi, A.; Parrain, J.-L.; Thibonnet, J.; Commeiras, L. *Org. Process Res. Dev.*, **2020,** *24*, 606.
183 Baldwin, J. E. *J. Chem. Soc., Chem. Commun.*, **1976,** 734.
184 Marcoux, J.-F.; Marcotte, F.-A.; Wu, J.; Dormer, P. G.; Davies, I. W.; Hughes, D.; Reider, P. J. *J. Org. Chem.*, **2001,** *66*, 4194.
185 Stewart, G. W.; Brands, K. M. J.; Brewer, S. E.; Cowden, C. J.; Davies, A. J.; Edwards, J. S.; Gibson, A. W.; Hamilton, S. E.; Katz, J. D.; Keen, S. P.; Mullens, P. R.; Scott, J. P.; Wallace, D. J.; Wise, C. S. *Org. Process Res. Dev.*, **2010,** *14*, 849.
186 Davies, I. W.; Marcoux, J.-F.; Corley, E. G.; Journet, M.; Cai, D.-W.; Palucki, M.; Wu, J.; Larsen, R. D.; Rossen, K.; Pye, P. J.; DiMichele, L.; Dormer, P.; Reider, P. J. *J. Org. Chem.*, **2000,** *65*, 8415.
187 Pearson, T. J.; Shimazumi, R.; Driscoll, J. L.; Dherange, B. D.; Park, D.-I.; Levin, M. D. *Science*, **2023,** *381*, 1474.
188 Haffner, C. *Tetrahedron Lett.*, **1994,** *35*, 1349.
189 Doering, W. von E.; Odum, R. A. *Tetrahedron*, **1966,** *22*, 81.

190 Bruce, A. *PNAS*, **2018,** *115*, 10836.
191 Glinkerman, C. M.; Boger, D. L. *J. Am. Chem. Soc.*, **2016,** *138*, 12408.
192 Chen, Z.; Zhu, J.; Xie, H.; Li, S.; Wu, Y.; Gong, Y. *Org. Lett.*, **2010,** *12*, 4376.
193 Chen, Z.; Zhu, J.; Xie, H.; Li, S.; Wu, Y.; Gong, Y. *Org. Biomol. Chem.*, **2011,** *9*, 5682.
194 Johnson, P. *Org. Process Res. Dev.*, **2019,** *23*, 2243.
195 Gatti Mcarthur, S.; Goetschi, E.; Wichmann, J.; Woltering, T. J. WO2007/110337, A1, **2007**.
196 Chen, C.-S.; Chen, Y.-L.; Chen, Y.-C.; Jadhav, A. WO2018/026811, A2, **2018**.
197 Wang, Z.; You, C.; Wang, C.; Weng, Z.; *J. Org. Chem.*, **2019,** *84*, 14926.
198 Flick, A. C.; Leverett, C. A.; Ding, H. X.; McInturff, E.; Fink, S. J.; Mahapatra, S.; Carney, D. W.; Lindsey, E. A.; DeForest, J. C.; France, S. P.; Berritt, S.; Bigi-Botterill, S. V.; Gibson, T. S.; Liu, Y.; O'Donnell, C. J. *J. Med. Chem.*, **2021,** *64*, 3604.
199 Fier, P. S.; Kim, S.; Cohen, R. D. *J. Am. Chem. Soc.*, **2020,** *142*, 8614.
200 Guillemard, L.; Kaplaneris, N.; Ackermann, L.; Johansson, M. J. *Nature Rev. Chem.*, **2021,** *5*, 522.
201 Roflumilast, sold under the brand name Daxas among others, is a medication used for the treatment of chronic obstructive pulmonary disease, plaque psoriasis, seborrheic dermatitis, and atopic dermatitis. It acts as a selective, long-acting inhibitor of the enzyme phosphodiesterase-4 (PDE-4). It has anti-inflammatory effects.
202 a) Siesel, D. US patent 8,124,781 B2, **2012**. b) Siesel, D. U.S. Patent 8,461,342 B2, **2013**. c) Miller, M. T.; McCartney, J.; Hadida-Ruah, S. S.; Zhou, J. U.S. Patent 9,321,725 B2, **2016**.
203 Doiron, J. E.; Le, C. A.; Ody, B. K.; Brace, J. B.; Post, S. J.; Thacker, N. L.; Hill, H. M.; Breton, G. W.; Mulder, M. J.; Chang, S.; Bridges, T. M.; Tang, L.; Wang, W.; Rowe, S. M.; Aller, S. G.; Turlington, M. *Chem. Eur. J.*, **2019,** *25*, 3662.
204 Hughes, D. L. *Org. Process. Res. Dev.*, **2019,** *23*, 2302.
205 Bhirud, S. B.; Kadam, S. M.; Kansagra, B. P.; Kansagra, B. P.; Bhadane, S, N.; Kale, S. K.; Patil, U. D. WO2017056031, A1, **2017**.
206 Fier, P. S. *J. Am. Chem. Soc.*, **2017,** *139*, 9499.
207 Bull, J. A.; Mousseau, J. J.; Pelletier, G.; Charette, A. B. *Chem. Rev.*, **2012,** *112*, 2642.
208 Angelaud, R.; Reynolds, M.; Venkatramani, C.; Savage, S.; Trafelet, H.; Landmesser, T.; Demel, P.; Levis, M.; Ruha, O.; Rueckert, B.; Jaeggi, H. *Org. Process Res. Dev.*, **2016,** *20*, 1509.
209 Bao, X. F.; Zhong, M. Y.; Wu, Z. X.; Zhang, Q. Y.; Wang, L. L.; Chen, G. L. *Rus. J. Gen. Chem.*, **2023,** *93*, 2694.
210 Zhao, Y.; Zhu, L.; Provencal, D. P.; Miller, T. A.; O'Bryan, C.; Langston, M.; Shen, M.; Bailey, D.; Sha, D.; Palmer, T.; Ho, T.; Li, M. *Org. Process Res. Dev.*, **2012,** *16*, 1652.
211 Wang, W.; Zhao, X.; Yuan, Q.; Shi, H.; Tian, Q.; Liu, Q.; Zhou, Z.; Wang, X.; Chen, Z.; Chen, L.; Tan, H.; Fang, B.; Jiang, L.; Liu, Y.; Sun, J.; Zeng, F.; Li, T. US 9,556,171 B2, **2017**.
212 a) Belema, M.; Meanwell, N. A. *J. Med. Chem.*, **2014,** *57*, 5057. b) Dore, A.; Asproni, B.; Scampuddu, A.; Pinna, G. A.; Christoffersen, C. T.; Langgard, M.; Kehler, J. *Eur. J. Med. Chem.*, **2014,** *84*, 181. c) Bae, E. J.; Choi, W. G.; Pagire, H. S.; Pagire, S. H.; Parameswaran, S.; Choi, J.-H.; Yoon, J.; Choi, W.-i.; Lee, J. H.;

Song, J. S.; Bae, M. A.; Kim, M.; Jeon, J.-H.; Lee, I.-K.; Kim, H.; Ahn, J. H. *J. Med. Chem.*, **2021,** *64*, 1037. d) Ivashchenko, A. A.; Ivanenkov, Y. A.; Aladinskiy, V. A.; Karapetian, R. N.; Koryakova, A. G.; Ryakhovskiy, A. A.; Mitkin, O. D.; Kravchenko, D. V.; Savchuk, N. P.; Zagribelnyy, B. A.; Ivashchenko, A. V. *Bioorg. Med. Chem.*, **2020,** *28*, article number 115716.

213 a) Liminal Biosciences, WO2022/167457, **2022,** A1. b) Hoffmann La Roche, US2013/59833, A1, **2013**. c) Enanta Pharmaceuticals, US2020/2314, A1, **2020**. d) Sichuan University, US2017/253614, A1, **2017**.

214 Tahlan, S.; Kumar, S.; Narasimhan, B. *BMC Chemistry*, **2019,** *13*, 101.

215 Ramalingam, S. S.; Blais, N.; Mazieres, J.; Reck, M.; Jones, C. M.; Juhasz, E.; Urgan, L.; Orlov, S.; Barlesi, F.; Kio, E.; Keiholz, U.; Qin, Q.; Qian, J.; Nickner, C.; Dziubinski, J.; Xiong, H.; Ansell, P.; McKee, M.; Giranda, V.; Gorbunova, V. *Clin. Cancer Res.*, **2017,** *23*, 1937.

216 Malamas, M. S.; Manas, E. S.; McDevitt, R. E.; Gunawan, I.; Xu, Z. B.; Collini, M. D.; Miller, C. P.; Dinh, T.; Henderson, R. A.; Keith Jr., J. C.; Harris, H. A. *J. Med. Chem.*, **2004,** *47*, 5021.

217 Guru, M. M.; Ali, M. A.; Punniyamurthy, T. *J. Org. Chem.*, **2011,** *76*, 5295.

218 Sexton, W. A. *J. Chem. Soc.*, **1942,** 303.

219 Davoll, J. *J. Chem. Soc.*, **1960,** 308.

220 Israel, M.; Jones, L. C.; Modest. E.J. *Tetrahedron Lett.*, **1968,** *9*, 4811.

221 Remond, G.; Portevin, B.; Bonnet, J.; Canet, E.; Regoli, D.; De Nanteuil, G. *Eur. J. Med. Chem.*, **1997,** *32*, 843.

222 Mohan, R. D.; Krishna, R. P.; Venkat, R. B. WO2010128516, A2, **2010.**

223 Yang, F.; Wu, C.; Li, Z.; Tian, G.; Wu, J.; Zhu, F.; Zhang, J.; He, Y.; Shen, J. *Org. Process Res. Dev.*, **2016,** *20*, 1576.

224 Qiu, A. Y.; Zhang, C. X.; Ye, Z. F. CN108129396, A, **2018.**

225 Davis, R. B.; Pizzini, L. C. *J. Org. Chem.*, **1960,** *25*, 1884.

226 Li, S.; Liu, K. CN106674144, B, **2021.**

227 Zhou, Y.; Mukherjee, S.; Huang, D.; Chakraborty, M.; Gu, C.; Zong, G.; Stashko, M. A.; Pearce, K. H.; Shears, S. B.; Chakraborty, A.; Wang, H.; Wang, X. *J. Med. Chem.*, **2022,** *65*, 6869.

228 Beier, P.; Pastyrikova, T. *Beil. J. Org. Chem.*, **2013,** *9*, 411.

229 Ando, K.; Shimazu, Y.; Seki, N.; Yamataka, H. *J. Org. Chem.*, **2011,** *76*, 3937.

230 Zhao, X. Z.; Hymel, D.; Burke Jr., T. R. *Bioorg. Med. Chem. Lett.*, **2016,** *26*, 5009.

231 Burkhard, J. A.; Tchitchanov, B. H.; Carreira, E. M. *Angew. Chem. Int. Ed.*, **2011,** *50*, 5379.

232 Liu, J.; Fitzgerald, A. E.; Lebsack, A. D.; Mani, N. S. *Org. Process Res. Dev.*, **2011,** *15*, 382.

233 Shiraishi, N.; Hoshii, H.; Hamaguchi, W.; Honjo, E.; Takuwa, T.; Kondo, Y.; Goto, T. US2015/111876, A1, **2015.**

234 Cooper, R. D. G.; Jose, F. L. *J. Am. Chem. Soc.*, **1970,** *92*, 2575.

235 Yoshioka, M. *Pure Appl. Chem.*, **1987,** *59*, 1041.

236 Scott, A. I.; Shankaranarayan, R.; Chung, S.-K. *Heterocycles*, **1990,** *30*, 909.

237 Wang, W.; Wu, J. CN105777780 A, **2016.**

238 Li, S.; Guo, S.; He, D.; Shao, Q.; Liu, Y. CN111763221 A, **2020.**

239 Zhu, C.; Zeng, H.; Liu, C.; Cai, Y.; Fang, X.; Jiang, H.; *Org. Lett.*, **2020,** *22*, 809.

240 Li, F.; Nie, J.; Sun, L.; Zheng, Y.; Ma, J.-A. *Angew. Chem., Int. Ed.*, **2013,** *52*, 6255.
241 Britton, J.; Jamison, T. F. *Eur. J. Org. Chem.*, **2017,** *2017*, 6566.
242 Pfeiffer, W.-D.; Dilk, E.; Bulka, E. *Synthesis*, **1977,** 196.
243 Brandt, T. A.; Caron, S.; Damon, D. B.; DiBrino, J.; Ghosh, A.; Griffith, D. A.; Kedia, S.; Ragan, J. A.; Rose, P. R.; Vanderplas, B. C.; Wei, L. L. *Tetrahedron*, **2009,** *65*, 3292.
244 Asano, T.; Yamazaki, H.; Kasahara, C.; Kubota, H.; Kontani, T.; Harayama, Y.; Ohno, K.; Mizuhara, H.; Yokomoto, M.; Misumi, K.; Kinoshita, T.; Ohta, M.; Takeuchi, M. *J. Med. Chem.*, **2012,** *55*, 7772.
245 Yoshida, S.; Hayashi, Y.; Obitsu, K.; Nakamura, A.; Kikuchi, T.; Sawada, T.; Kimura, T.; Takahashi, T.; Mukuta, T. *Org. Process Res. Dev.*, **2012,** *16*, 1818.
246 Roth, M.; Dubs, P.; Gotschi, E.; Eschenmoser, A. *Helv. Chim. Acta*, **1970,** *54*, 710.
247 Broadhurst, M. J.; Grigg, R. *J. Chem. Soc., Perkin 1*, **1972,** 1124.
248 Ohigashi, A.; Temmaru, K.; Hashimoto, N. *Org. Process Res. Dev.*, **2006,** *10*, 159.
249 Heindel, N. D.; Chun, M. C. *Chem. Commun.*, **1971,** 664.
250 Culbertson, T. P. *J. Heterocycl. Chem.*, **1979,** *16*, 1423.
251 Humphrey, G. R.; Pye, P. J.; Zhong, Y.-L.; Angelaud, R.; Askin, D.; Belyk, K. M.; Maligres, P. E.; Mancheno, D. E.; Miller, R. A.; Reamer, R. A.; Weissman, S. A. *Org. Process Res. Dev.*, **2011,** *15*, 73.
252 Summa, V.; Petrocchi, A.; Matassa, V. G.; Taliani, M.; Laufer, R.; Franasco, R. D.; Altamura, S.; Pace, P. *J. Med. Chem.*, **2004,** *47*, 5336.
253 Stansfield, I.; Avolio, S.; Colarusso, S.; Gennari, N.; Narjes, F.; Pacini, B.; Ponzi, S.; Harper, S. *Bioorg. Med. Chem. Lett.*, **2004,** *14*, 5085.
254 Koch, U.; Attenni, B.; Malancona, S.; Colarusso, S.; Conte, I.; Di Filippo, M.; Harper, S.; Pacini, B.; Giomini, C.; Thomas, S.; Incitti, I.; Tomei, L.; DeFrancesco, R.; Altamura, S.; Matassa, V. G.; Narjes, F. *J. Med. Chem.*, **2006,** *49*, 1693.
255 Pacini, B.; Avolio, S.; Ercolani, C.; Koch, U.; Migliaccio, G.; Narjes, F.; Pacini, L.; Tomei, L.; Harper, S. *Bioorg. Med. Chem. Lett.*, **2009,** *19*, 6245.
256 Slavish, P. J.; Cuypers, M. G.; Rimmer, M. A.; Abdolvahabi, A.; Jeevan, T.; Kumar, G.; Jarusiewicz, J. A.; Vaithiyalingam, S.; Jones, J. C.; Bowling, J. J.; Price, J. E.; DuBois, R. M.; Min, J.; Webby R. J.; Rankovic, Z.; White, S. W. *Eur. J. Med. Chem.*, **2023,** *247*, art. no. 115035.
257 a) Mao, Y.; Tian, W.; Huang, Z.; An, J. *J. Heterocycl. Chem.*, **2014,** *51*, 594. b) Jurok, R.; Cibulka, R.; Dvorakova, H.; Hampl, F.; Hodacova, J. *Eur. J. Org. Chem.*, **2010,** *27*, 5217.
258 Feng, J.; Zhang, Z.; Wallace, M. B.; Stafford, J. A. Kaldor, S. W.; Kassel, D. B.; Navre, M.; Shi, L.; Skene, R. J.; Asakawa, T.; Takeuchi, K.; Xu, R.; Webb, D. R.; Gwaltney, II S. L. *J. Med. Chem.*, **2007,** *50*, 2297.
259 Kelly, R. C.; Koztecki, L. H. WO 2008067465, A1, **2008**.
260 Yanadaf, R.; Yoneda, Y.; Yazaki, M.; Mimura, N.; Taga, T.; Yoned, F.; Yanada, K. *Tetrahydron Asy.*, **1997,** *8*, 2319.
261 Löffler, M.; Carrey, E. A.; & Zameitat, E.; *Nucleosides, Nucleotides and Nucleic Acids*, **2016,** *35*, 566.
262 Shi, C. CN109761916, **2019,** A.
263 Nath, M.; Vats, M.; Roy, P. *Eur. J. Med. Chem.*, **2013,** *59*, 310.
264 Basnak, L.; Farkas, J. *Collect. Czech. Chem. Commun.*, **1976,** *41*, 311.

265 Graham, M. A. ; Askey, H.; Campbell, A. D.; Chan, L.; Cooper, K. G.; Cui, Z.; Dalgleish, A.; Dave, D.; Ensor, G.; Galan, R.; Hamilton, P.; Heffernan, C.; Jackson, L. V.; Jing, D.; Jones, M. F.; Liu, P.; Mulholland, K. R.; Pervez, M.; Popadynec, M.; Randles, E.; Tomasi, S.; Wang, S. *Org. Process Res. Dev.*, **2021,** *25*, 43.

266 Duffy, K. J.; Fitch, D. M.; Jin, J.; Liu, R.; Shaw, A. N.; Wiggall, K. WO2007/150011, A2, **2007**.

267 a) Bose, A. K.; Garrat, S. *J. Am. Chem. Soc.*, **1962,** *84*, 1310. b) Bose, A. K.; Garratt, S. *Tetrahedron*, **1963,** *19*, 85.

268 O'Brien, A. G.; Ricci, E. M. *Org. Proc. Res. Dev.*, **2018,** *22*, 399.

269 Oda, M.; Furuya, T.; Hasebe, M.; Kuroki, N. WO 2007/072999 A1, **2007**.

270 Zaragoza, F.; Gantenbein, A. *Org. Process Res. Dev.*, **2017,** *21*, 448.

271 VerSek, B.; Ogorevc, B.; Stanovnik, B.; Tisler, M. *Monatsh. Chem.*, **1983,** *114*, 789.

272 a) Siu, M.; Pastor, R.; Liu, W.; Barrett, K.; Berry, M.; Blair, W. S.; Chang, C.; Chen, J. Z.; Eigenbrot, C.; Ghilardi, N.; Gibbons, P.; He, H.; Hurley, C. A.; Kenny, J. R. S.; Khojasteh, C.; Le, H.; Lee, L.; Lyssikatos, J. P.; Magnuson, S.; Pulk, R.; Tsui, V.; Ultsch, M.; Xiao, Y.; Zhu, B.-Y.; Sampath, D. *Bioorg. Med. Chem. Lett.*, **2013,** *23*, 5014. b) Sabourault, N. L.; Faveri, C.; Sheikh, A. WO 2015117981 A1, **2015**.

273 Levy, D. V.; Sclafani, J. A.; Bakale, R. P. *Org. Process Res. Dev.*, **2016,** *20*, 2085.

274 Du, W.; Zhang, F.; Yan, P.; Lai, Q.; Li, J.; Zhu, D.; Ye L. *Chemistry Select*, **2021,** *6*, 4212.

275 Ishimoto, K.; Fukuda, N.; Nagata, T.; Sawai, Y.; Ikemoto, T. *Org. Process Res. Dev.*, **2014,** *18*, 122.

276 Bell, B. M.; Fanwick, P. E.; Graupner, P. R.; Roth, G. A. *Org. Process Res. Dev.*, **2006,** *10*, 1167.

277 Humphrey, G. R.; Pye, P. J.; Zhong, Y.-L.; Angelaud, R.; Askin, D.; Belyk, K. M.; Maligres, P. E.; Mancheno, D. E.; Miller, R. A.; Reamer, R. A.; Weissman, S. A. *Org. Process Res. Dev.*, **2011,** *15*, 73.

278 William, O.; Wernerr, A. *Can. J. Chem.*, **1987,** *65*, 166.

279 Miller, T.; Papaioannou, N.; WO2018/144620, A1, **2018**.

280 Miyake, A.; Yoshimura, Y.; Yamaoka, M.; Nishimura, T.; Hashimoto, N.; Imada, A. *J. Antibiot.*, **1992,** *45*, 709.

281 Tatsuta, K.; Miura, S.; Gunji, H.; Tamai, T.; Yoshida, R.; Inagaki, T.; Kurita, Y. *Bull. Chem. Soc. Jpn.*, **1994,** *76*, 1701.

282 Anderson, E. D.; Duerfeldt, A. S.; Zhu, K.; Glinkerman, C. M.; Boger, D. L. *Org. Lett.*, **2014,** *16*, 5084.

283 Duerfeldt, A. S.; Boger, D. L. *J. Am. Chem. Soc.*, **2014,** *136*, 2119.

284 Vuluga, D.; Legros, J.; Crousse, B.; Bonnet-Delpon, D. *Green Chem.*, **2009,** *11*, 156.

285 Ohsawa, A.; Kaihoh, T.; Itoh, T.; Okada, M.; Kawabata, C.; Yamaguchi, K.; Igeta, H. *Chem. Pharm. Bull.*, **1988,** *36*, 3838.

286 Salituro, F. G.; Harrison, B. L.; Baron, B. M.; Nyce, P. L.; Stewart, K. T.; McDonald, I. A. *J. Med. Chem.*, **1990,** *33*, 2944.

287 Salituro, F. G.; Harrison, B. L.; Baron, B. M.; Nyce, P. L.; Stewart, K. T.; Kehne, J. H.; White, H. S.; McDonald, I. *J. Med. Chem.*, **1992,** *35*, 1791.

288 Baqi, Y.; Pillaiyar, T.; Abdelrahman, A.; Kaufmann, O.; Alshaibani, S.; Rafehi, M.; Ghasimi, S.; Akkari, R.; Ritter, K.; Simon, K.; Spinrath, A.; Kostenis, E.; Zhao, Q.; Köse, M.; Namasivayam, V.; Müller, C. E. *J. Med. Chem.*, **2018,** *61*, 8136.

289 Nagasaka, T.; Ohki, S. *Chem. Pharm. Bull.*, **1977,** *25*, 3023.

290 Thede, K.; Mengel, A.; Christ, C.; Kuhnke, J.; Johannes, S. A. L.;Buchgraber, P.; Klar, U.; Sack, U.; Kaulfuss, S.; Fernandez-Montalvan, A. E.; Werbeck, N.; Monning, U.; Nowak-Reppel, K.; Wittrock, S.; Mckinney, D.; Serrano-Wu, M. H.; Lemke, C.; Fitzgerald, M.; Nasveschuk, C.; Lazarski, K.; Ferrara, S. J.; Furst, L.; Wei, G.; Mccarren, P. R.; Harvey, R. A. WO2019/096922, A1, **2019**.

291 a) Wang, T.; Ueda, Y.; Zhang, Z.; Yin, Z.; Matiskella, J.; Pearce, B. C.; Yang, Z.; Zheng, M.; Parker, D. D.; Yamanaka, G. A.; Gong, Y.-F.; Ho, H.-T.; Colonno, R. J.; Langley, D. R.; Lin, P.-F.; Meanwell, N. A.; Kadow, J. F. *J. Med. Chem.*, **2018,** *61*, 6308. b) Ueda, Y.; Connolly, T.; Kadow, J.; Meanwel, N.; Wang, T.; Chen, C. P.; Yeung, K.S.; Zhang, Z.; Leahy, D.; Pack, S.; Soundararajan, N.; Sirard, P.; Levesque, K. US20050209246, A1, **2005**.

292 Zhang, Z.; Yang, Z.; Meanwell, N. A.; Kadow, J. F.; Wang, T. *J. Org. Chem.*, **2002,** *67*, 2345.

293 Bartoli, G.; Palmieri, G.; Bosco, M.; Dalpozz, R. *Tetrahedron Lett.*, **1989,** *30*, 2129.

294 Bosco, M.; Dalpozzo, R.; Bartoli, G.; Palmieri, G.; Petrini, M. *J. Chem. Soc., Perkin Trans. 2*, **1991,** 657.

295 Bundy, G. L.; Banitt, L. S.; Dobrowolski, P. J.; Palmer, J. R.; Schwartz, T. M.; Zimmermann, D. C.; Lipton, M. F.; Mauragis, M. A.; Veley, M. F.; Appell, R. B.; Clouse, R. C.; Daugs, E. D. *Org. Process Res. Dev.*, **2001,** *5*, 144.

296 Agami, C.; Couty, F.; Evano, G. *Tetrahedron Lett.*, **1999,** *40*, 3709.

297 Sharma, V.; Kumar, V. *Med. Chem. Res.*, **2014,** *23*, 3593.

298 Singh, G. S.; Mmatli, E. E. *Eur. J. Med. Chem.*, **2011,** *46*, 5237.

299 Martinez, T.; Alahyen, I.; Lemiere, G.; Mouries-Mansuy, V.; Fensterbank, L. *Org. Process Res. Dev.*, **2020,** *24*, 817.

300 Willemsens, B.; Vervest, I.; Ormerod, D.; Aelterman, W.; Fannes, C.; Mertens, N.; Marko, I. E.; Lemaire, S. *Org. Process Res. Dev.*, **2006,** *10*, 1275.

301 Genung, N. E.; Wei, L.; Aspnes, G. E.; *Org. Lett.*, **2014,** *16*, 3114.

302 Rodgers, J. D.; Johnson, B. L.; Wang, H.; Greenberg, R. A.; Erickson-Viitanen, S.; Klabe, R. M.; Cordova, B. C.; Rayner, M. M.; Lam, G. N.; Chang, C.-H. *Bioorg. Med. Chem. Lett.*, **1996,** *6*, 2919.

303 Sun, J.-H.; Teleha, C. A.; Yan, J.-S.; Rodgers, J. D.; Nugiel, D. A. *J. Org. Chem.*, **1997,** *62*, 5627.

304 Jacobson, P.; Huber, L. *Ber.*, **1908,** *41*, 660.

305 R. Huisgen and H. Nakaten, *Ann.*, **1954,** *586*, 84.

306 Huisgen, R.; Bast, K. *Organic Syntheses*; Wiley: New York, **1973,** *Coll. Vol. V*, 650.

307 Le Fouler, V.; Chen, Y.; Gandon, V.; Bizet, V.; Salome, C.; Fessard, T.; Liu, F.; Houk, K. N.; Blanchard, N. *J. Am. Chem. Soc.*, **2019,** *141*, 15901.

308 Le Fouler, V.; Brach, N.; Bizet, V.; Lanz, M.; Gallou, F.; Bailly, C.; Hoehn, P.; Parmentier, M.; Blanchard. N. *Org. Process Res. Dev.*, **2020,** *24*, 776.

309 Follmann, M.; Ackerstaff, J.; Redlich, G.; Wunder, F.; Lang, D.; Kern, A.; Fey, P.; Griebenow, N.; Kroh, W.; Becker-Pelster, E.-M.; Kretschmer, A.; Geiss, V.; Li, V.; Straub, A.; Mittendorf, J.; Jautelat, R.; Schirok, H.; Schlemmer, K.-H.; Lustig, K.; Gerisch, M.; Knorr, A.; Tinel, H.; Mondritzki, T.; Trübel, H.; Sandner, P.; Stasch, J.-P. *J. Med. Chem.*, **2017,** *60*, 5146.

310 Ramharter, J.; Kulhanek, M.; Dettling, M.; Gmaschitz, G.; Karolyi-Oezguer, J.; Weinstabl, H.; Gollner, A. *Org. Process Res. Dev.*, **2022,** *26*, 2526.

311 Kurth, M. J.; Olmstead, M. M.; Haddadin, M. J. *J. Org. Chem.*, **2005,** *70*, 1060.

312 Lv, Z.; Wang, B.; Hu, Z.; Zhou. Y.; Yu, W.; Chang, J. *J. Org. Chem.*, **2016,** *81*, 9924.

313 Rasal, K. B.; Yadav, G. D. *Org. Process Res. Dev.*, **2016,** *20*, 2067.

314 Ma, J.; Han, B.; Song, J.; Hu, J.; Lu, W.; Yang, D.; Zhang, Z.; Jiang, T.; Hou, M. *Green Chem.*, **2013,** *15*, 1485.

315 a) Koepfly, J. B.; Mead, J. F.; Brockman, J. A., Jr. *J. Am. Chem. Soc.* **1947,** *69*, 1837. b) Jiang, S.; Zeng, Q.; Gettayacamin, M.; Tungtaeng, A.; Wannaying, S.; Lim, A.; Hansukjariya, P.; Okunji, C. O.; Zhu, S.; Fang, D. *Antimicrob. Agents Chemother.*, **2005,** *49*, 1169.

316 Punthasee, P.; Vanitcha, A.; Wacharasindhu, S. *Tetrahedron Lett.*, **2010,** *51*, 1713.

317 Chandregowda, V.; Rao, G. V.; Reddy, G. C. *Org. Process Res. Dev.*, **2007,** *11*, 813.

318 Fischer, U.; Mohler, H.; Schneider, F.; Widmer, U.; *Helv. Chim. Acta*, **1990,** *73*, 763.

319 Kawasaki, H.; Abe, H.; Hayakawa, K.; Iida, T.; Kikuchi, S.; Yamaguchi, T.; Nanayama, T.; Kurachi, H.; Tamaru, M.; Hori, Y.; Takahashi, M.; Yoshida, T. WO **2005,** 121142 A1.

320 Zhou, J.; Liu, P.; Lin, Q.; Metcalf, B. W.; Meloni, D.; Pan, Y.; Xia, M.; Li, M.; Yue, T.-Y.; Rodgers, J. D.; Wang, H. WO2010/083382, A2, **2010**.

321 Rodgers, J. D.; Shepard, S.; Li, Y.-L.; Zhou, J.; Liu, P.; Meloni, D.; Xia, M. WO2009/114512, A2, **2009**.

322 Qi, L. X.; Meng, L. B.; Yang, Y. M.; Zhang, M. F.; Ju, L. X. CN109651424, A, **2019**.

323 Bagdanoff, J. T.; Chen, Z.; Acker, M.; Chen, Y.-N.; Chan, H.; Dore, M.; Firestone, B.; Fodor, M.; Fortanet, J.; Hentemann, M.; Kato, M.; Koenig, R.; LaBonte, L. R.; Liu, S.; Mohseni, M.; Ntaganda, R.; Sarver, P.; Smith, T.; Sendzik, M.; Stams, T.; Spence, S.; Towler, C.; Wang, H.; Wang, P.; Williams, S. L.; LaMarche, M. J. *J. Med. Chem.*, **2019,** *62*, 1781.

324 Bagdanoff, J. T.; .; Chen, Z.; Dore, M.; Fortanet, J.; Kato, M.; Lamarche, M. J.; SMITH, T. D.; Williams, S. WO2016/203404, A1, **2016.**

325 Purushottamachar, P.; Murigi, F. N.; Njar, V. C. O. *Org. Process Res. Dev.*, **2016,** *20*, 1654.

326 Njar, V. C. O.; Purushottamachar, P. WO2014/165815, A2, **2014.**

327 Winstein, S.; Adams, R. *J. Am. Chem. Soc.*, **1948,** *70*, 838.

328 Shoppee, C. W.; Summers, G. H. R. *J. Chem. Soc.*, **1952,** 3361.

329 Freiberg, L. A. *J. Org. Chem.*, **1965,** *30*, 2476.

330 Hughes, D. L. *Org. Process Res. Dev.*, **2021,** *25*, 1089.

331 Vitalievich, K. A.; Stepanovna, S. T.; Vadimovich, L. N.; Vladimirovich, D. D.; Makhailovich, K. S.; Viktorovna, S. G.; Anatolievich, S. A.; Valerievna, F. V.; Maximovna, N. V.; Viktorovna, D. M.; Valerievna, E. O.; Vasilievich, S. V. RU 0002532902, November 20, **2014.**

332 Hessler, E. J.; Van Rheenen, V. H. US1980/4216159, A, **1980.**

333 Koch, G.; Jeck, R.; Hartmann, O.; Kusters, E. *Org. Process Res. Dev.*, **2001,** *5*, 211.

334 Liu, J. P.; Mander, L. N.; Willis, A.C. *Tetrahedron*, **1998,** *54*, 11637.

335 Quílez del Moral, J. F.; Martínez, C. R.; Pérez del Pulgar, H.; González, J. E. M.; Fernández, I.; López-Pérez, J. L.; Fernández-Arteaga, A.; Barrero, A. F. *J. Org. Chem.*, **2021,** *86*, 3344.

336 Roy, P.; Maturano, J.; Hasdemir, H.; Lopez, A.; Xu, F.; Hellman, J.; Tajkhorshid, E.; Sarlah, D.; Das, A. *J. Nat. Prod.*, **2024,** 87, 639.

337 Liu, J. S.; Zhu, Y. L.; Yu, C. M.; Zhou, Y. Z.; Han, Y. Y.; Wu, F. W.; Qi, B. F. *Can. J. Chem.*, **1986,** *64*, 837.

338 Morita H.; Kobayashi, J. *J. Org. Chem.*, **2002, 67**, 5378.

339 Morita, H.; Arisaka, M.; Yoshida, N.; Kobayashi, J. *J. Org. Chem.*, **2000,** *65*, 6241.

340 Grierson, D. *Org. React.*, **1990,** *39*, 85.

341 Li, H.; Wang, X.; Wang, J. CN110950859, A, **2020**.

342 Desmaele, D.; Mekouar, K.; and d'Angelo, J. *J. Org. Chem.*, **1997,** *62*, 3890.

5

Mechanism Problems from Reactions Give Unexpected, Undesired, or Side Products

When you first glance at the title of this chapter, a question might immediately come to mind: Why should we focus on unexpected, undesired by-products, side products, or even impurities? After all, in our daily work, most reactions proceed smoothly, yielding the desired or expected products. However, due to the complexity of each specific case, unexpected results, undesired products, and impurities are often encountered. Identifying the structures of these side products is not merely a matter of scientific curiosity; understanding their formation mechanisms is also crucial for addressing the underlying issues. It's important to recognize that the formation of these unexpected products is not arbitrary—it follows logical chemical principles.

From a quality assurance perspective, controlling the percentage of certain impurities below a specified threshold is essential. This is particularly critical in process chemistry and the production of active pharmaceutical ingredients (APIs). By understanding how these side products are formed, we can develop more effective strategies to control or even eliminate them. Additionally, reducing or suppressing the formation of undesired products or side products directly increases the yield of the desired product. Yield, or overall yield, is a key factor in any synthesis or API production process. Higher yields not only mean more product but also simplify the workup and purification processes, reduce waste, and lead to greater financial returns.

This chapter compiles 32 examples that address the mechanisms behind the formation of undesired products, side products, and impurities. Notably, the majority of these examples are drawn from articles published in the journal *Organic Process Research & Development*, highlighting the significance of controlling side products and impurities in the field of process chemistry. This emphasis underscores their critical role in ensuring efficient and high-quality chemical processes.

Overcoming Synthetic Challenges in Medicinal Chemistry: Mechanistic Insights and Solutions, First Edition. Tongshuang Li.

5.1 Formation of Noncyclic Alkane and Derivatives

5.1.1 Alcohol

Problem 122: An Unexpected Impurity from a Grignard Reaction

The addition of Grignard reagents to benzoyl groups can yield crucial structural motifs found in APIs, such as lusutrombopag, [1] aprepitant, [2] and lorlatinib [3]. Advancements in this reaction methodology are expected to significantly enhance the production processes of these drug substances. In a study by Hosoya et al., a Grignard reaction was conducted using methyl 4-formylbenzoate (**1**) and methyl magnesium bromide [4, 5]. The target product (**2**) was obtained in 61% yield, while an unexpected impurity (**3**) was also detected, accounting for 10% of the yield.

Please provide a mechanism for the formation of **3**.

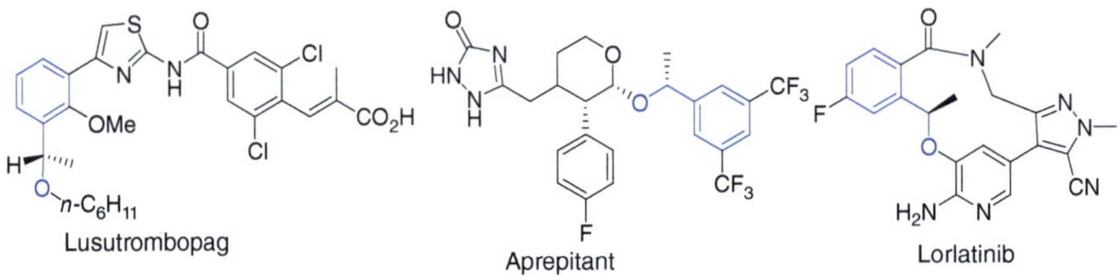

Lusutrombopag Aprepitant Lorlatinib

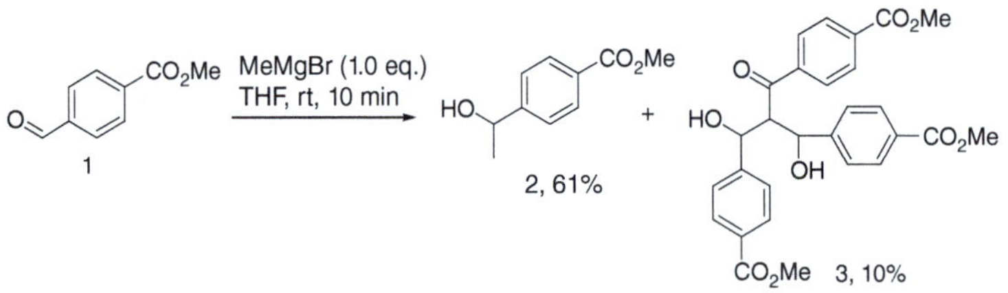

1 → MeMgBr (1.0 eq.), THF, rt, 10 min → 2, 61% + 3, 10%

Experimental Procedure (from *Org. Process Res. Dev.*, **2020**, *24*, 405)

Dimethyl 4,4′-(1,3-dihydroxy-2-(4-(methoxycarbonyl)benzoyl)propane-1,3-diyl) dibenzoate (3). Two solutions were fed by the Vapourtec V-3; solution A was a solution of **1** in 10 volumes of THF (0.559 mmol/mL) and solution B was MeMgBr in THF (1.07 mmol/mL). Flow rate of solution A was set as 1.09 mL/min, and that of solution B was set as 0.57 mL/min so that the equivalent of MeMgBr would be adjusted to 1.00 equiv. Two solutions were poured into a four-neck flask containing THF (10 mL) with a stirring bar at 25 °C for 11 minutes. Input amount of the starting material **1** was defined as 1.10 g (0.559 mmol/mL × 1.09 mL/min × 11 min × 164.16 mg/mmol). The reaction mixture was stirred for 90 minutes and poured into the mixture of ethyl acetate (80 mL) and aqueous solution of citric acid prepared from citric acid monohydrate (2.0 g) and water (38 mL) at 0 °C. The organic phase was separated, and the organic phase was washed with water (20 mL). The organic phase was evaporated, and the residue was purified by flash column chromatography on silica gel (eluent: *n*-hexane/ethyl acetate 75/25 (v/v) to 60/40 (v/v)) to afford **3** (74.2 mg, 95.2 area%, containing ethyl acetate (1 wt%) and *n*-hexane (2 wt%) from ^{1}HNMR). ^{1}HNMR (400 MHz, $CDCl_3$) δ 7.93–7.90 (6H, m), 7.68–7.66 (2H, m), 7.41–7.39 (4H, m), 5.02 (2H, dd, J = 6.3 and 6.3 Hz), 4.08 (1H, t, J = 6.2 Hz), 3.89 (3H, s), 3.87 (6H, s), 3.78 (2H, d, J = 6.5 Hz). ^{13}CNMR

(100 MHz, $CDCl_3$) δ 205.77, 166.76, 166.13, 146.82, 141.63, 134.17, 129.99, 129.95, 129.69, 128.21, 126.01, 74.17, 59.25, 52.27, 52.25. MS (ESI+) m/z 529 $[M + Na]^+$.

Methyl 4-(1-hydroxyethyl)benzoate (2). Input amount of the starting material: 1000.2 mg. Isolated amount of recovery **1** and **2** were 80.2 mg (8%, 92 area%, containing 8 area% of methyl 4-(hydroxymethyl)benzoate **5**) and 1002.0 mg (91%, 99.3 area%), respectively. ^{1}H NMR (400 MHz, $CDCl_3$) δ 7.99–7.96 (2H, m), 7.42–7.40 (2H, m), 4.92 (1H, qd, J = 6.5 and 3.0 Hz), 3.89 (3H, s), 2.48 (1H, d, J = 3.1 Hz), 1.48 (3H, d, J = 6.6 Hz). ^{13}C NMR (100 MHz, $CDCl_3$) δ 167.13, 151.17, 129.90, 129.19, 125.38, 69.97, 52.16, 25.34. MS (ESI+) m/z 163 $[M—OH]^+$.

The Mechanism

During the addition of MeMgBr, the starting material **1** and the magnesium salt of the target product **2** (denoted as **A**) coexist in the reaction mixture. The aldehyde group in **1** coordinates with the magnesium in **A**, facilitating an Oppenauer oxidation process. This results in the oxidation of the magnesium salt of **2** by **1**, yielding the ketone **4** and benzyl alcohol **5**. A similar magnesium salt-promoted tandem nucleophilic addition-Oppenauer oxidation of alcohols by aldehydes has been recently reported [5]. Subsequently, the ketone [4] undergoes two aldol reactions to form the trimerized by-product **3**. Both the ketone **4** and benzyl alcohol **5** were identified by high-performance liquid chromatography (HPLC), and their structures were confirmed by comparison with authentic samples.

*base could be MeMgBr, species A and/or B

Continuous flow synthesis avoided the side product and the yield of **2** was improved to 91% [4].

Problem 123: Formation of a Dimer Impurity in the Development of Delafloxacin

Delafloxacin is a 6-fluoroquinolone antibiotic with excellent antibacterial activity against Gram-positive organisms, including both methicillin-susceptible *Staphylococcus aureus* and methicillin-resistant *S. aureus*. It was approved by the U.S. Food and Drug Administration in June 2017. A key step in the synthesis of delafloxacin is a selective chlorination on the 8-position of the functionalized quinolone **1**. A new impurity up to 0.43 area% by HPLC was detected, and this new impurity is difficult to purge during the final salt formation. The structure of the new impurity was identified as a dimer **4** by comparison with synthetic standard [6].

Please propose a mechanism for the formation of **4**.

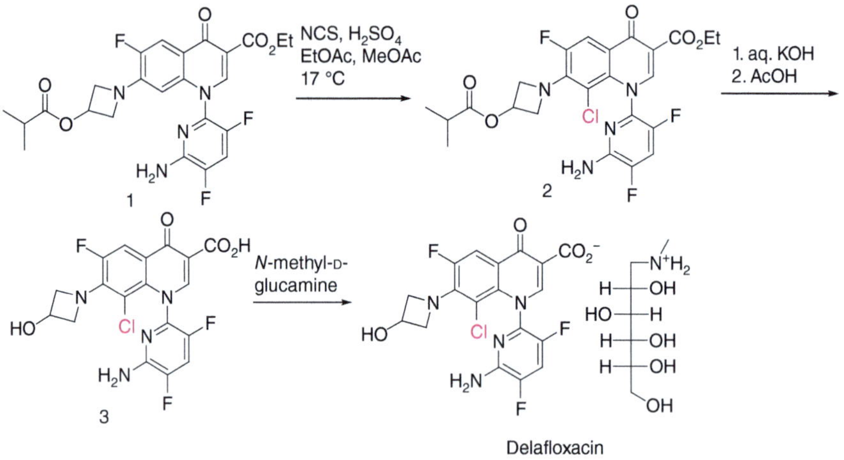

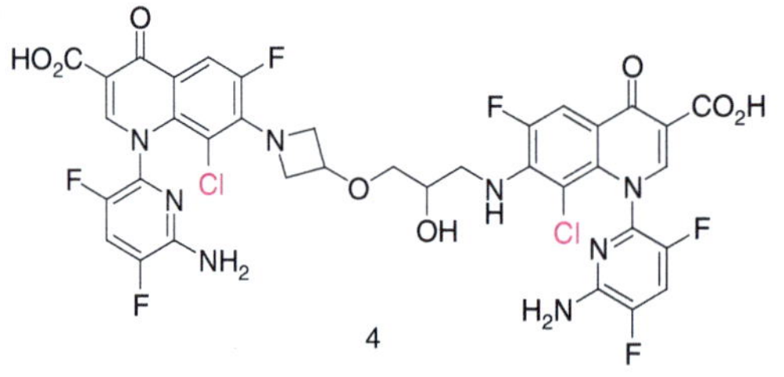

NCS, H_2SO_4
EtOAc, MeOAc
17 °C

1. aq. KOH
2. AcOH

N-methyl-D-glucamine

1 2 3 Delafloxacin 4

Experimental Procedure (from *Org. Process Res. Dev.*, **2009**, *13*, 54)
The dimer **4** was synthesized from **5**, **6**, and **7** in seven steps [6].

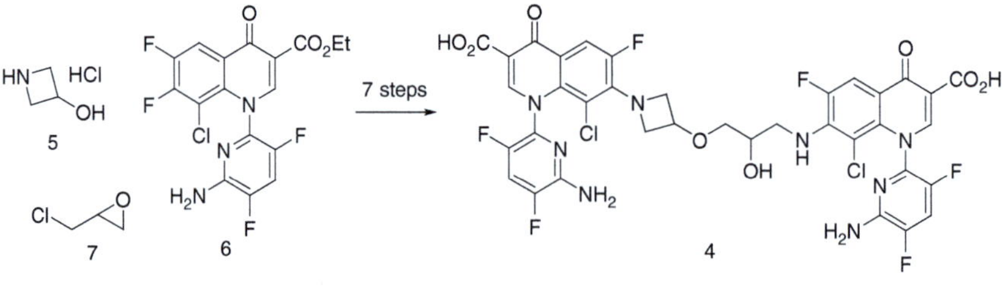

7 steps

5 6 7 4

Compound **4**. Yellow crystals. Mp 198–205 °C dec; ^{1}HNMR (300 MHz, DMSO-d_6): δ 3.28–3.45 (2H, m), 3.45–3.78 (2H, m), 3.79–3.88 (1H, m), 4.16–4.33 (3H, m), 4.61–4.75 (2H, m), 5.25 (1H, br s), 6.23–6.35 (1H, m), 6.76 (4H, s), 7.79 (1H, d, J = 13.7 Hz), 7.90 (1H, d, J = 13.8 Hz), 7.93 (2H, dd, J = 9.7, 2.4 Hz), 8.70 (1H, s), 8.71 (1H, s), 14.59 (2H, br s); ^{13}CNMR (75 MHz, $CDCl_3$): δ 48.1 (d, J_F = 11 Hz), 63.8, 68.4, 69.0 (d, J_F = 5 Hz), 70.6 (d, J_F = 6 Hz), 104.5 (d, J_F = 6 Hz), 105.9 (d, J_F = 7 Hz), 107.8, 108.2, 109.8 (d, J_F = 23 Hz), 110.8 (d, J_F = 23 Hz), 113.4 (d, J_F = 23 Hz), 113.7 (d, J_F = 23 Hz), 115.8 (d, J_F = 8 Hz), 116.6 (d, J_F = 8 Hz), 133.3 (dd, J_F = 14, 3 Hz), 133.5 (dd, J_F = 14, 4 Hz), 134.8, 135.9, 141.0 (d, J_F = 12 Hz), 142.1 (d, J_F = 12 Hz), 142.8 (dd, J_F = 249, 5 Hz), 143.3 (dd, J_F = 249, 5 Hz), 145.1 (dd, J_F = 259, 5 Hz), 145.4 (dd, J_F = 260, 5 Hz), 145.6 (2×, d, J_F = 15 Hz), 149.5 (d, J_F = 248 Hz), 150.1 (2×), 150.2 (d, J_F = 249 Hz), 164.7, 164.8, 175.8 (d, J_F = 3 Hz), 175.9 (d, J_F = 3 Hz); IR: (KBr) 1727, 1622, 1489, 1439 cm^{-1}; ES-HRMS m/z: (M^+ + H) calcd for $C_{36}H_{25}Cl_2F_6N_8O_8$ 881.1071, found 881.1090.

The Mechanism

The formation of impurity **4** may be initiated by an acid-catalyzed activation of the azetidine ring, followed by an isobutyric ester-induced ring opening to generate species **A**. This reactive intermediate **A** could then undergo cleavage by chloride ions (Cl^-) to form intermediate **B**. The chloride ions may be generated from *N*-chlorosuccinimide (NCS) upon quenching the reaction with Na_2SO_3. Subsequently, the chlorine in intermediate **B** could be displaced by the hydroxyl group of the ethyl ester of compound **3**, resulting in the formation of dimer **C**. Finally, hydrolysis of dimer **C** by aqueous potassium hydroxide (KOH) would yield impurity **4**.

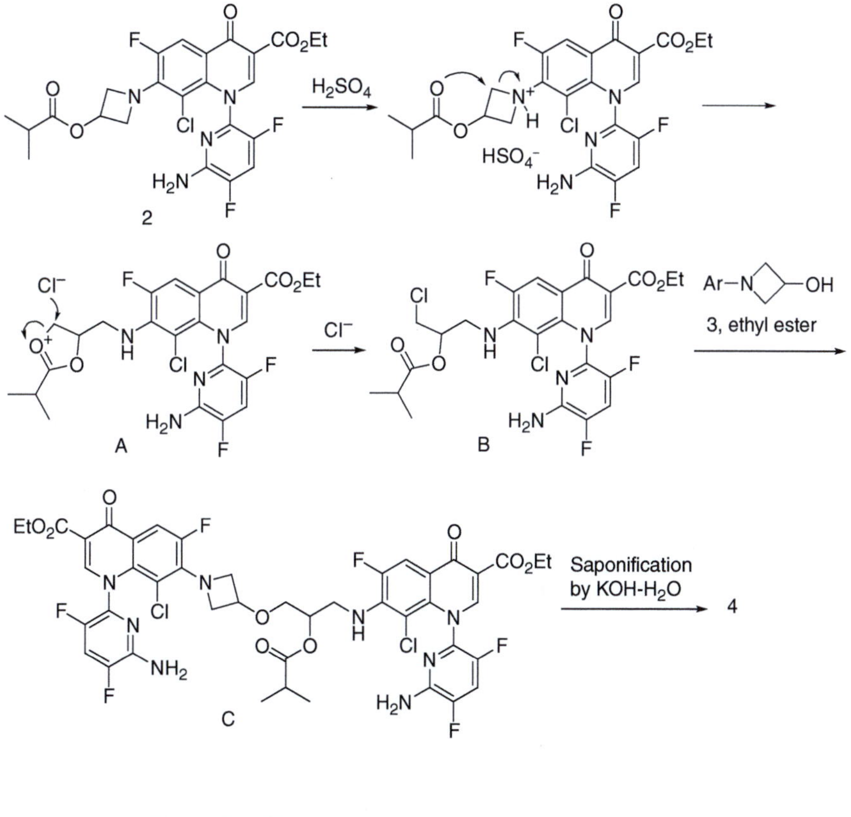

The mechanism for quenching NCS by Na_2SO_3 is as follows. During this workup, Cl^- (NaCl) is produced.

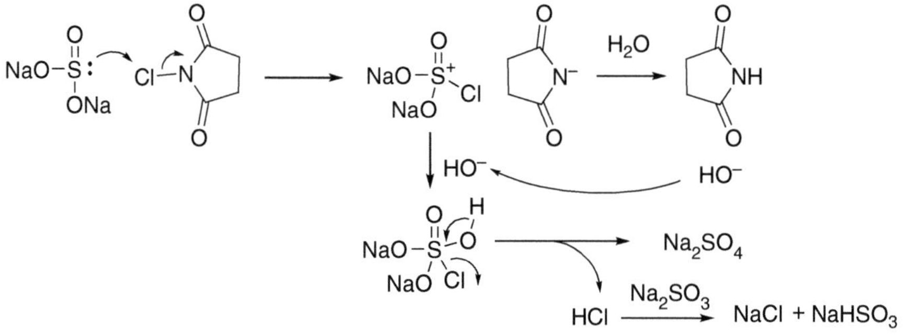

There is another possibility for the formation of impurity **4**. Species **A** could be attacked by ethyl ester of **3** before the Na_2SO_3 quenching, which would also lead to intermediate **C**.

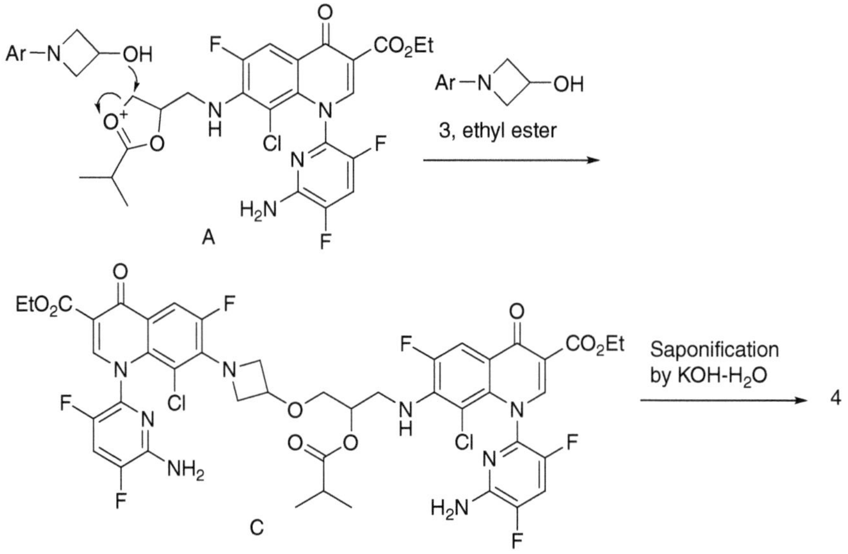

Problem 124: Formation of an Amidino-Diol Side Product

WAY-315193 (**1**) was identified as a progesterone receptor modulator. In a synthesis of **1**, treatment of the diol **2** by *p*-toluenesulfonyl chloride followed by methylamine provided **1** in 34% yield. Meanwhile, a side amidino-diol product **3** was also obtained [7].

Please suggest a mechanism for the formation of **3**.

1. TsCl, Et_3N
Bu_2SnO(cat)
DCM
2. $MeNH_2$, EtOH

2 → 1 (WAY-315193) + 3

Experimental Procedure (from *Org. Process Res. Dev.* **2009**, *13*, 880)

(2*S*,3*S*)-3-((*E*)-7-Fluoro-3,3-dimethyl-2-(methylimino)indolin-1-yl)-3-(3-fluorophenyl)propane-1,2-diol (3). To a solution of diol **2** (11.4 g, 32.9 mmol) in dichloromethane (70 mL), were added Bu_2SnO (0.164 g, 0.66 mmol, 2 mol%) and triethylamine (13.7 mL, 98.7 mmol, 3 equiv). A solution of *p*-toluenesulfonyl chloride (6.27 g, 32.9 mmol) in dichloromethane (30 mL) was added over 15 min at 20 °C. The reaction mixture was stirred at ambient temperature for 3 h, at which point HPLC analysis indicated 80:20 areas ratio of monotosylate : diol. Diisopropylethylamine (5 mL, 28.7 mmol) was added, the reaction mixture was kept at room temperature for 18 h, diluted with acetonitrile (150 mL), and stirred at 40 °C for 7.5 h. Methylamine (41 mL of 8 M EtOH solution, 328 mmol) was added at 10 °C. The reaction mixture was allowed to warm up to room temperature, stirred for 24 h, and then concentrated in vacuum. Methyl tertiary-butyl ether (MTBE) (100 mL) was added, followed by 2 N HCl (100 mL). The phases were separated. The aqueous phase was washed with 50 mL of MTBE. The acidic aqueous phase was cooled to 10 °C. MTBE (100 mL) was added, followed by 5 N NaOH (40 mL). The precipitated solids were filtered to afford 4.11 g of crude product. The solids were slurried in 12 mL of acetonitrile at 81 °C, cooled to room temperature, and filtered to afford 2.89 g of the title product. The remaining biphasic aqueous NaOH/MTBE system was further processed to afford a second crop of product. The MTBE phase was separated, the aqueous phase was extracted with 50 mL of MTBE. The combined MTBE phase was dried over Na_2SO_4 and concentrated to yield 2.81 g of crude product, which was slurried in 5 mL of acetonitrile at 81 °C, cooled to room temperature, and filtered to afford 1.608 g of the title product (38% combined yield). ^{1}H NMR (400 MHz, CD_3CN, 55 °C) δ: 7.30–7.19 (m, 3H), 6.98–6.90 (m, 2H), 6.86–6.76 (m, 2H), 5.67 (br, 1H), 4.40 (br, 1H), 3.58 (dd, $J = 11, 5$ Hz, 1H), 3.46–3.38 (m, 1H), 3.43 (s, 3H), 1.66 (s, 3H), 1.59 (s, 3H). For detailed 2D NMR studies, see Supporting Information. MS (positive ESI, for M + H): m/z 361.

The Mechanism

Initially, the monotosylate **A** is formed. Subsequently, the carbon bearing the TsO group is attacked by the adjacent hydroxyl group, leading to the formation of epoxide **B** (path A). This epoxide is then opened by methylamine to produce the desired product **1**. Alternatively, the carbon with the TsO group in **A** can undergo intramolecular attack by the amide oxygen, generating the cationic intermediate **C** (path B). Methylamine can react with **C** at the imidate carbon, displacing the imidate oxygen, opening the six-membered ring, and yielding the corresponding amidino-diol **3**. Additionally, the reaction of C with water would regenerate the starting amido-diol **2**.

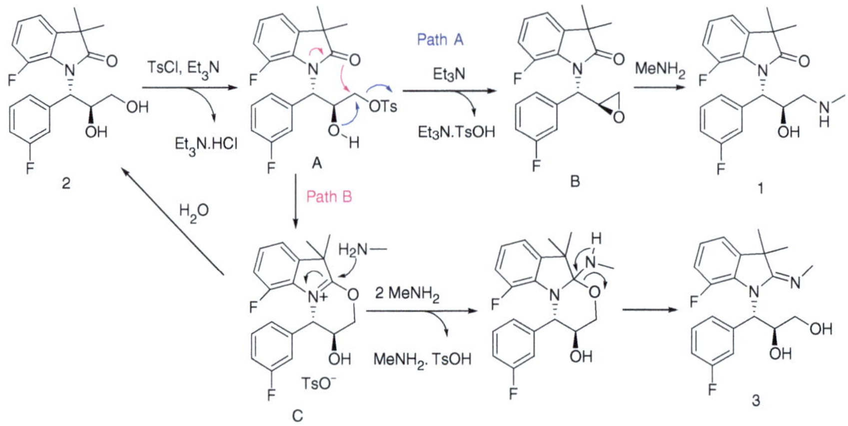

5.1.2 Amine

Problem 125: Formation of Ketoamine During Semmler–Wolff Aromatization
Dasabuvir (ABT-133) is an antiviral medication for the treatment of hepatitis C. Sulfonamide **1** is an intermediate for the preparation of ABT-333. In a process study of the synthesis of **1**, the Semmler–Wolff reaction was employed for the conversion of oxime **2** to naphthylamine **3** [8]. The reaction is shown as below. Except the desired product **3**, a side product **4** was also obtained in around 14% yield.

Please provide a mechanism for the formation of **4**.

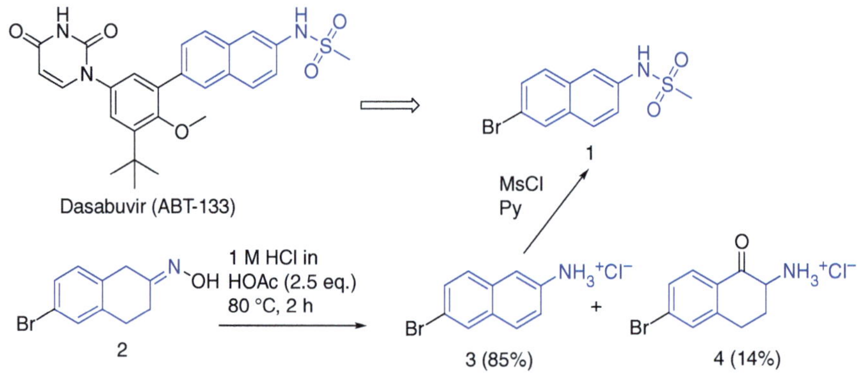

Experimental Procedure (from *Org. Process Res. Dev.*, **2014**, *18*, 1696)
6-Bromo-2-naphthylamine hydrochloride salt (3) and 6-bromo-1-oxo-1,2,3,4-tetrahydronaphthalen-2-aminium chloride (4). A mixture of oxime of 6-bromo-2-tetralone **2** (97.3 kg, 405.5 mol) and 1 M HCl in glacial acetic acid (1122.0 kg, 1030.0 mol) was heated to 75 ± 5 °C. After heating at 75 ± 5 °C for 3.5 h, HPLC analysis indicated that **2** was <0.5 peak area %. The reaction mixture was cooled to 50 °C, filtered, and washed sequentially with 220 kg of acetic acid and MTBE (700 kg).

After being dried under vacuum at 50 °C for 5 h, 79.9 kg of the product was obtained as off-white solids (76%). By-product ketoamine **4** (formed in ~14 mol% in the reaction) partially precipitated overnight from the mother liquors as hydrochloride salt and was filtered and washed with MTBE to provide an analytical sample for characterization. HPLC (method B) retention times: **3** 6.79 min, **4** 2.12 min.

Analytical data for **3**: ^{1}H NMR (400 MHz, DMSO-d_6): δ 8.21 (d, J = 2 Hz, 1H), 7.98 (d, J = 8.8 Hz, 1H), 7.90 (d, J = 8.8 Hz, 1H), 7.85 (d, J = 1.6 Hz, 1H), 7.65 (dd, J = 2, 8.8 Hz, 1H), 7.5 (dd, J = 2, 8.8 Hz, 1H). ^{13}C NMR (100 MHz, DMSO-d_6): δ 132.0, 131.2, 130.9, 129.5, 129.2 (two peaks overlapping), 128.4, 122.0, 119.5, 118.8. Mp 281 °C (decomp.). Anal. Calcd for $C_{10}H_9BrClN$ (HCl salt): C, 46.46; H, 3.51; N 5.42. Found: C, 46.12; H, 3.19; N, 5.39. HRMS calcd for $C_{10}H_8BrN$ 220.9840, obsd 220.9839.

Analytical data for **4**: ^{1}H NMR (400 MHz, DMSO-d_6): δ 8.77 (s, br, 3H), 7.81 (d, J = 8 Hz, 1H), 7.67 (d, J = 4, 1H), 7.60 (dd, J = 4, 12 Hz, 1H), 4.37 (d, J = 12 Hz, 1H), 3.18 (ddd, J = 4, 12, 16 Hz, 1H), 3.04 (ddd, J = 2, 4, 20 Hz, 1H), 2.52 (m, 1H), 2.16 (ddd, J = 4, 8, 26 Hz, 1H). ^{13}C NMR (100 MHz, DMSO-d_6): δ 191.6, 145.8, 131.3, 129.7, 129.0, 128.3, 128.2, 54.4, 27.3, 26.8. Mp 254.4 °C. Anal. Calcd for $C_{10}H_{11.5}BrCl_{1.5}NO$ (1.5 × HCl salt): C, 40.74; H, 3.93; N, 4.75; Cl, 18.04. Found: C, 38.11; H, 3.74; N, 4.31; Cl, 18.91. HRMS calcd for $C_{10}H_{10}BrNO$ 238.9946, obsd 238.9945.

The Mechanism

Under acidic condition, oxime **2** is protonated to give species **2a**, which can proceed via two tautomerization pathways. Following pathway (**a**), tautomer **2b** loses water to give **2c**, which then tautomerizes/aromatizes to provide desired product **3**. Alternatively, through pathway (**b**), **2a** can tautomerize to **2d**, which can then undergo hydration while cleaving the N—O bond to afford **2e**, which undergoes a double tautomerization to generate side product **4**.

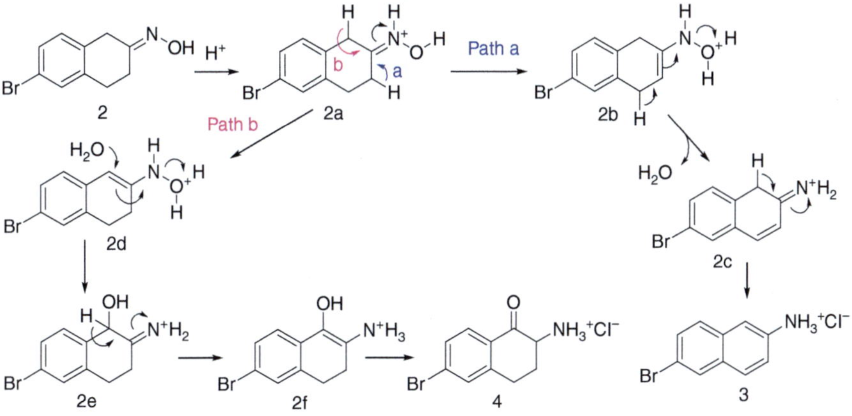

According to this mechanism, water is added in pathway (**b**). Therefore, removal of water from the reaction should have a positive impact on the formation of **3**. Thus, by adding 1.2 eq of acetic anhydride as a scavenger to remove water from the reaction, the yield of **3** was improved to 94% and the yield of **4** was reduced to 2.4% [8].

In another report when oxime **5** was submitted to the Semmler-Wolff aromatization conditions (Ac_2O, AcCl, refluxing), 4-diacetylamino-3-acetoxy-1,2-dihydrophenanthrene **6** was obtained in 27% yield [9].

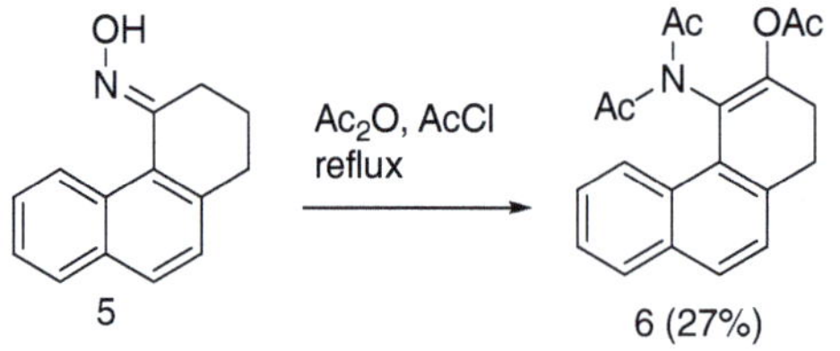

The mechanism for the formation of **6** should be as follows.

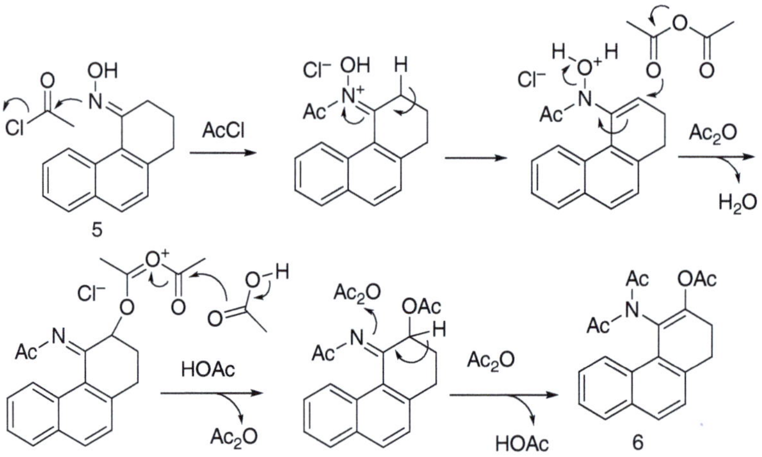

Problem 126: Formation of an Aniline Impurity by Smiles Rearrangement

In the synthesis of a novel renin inhibitor (**5**) for the treatment of hypertension, one synthetic step involved the alkylation of phenol **1** with bromide **2**. In addition to the desired product **3**, an impurity **4** was also identified. The percentage of **4** was increased over the reaction time [10].

Please explain the mechanism for the formation of **4**.

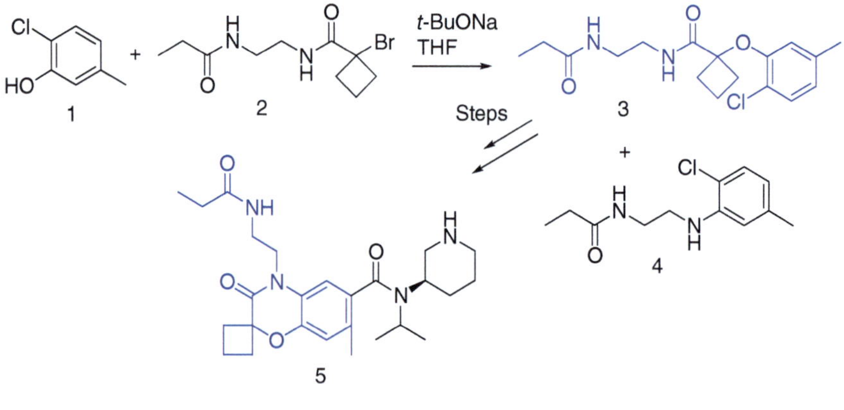

Experimental Procedure (from *Org. Process Res. Dev.*, **2019**, *23*, 499*)

1-(2-Chloro-5-methylphenoxy)-*N*-(2-propionamidoethyl)cyclobutane-1-carboxamide (3). To a solution of *t*BuONa (14.2 kg, 147.3 mol, 1.2 equiv) in toluene (523.0 kg) and 1-BuOH (187.0 kg) was added **1** (17.5 kg, 122.7 mol, 1.0 equiv) with stirring under a nitrogen atmosphere. The mixture was stirred at 20 °C for 1 h. To the mixture was added **2** (34.0 kg, 122.7 mol, 1.0 equiv) over 40 min and stirred at the same temperature. The mixture was heated to 45 °C, and the reaction was monitored by HPLC until completion (2.5 h). After cooling to 25 °C, water (250.0 kg) was added with stirring for 15 min. The separated organic phase was washed with water (84.0 kg) and concentrated to give a suspension (245.0 kg). The suspension was heated to 45 °C and stirred at the same temperature for 30 min and then cooled to 0 °C over 2 h. *n*-Heptane (84.0 kg) was added dropwise to the cooled suspension, and the resulting suspension was stirred at the same temperature for 2.5 h. The precipitate was filtered, washed with a mixture of toluene and *n*-heptane twice, and dried to give **3** (36.4 kg, 107.6 mol, 87.6% yield) as a white solid. ^{1}H NMR (400 MHz, $CDCl_3$) δ 7.22 (d, J = 8.0 Hz, 1H), 6.71 (d, J = 8.0, 1.2 Hz, 1H), 6.59 (t, J = 5.6 Hz, 1H), 6.29 (d, J = 1.2 Hz, 1H), 5.81 (t, J = 5.6 Hz, 1H), 3.47–3.31 (m, 2H), 3. 25–3.19 (m, 2H), 2.73–2.65 (m, 2H), 2.41–2.32 (m, 2H), 2.21 (s, 3H), 2.03 (q, J = 8.0 Hz, 2H), 2.00–1.84 (m, 2H), 1.03 (t, J = 8.0 Hz, 3H); ^{13}C NMR (100 MHz, $CDCl_3$) δ 174.2, 173.8, 150.1, 137.8, 130.2, 123.3, 120.7, 116.5, 81.9, 40.0, 39.1, 31.7 (2C), 29.4, 21.3, 13.7, 9.7; HRMS (ESI) calcd for $C_{17}H_{23}ClN_2O_3$ m/z 339.1470 $[M + H]^+$, found 339.1471.

The formation of the impurity* ***4* *was discussed in the article [10]. However, the isolation and identification of* ***4*** *was not disclosed.*

The Mechanism

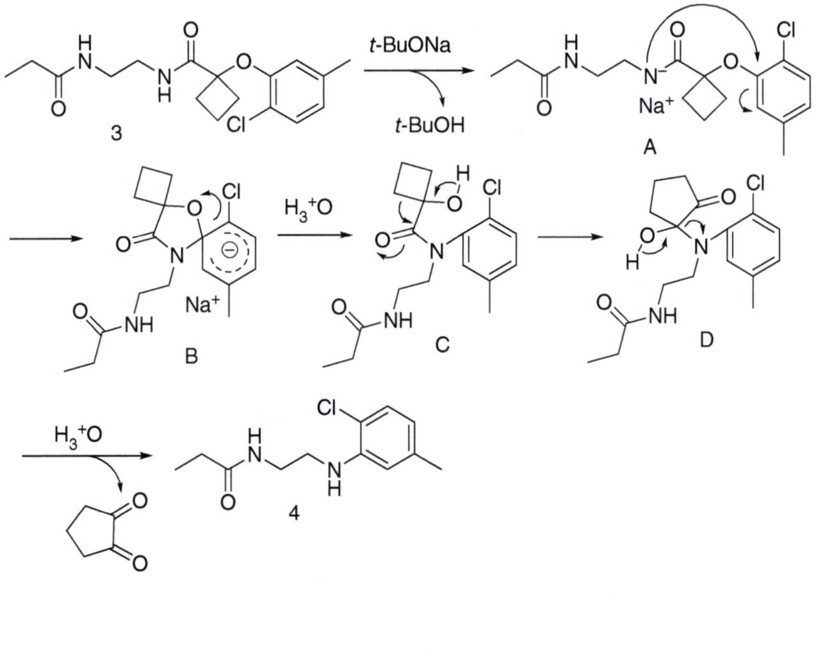

The formation of aniline **4** can be rationalized through a Smiles rearrangement, as outlined below. Upon generation of the desired alkylation product **3**, the amide NH group is deprotonated by the excess base (initially, 1.2 equivalents of *t*-BuONa were added), yielding the sodium salt **A**. The nitrogen anion in **A** then acts as a nucleophile, attacking the phenyl carbon adjacent to the ether linkage, resulting in the formation of the five-membered spiro anion **B**. Under acidic aqueous conditions, cleavage of the C—O bond produces the cyclobutanol amide **C**. This cyclobutanol amide, activated by the acidic environment, undergoes a 1,2-alkyl migration accompanied by ring expansion to form the amino cyclopentanol **D**. Subsequent hydrolysis of **D** under acidic aqueous conditions leads to the formation of the impurity **4**, along with the co-production of cyclopentane-1,2-dione.

5.2 Formation of Cyclic Alkane and Derivatives

5.2.1 Cyclopropane

Problem 127: Formation of a Cyclopropane Impurity

GW641597X (**1**) was under development as a peroxisome proliferator-activated receptor (PPAR)-α agonist for the treatment of dyslipidemia. During the synthesis of compound **1**, a series of steps were employed: deacetylation of acetate **2** using *t*-BuOK, followed by alkylation of phenol **3** with chloromethyl oxadiazole **4**, and finally hydrolysis of the ethyl ester to yield the desired product **2** in 76% yield. However, this process also generated a cyclopropane impurity **5** in approximately 1% yield [11]. To improve the reaction, a milder base, K_2CO_3, was used in ethanol, which resulted in the smooth hydrolysis of **2** and subsequent coupling with alkyl chloride **4** was intrinsically cleaner.

Please suggest a mechanism for the formation of **5**.

K_2CO_3, EtOH, reflux, 2 h

1. **4**, Reflux, 19 h
2. NaOH, 40–45 °C, 2 h

2 3

1 (GW641597X) (71%)

5 (~1%)

Experimental Procedure (from *Org. Process Res. Dev.*, **2020**, *24*, 371)

2-(4-((3-(4-(*tert*-Butyl)phenyl)-1,2,4-oxadiazol-5-yl)-methoxy)-2-methylphenoxy)-2-methylpropanoic Acid (1). Crude (GW641597X) via ethyl 2-(4-((3-(4-(*tert*-butyl)-phenyl)-1,2,4-oxadiazol-5-yl)methoxy)-2-methylphenoxy)-2-methylpropanoic acid. Potassium carbonate (10.9 kg, 79.9 mol) was added to a stirred solution of crude acetate **2** (ca. 25 kg) in absolute ethanol (125.0 kg), followed by an ethanol (10.0 kg) wash, and the suspension was heated to reflux (76 °C) and stirred at reflux for 2 h. A solution of crude alkyl chloride **4** (ca. 20.9 kg) in tetrahydrofuran (THF) (67.2 kg) was then added, with a 16.0 kg ethanol line wash, followed by three "put and take" distillations with ethanol (3 × 53.9 kg) to remove THF. The resulting solution was then stirred at reflux for ca. 19 h. The reaction mixture was cooled to 43 °C, treated with aqueous sodium hydroxide (20.5 kg in 27.0 kg of water), and stirred at 40–45 °C for ca. 2 h. The mixture was quenched with aqueous hydrochloric acid (30.7 kg of concentrated hydrochloric acid in 25.6 kg of water) at 40–45 °C and cooled to 25 °C before being stirred for 2 h at that temperature. Crude GW641597X was collected by centrifugation, washed once with ethanol (54.0 kg), followed by twice with water (2 × 136.4 kg), and dried at 60–65 °C to give the title compound **1** as a cream solid (22.1 kg, 71%).

2-(4-((3-(4-(*tert*-Butyl)phenyl)-1,2,4-oxadiazol-5-yl)-methoxy)-2-methylphenoxy)-2-methylpropanoic Acid (1). Final or API Grade (GW641597X). Crude **1** (GW641597X) (22.1 kg, 52 mol) was diluted with denatured ethanol (122.2 kg), stirred, and heated to reflux (76 °C) until dissolved. The hot solution was filtered into a separate vessel, followed by a 17.5 kg denatured ethanol wash, before being cooled to 20 °C over ca. 2 h. The slurry was stirred at ca. 18 °C for 2 h before being filtered, and the cake was washed with 30.3 kg of denatured ethanol. The product was then dried at 55–60 °C under vacuum and sieved through a 0.024 inch sieve screen to give API grade **1** (GW641597X) as a white solid (18.3 kg, 81.9%). ^{1}HNMR (400 MHz, dimethyl sulfoxide (DMSO)-d_6): δ 12.95 (br s, 1H), 7.95 (d, J = 8.4 Hz, 2H), 7.57 (d, J = 8.4 Hz, 2H), 6.94 (d, J = 3.1 Hz, 1H), 6.82 (dd, J = 8.8, 3.1 Hz, 1H), 6.73 (d, J = 8. Hz, 1H), 5.48 (s, 2H), 2.15 (s, 3H), 1.46 (s, 6H), 1.30 (s, 9H); ^{13}CNMR (101 MHz, DMSO-d_6): δ 175.7, 175.3, 167.6, 154.5, 151.8, 148.4, 130.5, 126.9 (2C), 126.1 (2C), 123.1, 117.9, 117.4, 112.2, 78.7, 61.2, 34.7, 30.8 (3C), 25.0 (2C), 16.6; high-resolution mass spectrometry (HRMS) $[M + H]^+$: calcd for $C_{24}H_{29}N_2O_5$: 425.2076; found: 425.2057.

1,2,3-Tris(3-(4-(*tert*-butyl)phenyl)-1,2,4-oxadiazol-5-yl)-cyclopropane (5). Sodium hydride (0.19 g, 4.8 mmol) was added in one portion to a stirred solution of alkyl chloride **4** (1.0 g, 4.0 mmol) in THF (20 mL) and dimethylformamide (10 mL) at 20 °C under nitrogen for 15.5 h. The reaction was quenched into water (150 mL), and the off-brown precipitate was collected by filtration, washed twice with water (20 mL), and dried. The crude product was slurried in methanol (20 mL) at 50 °C for ca. 2 h and then filtered and washed twice with methanol (5 mL) to yield the title compound **5** as a cream solid (0.58 g, 68%). ^{1}H NMR (400 MHz, $CDCl_3$): δ 8.02 (d, J = 8.6 Hz, 2H), 7.90 (d, J = 8.5 Hz, 4H), 7.54 (d, J = 8.6 Hz, 2H), 7.44 (d, J = 8.5 Hz, 4H), 4.16 (t, J = 5.9 Hz, 1H), 3.75 (d, J = 5.9 Hz, 2H), 1.38 (s, 9H), 1.34 (s, 18H); HRMS $[M + H]^+$: calcd for $C_{39}H_{43}N_6O_3$: 643.3396; found: 643.3422.

The Mechanism

The formation of **5** should be a result of trimerization of the chloromethyl oxadiazole **4** under the strong base conditions. In the presence of *t*-BuOK, the α-proton in **4** could be trapped to form anion **4a**. Then it would attack another chloromethyl group followed by elimination of HCl to generate the alkene dimer **4b**. Anion **4a** attacks the alkene and cyclizes to give **5**. Indeed, treatment of **4** with NaH provided **5** in 68% yield [11]. Similar cyclopropanation was reported previously in the literature [12, 13, 14].

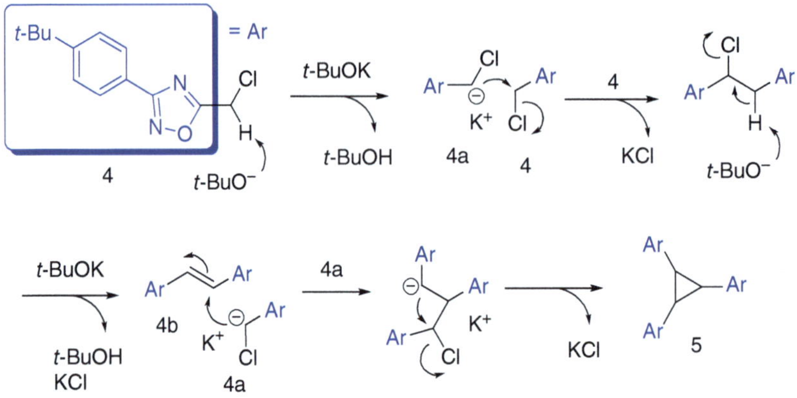

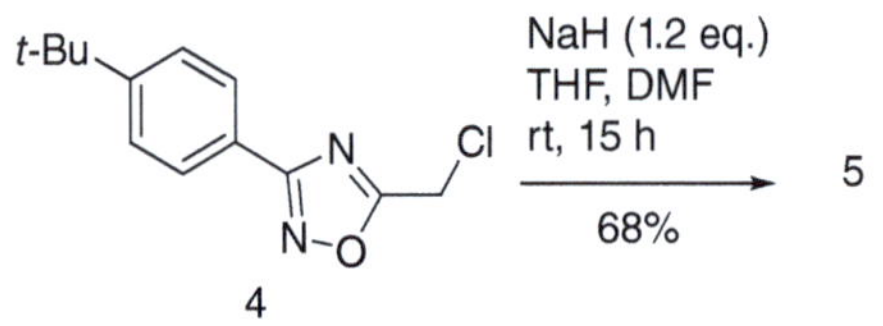

5.3 Formation of Aromatic Compounds

5.3.1 Benzene

Problem 128: Formation of an Undesired Substituted Benzene Side Product
Dolutegravir sodium (**1**), an inhibitor of HIV integrase, is a drug for the treatment of HIV/AIDS infection. In a process study, the diethyl 3-(benzyloxy)-4-oxo-4*H*-pyran-2,5-dicarboxylate (**4**) was designed as a key intermediate. The initial synthesis of **4**, as illustrated below, gave the desired product **4** in low yield (35%) and a side product **5** in considerable amount (8%) [15].

Please suggest a mechanism for the formation of **5** and propose a solution for solving the selectivity and low yield problems.

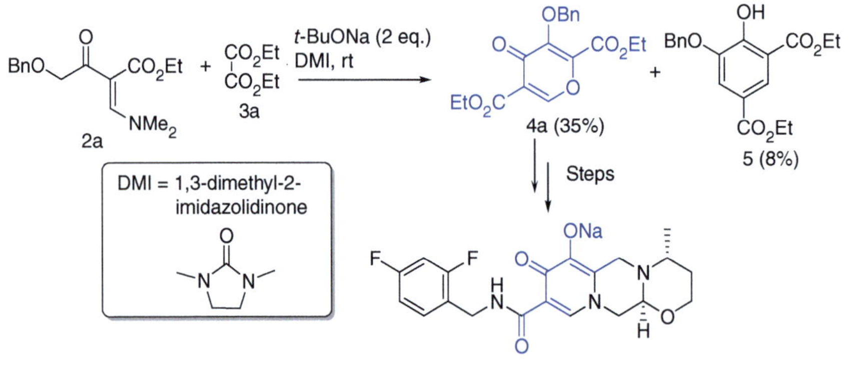

DMI = 1,3-dimethyl-2-imidazolidinone

Dolutegravir sodium (1)

Experimental Procedure (from *Org. Process Res. Dev.* **2019**, *23*, 565*)

2b + CO_2Me–CO_2Me (3b) → MeONa (1.5 eq.), DMI, rt, 4 h, 65% → 4c

Dimethyl 3-(benzyloxy)-4-oxo-4H-pyran-2,5-dicarboxylate (4c). To 1,3-dimethyl-2-imidazolidinone (1000 mL) was added 28% methanolic solution of sodium methoxide (438 mL, 2.16 mol, 1.4 eq) and cooled to the ice bath temperature. To the solution was added dimethyl oxalate (**3b**, 509 g, 4.31 mol, 2.8 eq) portionwise and stirred at room temperature for 30 min. To this was added crude oil of **2b** (430.7 g, 1.55 mmol, 1.0 eq) and stirred at room temperature for 2.5 h. Methanolic solution of sodium methoxide (29 mL, 0.14 mol) was added and stirred for an additional 1.5 h. The reaction mixture was added to the mixture of ethyl acetate (1600 mL), 2 M aqueous hydrochloric acid (2000 mL), and cold water (798 g) and extracted. The organic layer was washed with 5% aqueous sodium hydrogen carbonate (2000 mL) and 2% aqueous sodium chloride (2000 mL). The aqueous layer was extracted with ethyl acetate (2000 mL), and the combined organic layer was dried over magnesium sulfate and concentrated. To the resultant brown oil was added diisopropyl ether (600 mL) and *n*-hexane (400 mL) with stirring, and the precipitated solid was collected by filtration, washed with diisopropyl ether (200 mL × 3), and dried to give **4c** (314.82 g) as a pale yellow solid. Additional **4c** (7.55 g) was obtained from the concentrated mother liquor by column chromatography (silica gel, chloroform/ethyl acetate = 99:1), yield 65%. ^{1}H NMR (400 MHz, $CDCl_3$) δ 8.30 (s, 1H), 7.49–7.45 (m, 2H), 7.40–7.30 (m, 3H), 5.34 (s, 2H), 3.93 (s, 3H), 3.89 (s, 3H). ^{13}C NMR (100 MHz, $CDCl_3$) δ 172.4, 162.9, 160.2, 159.7, 150.2, 145.2, 135.9, 129.0, 128.6, 128.5, 120.6, 75.1, 53.2, 52.7.

The mechanism for the formation of the side product* **5 *was discussed in the article* [15]. *However, the isolation and identification of* **5** *was not disclosed.*

The Mechanism

Under basic (*t*-BuONa) condition, the C-4 in starting material **2a** is deprotonated and an anion **A** is generated. The diethyl oxalate **3a** is first attacked by **A**, followed by cyclization to provide the desired product **4a**. Meanwhile, the anion **A** could be dimerized by replacing the leaving group NMe_2 in **2a** followed by cyclization and elimination of Me_2NH to yield the side product **5**.

Under basic conditions (*t*-BuONa), the C-4 position in the starting material **2a** undergoes deprotonation, generating an anion intermediate **A**. This anion **A** then nucleophilically attacks diethyl oxalate **3a**, leading to cyclization and formation of the desired product **4a**. Concurrently, anion **A** can also undergo dimerization by displacing the leaving group NMe_2 in **2a**, followed by cyclization and elimination of Me_2NH, resulting in the formation of the side product **5**.

Based on this mechanism, the dimerization of **2a** and the following cyclization are just catalyzed by *t*-BuONa. One equivalent of *t*-BuONa is consumed only at the final elimination stage. To suppress the side reaction, increasing both the amount and the reactivity of the oxalate may be beneficial. For example, dimethyl oxalate **3b** is more active than diethyl oxalate **3a**. The order of reagent addition is also crucial, such as adding **2b** to a pre-mixed solution of **3b** and base. To exclude the possibility of transesterification, MeONa could be a substituent of *t*-BuONa and methyl ester **2b** is a replacement of ethyl ester **2a**. With those considerations in mind, the Japanese chemists performed the following reaction and the desired product **4c** was obtained in high yield [15].

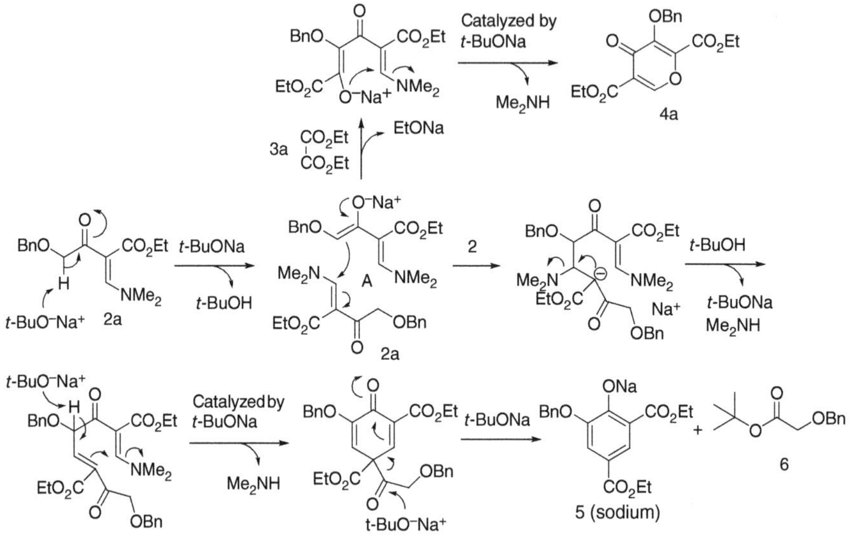

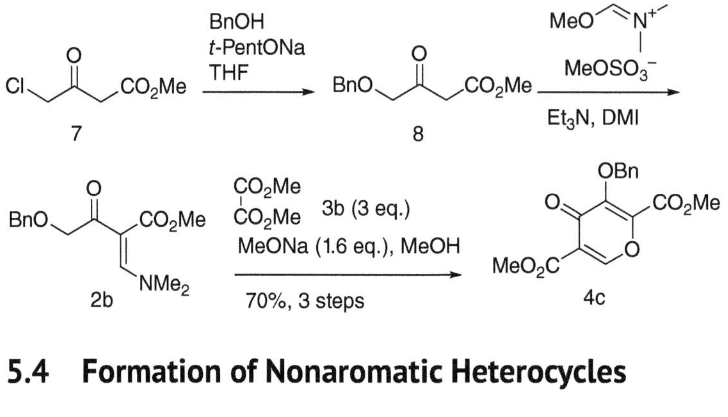

5.4 Formation of Nonaromatic Heterocycles

5.4.1 Aziridine

Problem 129: Formation of an Aziridine-a Neighboring Group Participated Cyclization

Diamine **2** was developed as a crucial intermediate for the synthesis of the direct oral anticoagulant (DOAC) edoxaban (**1**). To achieve this synthesis, researchers at Daiichi Sankyo chose to utilize compound **3** as a starting material. This compound features an **N**-protecting group linked to a tether with a terminal nitrogen, which acts as a nucleophile. The nucleophilic nitrogen was intended to displace an adjacent mesylate group through an intramolecular cyclization. Subsequent removal of the tether was expected to yield the desired cis-diamine. However, when the reaction was conducted in acetonitrile at 80 °C in the presence of trimethylamine, the anticipated product **4** was not obtained. Instead, compounds **5** and **6** were identified as the primary products [16].

Due to the inherent instability of the strained ring system, aziridine **5** could not be isolated in pure form. Its structure was tentatively characterized based on liquid chromatography mass spectrometry (LCMS) and NMR data obtained from the crude reaction mixture. As the reaction progressed, the formation of compound **6** increased, suggesting its role as a major product under the given conditions.

2 → Edoxaban (1)

3 → 4, 5, 6 (Et_3N, CH_3CN, 80 °C)

	4	5	6
Reaction time 1 h	Not detected	78%	9%
Reaction time 8 h	Not detected	Not detected	71%

When an ethyl ester **7** was employed as starting material under the cyclization conditions, the corresponding aziridine **8** was also formed. However, **8** remained unchanged and cyclic sulfamides **9** and **10** were not observed. In addition, **8** was stable enough to be isolated by silica gel column chromatography.

7 → (Et_3N, CH_3CN, 80 °C, 4 h) 8 (94%), 9 (not detected), 10 (not detected)

The goal of this research was to develop a process to produce isolable intermediates, such as mono-Boc-protected diamine **2**. Based on the findings of the Cbz-protected analog **3**, switch the Cbz to Boc-protected substrate **11**, the key reaction proceeded smoothly to produce **12** in 86% yield with excellent regio- and stereoselectivity. Further desulfonylation of **12** to **2** was achieved by employing pyridine-H_2O. After optimization, it was shown that the four-step cyclization sequence and desulfonylation reactions (**11** to **2**) could be conducted in one pot and **2** was isolated in 85% yield as the oxalate salt with >99% purity [16].

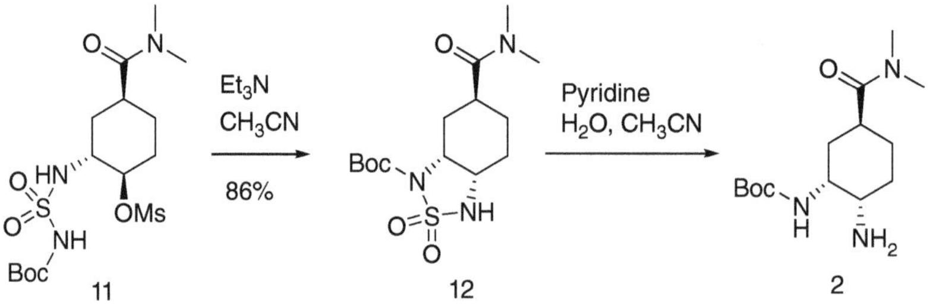

Et_3N, CH_3CN, 86% (**11** → **12**); Pyridine, H_2O, CH_3CN (**12** → **2**)

Please provide mechanism for the transformation from **3** to **5** and **6** (or from **11** to **12**).

Experimental Procedure (from *Org. Process Res. Dev.*, **2019**, *23*, 524*)
***tert*-Butyl{(1*R*,2*S*,5*S*)-2-amino-5-[(dimethylamino)-carbonyl]cyclohexyl} carbamate Oxalate (2) from 11.** Mesylate **11** (1430 kg, 3.224 kmol) was charged to a reactor, followed by the addition of acetonitrile (6435 L) and triethylamine (503 L, 3.579 kmol). The mixture was stirred at 70 °C for 2 h. After the reaction completion, water (358 L) and pyridine (1287 L, 15.98 kmol) were added, and the mixture was stirred at 75 °C for 5 h. NaCl aqueous solution (20%, 1430 L), water (200 L), toluene (14 300 L), and NaOH aqueous solution (25%, 2145 L) were added, and the mixture was stirred at 50 °C; then the layers were separated. The organic layer was washed with NaCl aqueous solution (20%, 1430 L) and NaOH aqueous solution (25%, 286 L). The organic layer was concentrated under reduced pressure to a volume of 4290 L, and then acetonitrile (8580 L) was added and filtered to give the solution of **2**. Oxalic acid (406 kg, 4.510 kmol), water (601 L), and acetonitrile (10 100 L) were charged to another reactor, followed by the addition of **2** solution at 35 °C for 2 h. The mixture was stirred at 50 °C for 2 h, then cooled to 25 °C, and filtrated. The cake was washed with an aqueous acetonitrile solution (93%, 5434 L) to give the crude crystalline **2** as a hydrate. Then the crude **2** was charged into acetonitrile (10 010 L), and the mixture was concentrated under reduced pressure to remove most of the water. The slurry was filtrated at 30 °C and then dried under full vacuum at 60 °C to give **2** (1010.8 kg, 2.692 kmol, 83.5% yield) as a white crystalline solid.

Compound **2**, ^{1}HNMR ($CDCl_3$) 4.10 (br, 1H), 3.32 (d, J = 12.2 Hz, 1H), 2.96 (m, 3H), 2.80 (t, J = 12.4 Hz, 1H), 2.77 (s, 3H), 1.72–1.83 (m, 3H), 1.63 (t, J = 22.7 Hz, 1H), 1.37–1.49 (m, 2H), 1.30 (s, 9H).

NMR spectra of aziridine **8**: ^{1}HNMR ($CDCl_3$) 7.35–7.39 (m, 5H), 5.22 (s, 2H), 4.16 (q, J = 7.0 Hz, 2H), 3.21–3.23 (m, 1H), 3.10–3.13 (m, 1H), 2.41–2.45 (m, 1H), 1.91–1.98 (m, 2H), 1.80–1.86 (m, 1H), 1.72–1.75 (m, 1H), 1.34–1.45 (m, 1H), 1.26 (t, J = 7.0 Hz, 3H).

The mechanism for the formation of **5 and **6** was discussed in the article [16]. It is stated that the structure of **6** was assigned by $^1H-^1H$ COSY spectra. However, the isolation and identification of **6** was not disclosed.*

The Mechanism

Deprotonation of the sulfonamide from the cyclohexane side generates anion **A**. Then cyclization occurs to form aziridine **5**. The aziridine ring is opened by the pendant dimethylamide group to afford **B**. After proton transfer, the anion **C** recyclizes to form **6**.

Under the cyclization conditions, the corresponding aziridine **8** was produced from ethyl ester **7**. The corresponding cyclic sulfamides **9** and **10** were not observed. The result supports the participation of the dimethylamide group in the formation of **6**.

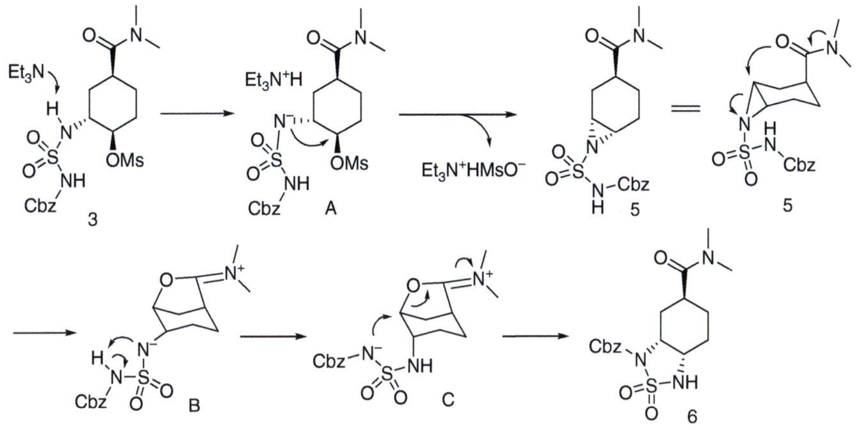

5.4.2 Pyrrolidine

Problem 130: Formation of Ring-contracted Side Product During Fluorination

Compound **1** was identified as a highly potent and selective inhibitor of estrogen-related receptor 1 (ERR1) for the treatment of hyperglycemia in patients with type 2 diabetes mellitus. In a process synthesis of **1**, fluorination of intermediate **2** with Deoxo-Fluor provided the desired product **3** and a ring-contracted side product **4** [17]. Reducing the amount of Deoxo-Fluor from 1.35 equiv to 1.1 equiv, led to a ~4.3/1 **3**/**4** mixture with <1% of starting **2**. Triturating the crude **3**/**4** in EtOAc provided >98% purity **3** with an isolated yield 55%. The structure of the ring-contracted side product **4** was elucidated by HNMR and confirmed by X-ray diffraction studies.

Please suggest a mechanism for the formation of **4**.

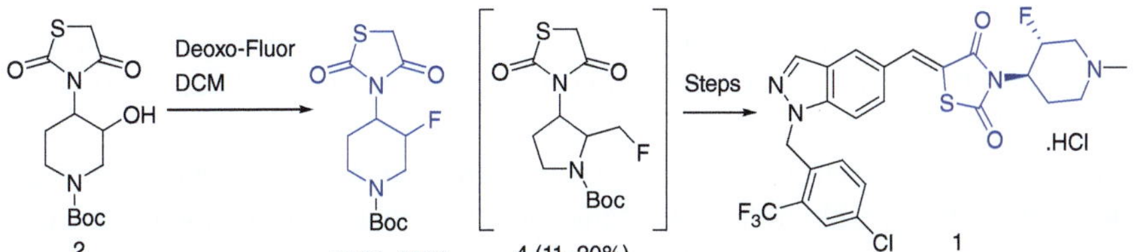

Experimental Procedure (from *Org. Process Res. Dev.* **2014**, *18*, 321*)

***tert*-Butyl 4-(2,4-dioxothiazolidin-3-yl)-3-fluoropiperidine-1-carboxylate (3).** To the cold solution of compound **2** (700.0 g, 2.18 mol) in DCM (11.0 L) at 0 °C, was added dropwise a solution of bis(2-methoxyethylaminosulfur trifluoride) (426.0 mL, 2.31 mol) in DCM (1.0 L) over 1 h. The resulting mixture was stirred at 0 °C for 2 h and then stirred at 20 °C for 20 h. The progress of the reaction was monitored by HPLC and LCMS. The reaction was cooled to 0 °C with fast agitation, and a solution of 2 N NaOH (2.0 L) was carefully added. (Caution: Addition of the first 500 mL of 2 N NaOH solution is very exothermic; the internal temperature was maintained between 0 and 10 °C by adjusting the addition rate.) The resulting suspension was allowed to stir for 1 h at 0 to 10 °C. After phase separation, the organic phase was washed with saturated $NaHCO_3$ (2.0 L) and D.I. H_2O (2.0 L) and then was dried over Na_2SO_4. The solvent was concentrated to dryness under house vacuum to give the crude **3** as an oily material. This crude **3** was triturated with EtOAc (1.0 L) at 25 °C for 30 min, the solid was collected by filtration, and the filter cake was washed with heptanes (300 mL × 3) and dried in a vacuum oven at 50 °C for 20 h to afford 382.0 g (55% isolated yield; 94.1% of **3**, 3.9% of **4**, 1.9% of **2**; HPLC area%) of compound **3** as a slightly yellow solid. The filtrate was concentrated to give 116.0 g of an oily, yellowish solid, which was recrystallized in EtOAc (116 mL) to afford an additional 81.5 g (12% isolated yield; 98.0% of **3**, 2.0% of **4**; HPLC area%) of compound **3** as a yellowish solid. Mp = 151–152 °C. ^{1}HNMR (300 MHz, $CDCl_3$) δ 1.47 (s, 9 H, 3 CH_3), 1.75–1.65 (m, 1 H), 2.38 (ddd, J = 4.40, 4.65, 12.71 Hz, 1 H), 2.62–2.83 (m, 2H), 3.97 (s, 2H), 4.08–4.26 (m, 1H), 4.28–4.38(m, 1H), 4.41–4.66 (m, 1H), 5.15 (ddd, J = 5.6, 10.27, 50.13 Hz, 1H). ^{13}CNMR (400 MHz, $CDCl_3$) δ 171.65, 171.36, 154.10, 84.00, 80.70, 57.07, 47.03, 42.57, 33.27, 28.37 (3 C), 26.50. LC/MS m/z 319.4 ($(MH)^+$ not detected), 341.1 $(MNa)^+$, and 219.1 $[M—Boc]^+$. Calcd for $C_{13}H_{19}FN_2O_4S$ (MW = 318.36): C, 49.09; H, 6.02; N, 8.80; F, 5.97; S, 10.07. Found: C, 49.15; H, 6.33; N, 8.78; F, 5.93; S, 10.37.

The mechanism for the formation of* ***4* *was discussed in the article* [17]. *It is stated that the structure of* ***4*** *was confirmed by* ^{1}*HNMR and by X-ray diffraction. However, the isolation and identification of* ***4*** *was not disclosed.*

The Mechanism

Direct reaction of the hydroxyl group in compound **2** with Deoxo-Fluor generates intermediate **A**. In **pathway a**, fluoride (F^-) then attacks the activated C-3 leading to the desired product **3**. Alternatively, the nearby oxygen of the thiazolidine ring can participate in an intramolecular attack, forming cyclic intermediate **B**. Subsequent attack by F^- at C-3 in intermediate **B** also yields product **3**. However, the Boc-protected amine on the piperidine ring can further react by attacking C-3 to form aziridium intermediate **C**. Finally, F^- attacks C-2 of the aziridine ring, resulting in the ring-contracted product **4**.

The formation of aziridium intermediate **C** does not necessarily proceed via intermediate **B**. It may also be generated directly from intermediate **A** through *N*-Boc attacking the sulfonate-bearing carbon.

Ring contraction caused by neighboring group participation was also reported during a diethylaminosulfur trifluoride (DAST)-mediated fluorination in the following case [18]. When compound **5** was treated with DAST, a mixture of **6** and **7** was obtained.

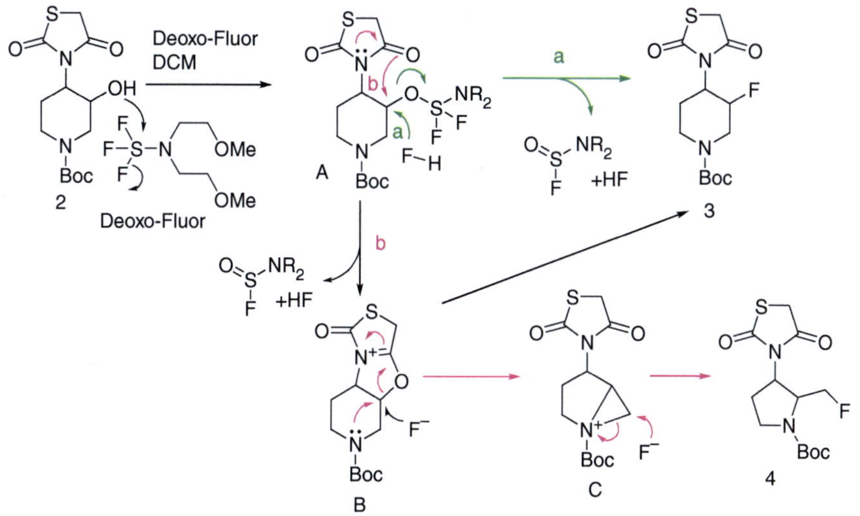

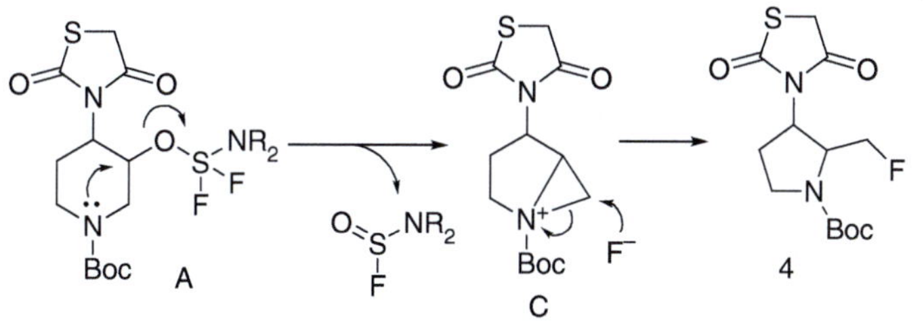

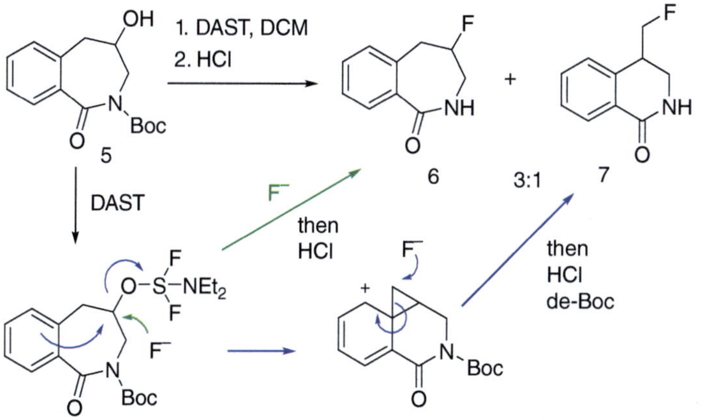

It can also be predicted that the following side product **8** is almost impossible to produce since the nitrogen is not a nucleophile due to the conjugation to an extra carbonyl.

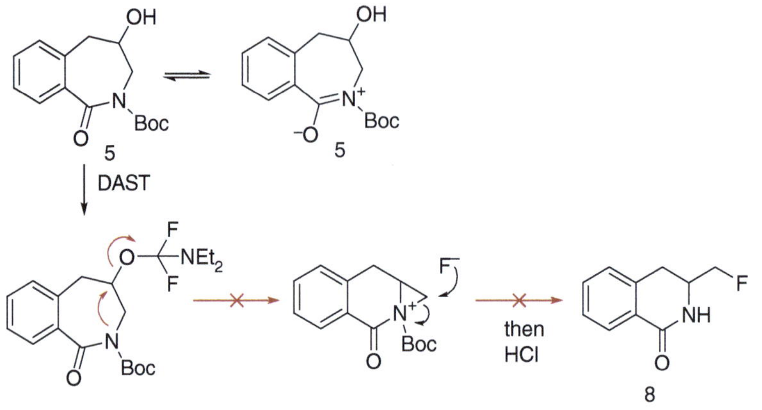

OH N Boc O 5 ⇌ OH N+ Boc -O 5

DAST

F O NEt$_2$ F N Boc O → (×) N+ Boc O F- → (×) then HCl F NH O 8

5.4.3 Piperidine

Problem 131: An Unusual DAST-Mediated Rearrangement

During attempts to synthesize 3-(3-fluoropiperidin-4-yl)-2-phenyl-1H-indole (**3b**) as part of a medicinal chemistry program, [19] chemists from Merck discovered an unusual DAST-mediated rearrangement [20]. Treatment of compound **1** with DAST at low temperature did not yield the desired Cbz-protected product **3a**; instead, while the rearranged product **2a** was afforded in 99% yield. After removal of the Cbz group by transfer hydrogenolysis, the 4-fluoro-3-indolyl isomer **2b** was obtained in excellent yield. To support the structural assignment of **2b**, a genuine sample of the 3-fluoro-4-indolyl substituted piperidine **3b** was needed for comparison. Reaction of 2-phenyl-1H-indole **4** with 3-fluoro-4-oxo-piperidine-1-carboxylic acid *tert*-butyl ester **5** furnished tetrahydropyridine **6** as a single isomer. Ionic reduction with trifluoroacetic acid/triethylsilane gave the trans-fluoropiperidine **3b**. The structure of **2b** and **3b** were determined by double quantum filtered correlation spectroscopy experiments.

Please provide a mechanism for the transformation from **1** to **2a**.

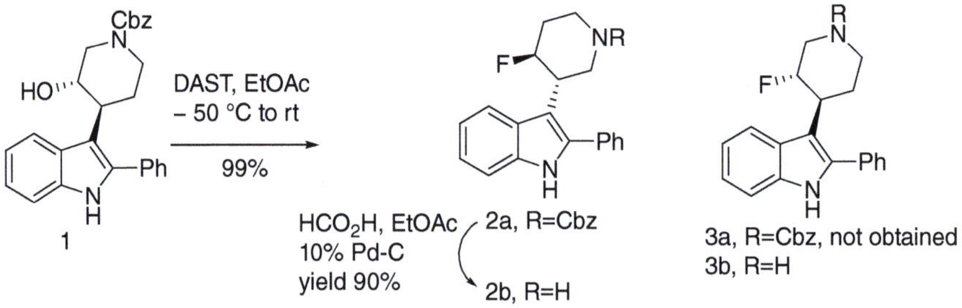

Cbz N HO Ph N H 1 — DAST, EtOAc - 50 °C to rt, 99% → F NR Ph N H; HCO$_2$H, EtOAc 10% Pd-C yield 90%; 2a, R=Cbz → 2b, R=H

R N F Ph N H 3a, R=Cbz, not obtained 3b, R=H

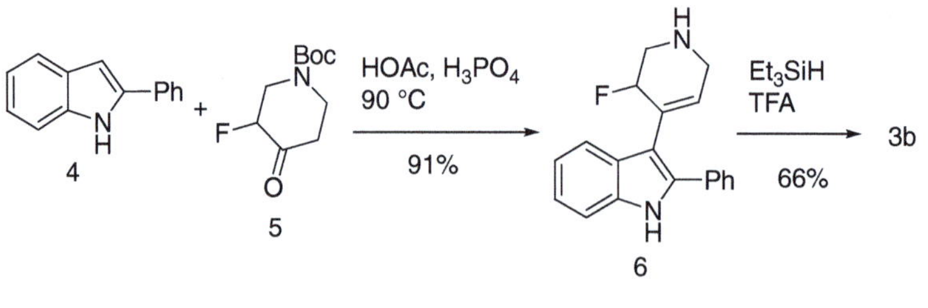

Ph N H 4 + Boc N F O 5 — HOAc, H$_3$PO$_4$ 90 °C, 91% → H N F Ph N H 6 — Et$_3$SiH TFA, 66% → 3b

Experimental Procedure (from *J. Org. Chem.*, **2000**, *65*, 4984)

(3RS,4RS)-4-Fluoro-3-(2-phenyl-1H-indol-3-yl)-piperidine-1-carboxylic Acid Benzyl Ester (2a). A cooled (−50 °C) solution of racemic alcohol **1** (1.0 g, 2.3 mmol) in anhydrous ethyl acetate (20 mL) was treated with DAST (350 μL, 2.6 mmol). Stirring at −50 °C was continued for 1 h before the solution was allowed to warm to ambient temperature over 3 h. The reaction mixture was then poured into saturated aqueous sodium hydrogencarbonate (100 mL), and the product was extracted into ethyl acetate (100 mL). The organic phase was washed with water (100 mL) and brine (100 mL) and then dried over anhydrous sodium sulfate. This solution was filtered, treated with decolorizing charcoal (0.5 g), and filtered again. The resulting pale yellow filtrate was then treated with Raney nickel (1 mL of a 50% slurry in water). After standing at ambient temperature for 2 h, the solution was filtered and evaporated to dryness to afford the title compound (995 mg, 99%) as a white solid: mp 199–201 °C (EtOH); ^{1}H NMR (360 MHz, $CDCl_3$) δ 1.70–1.85 (m, 1H), 2.20–2.30 (m, 1H), 2.90–3.10 (m, 1H), 3.20–3.30 (m, 1H), 3.40–3.50 (m, 1H), 4.20–4.50 (m, 2H), 5.10 (s, 2H), 5.24 (dtd, J = 54, 11 and 5 Hz, 1H), 7.12 (t, J = 7 Hz, 1H), 7.18 (t, J = 7 Hz, 1H), 7.20–7.43 (m, 9H), 7.52–7.56 (m, 2H), 7.67 (d, J = 8 Hz, 1H), 8.15 (s, 1H); MS (ES$^+$) m/z 429 $(M + H)^+$. Anal. Calcd for $C_{27}H_{25}FN_2O_2$: C, 75.68; H, 5.88; N, 6.54. Found: C, 75.45; H, 5.80; N, 6.34.

(3RS,4RS)-3-(4-Fluoro-piperidin-3-yl)-2-phenyl-1H-indole (2b). A solution of racemic fluoropiperidine **2a** (975 mg, 2.3 mmol) in ethyl acetate (50 mL) was treated with 99% formic acid (5 mL) and then with 10% palladium on activated carbon (120 mg), and the resulting suspension was stirred at ambient temperature for 4 h. The reaction mixture was filtered and then preadsorbed directly onto silica. Purification by flash chromatography eluting with dichloromethane/methanol/concentrated ammonia (93:7:1) gave the title compound (603 mg, 90%) as white needles: mp 134–136 °C (MeOH); ^{1}H NMR (400 MHz, d_6-DMSO) δ 1.42–1.58 (m, 1H, piperidine H-5$_{ax}$), 2.08–2.18 (m, 1H, piperidine H-5$_{eq}$), 2.72 (dd, J = 12.5 and 12.5 Hz, 1H, piperidine H-6$_{ax}$), 2.90–2.94 (m, 1H, piperidine H-2), 2.98–3.15 (m, 3H, piperidine H-6$_{eq}$, piperidine H-2′ and piperidine H-3$_{ax}$), 5.20 (1H, dddd, J = 49, 10, 10, and 2 Hz, piperidine H-4$_{ax}$), 6.99 (dd, J = 8.0 and 7.5 Hz, 1H, indole H-5), 7.09 (dd, J = 8.0 and 7.5 Hz, 1H, indole H-6), 7.37 (d, J = 8.0 Hz, 1H, indole H-7), 7.41 (dd, J = 7.5 and 7.5 Hz, 1H, phenyl H-4), 7.52 (dd, J = 7.5 and 7.5 Hz, 2H, phenyl H-3), 7.59 (d, J = 7 Hz, 2H, phenyl H-2), 7.77 (d, J = 8.0 Hz, 1H, indole H-4), 11.21 (s, 1H, indole NH); MS (ES$^+$) m/z 295 $(M + H)^+$. Anal. Calcd for $C_{19}H_{19}FN_2.CH_4O$: C, 73.59; H, 7.10; N, 8.58. Found: C, 73.66; H, 7.05; N, 8.62.

(3RS,4RS)-3-(3-Fluoro-piperidin-4-yl)-2-phenyl-1H-indole (3b). A solution of allylic fluoride **6** (500 mg, 1.7 mmol) in trifluoroacetic acid (10 mL) was treated with triethylsilane (500 μL, 3.1 mmol) and then stirred at ambient temperature for 3 h. The reaction mixture was poured carefully into a saturated aqueous solution of sodium hydrogen carbonate (100 mL), and the product was extracted into ethyl acetate (100 mL). The organics were washed with water (75 mL) and then brine (75 mL) to give an oil. This oil was purified by flash chromatography eluting with dichloromethane/methanol/concentrated ammonia (95:4.5:0.5) to furnish the title compound (330 mg, 66%) as a white solid: oxalate salt, colorless needles, mp 206–208 °C (EtOH); ^{1}H NMR (400 MHz, d_6-DMSO, free base) δ 1.67–1.78 (m, 1 H, piperidine H-5$_{eq}$), 2.12–2.26 (m, 1 H, piperidine H-5$_{ax}$), 2.34–2.48 (m, 2H, piperidine H-6$_{ax}$ and piperidine H-2$_{ax}$), 2.90 (broad d, J = 12 Hz, 1H, piperidine H-6$_{eq}$), 3.03–3.13 (m, 1H, piperidine H-4$_{ax}$), 3.32 (obscured by H_2O, 1H, piperidine H-2$_{eq}$), 5.03 (dddd, J = 49, 10, 10, and 5 Hz, 1H, piperidine H-3$_{ax}$), 7.01 (dd, J = 8.0 and 7.5 Hz, 1H, indole H-5), 7.10 (dd, J = 8.0 and 7.5 Hz, 1H, indole H-6), 7.38 (d, J = 8.0 Hz, 1H, indole H-7), 7.40–7.43 (m, 1H, phenyl H-4), 7.49–7.55 (m, 2H, phenyl H-3), 7.56–7.60 (m, 2H, phenyl H-2), 7.80 (d, J = 8.0 Hz, 1H, indole H-4), 11.18 (s, 1H, indole NH); MS (ES$^+$) m/z 295 (M + H)$^+$. Anal. Calcd for $C_{19}H_{19}FN_2.C_2H_2O_4$: C, 65.62; H, 5.51; N, 7.29. Found: C, 66.01; H, 5.44; N, 7.20.

The Mechanism

The mechanism was proposed in the literature [20]. The rearrangement is induced by the indole nucleus as a neighboring group. Two possible conformations, **A** and **B**, of the intermediate allow trans-diaxial ring opening; pathway A gives rise to the observed fluoropiperidine **2a** while pathway B (not observed) produces the isomer **3a**. A possible reason for the regioselectivity is that conformer **B** is destabilized relative to **A** as a result of steric congestion caused by the group *endo* to the piperidine ring.

Similar observation was also reported in the following reaction [21]. When alcohol **7** was treated with 1 equivalent of DAST, compounds **8** and **9** were obtained in 34% and 17%, respectively. In this case, the 4-hydroxyphenyl is the neighboring group that participates in the formation of the spiro intermediate.

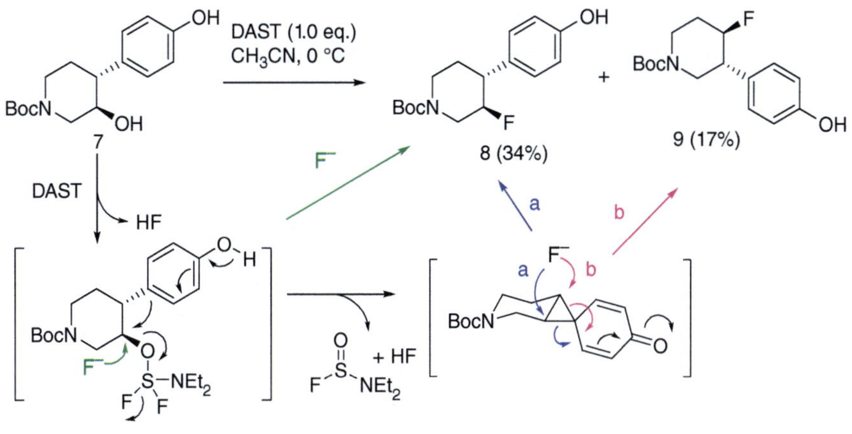

5.4.4 Tetrahydropyrazine

Problem 132: Formation of an Unexpected Tetrahydropyrazine Product

Lofexidine hydrochloride (**1**•HCl) is an antihypertensive medication currently used to treat physical and psychological symptoms of opioid dependence. Donnola's group attempted to develop a process chemistry for the production of **1** by following the literature-reported method [22] for constructing the dihydro-1H-imidazole core. Treatment of the ester **2** with ethylenediamine in the presence of Me_3Al in toluene, lofexidine **1** was not obtained, but the reaction quantitatively gave an unexpected product identified as (5-(2,6-dichlorophenoxy)-6-methyl-1,2,3,6-tetrahydropyrazine) (**3**) [23]. The synthetic procedure and proton-NMR were reported in a patent application [24]. Upon further investigation, Donnola's group found that formation of the imidazoline ring was more efficient when performed in a one-pot way, charging ethylenediamine, titanium isopropoxide (catalyst), toluene (used as the solvent), and ester intermediate **2** and allowing the reaction at reflux (110 °C). Applying this procedure, they discovered a quantitative conversion (>95%) of **2** to lofexidine **1** in a few hours.

Please suggest a mechanism for the formation of **3**.

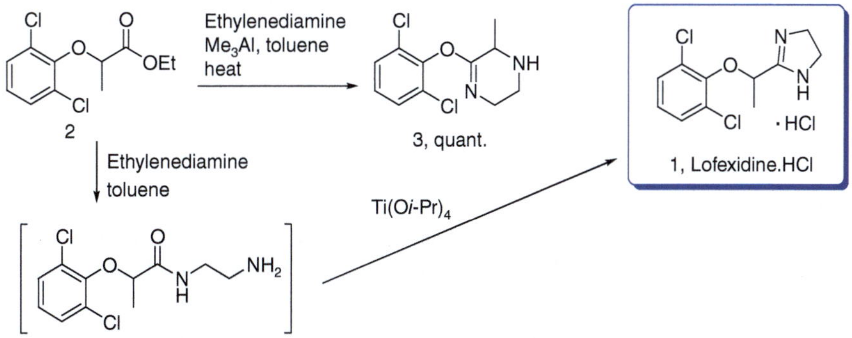

Experimental Procedure

Preparation of 5-(2,6-dichlorophenoxy)-6-methyl-1,2,3,6-tetrahydropyrazine (3) (from WO2020/254580, **2020**, A1)

Toluene (32 mL, 6.4 volumes) is loaded at room temperature into a 100-mL reactor rendered inert with N_2. The temperature is reduced to 0 °C, and 2M trimethylaluminum in toluene (15.50 g, 38.49 mmol) is added. While maintaining said temperature, a solution of ethylenediamine (2.3 g, 38.40 mmol) in toluene (13.4 ml, 2.7 volumes) is added, resulting in an exothermic reaction. At the end of the addition, the mass is brought to room temperature, and stirred at said temperature for 1 h. The solution is cooled again to 0 °C, and a solution of (+/−) ethyl 2-(2,6-dichlorophenoxy) propionate (5 g, 19 mmol) in toluene (20 mL, 4 volumes) is added slowly. The reaction mixture is stirred under reflux, 110 °C. When 95% conversion to 5-(2,6-dichlorophenoxy)-6-methyl-l,2,3,6-tetrahydropyrazine is observed with a residue of 5% (+/−) ethyl 2-(2,6-dichlorophenoxy)propionate (after about 3 h), the yellow solution is again cooled to 0 °C, MeOH (9 mL) is added very slowly and, when fume formation has ended, H_2O (20 mL, 4 volumes) is added, again at 0 °C, thus precipitating the aluminum salts, which are removed by vacuum filtration. The solution is diluted with ethyl acetate (40 mL, 8 volumes), the phases are separated, and the aqueous phase is washed with further ethyl acetate (40 mL, 8 volumes). The combined organic phases are washed with H_2O (40 mL, 8 volumes), and finally concentrated to residue by complete evaporation of the solvent, until a yellowish-white solid is obtained. Yield >86%. Lofexidine (**1**) is not obtained, but a different product, with 95% purity, of formula ^{1}HNMR (400 MHz, Chloroform-d) δ 7.42 (ddd, $J = 8.3$, 6.8, and 1.5 Hz, 2H Arom), 7.26 (t, $J = 8.1$ Hz, 1H Arom), 4.19–3.91 (m, 4H, CH, CH_2, CH), 3.82–3.46 (m, 2H, CH, NH), 1.21 (d, $J = 6.6$ Hz, 3H, CH_3).

Synthesis of Lofexidine (1). (from *Org. Process Res. Dev.* **2021**, *25*, 1816)

Titanium isopropoxide (2.57 kg, 9.00 moles) and toluene (9.5 L, 6 volumes vs. intermediate **2**) were charged at 25 °C into a 50-L Kilolab Hastelloy reactor blanketed by a N_2 flow. A solution of ethylenediamine (0.55 kg, 9.00 moles) in toluene (4 L, 2.5 volumes vs. intermediate **2**) was then added. The reaction mixture was stirred at 25 °C for 1 h, and a toluene solution of ethyl 2-(2,6-dichlorophenoxy)propionate (**2**) (7.08 kg, 6.000 moles) was added. The reaction mixture was heated to reflux (110 °C) and stirred for 18 h. The reaction was checked by ultra performance liquid chromatography (conversion: 98%). The mixture was then cooled to 25 °C, and the yellow-orange solution was added to a 30% tartaric acid solution (23.8 L, 15 volumes vs. intermediate **2**) previously prepared in a 50-L Kilolab glass-lined reactor (titanium isopropoxide reacted with tartaric acid in an aqueous environment, forming a soluble Ti species; in this case, no TiO_2 was formed, and Ti salts were completely soluble). The organic layer was discarded, and toluene was added to the aqueous acidic layer containing titanium salts and the product as the tartrate salt (8.9 L, 5.6 volumes vs. intermediate **2**). The mixture was basified by adding 30% sodium hydroxide until it reached a pH of 12. The layers were separated, and the aqueous layer was washed with toluene (14.3 L, 9 volumes vs. intermediate **2**). The mixed organic phases were washed with water (8.9 L, 5.6 volumes vs. intermediate **2**) and concentrated to a low volume. The obtained suspension was filtered at 25 °C, and the needles were dried at 40 °C to yield the crude lofexidine-free base as a yellowish solid. Yield >90% was calculated from the starting material (2,6-dichlorophenol) of the previous step. Purity >95% (Area %). The resulting product could be used for the precipitation of hydrochloride salt **1** or could be further purified by crystallization with heptane or hexane or by pulping at room temperature in methylisobutylketone or methylethylketone. ^{1}HNMR characterization of lofexidine-free base (400 MHz, chloroform-d) δ 7.31 (d, J = 8.1 Hz, 2H Arom), 7.01 (t, J = 8.1 Hz, 1H Arom), 5.16 (q + s (broad), J = 6.6 Hz, 2H, CH, NH), 3.65 (dd, J = 25.5 and 8.6 Hz, 4H, CH_2—CH_2), 1.59 (d, J = 6.6 Hz, 3H, CH_3). ^{13}CNMR of lofexidine-free base (400 MHz, chloroform-d) δ 167.9, 149.5, 129.6, 129.2, 125.3, 77.0, 51.7, 18.9.

The Mechanism

The literature discussed the mechanism for the formation of **3** [23]. The first reaction should be an amidation of the ester **2** by ethylenediamine to give intermediate **A**. Then a dyotropic rearrangement [25] occurs to provide ester **B**. Finally, the ester **B** undergoes cyclization leading to the unexpected product **3**.

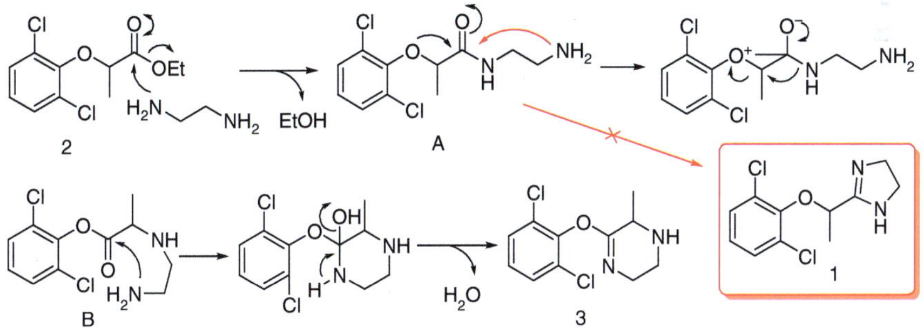

However, the above mechanism seems implausible. We can be almost certain that the amino amide **A** is the first intermediate. Due to the tautomerization, amide C—N single bond is a partially double bond, thus, much stronger than a simple C—N single bond and much stronger than an ester C—O single bond. Therefore, conversion of amide **A** to ester **B** via a dyotropic rearrangement is not thermodynamically favored. Secondly, even if the amino ester **B** were formed, intramolecular attack of the terminal amino group on the ester carbonyl could lead to its decomposition into 2,6-dichlorophenol (**C**) and 3-methylpiperazin-2-one (**D**), as the 2,6-dichlorophenoxy group is a good leaving group.

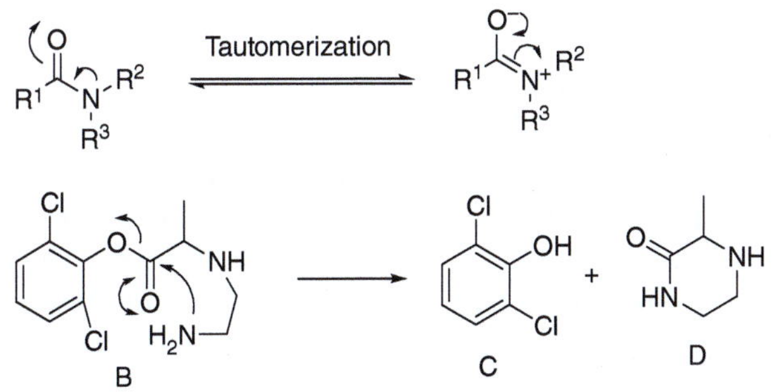

The ^{1}H NMR spectrum alone is insufficient to determine the structure of compound **3** (as described in patent [24]). Based on the structure of **3** assigned in both the article [23] and the patent, [24] the provided ^{1}H NMR data are incorrect for the signal at δ 7.42 (ddd, $J = 8.3$, 6.8, and 1.5 Hz, 2H Arom). This signal should instead appear as a doublet corresponding to the 4-H and 6-H protons on the phenyl ring. Without this correction, the ^{1}H NMR data do not align with the proposed structure of **3**. To conclusively confirm the structure of the unexpected product **3**, additional analytical techniques such as ^{13}C NMR, mass spectrometry, elemental analysis, or HRMS are required. A plausible mechanism for the formation of compound **3** can only be proposed once its structure has been definitively established.

5.4.5 Tetrahydropyrimidine

Problem 133: Formation of Impurities During a CDI-promoted Cyclization
LY2886721 (**1**) is a potent and selective inhibitor of beta-amyloid-cleaving enzyme that was advanced to phase 2 clinical trials as a potential treatment for Alzheimer's disease (AD). In a process synthesis of **1**, key intermediate **2** was converted to compound **3** in 75% yield by the following reactions. Two impurities **4** and **5** were also identified from the reaction [26]. The process was optimized by fine tuning the operation, controlling the equivalents of triethanolamine, BzNCS, and carbonyldiimidazole (CDI) [27].

Please provide mechanism for the formation of **3**, **4**, and **5**.

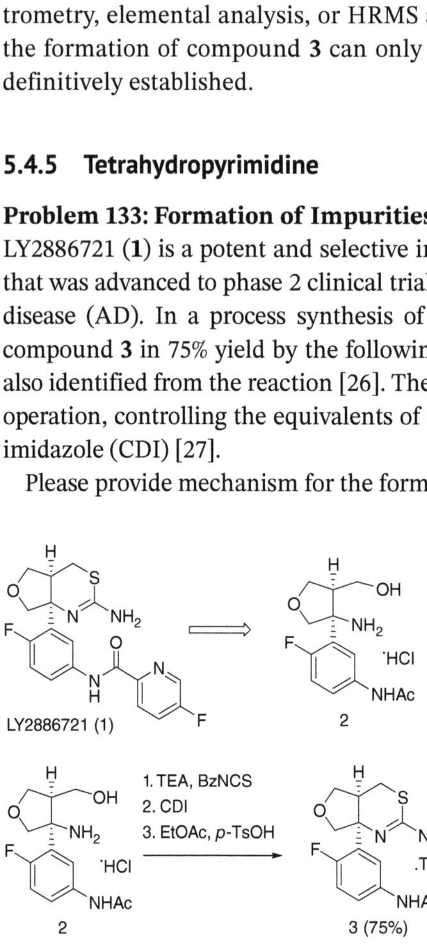

LY2886721 (1)

2

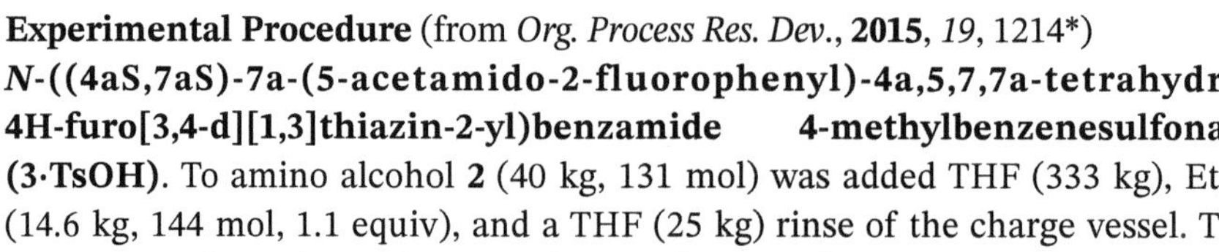

1. TEA, BzNCS
2. CDI
3. EtOAc, *p*-TsOH

3 (75%)

4

5

Experimental Procedure (from *Org. Process Res. Dev.*, **2015**, *19*, 1214*)
***N*-((4aS,7aS)-7a-(5-acetamido-2-fluorophenyl)-4a,5,7,7a-tetrahydro-4H-furo[3,4-d][1,3]thiazin-2-yl)benzamide 4-methylbenzenesulfonate (3·TsOH)**. To amino alcohol **2** (40 kg, 131 mol) was added THF (333 kg), Et_3N (14.6 kg, 144 mol, 1.1 equiv), and a THF (25 kg) rinse of the charge vessel. The mixture was heated at 50 °C as a solution of BzNCS (21.4 kg, 131 mol, 1.0 equiv)

in THF (66 kg) was added over 1.5 h, followed by a THF (5 kg) rinse of the charge vessel. After 1 h, the in-process control (IPC) showed <2.0% remaining **2**, or an additional charge of BzNCS (0.66 kg, 4.0 mol, 0.03 equiv) was added (required on only one batch). To this solution at 50 °C was added CDI (21.3 kg, 131 mol, 1.0 equiv) as a solid through the reactor inlet connected to a glove bag. After 60 min, the IPC showed <2.0% remaining alcohol intermediate **3a** or an additional charge of CDI (charge adjusted based on the area % remaining **3a**, i.e., if 3% **3a**, then add 3% of the original CDI charge) was added. For 3 of 8 batches, an additional charge of CDI was required. After the IPC was met, the mixture was heated at reflux for 8 h. The mixture was transferred to another reactor with a THF (40 kg) rinse, and the solvent was exchanged under vacuum using EtOAc (890 kg) added in two portions, until GC showed <2% THF remaining. 10% aqueous citric acid (200 kg) was added, and the layers were agitated and separated at 20 °C. An additional wash with 10% aqueous citric acid (200 kg) was completed. The organic layer was heated at 60 °C as an 8% solution of p-TsOH·H_2O (130 kg, 54.7 mol p-TsOH·H_2O, 0.42 equiv) in EtOAc was added. The solution was seeded with 0.2 kg of **3·TsOH** in 0.5 L of EtOAc. After 30–45 min, a slurry resulted, and an 8% solution of p-TsOH·H_2O (182 kg, 76.6 mol p-TsOH·H_2O, 0.58 equiv) in EtOAc was added over 3 h. After an additional 1 h at 60 °C, the mixture was cooled to 20 °C over 3 h and stirred for 1 h. The solids were collected by filtration on a 1-m^2 (1000 L) agitated filter and washed with EtOAc (2 × 72 kg) to afford the wet cake product. A total of 8 wet cake batches were collected and combined for drying in 5 batches to afford 478 kg (75% average yield for four 40 kg batches and four 42.5 kg batches) of aminothiazine **3·TsOH**. Purity = 99.8–99.9 HPLC area%. Assay = 98.7–99.6%. mp 144 °C. HPLC Method A t_R = 6.15 min. ^{1}H NMR ($CDCl_3$) 12.52 (br s, 1H), 11.91 (br s, 1H), 9.33 (s, 1H), 8.19–8.08 (m, 1H), 8.05 (d, J = 8.3 Hz, 2H), 7.67 (d, J = 7.8 Hz, 2H), 7.65–7.63 (m, 1H), 7.57 (t, J = 7.3 Hz, 1H), 7.34 (t, J = 7.8 Hz, 2H), 7.15 (d, J = 8.3 Hz, 2H), 7.09 (dd, J = 9.3, 12.2 Hz, 1H), 4.29 (s, 1H), 4.27–4.24 (m, 2H), 4.16 (dd, J = 6.6 and 9.0 Hz, 1H), 3.83–3.76 (m, 1H), 3.34 (dd, J = 2.9 and 13.7 Hz, 1H), 3.00 (dd, J = 4.4 and 14.2 Hz, 1H), 2.36 (s, 3H), 2.09 (s, 3H). ^{13}C NMR ($CDCl_3$) 171.38, 169.56, 166.71, 155.85, 153.95, 140.89, 136.30, 134.41, 129.72, 128.96, 128.91, 125.81, 123.34, 123.27, 122.12, 122.02, 120.39, 117.55, 117.36, 76.50, 76.47, 70.12, 68.07, 68.03, 39.34, 39.30, 25.18, 24.09, 21.32. HRMS (ESI): calcd for $C_{21}H_{21}FN_3O_3S$ [M + H^+]: 414.1282; found 414.1267.

The mechanisms for the formation of the side products **4 and **5** were discussed in the article* [26, 27]. *However, the isolation and identification of **4** and **5** were not disclosed.*

The Mechanism

The mechanisms for the formation of **3**, **4**, and **5** were discussed in the literature [26, 27] and can be summarized below. The impurity **4** is generated by excess of BzNCS, and the impurity **5** is generated by excess of CDI. Therefore, the equivalents of BzNCS and CDI, along with the addition rate, reaction temperature, solvent selection, concentration, and stirring speed are crucial parameters.

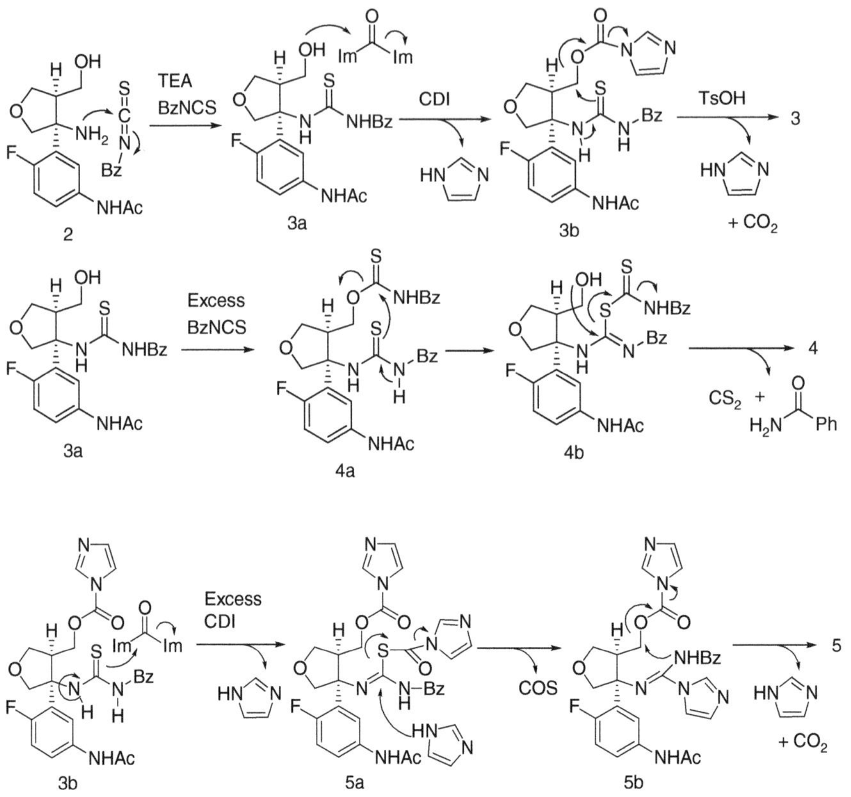

5.4.6 Oxazepane

Problem 134: An Unusual Stevens Rearrangement Catalyzed by Acid

During a total synthetic study of nucleoside antibiotic liposidomycin, Knapp's group wanted to run a thioglycosylation reactions of compound **1** with (methylthio) trimethylsilane catalyzed by trimethylsilyl trifluoromethanesulfonate (TMSOTf) in refluxing CCl_4. Only a trace amount of the expected thioglycoside **2** was observed, with the major product being the rearranged diacetate **3** [28].

Please provide a mechanism for the formation of **3**.

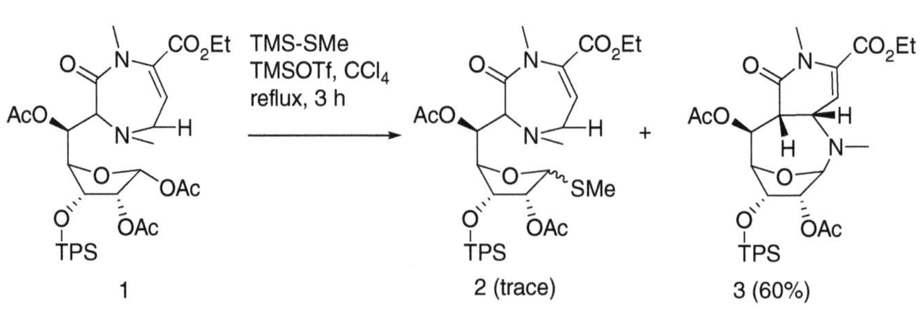

Experimental Procedure (from *Tetrahedron Lett.* **2002**, *43*, 5797)
Combination of the triacetate **1**, (methylthio)trimethylsilane (10 equiv), and TMSOTf (2.5 equiv) in dry carbon tetrachloride solution was heated at reflux for 3 h, whereupon **1** was consumed. Although a trace of the expected thioglycoside **2** was present, preparative thin-layer chromatography (TLC) of the reaction mixture gave a rearranged diacetate **3** (LC-FAB-MS *m/z* 665, MH^+ for M = C38H50N2O9Si) as the major product in 60% yield. ^1H and ^{13}C NMR assignments, nuclear Overhauser effect spectroscopy (NOESY), and heteronuclear multiple bond correlation (HMBC) for **3** can be found in the literature [28].

The Mechanism
This is a Stevens rearrangement catalyzed by Lewis acid. Usually, Stevens rearrangement is catalyzed by strong base. Loss of acetate from **1** promoted by TMSOTf and accompanied by least-motion *N*-participation on the furanose *β*-face would lead to the anomeric ammonium salt **A**. Elimination of trifluoromethanesulfonic acid assisted by *O*-silylation of the amide would give the iminium intermediate **B**, and then Mannich-like ring closure provides the product **3**.

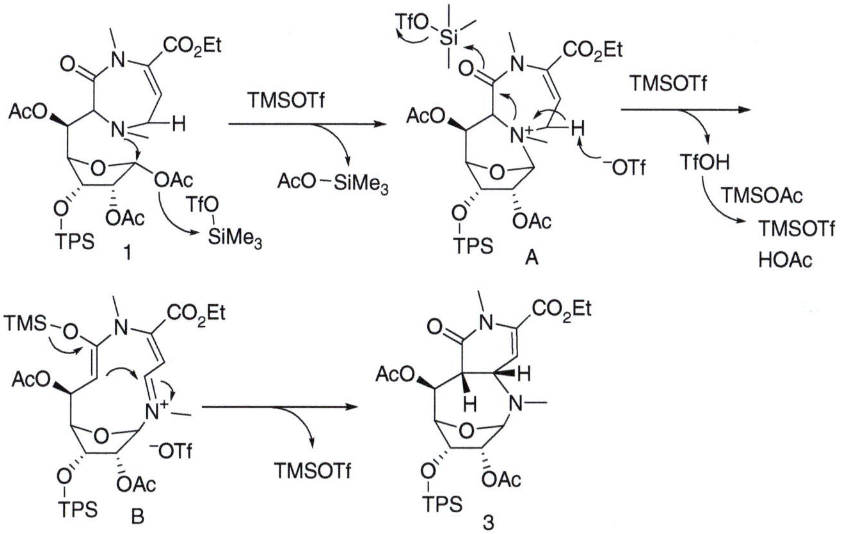

Based on the above mechanism, the nucleophilic ability of the tertiary amine could be decreased by protonation of the nitrogen as HCl salt. Since the TPS protecting group is not compatible with acidic HCl salt, switching the TPS to acetate would address this issue. The problem was finally solved by employing the hydrochloride salt of the tetraacetate donor **4**. Even with the initial protonation of the tertiary amine, the reaction of **4** with bis(trimethylsilyl)uracil **5** gave the required uracil nucleoside **6** in 18% isolated yield, which was sufficient for the constitutional synthesis [29].

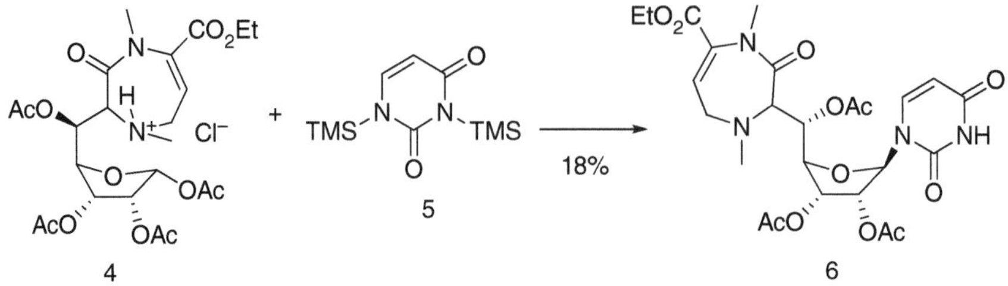

5.4.7 Tetrahydroquinoline

Problem 135: Formation of Unexpected Tetrahydrobenzoquinoline Side Product in a Hydrogenation

(S)-Enantiomer of 2-[2-(1-methyl-2-piperidyl)ethyl]cinnamanilide (MSA100, **5**) is an active 5-HT (5-hydroxytryptamine or serotonin) receptor antagonist. In a process study, [30] methylpyridinium *p*-toluenesulfonate **1** was chosen as a key intermediate. Catalytic hydrogenation of **1** generated the desired product **2** in 59% yield along with side products **3** and **4**.

Please suggest mechanism for the formation of **3** and **4**.

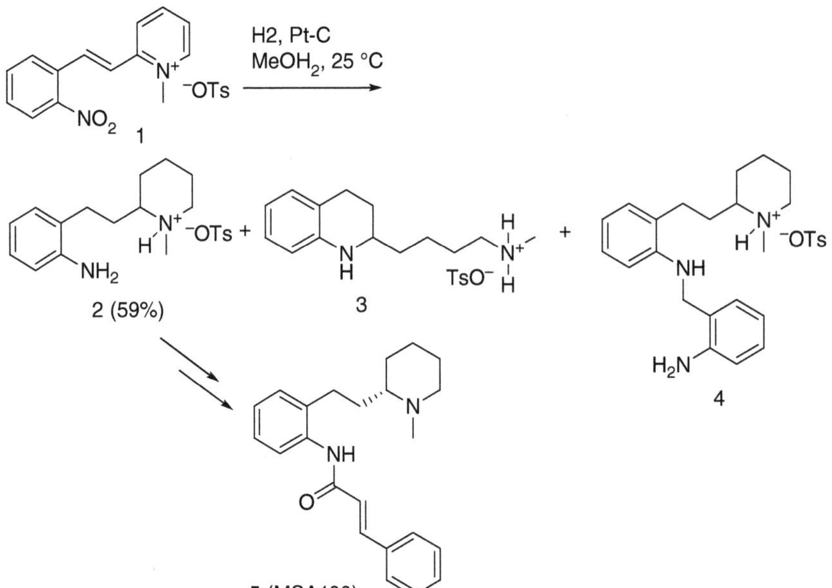

Experimental Procedure (from *Org. Process Res. Dev.* **2015**, *19*, 290*)

2-[2-(1-Methyl-2-piperidinyl)ethyl]-benzenamine 4-methylbenzenesulfonate (1:1) (2). To an inert hydrogenation vessel was charged **1** (43.9 g, 0.106 mol), 10%

Pt/C (1.87 g, 62.4% wet), and methanol (396 g) under N_2 atmosphere. The headspace was purged with N_2, and the vessel was charged with H_2 by pressurizing with H_2 to 65 psi, followed by depressurizing to 14.5 psi. The H_2 pressurization/depressurization cycle was repeated four times. After the final depressurization, the reactor's temperature, hydrogen pressure, and agitation speed were set at 30 °C, 75 psi, and 450 rpm, respectively. Hydrogenation was maintained under these conditions for an additional 6 h. The batch was cooled to 25 °C. The vessel was depressurized, purged with N_2 by pressurizing to 65 psi, and depressurized. The pressurizing–depressurizing cycle was repeated five times. The mixture was filtered over a pad of celite (8 g) and rinsed with methanol (2 × 45 g). The combined filtrate was concentrated at 35–45 °C under reduced pressure (80–160 mbar) until a final volume of ~150 mL was reached. 2-Propanol (353 g) was added and the mixture was concentrated at 35–45 °C under reduced pressure (80–160 mbar) until a final volume of ~150 mL was reached. The residue was heated to 60 °C and isopropyl acetate (44 g) was added while maintaining the temperature at 55–65 °C. The mixture was cooled to 40 °C over a period of 20 min and seeded with **2** (160 mg). The mixture was cooled to 20 °C over a period of 1 h and stirred for 4 h. The precipitate was filtered, rinsed with a solution of 2-propanol/isopropyl acetate (1:2 v/v) (2 × 42 g), and dried under reduced pressure (15–40 mbar) at 60 °C for 16 h to afford **2** (26.3 g, 64% yield) as a solid: mp 133–135 °C; HPLC for **2** (t_R = 5.20 min) 98.8% purity: Waters YMC ODS-AQ S-3 120 A, 150 × 3.0 mm, flow rate = 0.8 mL/min, 25 °C, gradient elution from 90:10 A–B to 40:60 A–B over 22 min; A = 10 mM NH4OAc in water; B = acetonitrile; UV λ = 240 nm.

* *The mechanisms for the formation of the side products* ***3*** *and* ***4*** *were discussed in the article* [30]. *However, the isolation and identification of* ***3*** *and* ***4*** *were not disclosed.*

The Mechanism

Complete hydrogenation of nitro, double bond, and the pyridinium ring will lead to the desired product **2** (pathway A). Alternatively, since the hydrogenation of the double bond and the pyridinium ring to piperidine are much slower than that of the nitro/nitroso group to aniline **B**, through pathway B, a cyclization of aniline **B** gives a spiro intermediate **C**. After subsequent three transformations from **B**, hydrogenation to **D**, rearrangement to **E**, and hydrogenation of **E**, side product **3** is provided. Under another scenario, a thermal [2 + 2] cycloaddition reaction between nitroso intermediate **A** and starting material **1** via pathway C is conceivable. The four-membered ring intermediate **F** thus formed undergoes a retro [2 + 2] cycloaddition to afford an imine intermediate **H** with concomitant generation of aldehyde **G**. Further hydrogenation of **H** generates side product **4**.

Employing the improved conditions (30 °C, 75 psi, and 450 rpm), **2** was generated in 71% yield containing no detectable amount of **4** and 14% of **3**. Side product **3** was easily removed from **2** by crystallization.

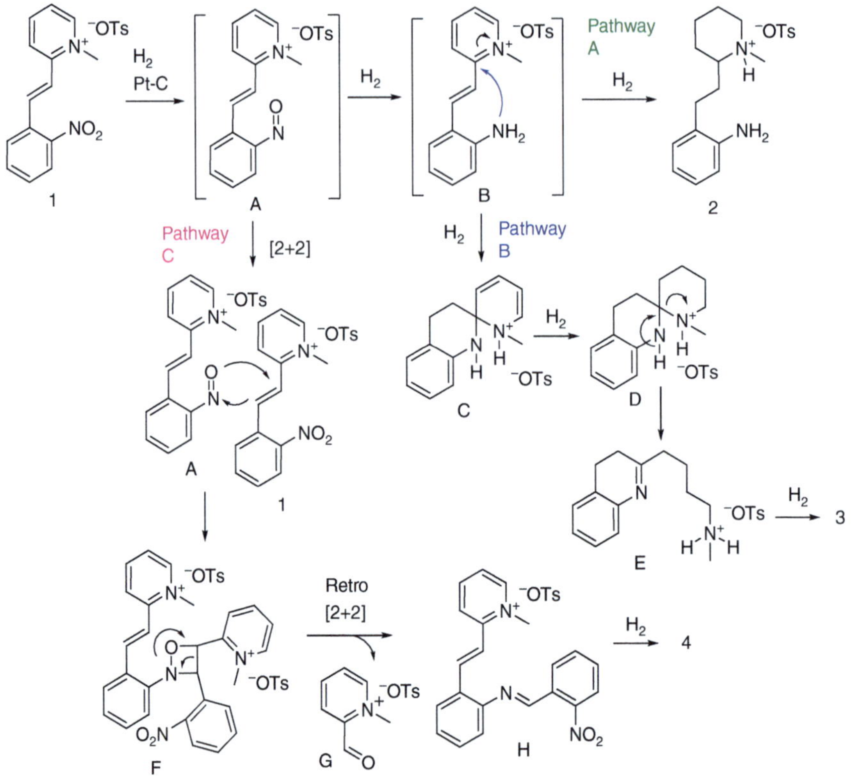

5.5 Formation of Monocyclic Aromatic Heterocycles

5.5.1 Pyrrole

Problem 136: Formation of an Amino Pyrrole Impurity

Heating the furan precursor **1** with ammonium acetate in *N*-methylpyrrolidone at 100–105 °C yielded the pyrrole product **2**. However, chemists at Pfizer discovered that an impurity was also generated during the reaction at a level of 0.7%. Even after recrystallization and Darco treatment, the impurity persisted at 0.5%. Through comparison with a synthetic standard, the impurity was identified as 2-aminopyrrole **3** [31]. The electron-deficient pyrrole **2** was further derivatized via diazotization using **p**-nitrobenzene diazonium chloride. Subsequent reduction of the diazo intermediates **4** and **5** produced 2- and 1-aminopyrroles (**3** and **6**, respectively). Understanding the formation mechanism of this impurity is essential for developing strategies to suppress its generation.

Please suggest a mechanism for the formation of impurity **3**.

Experimental Procedure (from *Org. Process Res. Dev.*, **2002**, *6*, 64)

4-Oxo-1,4,5,6,7,8-hexahydro-cyclohepta[b]pyrrole-3-carboxylic acid (2-fluoro-4-methoxy-phenyl)amide (2). Under nitrogen atmosphere, a mixture of **1** (735 g, 2.32 mol) and ammonium acetate (1249 g, 16.22 mol) in *N*-methylpyrrolidinone

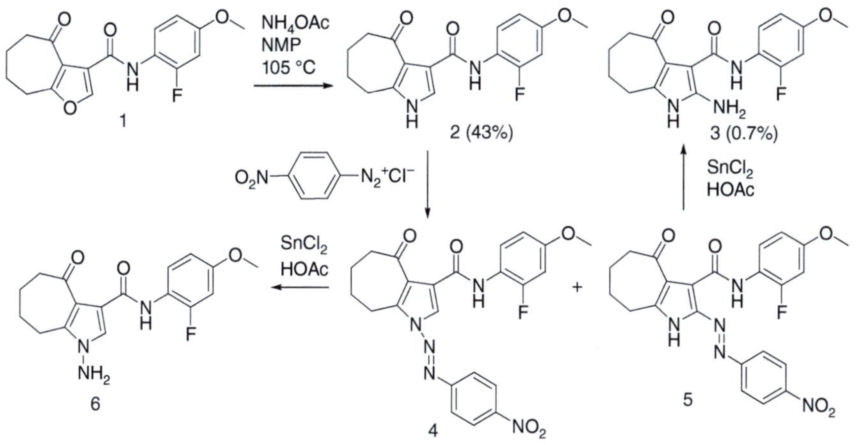

(1.47 L) was heated to 100–105 °C for 24 h. The reaction was cooled to 25 °C, and water (13.2 L) was added. The resulting mixture was stirred for 17 h at 25 °C, and filtered. The filter cake was rinsed with water and dried under vacuum at 40–45 °C to give 729 g of crude product as dark brown solids (purity 94.8% by HPLC). The crude product was stirred in methanol (33.7 L), and filtered through celite. To the filtrate was added 116 L of water, and the resulting slurry was stirred for 24 h. The mixture was filtered, and the solids were dried under vacuum at 40–45 °C to give 567 g of dark colored solids (purity 98.5% by HPLC). This crude product was stirred in acetone (17.0 L), and Darco G-60 (567 g) was added. After stirring for 30 min at 20–25 °C, the mixture was filtered. The filtrate was concentrated under vacuum to ~0.85 L and crystallization occurred. The mixture was granulated for 30 min and then filtered. The filter cake was rinsed with acetone (120 mL) and dried under vacuum at 50–60 °C to yield 320 g (1.01 mol, 43%) of **2** as off-white solids (purity 99.3% by HPLC). ^{1}H NMR * δ 12.40 (s, 1H), 12.04 (s, 1H), 8.18 (t, $J = 8.8$ Hz), 6.91 (s, 1H), 6.89 (dd, 1H, $J = 2.4$ and 12.8 Hz), 6.72 (dd, 1H, $J = 2.4$ and 8.8 Hz); ^{13}C NMR δ 200.12, 161.12, 155.88, 155.78, 154.78, 152.34, 146.57, 126.16, 123.62, 120.21, 117.84, 109.55, 101.76, 101.54, 55.56, 40.93, 25.34, 23.11, 20.69. MS m/z 317 (M + 1), 177, 149.

2-Amino-4-oxo-1,4,5,6,7,8-hexahydro-cyclohepta[*b*]pyrrole-3-carboxylic acid (2-fluoro-4-methoxy-phenyl)amide (3). To 4-nitroaniline (1.57 g, 11.4 mmol) in 6 N HCl (8.55 mL) at 0 °C was added dropwise with 30% $NaNO_2$ solution in water (wt/wt). After 10 min at 0 °C, **2** (1.20 g, 3.8 mmol) with NaOAc (2.50 g) in AcOH (50 mL) was added. The resulting reaction mixture was stirred at room temperature for 3 h and worked up by extraction with ethyl acetate. The 2-azo (47 mg) and *N*-azo (3.0 mg) intermediates were obtained by silica gel column chromatography: 0.87 g of **2** was recovered. The 2-azo intermediate (**5**, 45 mg) was heated with $SnCl_2$ (100 mg) in AcOH (2 mL) at 80 °C for 1 h to give the desired product **3** (17.5 mg) after preparative TLC purification: ^{1}H NMR * δ 12.29 (s, 1H), 8.14 (t, $J = 9.2$ Hz, 1H), 6.88 (dd, $J = 12.8$ and 2.8 Hz, 1H), 6.72 (dd, $J = 9.2$ and 2.8 Hz), 6.48 (s, 2H), 2.83–2.85 (s, 2H), 2.62–2.65 (s, 2H), 1.69–1.76 (s, 2H); ^{13}C NMR δ 198.28, 164.45, 155.07, 154.99, 154.33, 152.39, 148.24, 141.83, 123.38, 120.72, 116.39, 109.33, 101.57, 101.38, 55.52, 40.84, 25.22, 22.98, 20.70; MS m/z 331** (M + 1), 191, 163, 135.

*The ^{1}HNMR spectra of **2** and **3** were not fully presented here. "6.72 (dd, J = 9.2 and 2.8 Hz)" did not indicate the number of protons. Several proton signals, such as OMe, were missing.*

***The molecule weight of **3** is 331 and the mass spectrum should show m/z 332 (M + H$^+$). There might be a typo here.*

The Mechanism

Since the initial procedure for that reaction was not deoxygenated, the furan unit could be oxidized by oxygen from air to form 1,2-dioxetane **A** or **B** via radical reaction.

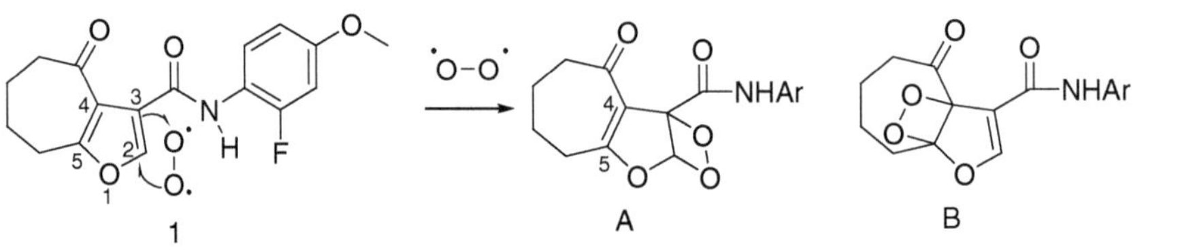

The dioxetane **A** is attacked by NH_3 (from ammonium acetate) to give the hydroperoxyl alcohol **C**. After the elimination of hydrogen peroxide and ring opening, diketone imine **D** is formed, which in turn cyclizes to hydroxyl pyrrole **E**. A second NH_3 is added to the hydroxyl pyrrole followed by elimination of water to produce the impurity **3**.

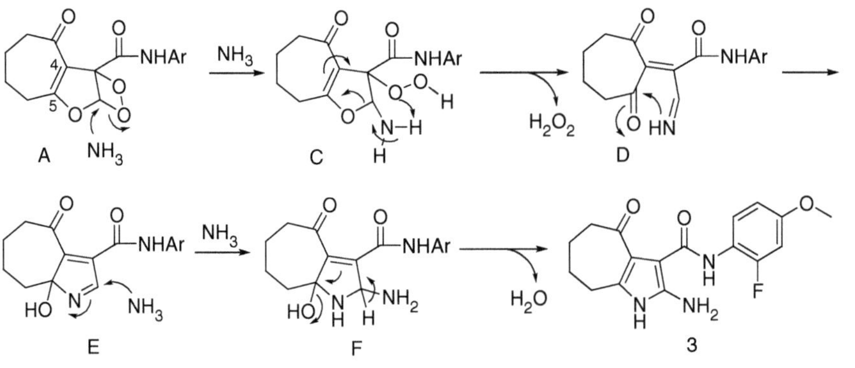

Starting from possible intermediate **B**, the impurity **3** could be formed analogously by following mechanism.

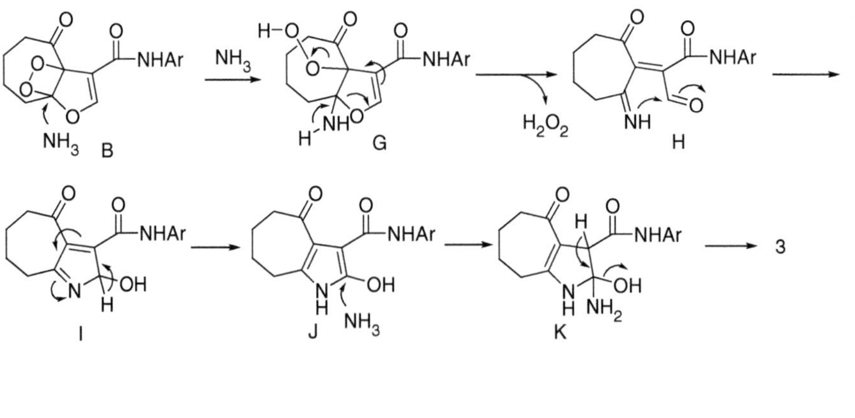

Once hydrogen peroxide is generated, it could oxidize the furan to the C2—C3 epoxide or the C4—C5 epoxide and those would also lead to the impurity **3**. That mechanism was proposed in the literature [31].

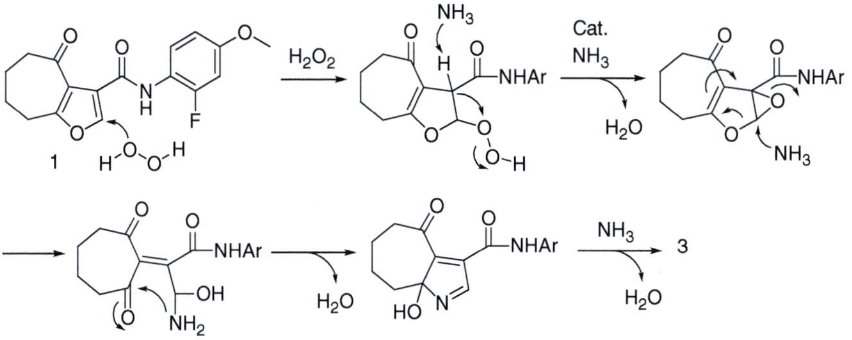

Reaction conditions free of oxidants helped prevent the formation of the 2-aminopyrrole impurity **3**; nitrogen sparging of the reaction mixture prior to heating eliminated the impurity formation.

5.5.2 Pyridine

Problem 137: Pyridyl Walks to Hydroxyl from Sulfone During an Alkylation
SB-462795 is a potent cathepsin K inhibitor under development for the treatment of osteoporosis and osteoarthritis. One of the key building blocks for the synthesis of SB-462795 is diene **3**, which is prepared from sulfonamide **1** and epoxide **2**. In the reaction, except the desired product **3**, an impurity **4** was also isolated and identified, in which the pyridyl group moved to hydroxyl and eliminated SO_2 [32]. The amount of **4** increased to 3–5% over prolonged reaction times. Treatment of **3** with Cs_2CO_3 in acetonitrile at 80 °C provided **4** in 70% yield.

Please suggest a mechanism for the formation of **4**.

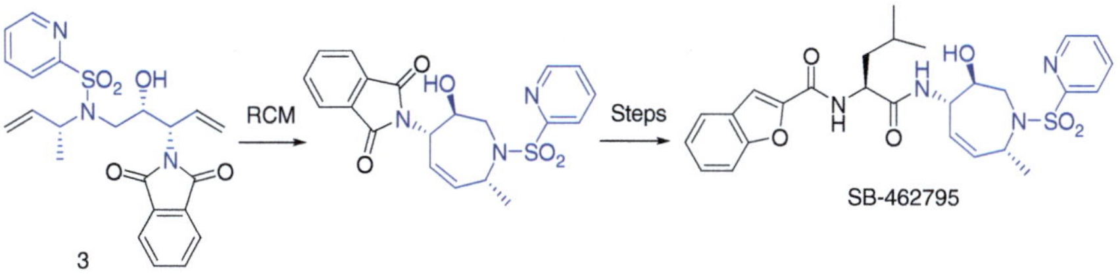

Experimental Procedure (from *Org. Process Res. Dev.*, **2008**, *12*, 226)
1,3,4,5-Tetradeoxy-3-(1,3-dioxo-1,3-dihydro-2H-isoindol-2-yl)-1-[[(1R)-1-methyl-2-propen-1-yl](2-pyridinylsulfonyl)amino]-L-threo-pent-4-enitol (3). In a glass-lined reactor, a slurry of (2*R*,3*S*)-3-(1,3-dioxoisoindolin-2-yl)-2-hydroxypent-4-en-1-yl 4-methylbenzenesulfonate (8.5 kg, 21.2 mol) and toluene (34 L) was heated to 56 °C under nitrogen, and DBU (3.3 kg, 21.7 mol) was added via pressure vessel to maintain the temperature between 55 and 65 °C. The starting material dissolved during the

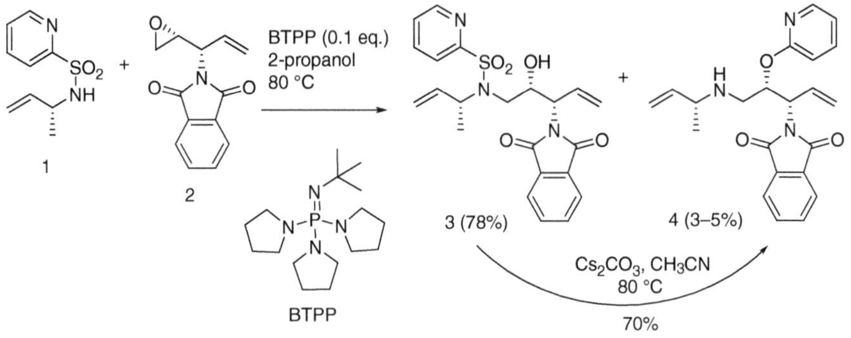

addition and the mixture was aged at this temperature for 60 min. Upon completion of the reaction, the mixture was cooled to 25 °C and 10% w/w aq. citric acid (0.42 L water with 45 g citric acid monohydrate) was added. After stirring for 10 min, water (25.5 L) was added, the mixture stirred and the layers allowed to separate. The organic layer was then washed twice with water (2 × 8.5 L). The solution of crude **2** was then transferred to a second reactor via an inline filter to remove a small layer of film, and the transfer line then rinsed with toluene (8.5 L). To this second reactor was charged **1** (4.5 kg, 21.2 mol) and the solvent partially removed by vacuum distillation to reach 15 L total volume (36.0 L toluene removed). 2-Propanol (25.5 L) was charged and the system vacuum distilled to achieve 15 L total volume (25.5 L removed). The total solvent volume was adjusted to 3.4 volumes using 2-propanol (15.0 L added), (tert-Butylimino) tris(pyrrolidino)phosphorane (BTPP) (0.7 kg, 2.12 mol) was added and the mixture heated to 80 °C over ~30 min. After aging for 3.5 h, the reaction was complete, and heptane was added (17.0 L) while maintaining the temperature 60 °C. The temperature was then adjusted to 53 °C and seeds of **3** (4.5 g in 2:1 heptane:2-propanol) were charged to the reactor. The slurry was held at 52 °C for 60 min, then cooled to ~10 °C at 0.5 °C/min. After holding for 30 min, the slurry was isolated by centrifugal filtration, washed with cold heptane/2-propanol (2:1 v/v, 2 × 8.5 L). The product was dried under vacuum at 45 °C to afford **3**: 7.25 kg (16.4 mol, 78%), >99.9% pure by HPLC. ^{1}H NMR ($CDCl_3$, 400 MHz) δ 8.62 (dm, 1H, J = 4.0 Hz), 8.02 (d, 1H, J = 7.9 Hz), 7.93 (td, 1H, J = 7.7, 1.6 Hz), 7.83 (m, 2H), 7.70 (m, 2H), 7.50 (dm, 1H, J = 0.9 Hz), 6.29 (m, 1H), 5.81 (m, 1H), 5.39 (d, 1H, J = 17.1 Hz), 5.30 (dd, 1H, J = 10.2, 0.5 Hz), 5.13 (m, 2H), 4.89 (bs, 1H), 4.68 (t, 1H, J = 9.0 Hz), 4.60 (dt, 1H, J = 9.2, 2.6 Hz), 4.39 (m, 1H), 3.56 (dd, 1H, J = 15.1, 2.6 Hz), 3.45 (dd, 1H, J = 15.3, 9.2 Hz), 1.30 (d, 3H, J = 6.9 Hz). ^{13}C NMR ($CDCl_3$, 100 MHz) δ 168.6, 158.6, 150.1, 138.9, 137.9, 134.3, 132.7, 132.4, 127.2, 123.7, 123.4, 121.3, 117.7, 69.2, 58.1, 56.2, 49.4, 18.1.

1,3,4,5-Tetradeoxy-3-(1,3-dioxo-1,3-dihydro-2H-isoindol-2-yl)-1-{[(1R)-1-methyl-2-propen-1-yl]amino}-2-O-2-pyridinyl-Lthreo-pent-4-enitol (4). Diene **3** (1.0 g, 2.27 mmol) and Cs_2CO_3 (0.775 g, 2.38 mmol) were heated in acetonitrile (5 mL) to 80 °C and stirred overnight. Upon cooling, the slurry was filtered through celite to remove solid residues, and the solvent removed under vacuum. The product was purified by flash column chromatography: 0.60 g, 70% yield. ES-MS: m/z: 378 (M + H^{+}). ^{1}H NMR ($CDCl_3$, 300 MHz) δ 7.88 (ddd, 1H, J = 5.0, 2.1, 0.8 Hz), 7.70 (m, 2H), 7.63 (m, 2H), 7.33 (ddd, 1H, J = 9.1, 7.1, 2.0 Hz), 6.61 (ddd, 1H, J = 7.2, 5.0, 0.9 Hz), 6.54 (dm, 2H, J = 9.2 Hz), 6.45 (m, 1H), 6.05 (m, 1H), 5.67 (m, 1H), 5.48

(dm, 1H, $J = 17$ Hz), 5.32 (dd, 1H, $J = 9.8$, 1.4 Hz), 5.16 (t, 1H, $J = 9.4$ Hz), 5.01 (m, 2H), 3.13 (m, 2H), 2.81 (dd, 1H, $J = 13$, 6.2 Hz), 1.13 (d, 3H, $J = 6.5$ Hz). ^{13}C NMR ($CDCl_3$, 100 MHz) δ 167.83, 163.11, 146.53, 142.64, 138.44, 133.53, 131.83, 131.69, 122.83, 120.73, 116.71, 114.10, 111.01, 72.99, 56.47, 56.41, 47.87, 21.32.

The Mechanism

As the reaction proceeded, it is likely that the concentration of sulfonamide **1** gradually decreased. Concurrently, the deprotonation of the hydroxyl group in product **3** by BTPP emerged as a competing side reaction. The resulting alkoxide could then attack the pyridine ring, triggering an irreversible loss of SO_2 and migration of the pyridyl group to the hydroxyl, ultimately forming impurity **4**. This proposed mechanism is corroborated by experimental evidence, where treating **3** with Cs_2CO_3 in acetonitrile resulted in the efficient conversion of **3** to **4** in high yield.

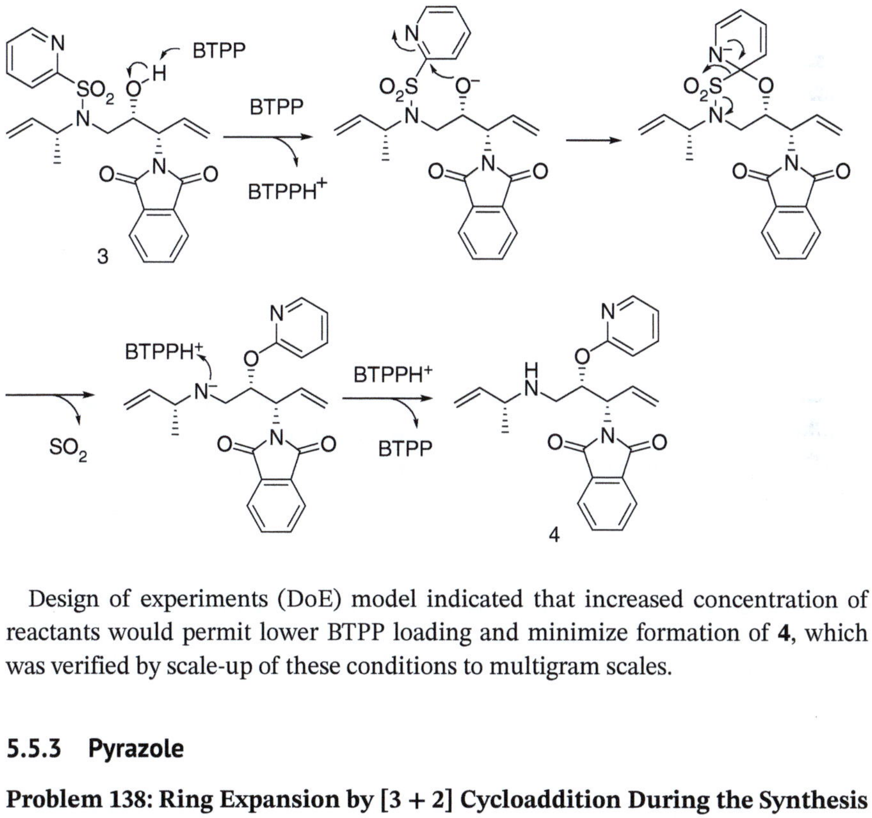

Design of experiments (DoE) model indicated that increased concentration of reactants would permit lower BTPP loading and minimize formation of **4**, which was verified by scale-up of these conditions to multigram scales.

5.5.3 Pyrazole

Problem 138: Ring Expansion by [3 + 2] Cycloaddition During the Synthesis of Pyrazoles

The pyrazole motif is found in many marketed drugs, such as pazopanib **1** and crizotinib **2**. In 2004, Jiang and Li reported a synthesis of pyrazoles by [3 + 2] cycloaddition of α-diazocarbonyl compounds with alkynes catalyzed by indium trichloride. [33]. Later on, Legros' group discovered that this reaction can be carried out by just heating the mixture of α-diazocarbonyl compounds and alkynes in neat [34]. Both articles reported that in the cases of diazo-indanone **3a** and diazo-tetralone **3b** as substrates, ring expansion products **5a** and **5b** were obtained in excellent yield [33].

Please provide a mechanism for the ring expansion reaction.

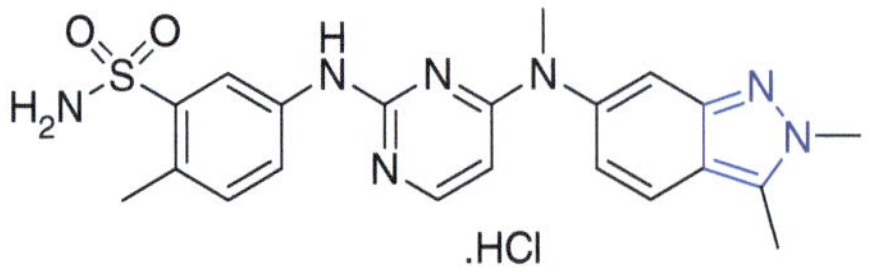

1, pazopanib

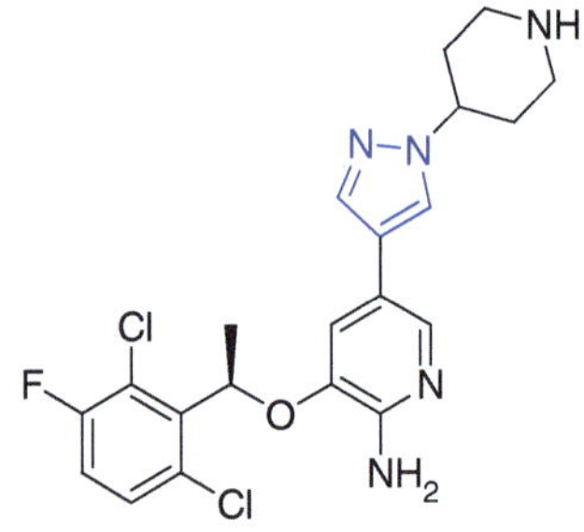

2, crizotinib

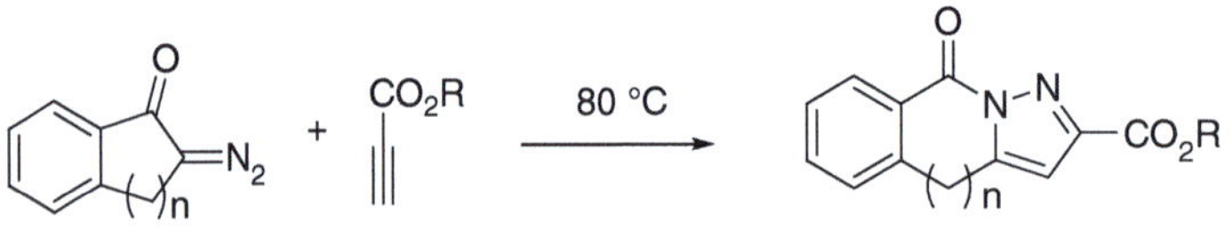

3a, n = 1
3b, n = 2

4

80 °C

5a, n = 1, R = Et, 90%
5b, n = 2, R = Et, 88%

Experimental Procedure (from *Green Chem.*, **2009**, *11*, 156, Supplementary Information)

General Procedure for the 1,3-dipolar Cycloaddition of Diazo Compounds 3a-f to Alkynes: A 5 mL round bottom flask equipped with a reflux condenser was charged with the diazo compound and the alkyne (1 mmol of the heaviest reagent/ 1.1 mmol of the most volatile one) and was heated at 80 °C. After completion of the reaction (see Table 2 in the article), the excess of reagent was evaporated under vacuum to afford the corresponding pyrazole as pure solid.

Ethyl 4,9-dihydro-9-oxopyrazolo[1,5-b]isoquinoline-2-carboxylate (5a, R=Et): 90% yield, yellow crystals; mp 171 °C. ^{1}HNMR ($CDCl_3$, 300 MHz), δ: 1.41 (t, J = 7.1, 3H); 4.43 (q, J = 7.1, 2H); 4.47 (s, 2H); 6.88 (s, 1H); 7.46 (d, J = 7.9, 1H); 7.52 (t, J = 7.9, 1H); 7.68 (t, J = 7.9, 1H); 8.43 (d, J = 7.9, 1H). ^{13}CNMR ($CDCl_3$, 75 MHz), δ: 14.2; 27.1; 61.6; 107.9; 125.4; 127.9; 128.4; 130.0; 134.3; 136.6; 142.7; 148.3; 157.6; 161.9. APCI m/z (rel.int.): 257 $[M + H]^+$ (100%). IR (ν, cm^{-1}): 1719.

Ethyl 10-oxo-5,10-dihydro-4H-pyrazolo[1,5-b][2]**benzazepine-2-carboxylate (5b, R=Et)**: 88% yield, red solid; mp: 115–116 °C. ^{1}HNMR (CDC13, 300 MHz), δ: 1.41 (t, J = 7.2, 3H); 3.15 (t, J = 5.2, 2H); 3.19 (t, J = 5.2, 2H); 4.42 (q, J = 7.2, 2H); 6.71 (s, 1H); 7.28 (d, J = 7.5, 1H); 7.43 (t, J = 7.5, 1H); 7.54 (t, J = 7.5, 1H); 8.25 (d, J = 7.5, 1H). ^{13}CNMR ($CDCl_3$, 75 MHz), δ: 14.0; 26.8; 32.6; 61.3; 109.6; 127.3; 129.1; 130.1; 133.8; 133.9; 140.7; 146.6; 147.0; 161.5; 163.5. APCI m/z (rel.int.): 271 $[M + H]^+$ (100%). IR (ν, cm^{-1}): 1705.

The Mechanism

Legros' group proposed a putative mechanism for the formation of the ring expanded product [34]. The 1,3-dipolar cycloaddition of **3a** with alkyne **4** affords the spiro pyrazole adduct **A**. Upon heating, the nitrogen in pyrazole attacks the carbonyl to form the tetracyclic species **B**, which rearranges to ring expanded product **5a**.

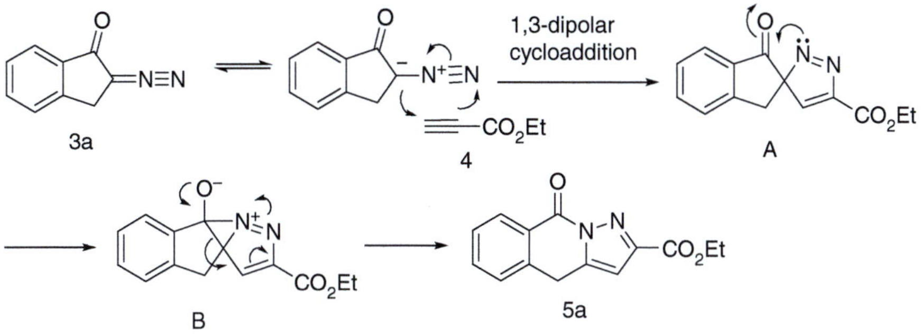

Alternatively, species **B** seems much tensioned, instable, and not easy to be formed. Jiang and Li discovered that reaction of methyl α-diazoarylacetates **7** and methyl propiolate **4** (R = Me) gave two pyrazole products **8** and **9** in excellent combined yields [33]. The possible mechanism for the formation of the minor product **9** proceeds by an initial 1,3-dipolar cycloaddition followed by a subsequent 1,3(5)-carboxylate shift of the initially formed 3,5-dicarboxylate-3-aryl-3H-pyrazole (**C**). The formation of the major product **8** could proceed by three steps: an initial 1,3-dipolar cycloaddition, a subsequent 1,5-aryl shift, and a 1,3-hydrogen shift of the 4H-pyrazole intermediate (**D**).

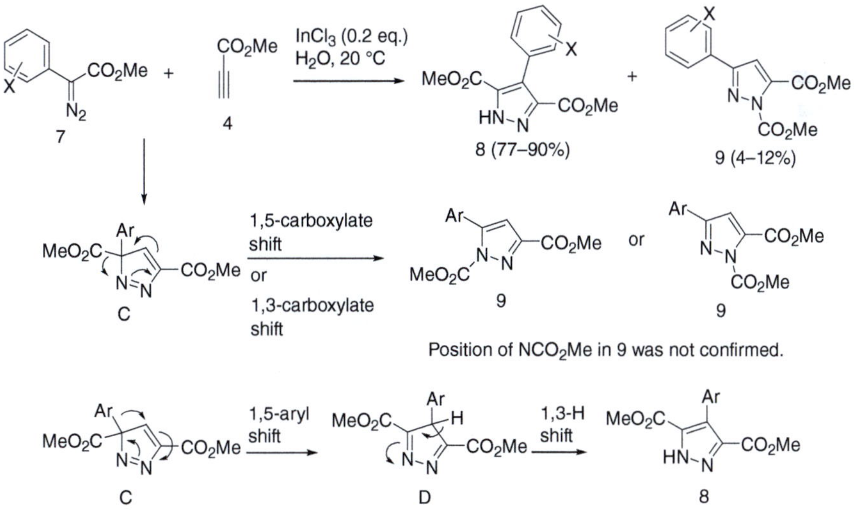

Position of NCO_2Me in 9 was not confirmed.

Based on above mechanism, a simple 1,2-carbonyl shift of the initially formed spiro pyrazole (**A**) seems a more reasonable mechanism for the formation of **5a**.

Judging by the ^1H and ^{13}C NMR data, can we exclude the possibility that the structure of the product is **5a′** as shown below?

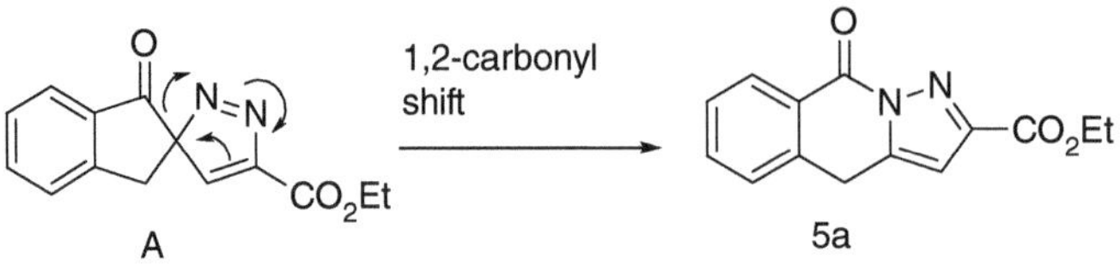

A ^{13}C DEPT NMR would address this uncertainty, since there are different numbers of CHs and quaternary carbons in **5a** and **5a′**.

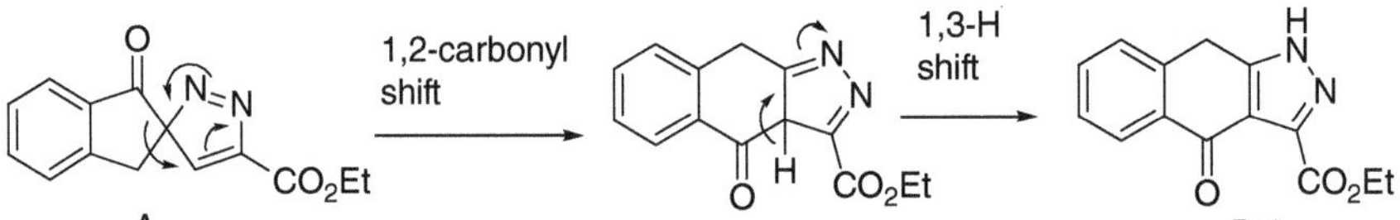

5.5.4 Oxazole

Problem 139: Formation of 2-(2,6-difluorophenyl)-4-(*p*-tolyl)oxazole as an impurity

Etoxazole (**1**) is a narrow spectrum systemic acaricide used to combat spider mites. *o*-Trifluoromethyl benzaldehyde oxime (FET-II-L, **2**) and mercaptobenzimidazole NK-12 (**3**) are more active than etoxazole (**1**) against mites in both greenhouse and field tests [35, 36].

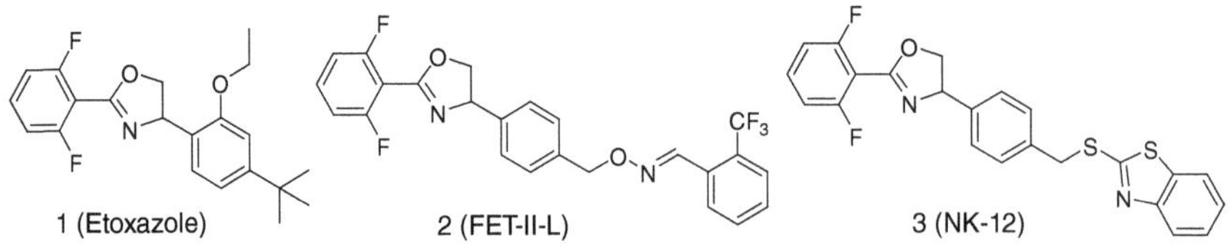

1 (Etoxazole) 2 (FET-II-L) 3 (NK-12)

In a process study for the synthesis of **2** and **3**, the key intermediate **6** was prepared from 2,6-difluorobenzonitrile (**4**) and 1-(chloromethyl)-4-vinylbenzene (**5**) as shown below. Except the desired product **6** was obtained in 86% yield, a side product **7** (6–8%) was also generated. To complete the synthesis of the target molecule **2**, **6** was treated with oxime **8** in the presence of sodium hydride, resulting in low yields of **2**. The main side product under these strongly alkaline conditions was **9**. By employing sodium hydroxide in methanol, **2** was obtained in 82% yield [37].

Please provide mechanisms for the formation of **6**, **7**, and **9**.

Experimental Procedure (from *Org. Process Res. Dev.*, **2020**, *24*, 216)

Preparation of N-(2-Bromo-1-(4-(chloromethyl)-phenyl)ethyl)-2,6-difluorobenzamide (6) via Ritter Reaction. To a solution of 2,6-difluorobenzonitrile (**4**, 15.3 g, 110 mmol), 1-(chloromethyl)-4-vinylbenzene (**5**, 15.2 g, 100 mmol), and NBS (19.6 g, 110 mmol) in 125 mL of CH_2Cl_2 was added H_2SO_4 (6.5 mL, 120 mmol) dropwise at 0 °C, and then the reaction mixture was stirred at 0 °C for 8 h until the reaction was complete, indicated by TLC (PE/EA = 4/1). The reaction was quenched by adding ice water (100 mL), and the aqueous phase was extracted with dichloromethane (50 mL × 3). The combined organic phase was washed with saturated Na_2CO_3 (100 mL) and brine (100 mL), dried over anhydrous sodium sulfate (about 3 g), and filtered. After removal of the organic solvent, the residue was purified by column chromatography on silica gel with PE/EA (5/1) as eluent to give **6** (33.4 g, 86 mmol, yield 86%) as a yellow solid. Mp 128–129 °C. ^{1}H NMR (400 MHz, DMSO-d_6) δ 9.43 (d, J = 8.4 Hz, 1H), 7.57–7.50 (m, 1H), 7.40 (d, J = 8.0 Hz, 2H), 7.31 (d, J = 8.0 Hz, 2H), 7.18 (t, J = 8.0 Hz, 2H), 5.29 (dd, J = 8.8, 4.4 Hz, 1H), 3.80 (dd, J = 10.4, 4.8 Hz, 1H), 3.63 (t, J = 10.0 Hz, 1H). ^{13}C NMR (100 MHz, DMSO-d_6) δ 159.9, 159.3 (dd, J = 248.9, 8.0 Hz), 140.4, 138.1, 132.2 (t, J = 9.9 Hz), 129.9, 127.7, 115.6 (t, J = 23.0 Hz), 112.4 (dd, J = 19.3 and 5.2 Hz), 55.0, 36.1, 34.5. HRMS (ESI): calcd for $C_{16}H_{14}BrClF_2N_2O$ $[M + H]^+$ 389.9889, found 389.9889.

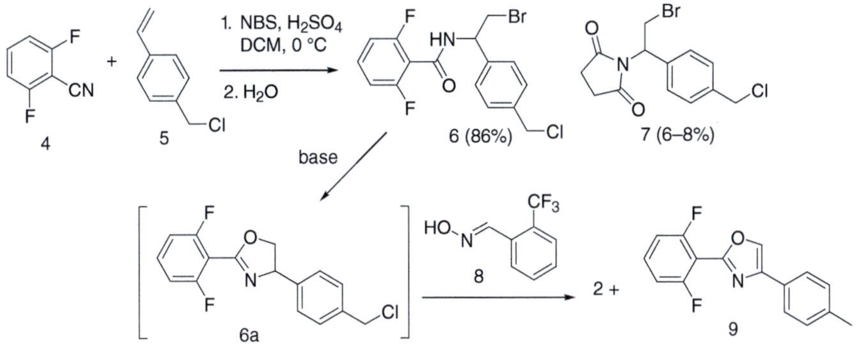

Formation of 1-(2-Bromo-1-(4-(chloromethyl)-phenyl)ethyl)pyrrolidine-2,5-dione (7). A small amount of **7** could also be separated from the above reaction mixture. ^{1}H NMR (400 MHz, DMSOd_6) δ 7.51–7.36 (m, 4H), 5.35 (dd, J = 10.4 and 5.6 Hz, 1H), 4.75 (s, 2H), 4.42 (t, J = 10.4 Hz, 1H), 4.19 (dd, J = 10.0, 5.6 Hz, 1H), 2.73 (s, 4H). ^{13}C NMR (100 MHz, DMSO-d_6) δ 178.1, 138.2, 137.2, 129.6, 128.4, 56.0, 44.9, 31.9, 28.3. HRMS (ESI): calcd for $C_{13}H_{14}BrClNO_2$ $[M + H]^+$ 329.9891, found 329.9892.

Preparation of Target Compound 2 (FET-II-L) via Stepwise Procedure. A solution of compound **6** (19.4 g, 0.05 mol, 1.0 equiv) and NaOH (2.4 g, 0.06 mol, 1.2 equiv) in methanol (100 mL) was stirred at room temperature for 12 h until full consumption of **6** indicated by TLC (PE/EA = 4/1). Compound **8** (9.6 g, 0.051 mol, 1.02 equiv) and another 1.2 equiv of NaOH (2.4 g, 0.06 mol) were added, and the mixture was stirred for another 48 h. After evaporation of the solvent, water (50 mL) and ethyl acetate (50 mL) were added, and the water layer was extracted with ethyl acetate (50 mL × 3). The combined organic layer was washed with 10% $NaHCO_3$(50 mL), water (50 mL), and brine (50 mL), then dried over anhydrous sodium sulfate (about 3 g), filtrated, and concentrated to give the title compound **2** (21.8 g, 94% crude yield, 87% HPLC purity) as a yellow oil. After recrystallization with petroleum ether and ethyl acetate (100 mL/10 mL), 19.2 g of product (83% yield) were obtained with 96% HPLC purity (quantitative method). White solid. Mp 52–53 °C. ^{1}H NMR (400 MHz, $CDCl_3$) δ 8.50 (s, 1H), 8.04 (d, J = 7.8 Hz, 1H), 7.66 (d, J = 7.7 Hz, 1H), 7.53 (t, J = 7.5 Hz, 1H), 7.49–7.41 (m, 4H), 7.37 (d, J = 8.1 Hz, 2H),

7.00 (t, $J = 8.3$ Hz, 2H), 5.50 (dd, $J = 10.2$ and 8.1 Hz, 1H), 5.24 (s, 2H), 4.84 (dd, $J = 10.3$, 8.5 Hz, 1H), 4.31 (t, $J = 8.2$ Hz, 1H). ^{13}C NMR (100 MHz, $CDCl_3$) δ 161.8 (dd, $J = 254$ Hz, 5 Hz), 158.2 (s), 146.3 (s), 142.3 (s), 137.3 (s), 133.0 (t, $J = 10.0$ Hz), 132.5 (s), 130.9 (s), 130.0 (s), 129.5 (s), 129.0 (s), 128.8 (s), 128.6 (s), 128.3 (q, $J = 31.1$ Hz), 127.9 (s), 127.4 (s), 126.4 (q, $J = 272$ Hz), 112.5 (dt, $J = 20.0$ and 2.0 Hz), 108.1 (t, $J = 18.0$ Hz), 77.0 (s), 75.4 (s), 70.6 (s). HRMS (ESI): calcd for $C_{24}H_{19}F_5N_2O_2$ $[M + H]^+$ 461.1283, found 461.1280.

Formation of 2-(2,6-Difluorophenyl)-4-(*p*-tolyl)-oxazole (9). A small amount of **9** could be separated from the reaction mixture. White solid. Mp: 97–98 °C. ^{1}H NMR (400 MHz, $CDCl_3$) δ 8.01 (s, 1H), 7.70 (d, $J = 8.0$ Hz, 2H), 7.44–7.30 (m, 1H), 7.21 (d, $J = 8.0$ Hz, 2H), 7.01 (t, $J = 8.4$ Hz, 2H), 2.36 (s, 3H). ^{13}C NMR (100 MHz, $CDCl_3$) δ 160.9 (dd, $J = 257.2$ and 5.6 Hz), 153.2, 142.2, 138.2, 134.0, 132.0 (t, $J = 10.4$ Hz), 129.5, 127.9, 125.7, 112.2 (dd, $J = 20.1$ and 5.1 Hz), 106.7 (t, $J = 16.3$ Hz), 21.3. HRMS (ESI) calcd for $C_{16}H_{11}F_2NO$ $[M + H]^+$ 272.0881, found 272.0868.

The Mechanism

The preparation of **6** from **4** and **5** is a Ritter reaction in which the cation is generated from 4-chloromethylstyrene (**5**) by NBS bromination. First, reaction of **5** with NBS gives bromonium ion **A**, which is readily transformed to benzyl cation **B**. The benzyl site of **B** is attacked by the nitrogen lone-pair electrons of difluorobenzonitrile (**4**) to form **C**, and subsequent hydrolysis gives **6**.

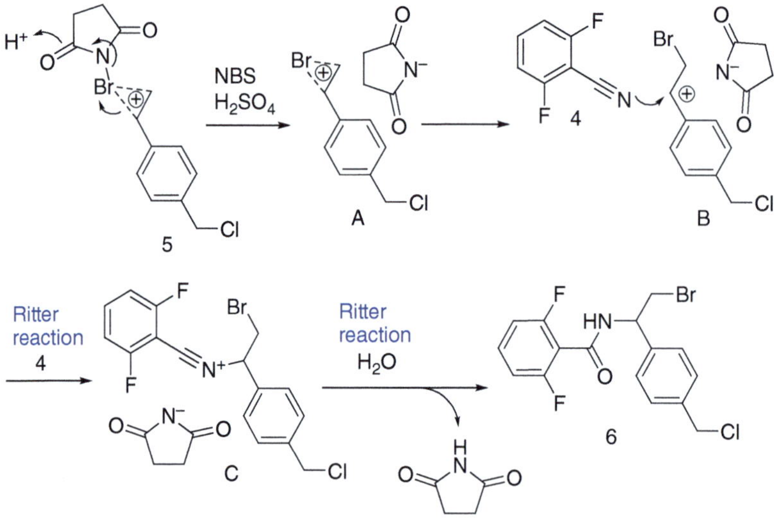

The succinimide anion that forms from NBS attacks benzyl cation **B** to give **7** as a side product.

Upon being treated with base, compound **6** will first cyclize to form an intermediate **D**, then the oxime **8** attacks the benzyl chloride to produce the product **2**. However, since the proton at the benzylic position of **D** is relatively acidic, under strong base conditions, such as NaH, it is easily deprotonated to give anion **E**.

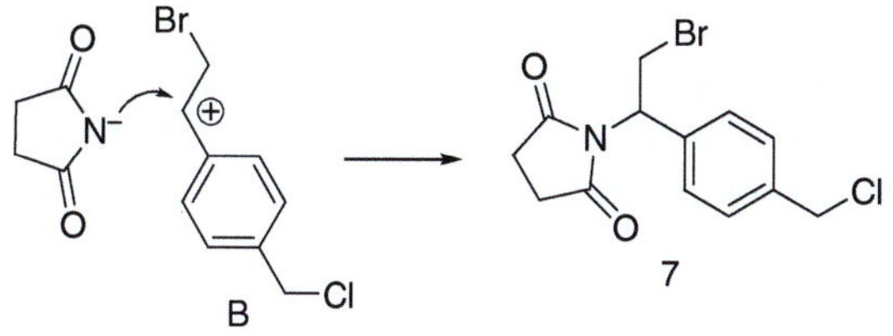

Subsequent elimination of Cl through the benzene ring leads to the formation of **F**, which then rearranges to the more stable, undesired product **9** via a [1,7]-H shift. Weaker (NaOH is weaker than NaH in basicity) base would suppress this side reaction.

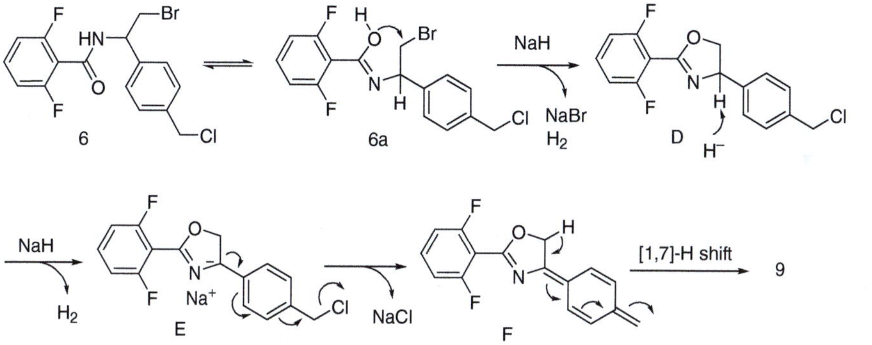

Problem 140: Formation of an Oxazoline Impurity in the Synthesis of a M1 Agonist

M1-type muscarinic acetylcholine receptors play a role in cognitive processing. The muscarinic receptor (M1) agonist is a potential treatment for AD. The aminoindanol-derived compound **1** is a clinical candidate for this purpose. In the last step of the synthesis of **1**, treatment of aniline **2** with thioimidate **3** in THF in the presence of K_2CO_3 and 4-(dimethylamino)pyridine (DMAP) provided the desired product **1**. An oxazoline impurity **4** was also identified [38].

Please propose a mechanism for the formation of **4**.

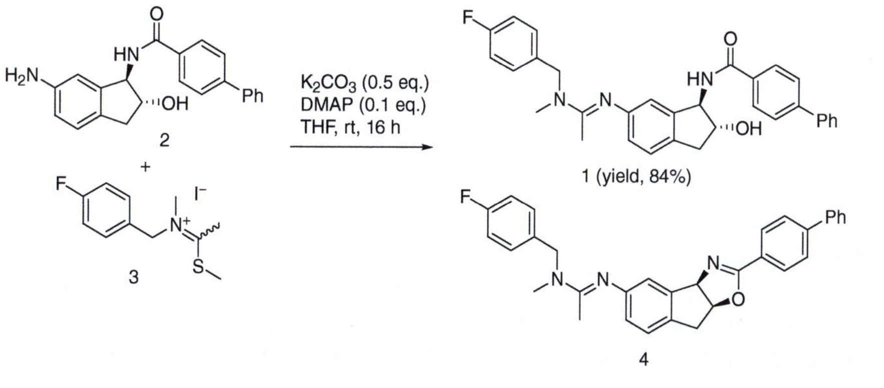

Experimental Procedure (from *Org. Process Res. Dev.*, **2009**, *13*, 198*)
***N*-[(1R,2R)-6-[[1-[[(4-Fluorophenyl)methyl]methylamino]-ethylidene]amino]-2,3-dihydro-2-hydroxy-1H-inden-1-yl]-[1,1′-biphenyl]-4-carboxamide, Hemihydrate (1). Kilolab procedure**. To a slurry of aniline **2** (0.969 kg, 2.8 mol) in THF (9.7 L) at room temperature was added thioimidate **3** (0.954 kg, 2.8 mol) and DMAP (34.5 g, 0.28 mol). When HPLC indicated that thioimidate **3** was <2 area % (20–24 h), the mixture was transferred to a rotary evaporator, and the solvent was removed *in vacuo*. The resulting foam was dissolved in CH_2Cl_2 (12.5 L), and the organic phase was washed with 1.0 N HCl (aq) (1 × 4 and 1 × 3 L), 1.0 M NaOH (aq) (1 × 2.44 L), and sat. NaCl (aq) (1 × 4 L). The organic phase was separated, dried (Na_2SO_4, 2.5 kg), and filtered, and the solvent was removed *in vacuo* to yield crude amidine **1** as a solid (1.8 kg, 93.5 area% by HPLC). The solid was dissolved in acetonitrile (9 L) while heating to 35–40 °C. After 0.5 h, seed crystals were added, and a thick, white slurry formed. The mixture was cooled to −10 to −15 °C in an ice/acetone bath and stirred for 1–2 h. The slurry was filtered, and the resulting solid was rinsed with cold acetonitrile (2 × 0.5 L) and dried *in vacuo* to provide amidine **1** as a partial CH_3CN solvate (1.10 kg, 77%, 98.6 area% by HPLC). Three CH_3CN solvate lots (2.86 kg) were combined and dissolved in 21.8 L of methanol. The solution was passed through a 0.45 μm carbon impregnated filter, and the tank and filter were rinsed with 24 L of methanol. To the solution was added 5.7 kg of H_2O over 35 min followed by 15 g of seed crystals. After 20 min, 1.15 kg of H_2O was added followed by 15 g of seed crystals. After 1 h, 1.15 kg of H_2O was added over 30 min followed by 15 g of seed crystals. After 10 min, 3.4 kg of H_2O was added over 1 h, and the slurry was stirred at room temperature for 1 h and at 0 °C for 45 min. The solid was collected by filtration, rinsed with a cold solution of 11.4 L of methanol and 2.9 L of H_2O, and dried under vacuum at 50 °C for 24 h. The solid was equilibrated in air for 4 h to afford amidine **1** as a white solid (hemihydrate), 2.19 kg, 76% recovery (typical laboratory recovery is 87%, but the wash of the carbon-impregnated filter was inadequate on scale), 99.3 area% by HPLC, 99.5 wt/wt% assay. $[\alpha]_D$ + 105.2 (MeOH, c = 1.0). IR ($CHCl_3$, cm^{-1}): 3489, 1643, 1606, 1600, 1509, 1482. ^{1}H NMR ($CDCl_3$): δ 7.90 (d, 2, J = 8.6), 7.69 (d, 2, J = 8.6), 7.63 (d, 2, J = 8.2), 7.48 (t, 2, J = 8.2, 7,6), 7.41 (d, 1, J = 7.3), 7.24 (dd, 2, J = 8.5 and 5.2), 7.14 (d, 1, J = 7.9), 7.04, (t, 2, J = 8.7), 6.72–6.63 (m, 3), 5.31 (t, 1, J = 5.6), 4.84 (br s, 1), 4.64 (dd, 2, J = 21.4, 15.6), 4.54 (dd, 1, J = 14.0, 7.9), 3.32 (dd, 1, J = 15.6, 7.9), 3.01 (s, 3), 2.95 (dd, 1, J = 15.7, 8.0), 1.97 (s, 3). ^{13}C NMR ($CDCl_3$): δ 15.3, 36.2, 38.3, 52.6, 65.2, 82.7, 115.7, 115.8, 116.7, 123.2, 125.9, 127.5, 127.6, 127.9, 128.4, 128.9, 129.2, 132.2, 133.8, 134.2, 139.7, 140.0, 145.1, 151.8, 157.6, 163.2, 169.7. MS (m/z): 508.2 (M + 1). EA calcd for $C_{32}H_{30}N_3O_2 \cdot ½ H_2O$: C, 74.40; H, 6.05; N, 8.13. Found: C, 74.50; H, 5.93; N, 8.17.

***The mechanisms for the formation of the impurity **4** and other impurities were discussed in the article. However, the isolation and identification of **4** were not disclosed.*

The Mechanism

Excess thioimidate **3** could be attacked by the alcohol of the desired product **1** to form the onium salt **A**, in which the masked hydroxyl is activated as a leaving group for ring closure, intramolecular S_N2, to oxazoline **4** with the stereochemistry inversion.

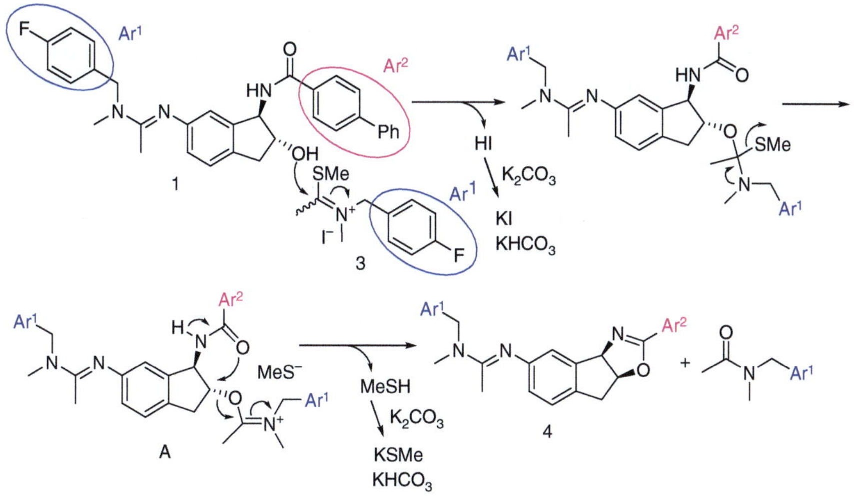

This side product was suppressed by fixing the **2**:**3** stoichiometry at 1:1.

5.5.5 Triazole

Problem 141: Formation of Dimethylcarbamoyl-OBt in a HBTU-promoted Amide Preparation

In a synthesis of cyclic Peptide Q at Ferring Pharmaceuticals, the last manufacturing upstream processing step was an amide formation by employing HBTU in DMF. A non-peptide-related impurity with an abundance of 0.09% was detected in the final API of Peptide Q [39]. For quality control and analytical purpose, determining its structure and understanding its formation mechanism are crucial. The structure of the impurity was identified as 1-[((dimethylamino)carbonyl]oxy)-1H-benzotriazole (**1**) or its isomer 1-(dimethylcarbamoyl)-1H-1,2,3-benzotriazol-3-ium-3-olate (**2**), both were abbreviated as dimethylcarbamoyl-OBt.

Please propose a mechanism for the formation of **1** or **2**.

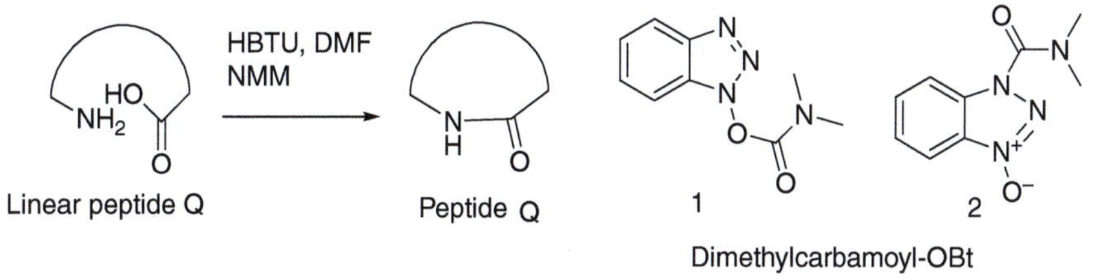

Linear peptide Q — HBTU, DMF, NMM → Peptide Q

1, 2 Dimethylcarbamoyl-OBt

Experimental Procedure (from *Org. Process Res. Dev.*, **2021**, *25*, 1923, Supporting Information*)

The Fate of HBTU Under Peptide Q Cyclization-like Conditions. 0.41 g (1.09 mmol) HBTU was dissolved in 3 mL DMF, and the derived HBTU/DMF solution was loaded in Syringe A. 5 mL DMF was loaded in Syringe B. In Syringe C, 0.88 mL DIPEA was loaded. The contents in Syringe A, B, and C were simultaneously slowly added at ambient temperature through a syringe pump to a reactor containing 7 mL DMF within 120 min. The solution in the reactor was stirred at ambient temperature, and its pH value was maintained above 8 by the addition of DIPEA, if necessary. After the addition, the solution in the reactor was stirred further at ambient temperature for 10 min before being poured to 135 mL 1% AcOH aqueous solution at 5 °C. The derived solution was stored overnight at 2–8 °C and filtered through a 2.4 μm filter. The content of the solution was analyzed by LC/MS.

LC/MS Analysis of the Dimethylcarbamoyl-OBt Impurity. The comparison of HBTU degradant and reference material 1-[((dimethylamino)carbonyl)oxy]-1Hbenzotriazole (dimethylcarbamoyl-OBt) by LCMS analysis is displayed in Figure S1. The results indicate a strikingly high similarity between the subject degradant from HBTU/DIPEA/DMF simulation reaction and the dimethylcarbamoyl-OBt reference. Both the retention time and the MS fragmentation strongly corroborate that HBTU was partially degraded into dimethylcarbamoyl-OBt adduct in DIEPA/DMF solution.

The subject non-peptide-related impurity in Peptide Q was isolated and subjected to NMR analysis by 1D ^{1}H - and ^{13}C NMR, COSY, NOESY, ^{13}C-HSQC, ^{13}C-HMBC, and ^{15}N-HMBC. The 1D ^{1}H NMR spectrum of the impurity identified four aromatic signals and two methyl signals as part of the impurity. ^{1}H NMR (600 MHz, DMSO-d_6) δ 8.10 (dd, J = 8.4 and 1.0 Hz, 1H), 7.79 (dd, J = 8.3 and 1.0 Hz, 1H), 7.63 (ddd, J = 8.2, 7.0, and 0.9 Hz, 1H), 7.48 (ddd, J = 8.1, 6.9, and 1.0 Hz, 1H), 3.21 (s, 3H), 3.00 (s, 3H). ^{13}C NMR (150.8 MHz, DMSO-d_6) δ 152.13, 143.09, 129.36, 128.89, 125.55, 120.20, 109.76, 38.08, 36.53.

The Mechanism

The formation of dimethylcarbamoyl-OBt (**1** or **2**) is presumably initiated by DMF as a nucleophile-attacking HBTU. One of the dimethylamino group in the formed species **A** migrates to the formimidate carbon to give the cation species **B**, which is possibly resonance with tautomers **C** and **D**. Water is added to those species to provide adduct **E;** simultaneously HPF_6 is released and trapped by NMM. Finally, elimination of DMF and dimethylamine from **E** leads to the impurity **2**.

Similar to the above mechanism, compound **1** could be generated by the following mechanism.

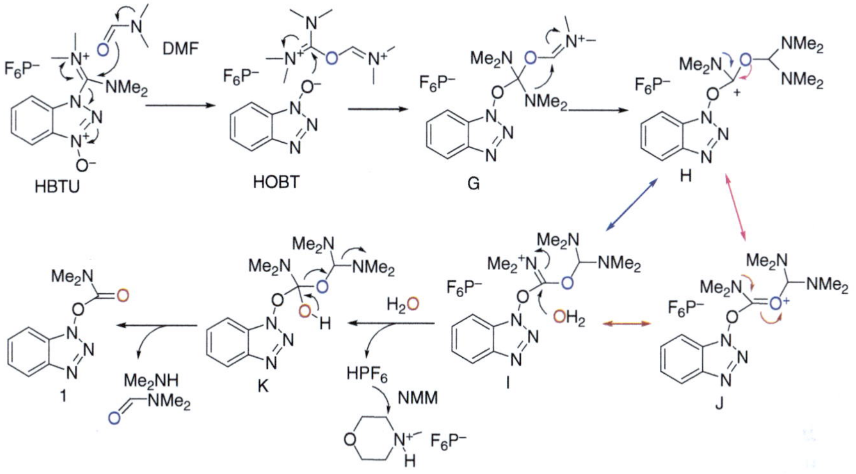

Compound **1** and compound **2** should be interconvertible via intermolecular carbamoyl transformation between N and O atoms.

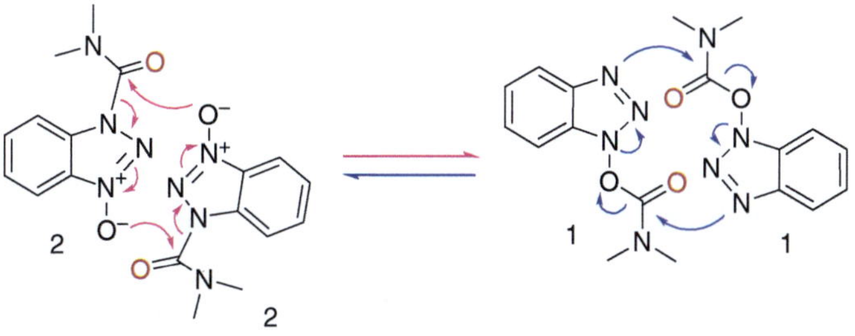

5.5.6 Tetrazole

Problem 142: Formation of a Tetrazole Impurity Caused by Acetone
Olmesartan medoxomil (OLM, **2**) is an antihypertensive prodrug used in the treatment and management of hypertension either alone or in combination with other drugs. In the process of deprotection of **1** by implementing the acetone-HCl system, an impurity **3** of 0.23 area% was identified [40]. It was proved challenging to purge out this impurity during the final recrystallization. The generation of **3** was suppressed (<0.1%) using statistical methods (DoE) via a definitive screening design.

Please propose a mechanism for the formation **3**.

Experimental Procedure (from *Org. Process Res. Dev.* **2021**, *25*, 1112, the spectra data were reported in the Supporting Information)

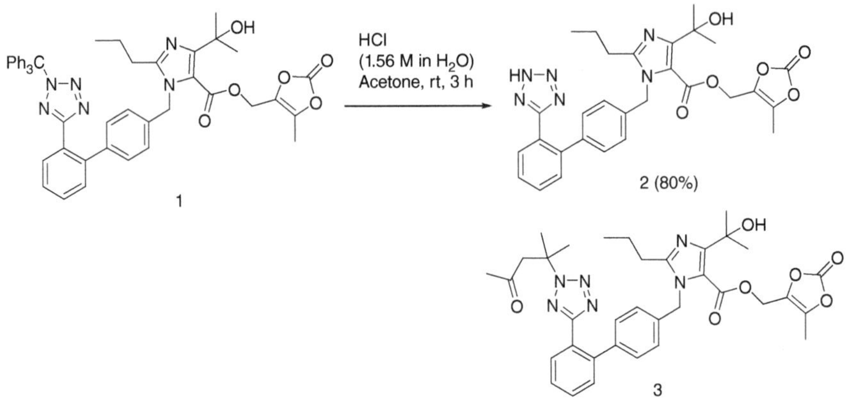

(5-Methyl-2-oxo-1,3-dioxolen-4-yl)methyl 4-(1-hydroxy-1-methylethyl)-2-propyl-1-[2′-(1H-tetrazol-5-yl)biphenyl-4-yl]methyl-1H-imidazole-5-carboxylate (2). A three-neck flask was charged with trityl-protected ester 1 (8.00 g, 9.92 mmol, 1.0 equiv), suspended in acetone (16 mL). Then while stirring, 1.56 M hydrochloric acid (16 mL) was added to the suspension. After being stirred at ambient temperature for 3 h, the complete conversion was indicated by TLC analysis (50% AcOEt/petroleum ether). The reaction mixture was diluted with H_2O (32 mL), cooled to 0 °C in an ice bath for another 0.5 h. Triphenylmethanol was filtered and washed with H_2O (2 × 5 mL). The combined filtrate was diluted with CH_2Cl_2 (20 mL), followed by careful neutralization with aqueous Na_2CO_3 solution (0.2 g/mL) to adjust the pH to 5.5. The organic layer was separated, and the aqueous layer was extracted with CH_2Cl_2 (3 × 10 mL). The combined organic layer was washed with H_2O (30 mL), dried over Na_2SO_4, and condensed *in vacuo* to afford an off-white solid. Recrystallization from acetone afforded olmesartan medoxomil as a pure white crystalline powder (4.38 g, 80.1% yield).

(5-Methyl-2-oxo-1,3-dioxol-4-yl)methyl 4-(2-hydroxypropan-2-yl)-1-((2′-(2-(2-methyl-4-oxopentan-2-yl)-2H-tetrazol-5-yl)-[1,1′-biphenyl]-4-yl)methyl)-2-propyl-1H-imidazole-5-carboxylate (3). To a three-neck flask, olmesartan medoxomil (**2**, 1.13 g, 2.02 mmol, 1.0 equiv) and mesityl oxide (2.98 g, 30.31 mmol, 15 equiv) were added. $BF_3{\cdot}Et_2O$ (0.32 g, 2.23 mmol, 1.1 equiv) was added dropwise to the reaction mixture. After being stirred at 50 °C for 5 h, the complete conversion was indicated by TLC analysis (50% AcOEt/petroleum ether). While stirring, H_2O (10 mL) was added dropwise to the reaction mixture, followed by extraction with ethyl acetate (3 × 10 mL). The organic layer was combined, washed with brine (10 mL), dried over Na_2SO_4, and condensed *in vacuo* to afford a yellow oil. The crude product was purified using short column chromatography (50% AcOEt/petroleum ether) to eliminate the mesityl oxide. The product was obtained as a light yellow oil (1.10 g, 82.9% yield). ^{1}HNMR (400 MHz, DMSO-d_6) δ 7.76 (dd, J = 7.6 and 1.5 Hz, 1H), 7.61 (td, J = 7.6 and 1.5 Hz, 1H), 7.53 (td, J = 7.5 and 1.4 Hz, 1H), 7.47 (dd, J = 7.6 and 1.4 Hz, 1H), 7.02 (d, J = 8.2 Hz, 2H), 6.85 (d, J = 8.0 Hz, 2H), 5.45–5.40 (s, 2H), 5.22 (s, 1H), 5.08 (s, 2H), 3.18 (s, 2H), 2.60 (t, J = 7.6 Hz, 2H), 2.08 (s, 3H), 1.95 (s, 3H), 1.63 (h, J = 7.4 Hz, 1H), 1.53 (s, 6H), 1.47 (s, 6H), 0.90 (t, J = 7.3 Hz, 3H). ^{13}CNMR (101 MHz, DMSO-d_6) δ 204.61, 163.55, 160.72, 157.66, 151.68, 150.95,

140.90, 140.44, 139.11, 136.01, 132.87, 130.52, 130.03, 129.14, 127.75, 126.06, 125.10, 116.14, 69.65, 63.65, 54.92, 54.17, 51.25, 48.02, 30.77, 29.67, 28.30, 26.97, 20.52, 13.69, 8.74. ^{15}N NMR (60.8 MHz, $CDCl_3$) δ 1.12, −44.72, −72.53, −74.67, −114.56, −207.50. MS m/z 657 (M + H^+).

The Mechanism

At the acidic condition, acetone dimerizes to form mesityl oxide by aldol condensation and dehydration [41]. Deprotection of the trityl in the starting material **1** yields the desired product **2**, which further reacts with the mesityl oxide via Michael-type addition to generate the impurity **3**. Indeed, treatment of **2** with mesityl oxide at the acetone-H_2O-HCl system provided **3** in 87% yield [40].

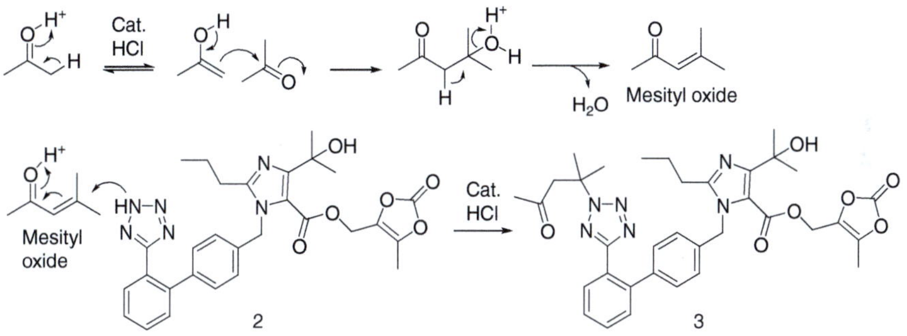

Mesityl oxide is genotoxic and could be generated from acetone in acidic environments, including acid salts. It could be further attacked by nucleophile via Michael-type addition. Notably, plenty of small-molecule drugs and intermediates have nitrogen-containing moieties or nucleophilic functional groups; therefore, more attention should be paid to impurities formed between mesityl oxide and reactive substrates.

5.5.7 Diazepine

Problem 143: Accidental Discovery of Librium

Chlordiazepoxide (**3**), trade name Librium, is a tranquilizer and approved for medical use in 1960. It was discovered accidently by Sternbach and Reeder in 1957. When they wanted to do an amination of 6-chloro-2-(chloromethyl)-4-phenylquinazoline 3-oxide (**1**) with methylamine, the expected product **2** was not obtained. Instead, the rearranged product **3** was obtained in high yield (82%) [42].

Could you please suggest a mechanism for its formation?

Experimental Procedure (from *J. Org. Chem.* **1961**, *26*, 1111)

7-Chloro-9-methylamino-5-phenyl-3H-1,4-benzodiazepine 4-oxide (3). Into 600 cc. of a cold 25% solution of methylamine in methanol was introduced with stirring 98 g of the hydrochloride of 6-chloro-2-chloromethyl-4-phenylquinazoline 3-oxide (**1**). The mixture was initially cooled to about 30° and then stirred at room temperature for 15 h. The precipitated chemically pure reaction product (50 g) was then filtered off. The mother liquor was concentrated *in vacuo* and the residue dissolved in methylene chloride. The methylene chloride solution was washed with water, dried with sodium sulfate, and concentrated *in vacuo*. The crystalline residue was triturated with a small amount of hot acetone to dissolve the more soluble impurities. The mixture was then cooled to 5° for 10 h and filtered, yielding 20.3 g of almost chemically pure material. The total yield was 70.3 g (82%). The product was recrystallized from a 15-fold volume of boiling ethanol, forming light yellow plates with a melting point of 236–236.5°. Anal. Calcd for $C_{16}H_{14}N_3OCl$: C, 64.11; H, 4.71. Found: C, 64.38; H, 4.66.

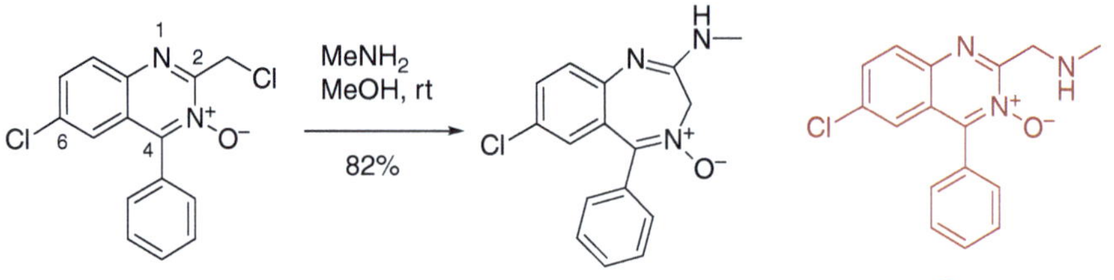

The NMR data were reported by Haran and Tuchagues [43]. ^{1}H NMR (90 MHz, DMSO-d_6) δ 8.11 (q, J = 4.4 Hz, 1H), 6.79–7.55 (m, 8H), 4.42 (brs, 2H), 2.85 (d, J = 4.7 Hz, 3H). ^{13}C NMR (62.86 MHz, DMSO-d_6) δ 152.80, 147.92, 140.54, 133.52, 130.55, 129.08, 128.81, 128.33, 127.86, 125.89, 124.45, 64.53, 27.96.

The Mechanism

Due to the electron-withdrawing property of the *N*-oxide, the C-2 is more electrophilic than the chloromethyl. Thus, methylamine would preferentially attack C-2 to give adduct **A**. Then ring opening releases imidamide oxime **B**. Recyclization of **B** along with lose HCl leads to Librium.

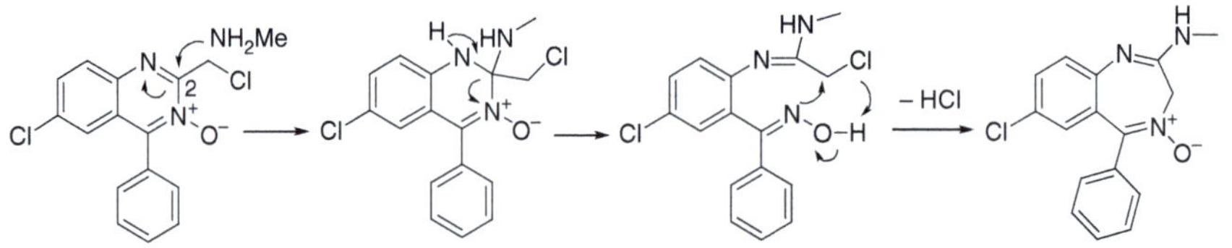

Secondary amines, such as dimethylamine and piperazine, provided only the normal products [42].

5.6 Formation of Bicyclic Aromatic Heterocycles

5.6.1 Benzofuran

Problem 144: Formation of Two Benzofuran Side Products During the Synthesis of Morin

Morin (**1**) is a natural product found in various plants of the Moraceae family. It is an antioxidant and is being considered as a promising chemo preventive anticancer agent that does not exhibit toxicity [44]. In a synthesis of morin, the oxidative cyclization of **2** by *tert*-butyl hydroperoxide (TBHP) provided desired product **3** in 39% yield. Meanwhile, two side products **4** and **5** were also obtained [45].

Please suggest mechanisms for the formation of **3**, **4**, and **5**.

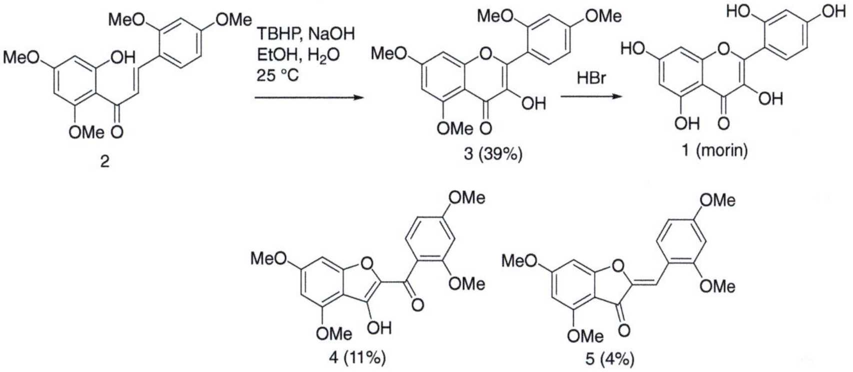

TBHP, NaOH
EtOH, H_2O
25 °C

HBr

2

3 (39%)

1 (morin)

4 (11%)

5 (4%)

Experimental Procedure (from *Org. Process Res. Dev.*, **2023**, *27*, 890)

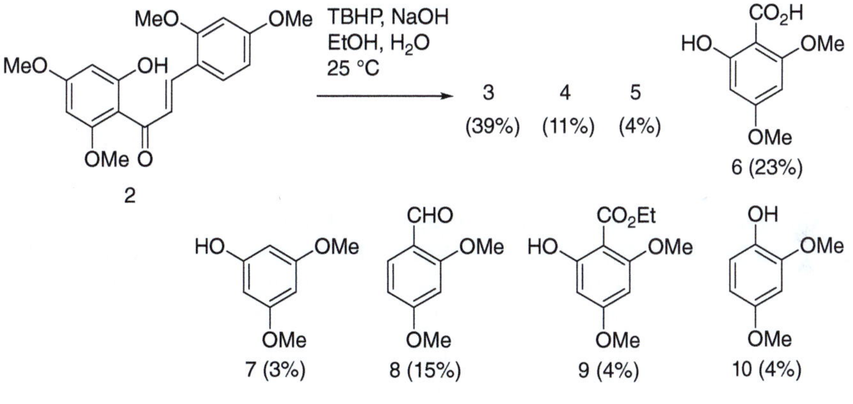

TBHP, NaOH
EtOH, H_2O
25 °C

2

3 (39%) 4 (11%) 5 (4%)

6 (23%)

7 (3%) 8 (15%) 9 (4%) 10 (4%)

2-(2,4-Dimethoxyphenyl)-3-hydroxy-5,7-dimethoxy-4H-chromen-4-one (3). In a 50 L reactor, a mixture of 3-(2,4-dimethoxyphenyl)-1-(2-hydroxy-4,6-dimethoxyphenyl)prop-2-en-1-one (**2**, 200 g, 0.581 mol, 1.00 equiv), ethanol (8 L), and NaOH (800 mL, 10.87 M, 15.0 equiv) was stirred at 24 °C for 0.5 h. Then, TBHP (70% aq., 800 mL, 5.81 mol, 10.0 equiv) was added, and the mixture was stirred at 24 °C for 23 h. The reaction was quenched with 1 M HCl and adjusted to pH 2–4. The aqueous phase was extracted with DCM (3 × 5 L). The organic phases were combined, and the solvent was removed. The remaining brown oily crude product was recrystallized from *n*-butanol and dried under vacuum at 50 °C. The dried yellow solid was added to aqueous NaOH (2 M, 1.1 L) and stirred at 24 °C overnight. The resulting suspension was filtered, and the filter cake was acidified by addition of 1 M HCl (13 L) to pH 4 and stirred for 1 h at 100 °C. The color of the suspension changed from yellow to white. The suspension was cooled to 25 °C and stirred for 12 h. After filtration of the suspension, the filter cake was dried under vacuum at 50 °C to give **3**

as a white solid (74.8 g, 0.209 mol, 36%). The product was obtained with a purity of 95% (HPLC analysis). In the synthesis on a 2 g scale, the following compounds were isolated via column chromatography: aurone **5** (4%), auronol **4** (11%), benzaldehyde **8** (15%), benzoic acid **6** (23%), benzoate **9** (4%), and phenols **7** (4%) and **10** (3%). **3**, R_f = 0.33, hexane/EtOAc (1:3). ^{1}H NMR (400 MHz, $CDCl_3$): δ 7.49 (d, J = 8.4 Hz, 1H, 6′-H), 6.76 (s, 1H, OH), 6.60 (dd, J = 8.4, 2.3 Hz, 1H, 5′-H), 6.57 (d, J = 2.3 Hz, 1H, 3′-H), 6.47 (d, J = 2.2 Hz, 1H, 8-H), 6.34 (d, J = 2.2 Hz, 1H, 6-H), 3.97 (s, 3H, 5-OCH_3), 3.86 (s, 6H, 4′-OCH_3, 7-OCH_3), 3.84 (s, 3H, 2′-OCH_3). ^{13}C NMR (101 MHz, $CDCl_3$): δ 172.09 (C-4), 164.23 (C-7), 162.80 (C-4′), 160.77 (C-5), 159.73 (8C-C-1O), 158.99 (C-2′), 142.80 (C-2), 138.74 (C-3), 132.02 (C-6′), 112.54 (C-1′), 106.94 (4C-C-5C), 104.99 (C-5′), 99.38 (C-3′), 95.68 (C-6), 92.70 (C-8), 56.51 (5-OCH_3), 56.03 (2′- OCH_3), 55.87 (7-OCH_3), 55.64 (4′-OCH_3). ^{1}H NMR (400 MHz, DMSO-d_6): δ 8.21 (s, 1H, OH), 7.36 (d, J = 8.4 Hz, 1H, 6′-H), 6.69 (d, J = 2.3 Hz, 1H, 3′-H), 6.63 (dd, J = 8.5, 2.3 Hz, 1H, 5′-H), 6.60 (d, J = 2.2 Hz, 1H, 8-H), 6.46 (d, J = 2.3 Hz, 1H, 6-H), 3.86 (s, 3H, 5-OCH_3), 3.84 (s, 3H, 7-OCH_3), 3.83 (s, 3H, 4′-OCH_3), 3.78 (s, 3H, 2′-OCH_3). ^{13}C NMR (101 MHz, DMSO-d_6): δ 171.15 (C-4), 163.47 (C-7), 162.06 (C-4′), 160.28 (C-5), 158.68 (8C-C-1O), 158.46 (C-2′), 143.11 (C-2), 138.83 (C-3), 131.88 (C-6′), 112.33 (C-1′), 106.95 (4C-C-5C), 105.05 (C-5′), 98.78 (C-3′), 95.54 (C-6), 92.68 (C-8), 56.19 (5-OCH_3), 55.90 (7-OCH_3), 55.79 (2′-OCH_3), 55.44 (4′-OCH_3). UV/vis (MeCN): λ_{max} (lg ε) = 333 (3.22), 292 (3.00), 244 (3.39), 204 (3.41). MS (ESI): m/z (%) = 359 $[M + H]^+$ (33), 381 $[M + Na]^+$ (37), 739 $[2 M + Na]^+$ (100).

5: R_f = 0.51, hexane/EtOAc (1:3). ^{1}H NMR (400 MHz, $CDCl_3$): δ 8.19 (d, J = 8.7 Hz, 1H, 6′-H), 7.26 (s, 1H, 2C=CH), 6.57 (dd, J = 8.8, 2.4 Hz, 1H, 5′-H), 6.44 (d, J = 2.4 Hz, 1H, 3′-H), 6.35 (d, J = 1.8 Hz, 1H, 7-H), 6.11 (d, J = 1.8 Hz, 1H, 5-H), 3.94 (s, 3H, 4-OCH_3), 3.89 (s, 3H, 6-OCH_3), 3.87 (s, 3H, 2′-OCH_3), 3.85 (s, 3H, 4′-OCH_3). ^{13}C NMR (101 MHz, $CDCl_3$): δ 180.69 (C-3), 168.72 (7C-C-1O), 168.54 (C-6), 162.34 (C-4′), 160.25 (C-2′), 159.40 (C-4), 146.95 (C-2), 132.94 (C-6′), 114.83 (C-1′), 105.83 (3C-C-4C), 105.65 (C-5′), 105.46 (2C=C), 98.07 (C-3′), 93.92 (C-5), 89.23 (C-7), 56.30 (4-OCH_3), 56.15 (6-OCH_3), 55.69 (2′-OCH_3), 55.56 (4′-OCH_3). UV/vis (MeCN): λ_{max} (lg ε) = 398 (3.45), 251 (3.30), 204 (3.55). MS (ESI): m/z (%) = 343 $[M + H]^+$ (45), 365 $[M + Na]^+$ (55), 707 $[2 M + Na]^+$ (100).

8: R_f = 0.44, hexane/EtOAc (2:1). ^{1}H NMR (400 MHz, $CDCl_3$): δ 10.27 (s, 1H), 7.79 (d, J = 8.7 Hz, 1H), 6.53 (ddd, J = 8.7, 2.2, and 0.7 Hz, 1H), 6.43 (d, J = 2.2 Hz, 1H), 3.89 (s, 3H), 3.86 (s, 3H). ^{13}C NMR (101 MHz, $CDCl_3$): δ 188.42, 166.30, 163.73, 130.82, 119.16, 105.88, 98.03, 55.73, 55.71. UV/vis (MeCN): λ_{max} (lg ε) = 308 (3.42), 271 (3.42), 223 (3.41). MS (ESI): m/z (%) = 167 $[M + H]^+$ (22), 189 $[M + Na]^+$ (100), 355 $[2 M + Na]^+$ (91).

6: R_f = 0.62, hexane/EtOAc (1:3). ^{1}H NMR (400 MHz, $CDCl_3$): δ 12.45 (s, 1H), 10.86 (s, 1H), 6.17 (d, J = 2.3 Hz, 1H), 6.02 (d, J = 2.3 Hz, 1H), 4.00 (s, 3H), 3.81 (s, 3H). ^{13}C NMR (101 MHz, $CDCl_3$): δ 170.87, 166.42, 165.75, 159.80, 95.20, 94.99, 91.47, 57.02, 55.81. UV/vis (MeCN): λ_{max} (lg ε) = 295 (2.66), 265 (3.06), 220 (3.08). MS (ESI): m/z (%) = 197 $[M - H]^-$ (100), 221 $[M + Na]^+$ (70), 419 $[2M + Na]^+$ (45).

10: R_f = 0.51, hexane/EtOAc (2:1). ^{1}H NMR (400 MHz, $CDCl_3$): δ 6.83 (d, J = 8.7 Hz, 1H), 6.49 (d, J = 2.8 Hz, 1H), 6.39 (dd, J = 8.7 and 2.8 Hz, 1H), 5.24 (s, 1H),

3.86 (s, 3H), 3.76 (s, 3H). ^{13}C NMR (101 MHz, $CDCl_3$): δ 153.68, 147.19, 139.92, 114.21, 104.38, 99.57, 56.02, 55.94. MS (ESI): m/z (%) = 155 $[M + H]^+$ (100), 177 $[M + Na]^+$ (32).

7: R_f = 0.42 hexane/EtOAc (2:1). ^{1}H NMR (400 MHz, $CDCl_3$): δ 6.08 (t, J = 2.2 Hz, 1H), 6.03 (d, J = 2.2 Hz, 2H), 5.41 (s, 1H), 3.74 (s, 6H). ^{13}C NMR (101 MHz, $CDCl_3$): δ 161.73, 157.51, 94.43, 93.33, 55.48. UV/vis (MeCN): λ_{max} (lg ε) = 266 (2.68), 233 (3.20). MS (ESI): m/z (%) = 155 $[M + H]^+$ (100).

9: Rf = 0.57, hexane/EtOAc (2:1). ^{1}H NMR (400 MHz, CDC13): δ 12.07 (s, 1H), 6.10 (d, J = 2.4 Hz, 1H), 5.95 (d, J = 2.4 Hz, 1H), 4.37 (q, J = 7.1 Hz, 2H), 3.80 (s, 3H), 3.80 (s, 3H), 1.39 (t, J = 7.1 Hz, 3H). ^{13}C NMR (101 MHz, CDC13): δ 171.32, 166.03, 165.34, 162.40, 97.01, 93.55, 91.74, 61.27, 56.13, 55.54, 14.34. MS (ESI): m/z (%) = 249 $[M + Na]^+$ (100).

(*E*)-2-((2,4-Dimethoxyphenyl)(hydroxy)methylene)-4,6-dimethoxybenzofuran-3(2H)-one (4). Aurone **5** (100 mg, 0.292 mmol, 1.0 equiv) was dissolved in ethanol (4 mL) and NaOH (0.4 mL, 10.87 N, 15.0 equiv). The reaction mixture was stirred at 24 °C for 0.5 h. TBHP 70% aq. (0.4 mL, 2.92 mmol, 10.0 equiv) was added, and the mixture was stirred for 23 h at 25 °C. The reaction was quenched with 1 N HCl and adjusted to pH 2. The aqueous phase was extracted with DCM (3 × 5 mL). The organic phases were combined and dried over sodium sulfate, and the solvent was removed. Auronol **4** was obtained as a colorless solid (8.4 mg, 8%) after silica gel column chromatography. R_f = 0.48, hexane/EtOAc (1:3). ^{1}H NMR (400 MHz, $CDCl_3$): δ 9.47 (s, 1H, OH), 7.19 (d, J = 8.7 Hz, 1H, 6′-H), 6.57–6.56 (m, 2H, 5′-H, 3′-H), 6.51 (d, J = 2.2 Hz, 1H, 7-H), 6.35 (d, J = 2.2 Hz, 1H, 5-H), 3.98 (s, 3H, 4-OCH_3), 3.85 (s, 3H, 6-OCH_3), 3.82 (s, 3H, 4′-OCH_3), 3.78 (s, 3H, 2′-OCH_3). ^{13}C NMR (101 MHz, $CDCl_3$): δ 163.00 (C-6), 162.59 (2C-C-1′C), 161.93 (C-3), 160.97 (C-4′), 158.79 (C-2′), 157.13 (C-4), 155.90 (7C-C-1O), 132.59 (C-6′), 113.11 (C-1′), 104.82 (C-5′), 100.50 (C-2), 99.25 (C-3′), 99.18 (3CC-4C), 95.47 (C-5), 94.39 (C-7), 57.00 (4-OCH_3), 55.99 (6-OCH_3), 55.89 (2′-OCH_3), 55.46 (4′-OCH_3). ^{1}H NMR (400 MHz, DMSO-d_6): δ 9.80 (s, 1H, OH), 7.04 (d, J = 8.3 Hz, 1H, 6′-H), 6.64 (d, J = 2.3 Hz, 1H, 7-H), 6.59 (d, J = 2.3 Hz, 2H, 5-H, 3′-H), 6.54 (dd, J = 8.3, 2.4 Hz, 1H, 5′-H), 3.97 (s, 3H, 4-OCH_3), 3.87 (s, 3H, 6-OCH_3), 3.79 (s, 3H, 4′-OCH_3), 3.69 (s, 3H, 2′-OCH_3). ^{13}C NMR (101 MHz, DMSO-d_6): δ 162.91 (C-6), 161.69 (2C-C-1′C), 161.23 (C-3), 160.21 (C-4′), 158.46 (C-2′), 157.44 (C-4), 155.12 (7C-C-1O), 132.58 (C-6′), 113.44 (C-1′), 104.59 (C-5′), 99.21 (C-2), 98.39 (C-3′), 98.31 (3C-C-4C), 95.36 (C-5), 93.93 (C-7), 56.98 (4-OCH_3), 56.05 (6-OCH_3), 55.37 (2′-OCH_3), 55.19 (4′-OCH_3). UV/vis (MeCN): λ_{max} (lg ε) = 316 (3.13), 215 (3.16), 205 (3.15). MS (ESI): m/z (%) = 359 $[M + H]^+$ (33), 381 $[M + Na]^+$ (77), 739 $[2\,M + Na]^+$ (100).

The Mechanism

The reaction begins the oxidation of the double bond in **2** with TBHP to give the epoxide **A**. Then the epoxide is attacked by the neighbor phenolic hydroxyl. The new ring could be formed by two paths. Path A is a *6-endo-tet* process [46] to provide the 3-hydroxychroman-4-one **B**. After the elimination of one molecule of water, the newly formed double bond in **C** is oxidized by TBHP again to generate epoxide **D**. Further rearrangement of **D**—possibly via diketone **E**—leads to the formation of product **3**.

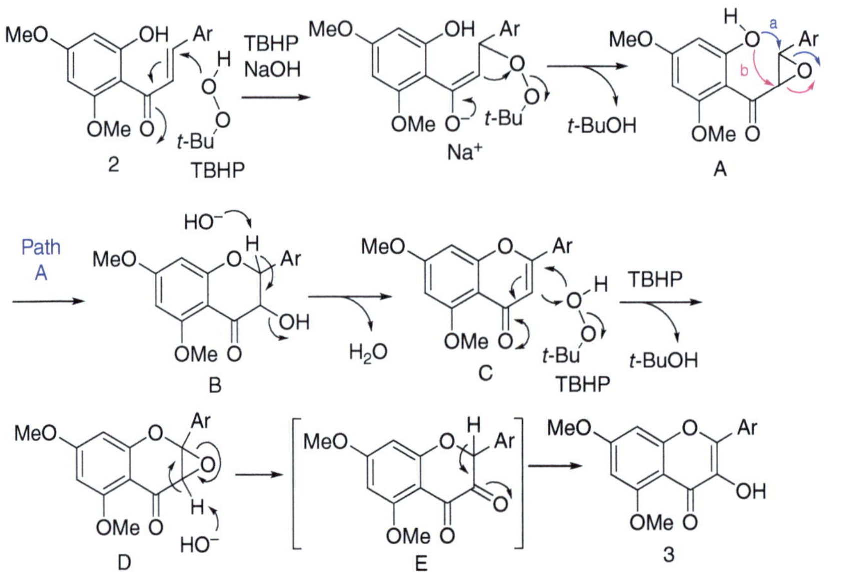

Another possible ring closure of **A** is a *5-endo-tet* process [46]. (Path B), which gives the benzofuran **F**. Elimination of one molecule of water from **F** produces the side product **5**.

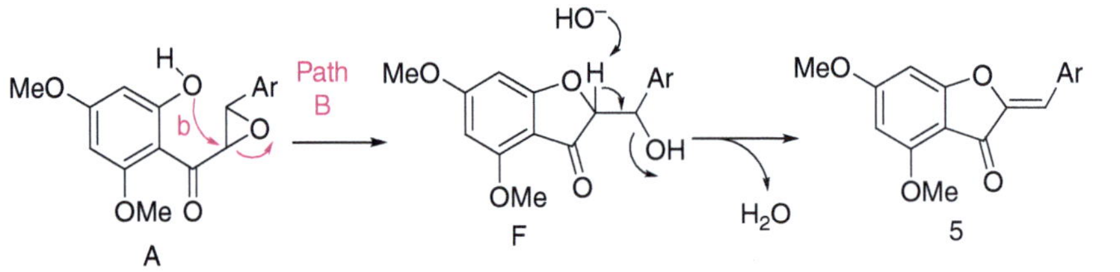

By the same manner for the formation of **3**, compound **5** is further oxidized by TBHP, via epoxide **G**, to generate side product **4**.

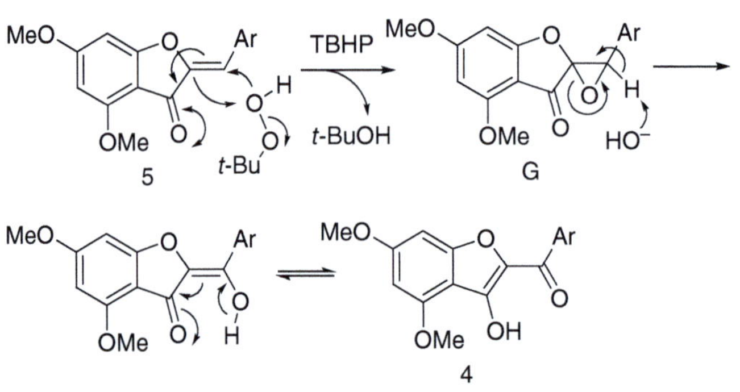

Indeed, when isolated aurone **5** was subjected to oxidative cyclization conditions (TBHP/NaOH, 24 h), auronol **4** (8%, 40% brsm) and flavonol tetramethylmorin (**3**, 8%) were obtained together with 80% of the unreacted starting material.

The transformation from **A** to **5** could be reversible. Addition of H_2O to **5** would give the benzofuran **F**. The partially deprotonated hydroxyl attacks the α-carbon of the carbonyl in the furan ring, leading to intermediate **A**. Compound **3** could be produced from **A** as we have proposed in the mechanism.

MeO Ar HO⁻ OMe O 5 H_2O MeO Ar O⁻ OMe O F MeO O⁻ Ar O OMe O A → 3

5.6.2 Benzoxazolinone

Problem 145: Formation of Undesired Benzoxazolinone from *N*,2-Dihydroxybenzamide

Compound **1** was identified as a potent 5-HT_4 partial agonist with favorable safety and pharmacokinetic profiles for treatment of AD and cognitive impairment [47]. The initial synthesis of **1** included the reaction as shown below. Due to the formation of undesired benzoxazolinone (**4**), the desired product benzo[*d*]isoxazol-3-ol (**3**) was obtained in moderate yield (43%) [48, 49].

Please suggest mechanism for the formation of **3** and **4**.

OH HN OH O O O 2 CDI THF 50 °C O–N OH O O 3 (43%) O O NH O O 4 O–N O N OH O O 1 (8 steps overall yield 3.3%)

Experimental Procedure (from *Org. Process Res. Dev.*, **2016**, *20*, 233*)

4-[(*R*)-(Tetrahydrofuran-3-yl)oxy]-benzo[d]isoxazol-3-ol (3). A solution of 2,*N*-dihydroxy-6-[(*R*)-(tetrahydrofuran-3-yl)oxy]-benzamide (**2**, 25 g, 0.105 mol) in 250 mL of THF was heated to 50 °C. Carbonyl diimidazole was added portionwise and the resulting mixture was stirred at 50 °C for 14 h. After cooling to room temperature, 100 mL of 2N HCl was added and the aqueous layer was extracted with ethyl acetate. The combined organic layers were then extracted three times with 10% aqueous potassium carbonate. The aqueous extracts were washed with ethyl acetate and then acidified to pH 2–3 with 2N HCl. The acidified aqueous layer was extracted with ethyl acetate. The ethyl acetate extracts were washed with brine, dried (Na_2SO_4), and concentrated to afford 20 g of product as a yellow solid (43% yield**). ^1H NMR (400 MHz, $CDCl_3$): δ 2.20 (m, 2H), 3.89 (m, 1H), 4.01 (m, 3H), 5.05 (m, 1H), 6.48 (d, J = 7.6 Hz, 1H). 6.92 (d, J = 7.6 Hz, 1H), 7.37 (t, J = 7.6 Hz, 1H).

** The mechanism for the formation of **3** and the side product **4** was discussed in the article* [49]. *However, the isolation and identification of **4** was not disclosed.*

*** I calculated the yield to be 86%.*

The Mechanism

Upon the addition of CDI to compound **2**, two molecules of imidazole are released, yielding intermediate **A**. Subsequently, a nucleophilic attack by the hydroxyl group on the dioxazole ring, coupled with the synergistic elimination of CO_2, leads to the formation of the desired product **3**. However, due to the electron-donating nature of the alkoxy group, a Lossen-type rearrangement also takes place, accompanied by the generation of CO_2, resulting in the formation of the isocyanate intermediate **B**. This intermediate readily undergoes cyclization to produce the undesired benzoxazolinone **4**.

To inhibit the Lossen rearrangement, a fluoro substituent—an electron-withdrawing group—was introduced at the C-6 position. This group could also serve as a leaving group for the subsequent introduction of the chiral tetrahydrofuran moiety. In practice, treatment of the fluoro compound **5** with CDI in refluxing THF provided the desired product **6** in 92% yield without any detectable Lossen rearrangement side product [49].

5.6.3 Quinoline

Problem 146: Formation of Quinoline-4-carboxylic Acids by Pfitzinger Reaction Conditions

Quinoline-4-carboxylic acid was a core structure in several interesting pharmaceutical compounds, such as antinorovirus compound **1**, [50] antimycobacterial compound **2**, [51] and PPARβ/δ inhibitor **3** [52]. Ivachtchenko et al. carried out a reaction in which *N,N*-diethyl-2,3-dioxoindoline-5-sulfonamide (**4**) was reacted with diethyl malonate under standard Pfitzinger reaction conditions. Instead of obtaining the expected product, 6-(*N,N*-diethylsulfamoyl)-2-hydroxyquinoline-4-carboxylic acid (**5**), they identified 6-(*N,N*-diethylsulfamoyl) quinoline-4-carboxylic acid (**6**) as the major product [53].

Please suggest a mechanism for the formation of **6**.

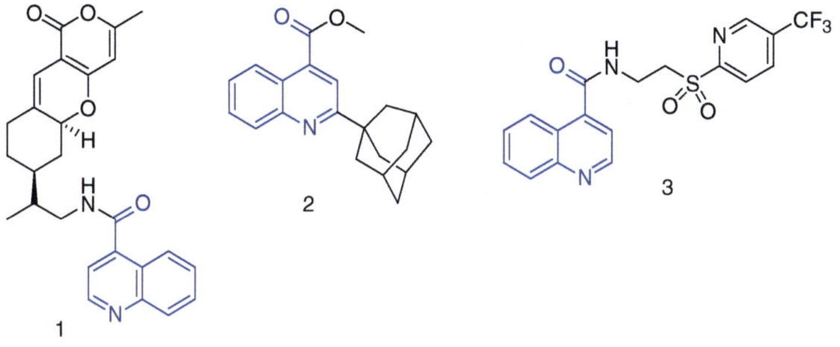

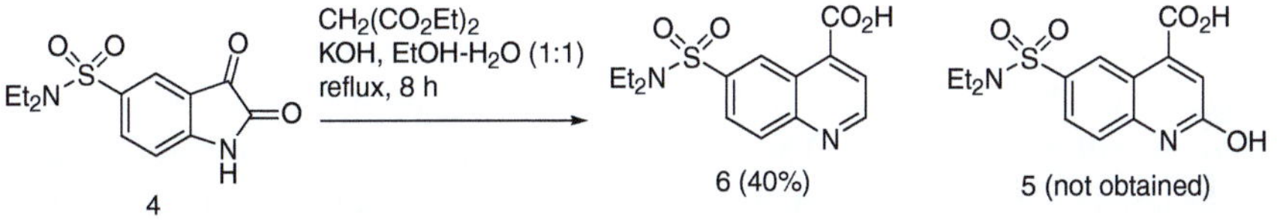

1 2 3

4 → 6 (40%); 5 (not obtained)

$CH_2(CO_2Et)_2$, KOH, EtOH-H_2O (1:1), reflux, 8 h

Experimental Procedure (from *Tetrahedron Lett.*, **2004**, *45*, 5473)

General Procedure for the Synthesis of 6-sulfamoylquinoline-4-carboxylic Acids. Method A: Using Conventional Heating. 5-Sulfamoylisatin **4** (5.0 mmol) was added to a solution of 5.64 g (100.7 mmol) of KOH in 32 mL of ethanol–water (1:1). The reaction mixture was heated at reflux for 8 h, then cooled to room temperature and acidified with 1N HCl until pH 1. The resulting mixture was extracted with ethyl acetate (3 × 50 mL), the organic extracts were combined and dried over anhydrous $MgSO_4$. The solvent was removed *in vacuo*, and the residue was purified by silica gel chromatography (eluent $CHCl_3$–CH_3OH, 19:1 v/v) or by recrystallization from ethyl acetate. Yield: 40% (**6**). Spectroscopic data of **6**: IR (KBr) 1705 cm^{-1} (νC=O); ^{1}H NMR (DMSO-d_6, 300 MHz (Varian), ppm) δ 1.05 (t, 6H, NCH_2CH_3), 3.22 (q, 4H, NCH_2CH_3), 8.11 (d, 1H, Ar), 8.13 (dd, 1H, Ar), 8.30 (d, 1H, Ar), 9.21 (d, 1H, Ar), 9.30 (d, 1H, Ar), 14.2 (br s, 1H, OH); ^{13}C NMR (DMSO-d_6 75 MHz (Varian), ppm) δ 14.48, 42.26, 124.7, 126.02, 126.65, 131.73, 136.87, 139.21, 149.62, 163.63, 167.12; *m/z* 309.0901 [M + H]. Anal. Calcd. For $C_{14}H_{16}N_2O_4S$: C, 54.53; H, 5.23; N, 9.08; S, 10.40. Found: C, 54.67; H, 5.46; N, 9.21; S, 10.51.

The Mechanism

If the reaction underwent an aldol condensation pathway, the mechanism should be as below and **5** should be the product. However, under the strong basic (KOH) condition, diethyl malonate would hydrolyze to malonic acid dianion. That makes the methylene not active enough to attack the carbonyl of the hydrolyzed intermediate **A**. Malonic acid dianion is not an electrophile to be attacked by the aniline in **A**. Therefore, neither aldol reaction nor amidation between **A** and malonic acid dianion would happen at this strong base conditions. Typical preparation of 2-hydroxyquinoline-4-carboxylic acids from isatins and malonic acid by Pfitzinger reaction is carried out at weak acidic conditions [54, 55].

While the EtO^- is active enough to attack the carbonyl of the starting material **4** to provide species **A**. Then intramolecular disproportionation occurs to release acetaldehyde and 3-hydroxyindolone **7**. Meanwhile, hydrolysis of **4** at the strong basic conditions will give aniline **C**. The acetaldehyde is active enough to readily react with the amine, after elimination of water, to provide vinylaniline **E**. Finally, vinylaniline **E** undergoes cyclization and dehydration to give product **6**. Diethyl malonate does not play any role in this reaction. 3-Hydroxyindolone **7** is the by-product of this reaction.

Upon the reaction of **4** with ^{13}C-ethanol under the described conditions, ^{13}C-labeled product **8** was obtained.

$CH_3{}^{13}CH_2OH$
KOH-H_2O
35%
4 → 8

Based on this mechanism, acetaldehyde is necessary for the formation of quinoline ring. Since for generation of one molecule of acetaldehyde, one molecule of the starting material **4** will be consumed, theoretically the maximum yield would be 50%. We can predict that if extra acetaldehyde, e.g., >1 equivalent, were added at the beginning of the reaction, the yield of product **6** would be improved and the formation of **7** would be suppressed. Other α-hydrogen-containing aldehydes would also react with isatins to give quinoline-4-carboxylic acids.

Problem 147: Formation of a Tricyclic Impurity During Alkylation of 8-hydroxyquinoline

The 3,4,8-trisubstituted quinoline derivatives **1a** and **1b** are reversible (H^+/K^+) ATPase inhibitors for the treatment of peptic ulcers, gastroesophageal reflux disease, and related disorders. In a process chemistry study [56], treatment of the 8-hydroxyquinoline **1c** with ethylene carbonate in *i*-PrOH in the presence of 1,4-diazabicyclo[2.2.2]octane (DABCO) gave the desired product **1b** (yield 66%) together with small quantities of the tricyclic compound **2**, the symmetrical carbonate **3**, and isopropyl carbonate **4**.

Please propose mechanisms for the formation of impurities **2**, **3**, and **4**.

Experimental Procedure (from *Org. Process Res. Dev.*, **1997**, *1*, 185)

3-Butyryl-8-(2-hydroxyethoxy)-4-[(2-methylphenyl)-amino]quinoline (1b).

Method A. Molten ethylene carbonate (2.6 kg, 29.5 mol) was heated to 90 °C and treated with the 8-hydroxyquinoline **1c** (260 g, 0.81 mol), added portionwise over ~5 min. The mixture was reheated to 90 °C and the resulting suspension treated with anhydrous K_2CO_3 (11.2 g, 81 mmol). The reaction mixture was heated at 90 °C for 3.5 h and subsequently allowed to cool to ~60 °C before water (2 dm^3) was added with stirring. The resulting slurry was poured into more water (2 dm^3) and stirring continued for 45 min. The solid was collected via suction filtration, washed with water, and dried at 50 °C to give a crude product (277 g), which was recrystallized from EtOAc (5 dm^3) to give **1b** (233 g, 71%) as yellow crystals: mp 115–118 °C (lit. mp 117–120 °C). The mother liquors from the above were concentrated *in vacuo*, and the residue was triturated with toluene (60 cm^3). The resulting solid was filtered off, washed with toluene, and dried at 50 °C to give yellow crystals (12.7 g). A sample (1.22 g) of the latter was subjected to chromatography on silica gel, eluting with EtOAc to give **6-butyryl-7-[(2-methylphenyl)imino]-2,3-dihydro-7H-pyrido[1,2,3-*de*]-1,4-benzoxazine (2)** (0.6 g) as bright yellow crystals: mp 178–180 °C (Found: C, 76.2; H, 6.5; N, 7.9. $C_{22}H_{22}N_2O_2$ requires C, 76.3; H, 6.4; N, 8.1);

1a, R=Me
1b, R=CH_2CH_2OH
1c, R=H

1c + ethylene carbonate; DABCO, *i*-PrOH, reflux, 10 h → 1b (66%), 2, 3, 4

1,4-diazabicyclo[2.2.2]octane (DABCO)

ν_{max} (KBr)/cm^{-1} 1686 (C=O), 1670 (C=N), 1620 (C=C), 1605 (Ar), and 1560 (Ar); δ_H (270 MHz; [$^2H_6$]DMSO) 0.70 (3H, t, *J* 7.3, $CH_3CH_2CH_2CO$), 1.20 (2H, sextet, *J* 7.3, $CH_3CH_2CH_2CO$), 2.13 (3H, s, $ArCH_3$), 2.33 (2H, t, *J* 7.3, $CH_3CH_2CH_2CO$), 4.15–4.19 (2H, m, -NCH_2CH_2O-), 4.37–4.41 (2H, m, -NCH_2CH_2O-), 6.44–6.47 (1H, m, Ar), 6.75–6.81 (1H, m, Ar), 6.91–7.13 (4H, m, Ar), 7.65–7.67 (1H, m, Ar), and 7.71 (1H, s, 5-H); δ_C(62.5 MHz; [$^2H_6$]DMSO) 13.0 ($CH_3CH_2CH_2CO$), 16.7 ($CH_3CH_2CH_2CO$), 17.4 ($ArCH_3$), 42.0 ($CH_3CH_2CH_2CO$), 47.5 (-NCH_2CH_2O-), 63.2 (-NCH_2CH_2O-), 115.9–150.9 (13C, ArC and 7-C), 141.1 (5-C), and 199.1 (CO); *m/z* (EI) 346 (M_+, 30), 317 ($M^+ - C_2H_5$, 100), 303 ($M^+ - C_3H_7$, 13), and 275 ($M^+ - C_3H_7CO$, 53).

Crude **1b** prepared as above (70 g) was chromatographed on silica gel (200 g) eluting with acetone followed by 5% MeOH in acetone to give the crude carbonate **3** (0.5 g) as the earlier eluting fraction, the bulk (0.3 g) of which was rechromatographed on silica gel (50 g) eluting with CH_2Cl_2 to give **bis[[[3-butyryl-4-[(2-methylphenyl)amino]quinol-8-yl]oxy]ethyl] carbonate (3)** (0.28 g) as a yellow foam: δ_H (400 MHz) 1.02 (3H, t, *J* 7.4, $CH_3CH_2CH_2CO$), 1.80 (2H, sextet, *J* 7.4, $CH_3CH_2CH_2CO$), 2.34 (3H, s, $ArCH_3$), 3.07 (2H, t, *J* 7.4, $CH_3CH_2CH_2CO$), 4.42–4.49 (2H, m, $ArOCH_2$-CH_2OCO), 4.70–4.75 (2H, m, $ArOCH_2CH_2OCO$), 6.85–7.30 (7H, m, Ar), 9.25 (1H, s, 2-H), and 11.80 (1H, s, NH); δ_C (100 MHz) 13.9 ($CH_3CH_2CH_2CO$), 18.2 ($CH_3CH_2CH_2$-CO), 18.5 ($ArCH_3$), 41.7 ($CH_3CH_2CH_2CO$), 65.9 ($ArOCH_2$-CH_2OCO), 66.5 ($ArOCH_2CH_2OCO$), 111.7–155.0 (16C, Ar C, OCO.O), and 203.1 (ArCO); *m/z* (FAB) 755 (MH^+, 8) and 347 [M(**32**)H^+, 100].

Method B. The 8-hydroxyquinoline **1c** (10 g, 31.2 mmol) was stirred with *i*-PrOH (60 cm^3) to give a thick slurry. Ethylene carbonate (13.21 g, 0.15 mol) was added and the mixture heated to give a clear solution. Following addition of DABCO (0.5 g,

4.5 mmol), the mixture was heated at reflux for 10 h. HPLC analysis of the reaction solution revealed mainly desired product **1b** together with small quantities of the tricyclic compound **2**, the symmetrical carbonate **3** (both identified by HPLC comparison with samples isolated above), and **[[3-butyryl-4-[(2-methylphenyl)amino]quinol-8-yl]oxy]ethyl 2-methylethyl carbonate (4)**, assigned on the basis of LC-ionspray-MS, *m*/*z* 451 (MH^+, 100), 409 ($MH^+ - C_3H_6$, 26), 365 ($MH^+ - C_3H_6 - CO_2$, 44), and 321 ($MH^+ - C_3H_6 - CO_2 - C_2H_4O$, 36). Water (40 cm^3) was added to the reaction mixture, and the product crystallized on cooling. The suspension was stirred for 18 h at ambient temperature and the product collected via suction filtration, washed with 40% aqueous *i*-PrOH (20 cm^3), and air-dried to give the 8-(2-hydroxyethoxy) quinoline **1b** (7.7 g, 66%) as a yellow solid, mp 116–118 °C, identical to the product previously reported and containing very low levels of impurities.

Method C. Ethylene carbonate (66.1 g, 0.75 mol) was taken up in *t*-BuOH (200 cm^3) at 40–50 °C, and the resulting solution was treated with the 8-hydroxyquinoline **1c** (50 g, 0.156 mol) and DABCO (2.5 g, 22.3 mmol). The solution was heated under reflux for 2 h. More *t*-BuOH (100 cm^3) and water (200 cm^3) were added with the temperature of the flask being maintained at >70 °C. The reaction mixture was subsequently cooled with rapid stirring to ~10 °C. The resulting precipitated product was collected via suction filtration, washed with a chilled *i*-PrOH (120 cm^3)/water (80 cm^3) mixture, and dried under reduced pressure to give **1b** (51.2 g, 90%) as a yellow crystalline solid: mp 118–120 °C; δ_H (400 MHz) 1.02 (3H, t, *J* 7.4, $CH_3CH_2CH_2CO$), 1.79 (2H, sextet, *J* 7.4, $CH_3CH_2CH_2CO$), 2.34 (3H, s, $ArCH_3$), 3.05 (2H, t, *J* 7.4, $CH_3CH_2CH_2CO$), 3.70 (1H, br s, OH), 4.04–4.07 (2H, m, $ArOCH_2CH_2OH$), 4.23–4.27 (2H, m, $ArOCH_2CH_2OH$), 6.90–7.30 (7H, m, Ar), 9.10 (1H, s, 2-H), and 11.97 (1H, s, NH).

The Mechanism

8-Hydroxyquinoline **1c** is activated by DABCO to form salt **A**, which in turn attacks ethylene carbonate to give carbonate **B**. At this stage, the major pathway is decarboxylation to provide the desired product **1b**.

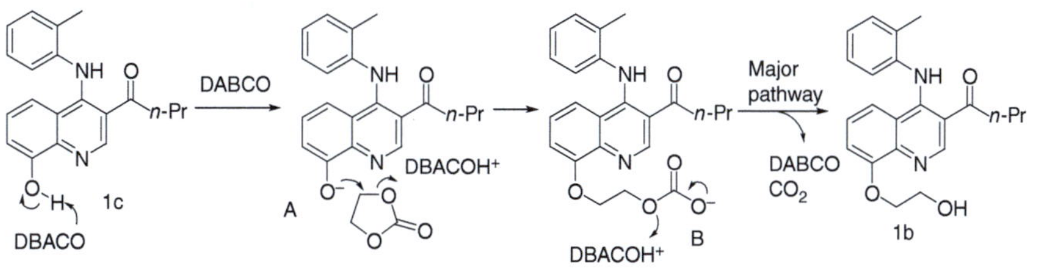

Meanwhile, the salt carbonate **B** may also exist in equilibrium with the hydrogen carbonate **C**, accompanied by the release of free base DABCO. Once **C** is generated, it could be attacked by alcohol **1b** to produce symmetric carbonate **3**. Hydrogen carbonate **C** can also be attacked by the solvent isopropanol, leading to the formation of carbonate **4**. Cyclization of **4** leads to tricyclic compound **2** and releases CO_2 and isopropanol. Tricyclic compound **2** could be also generated from carbonate **3** and from **C**.

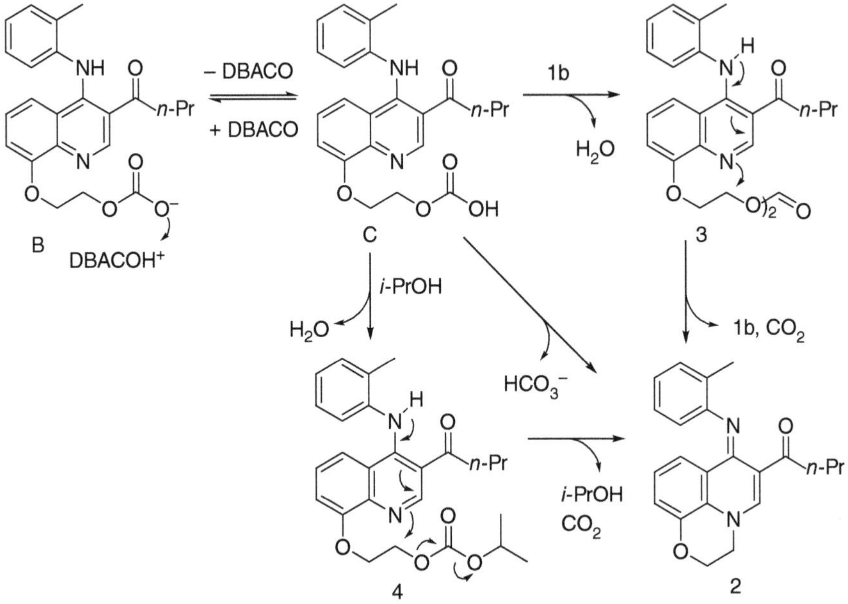

Since the formation of impurities **2** and **4** is attributed to the use of isopropanol as the solvent, replacing it with the more sterically hindered *t*-BuOH is expected to suppress side product formation. Indeed, when the reaction was carried out in *t*-BuOH, the desired product **1b** was obtained from **1c** in 97% yield.

5.6.4 Naphthyridine

Problem 148: Unexpected Formation of a Dibenzo[c,f]-2,7-naphthyridine-3,6-diamine

10,11-Dimethoxy-4-methyldibenzo[*c*,*f*]-2,7-naphthyridine-3,6-diamine (**1**) was identified as a potent compound with favorable biological profile. This compound was discovered accidently from a side product of the following reaction. The chemists from Wyeth wanted to prepare 4-((3-amino-2-methylphenyl)amino)-6,7-dimethoxyquinoline-3-carbonitrile (**4**) from 4-chloro-6,7-dimethoxyquinoline-3-carbonitrile (**2**) and 2-methylbenzene-1,3-diamine (**3**), an unexpected product **1** was also obtained as a side product. A reliable high yielding (80%) procedure for preparation of **1** was identified through the use of DoE [57].

Please suggest a mechanism for the formation of **1**.

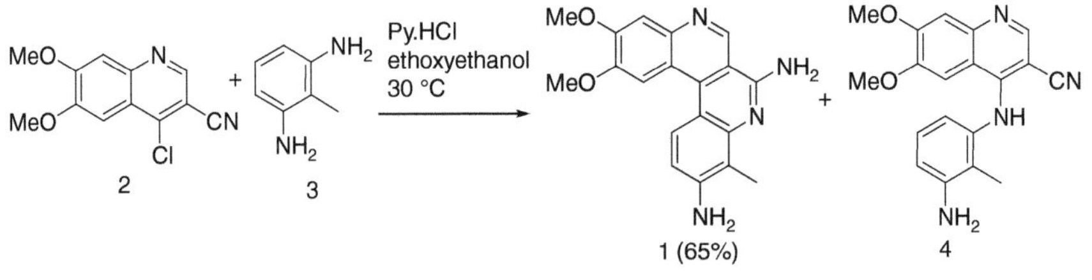

Experimental Procedure (from *Org. Process Res. Dev.*, **2007**, *11*, 450)
10,11-Dimethoxy-4-methyldibenzo[*c,f*]-2,7-naphthyridine-3,6-diamine (1)

Method A. A stirred mixture of 4-chloro-6,7-dimethoxyquinoline-3-carbonitrile **2** (25 mg, 0.10 mmol), 2,6-diaminotoluene **3** (49 mg, 0.40 mmol), and pyridine hydrochloride (58 mg, 0.50 mmol) (catalyst) in 2 mL of 2-ethoxyethanol was heated at 130 °C for 8 h. 10,11-Dimethoxy-4-methyldibenzo[*c,f*]-2,7-naphthyridine-3,6-diamine **1** was formed in 65% yield (based on LC/MS analysis).

Method B. A stirred mixture of 4-chloro-6,7-dimethoxyquinoline-3-carbonitrile **2** (25 mg, 0.10 mmol), 2,6-diaminotoluene **3** (24 mg, 0.20 mmol), and pyridine hydrochloride (12 mg, 0.10 mmol) in 2 mL of 2-ethoxyethanol was heated in a sealed tube at 210 °C in a microwave oven (Biotage, Initiator) for 1 h using a power limit of 300 W and a pressure inside the reaction vessel of less than 20 psi. 10,11-Dimethoxy-4-methyldibenzo[*c,f*]-2,7-naphthyridine-3,6-diamine **1** was formed in 80% yield (based on LC/MS analysis). The reaction mixture was concentrated under reduced pressure, diluted with ethyl acetate, and washed with brine. The organic layer was collected, dried over sodium sulfate, and concentrated. The crude product was purified by flash column chromatography using silica gel. Isolated yield: 72%. HPLC: R_t) 2.13 min; MS (ES^+): *m/z* 335.1 [M + H]. HRMS: 335.15122 [M + H]; 335.15026 [calculated]; ^{1}H NMR (DMSO-d_6): δ 9.4 (1H, s), 8.4 (1H, d), 8.2 (1H, s), 7.5 (1H, s), 6.9 (1H, br), 6.8 (1H, d), 5.4 (2H, br), 4.0 (3H, s), 3.9 (1H, s), 2.4 (3H, s); IR (cm^{-1}). Cyano group's absorption around 2200 cm^{-1} is not observed.

4-[(3-Amino-2-methylphenyl)amino]-6,7-dimethoxyquinoline-3-carbonitrile (4): ^{1}H NMR (DMSO-d_6): δ 9.3 (1H, s), 8.3 (1H, d), 7.8 (1H, s), 7.3 (1H, s), 6.9 (1H, m), 6.6 (1H, d), 6.5 (1H, d), 5.0 (2H, br), 4.0 (3H, s), 3.9 (1H, s), 1.9 (3H, s), HRMS: 335.14954 [M + H]; 335.15026 [calculated]; IR (cm^{-1}). Cyano group's absorption around 2200 cm^{-1}.

The Mechanism

This reaction can be regarded as a Friedel-Crafts acylation. Usually, the reaction of 4-chloropyridines with anilines yields diarylamines [58], as seen in the formation of product **4** in this case. However, due to the presence of a second amino group on the benzene ring, 2,6-diaminotoluene (**3**) exhibits high reactivity as a nucleophile in Friedel-Crafts acylation. At the same time, the 4-chloro-3-cyanoquinoline **2** is a good electrophile owing to the electron-withdrawing cyano group. The C4 position is further activated by pyridinium chloride, allowing nucleophilic attack by the C-3 position of compound **3** to form intermediate **A**. Subsequent cyclization between the amino and the cyano groups, followed by aromatization, leads to the final product **1**.

You can also draw the mechanism as shown below—it's basically the same as the one above.

Another possibility is that the amino group in **3** first attacks the cyano group in **2**, followed by the adjacent carbon acting as a nucleophile and attacking the quinoline moiety.

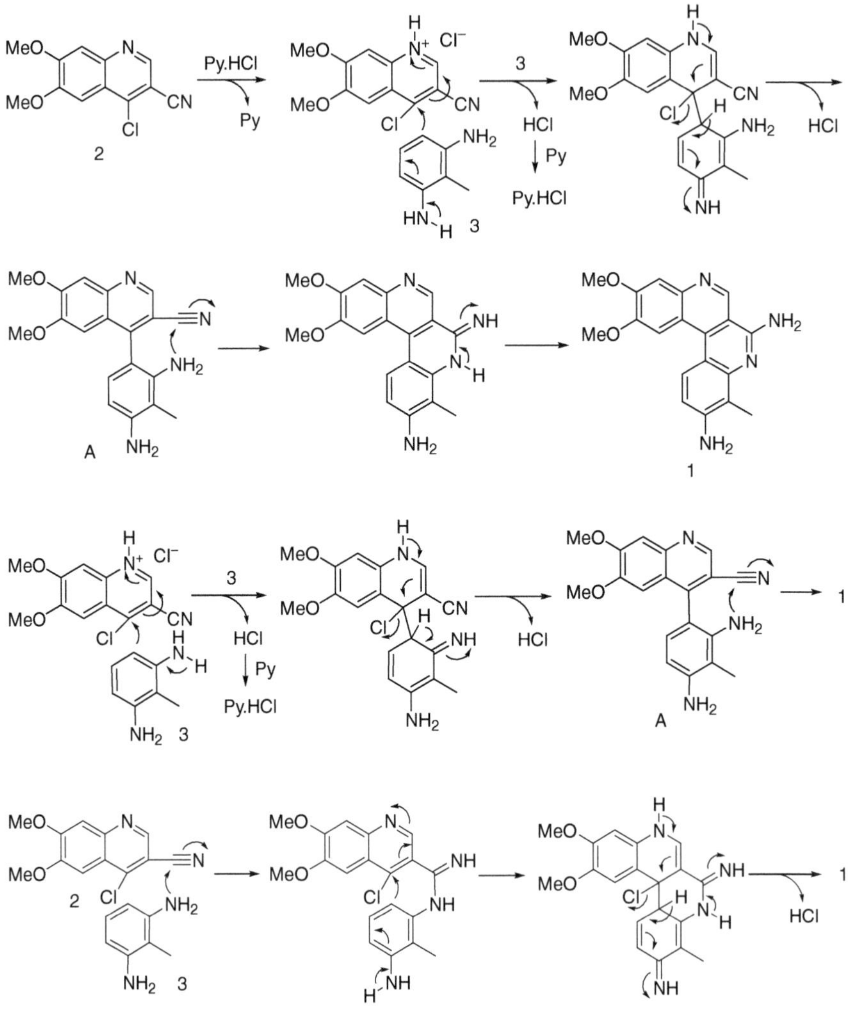

5.6.5 Pteridine

Problem 149: Formation of Two Pteridine Impurities During a Synthesis of Dexrazoxane

Dexrazoxane (**1**) has been used to protect the heart against the cardiotoxic side effects of chemotherapeutic drugs such as daunorubicin, doxorubicin, or other chemotherapeutic agents. In a process-related impurity study of dexrazoxane synthesis [59], two new impurities—present at levels ranging from 0.5% to 5%—were isolated by preparative HPLC. Their structures were identified as compounds **3** and **4** by IR, ^{1}H, ^{13}C, 2D NMR, and mass spectrometry. Their structures were finally confirmed by comparison with synthetic standards.

Please suggest mechanisms for the formation of **3** and **4**.

2 (tetra acid) $\xrightarrow[\text{160 °C, 17 h}]{\text{HCONH}_2\text{, toluene}}$ 1 (dexrazoxane) + 3, R^1=Me, R^2=H; 4, R^1=H, R^2=Me

Experimental Procedure (from *Org. Process Res. Dev.*, **2017**, *21*, 11)

5-(2-(3-Oxopiperazin-1-yl) propyl)-5,6-dihydropteridin-7(8H)-one (Impurity 3) M.p: 252.09 °C. ^{1}HNMR (DMSO-d_6, 400 MHz): δ_H 0.95 (3H, d, H4), 3.09–3.23 (2H, m, H2), 3.29–3.49 (5H, m, H14, H11, and H3), 3.94–4.04 (2H, m, H5), 8.14 (1H, s, H8), 8.02 (1H, s, H9). ^{13}CNMR (DMSO-d_6, 100 MHz): δ_C 50.54 (C2), 54.33 (C3), 11.28 (C4), 52.31 (C5), 166.81 (C6), 147.15 (C7), 128.95 (C8), 135.96 (C9), 147.07 (C10), 51.93 (C11 and C14), 172.41 (C12 and C13); HRMS (ESI) calcd for $C_{13}H_{17}O_3N_6$: 305. 13338 ([M + H]$^+$), found; 305.13566([M + H]$^+$). IR (KBr, cm^{-1}): 3248.13 (NH), 2970.38 & 2931.80 (CH), 1705.07 & 1689.64 (C=O), 1564.27 (NH bending), 1512.19 (C=N).

4-(2-(7-Oxo-7,8-dihydropteridin-5(6H)-yl)propyl)piperazine-2,6-dione (Impurity 4) M.p: 242.98 °C. ^{1}H NMR (DMSO-d_6, 400 MHz): δ_H 1.07 (3H, d, H4), 4.16 (2H, dd, H9, J = 6.4 Hz, 15.6 Hz), 3.73–3.82 (2H, m, H5), 3.31–3.42 (3H, m, H3, H5 and H8),2.76 (2H, dd, H2, J = 8.4 Hz, 12.8 Hz), 8.17 (1H, s, H13), 8.15 (1H, s, H14). ^{13}C NMR (DMSO-d_6, 100 MHz,): δ_C 45.12 (C2), 47.84 (C3), 13.53 (C4), 55.57 (C5 and C8), 171.82 (C6 and C7), 56.64 (C9), 167.33 (C10), 147.75 (C11), 147.39 (C12), 136.88 (C13), 128.95 (C14). HRMS (ESI) calcd for $C_{13}H_{17}O_3N_6$:305. 13334 ([M + H]$^+$), found; 305.13566 ([M + H]$^+$). IR (KBr, cm^{-1}): 3227.51 (NH), 2993.72 and 2920.23 (CH), 1703.14 (C=O), 1571.77 (NH bending), 1523.76 (C=N).

The Mechanism

Tautomerization of product **1** at ring B gives **A**, which attacks the formamide carbonyl group to form **B**. Then the condensation of **B** with another molecule of formamide gives **C**. An intramolecular cyclocondensation produces the corresponding

1 (dexrazoxane) ⇌ A $\xrightarrow{\text{HCONH}_2}$ B $\xrightarrow{\text{HCONH}_2}$ C ⟶ D $\xrightarrow{-3\ H_2O}$ 3

bicyclic ring **D** and finally aromatization occurs by dehydration to lead the impurity **3**. Following the same principle, tautomerization occurs at ring-A in **1** leads to the formation of impurity **4**.

The mechanism for the formation of the desired product 1 from 2 is shown below.

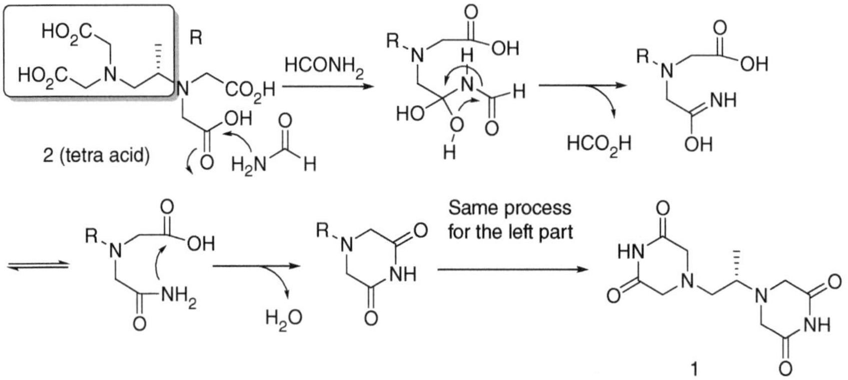

5.7 Miscellaneous

5.7.1 Steroid

Problem 150: Formation of Impurity F of Clobetasol Propionate

Clobetasol propionate (**1**) is a corticosteroid used to treat skin conditions such as eczema, contact dermatitis, seborrheic dermatitis, and psoriasis. Impurity F (**2**) is one of the impurities [60] during the production [61, 62]. During a stability study of clobetasol propionate (**1**), compound **2** was identified as a major degradant under acidic conditions at a level up to 1.39% [62]. As shown below, forced degradation experiments generated **2** in 2.7%. It has a relative retention time (RRT) on HPLC of 0.49 compared to clobetasol propionate.

Please propose a mechanism for the formation of impurity F (**2**) from clobetasol propionate under acidic conditions.

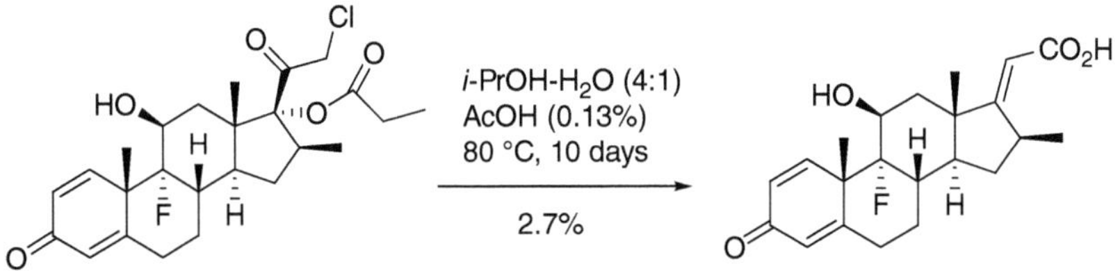

Clobetasol propionate (1) Impurity F (2)

Experimental Procedure (from https://doi.org/10.26434/chemrxiv.11809077.v1)
Preparation and Purification of RRT 0.49 Impurity from Forced Degradation. To prepare a suitable amount of pure RRT 0.49 impurity for NMR structural elucidation, a forced degradation of clobetasol propionate was designed to enrich and isolate RRT 0.49 impurity. Hence, a solution containing clobetasol propionate (5.0 g),

isopropanol (200 mL), water (50 mL), and acetic acid (0.33 mL) was heated at 80 °C for 10 days, during which period RRT 0.49 impurity grew to 2.7%. RRT 0.49 impurity was then isolated from a portion of the forced degradation sample solution by semipreparative HPLC, which was performed on a Shimadzu LC-8A system (Model XXI-4563) with a Waters Xterra RP18, 150 mm × 19 mm, 5 μm semipreparative HPLC column. The separation was carried out at room temperature, and the same mobile phase system as used in LC-PDA/UV-MSn analyses was utilized as per the gradient according to the following program: 0 min (40% B), 30 min (40% B), 35 min (55% B), 35.1 min (80% B), and 41 min (80% B). The HPLC separation was run at a flow rate of 9 mL/min with an injection volume of 1.8 mL and the detection wavelength for fraction collection was set at 240 nm. Approximately 20 mg of purified RRT 0.49 impurity was obtained with an HPLC purity of ~96%. ^{1}H NMR (DMSO-d$_6$, 400 MHz): δ_H 7.29 (1H, d, J = 10.2 Hz), 6.21 (1H, d, J = 10.2 Hz), 6.00 (1H, s), 5.50 (1H, brs), 5.33 (1H, s), 4.14 (H, brs), 2.63 (1H, td, J = 13.2 and 5.7 Hz), 2.47 (1H, m), 2.32 (1H, dd, J = 13.5 and 3.3 Hz), 2.05 (1H, m), 1.93 (1H, m), 1.86 (1H, d, J = 12.8 Hz), 1.86 (1H, m), 1.49 (3H, s), 1.42 (1H, d, J = 13.3 Hz), 1.34 (1H, m), 1.26 (1H, m), 1.23 (1H, d, J = 7.0 Hz), 1.18 (1H, m), 1.10 (3H, s). ^{13}C NMR (DMSO-d$_6$, 100 MHz): δ_C 185.2, 177.5, 167.4, 152.6, 129.1, 124.2, 109.6, 101.6 (d, J = 175.5 Hz), 70.4 (d, J = 36.5 Hz), 66.8, 47.9 (d, J = 22.0 Hz), 45.3, 40.5, 33.2 (d, J = 19.8 Hz), 32.9, 30.1, 27.0, 22.5 (d, J = 5.4 Hz), 21.3, 19.7, 5.9. HRMS (ESI) calcd for $C_{22}H_{28}FO_4$: 375.1972 ([M + H]$^+$), found: 375.1967 ([M + H]$^+$).

The Mechanism

Typical Favorskii rearrangement is catalyzed by base [63]. This is a case where impurity F (**2**) forms from clobetasol propionate (**1**) via an acid-catalyzed Favorskii rearrangement. In this particular circumstance the propionate is the leaving group rather than a typically halogen.

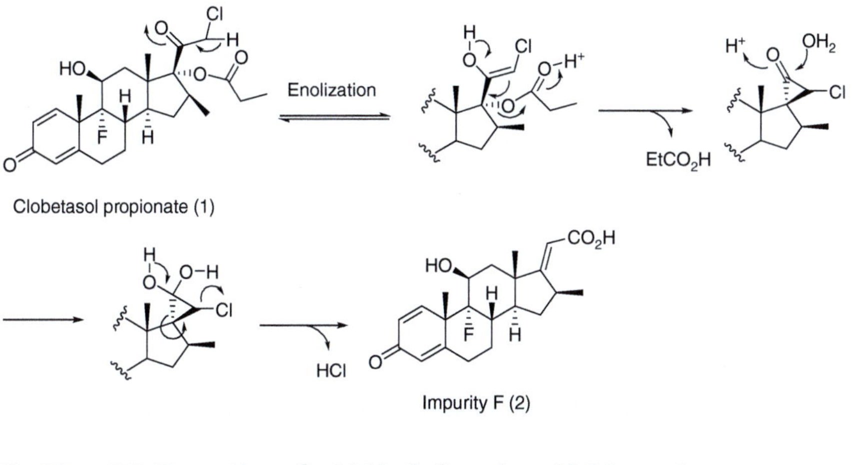

Enolization

$EtCO_2H$

Clobetasol propionate (1)

HCl

Impurity F (2)

Problem 151: Formation of a 24-Methylene Steroidal Impurity

In a homologation study of cholanic acid derivatives to 25-hydroxy vitamin D and congeners, demethoxycarbonylation of malonate **1** was carried out in DMSO with sodium chloride under reflux conditions. This reaction yielded the desired ester **2** (80% yield) along with trace amounts of 24-methylene derivative **3** [64].

Please suggest a mechanism for the formation of **3**.

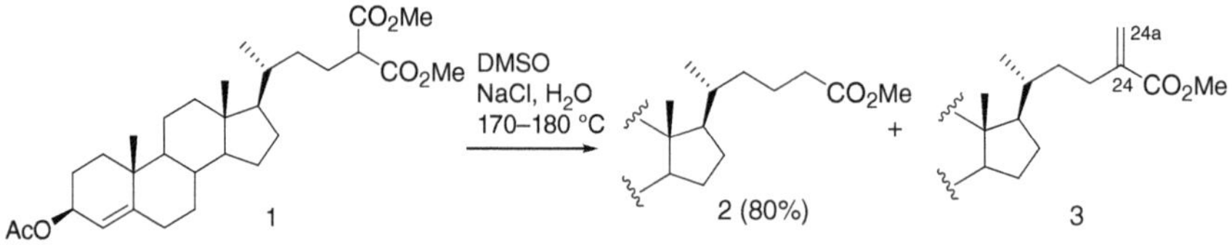

Experimental Procedure (from *Org. Process Res. Dev.*, **1998**, *2*, 290)

Methyl 3β-Acetoxy-23a-homo-5-cholen-24-oate (2). β A mixture of diester **1** (72 g, 143 mmol), sodium chloride (11 g, 188 mmol), and water (11 g, 611 mmol) in DMSO (1.2 L) was heated at 170–180 °C for 2.5 h. The mixture was poured into water (2 L) and left overnight. The precipitate was filtered off, rinsed with water, and dissolved in ethyl acetate. The solvent was removed and homoester **2** was obtained as yellow oil. Filtration over alumina (400 g) in toluene (2 L) gave homoester **2** (69 g) as a creamy white solid. Crystallization from methanol/benzene (1:1) gave colorless crystals (66 g, 80%), mp 110–112 °C; $[R]_D^{25})$ −45 (*c* 0.3) {ref 5, 110–112.5 °C, from methylene chloride-methanol; $[\alpha]_D^{25}$ − 45 (*c* 1, $CHCl_3$); IR 2940, 1730, 1375, 1261, 1049; ^1H NMR (200 MHz) δ 0.68 (3H, s, 18-CH_3), 0.93 (3H, d, *J*) 6 Hz, 21-CH_3), 1.02 (3H, s, 19-CH_3), 2.03 (3H, s, $COCH_3$), 3.7 (3H, s, $-COOCH_3$), 4.6 (1H, m, 3α-H), 5.37 (1H, br s, 4-H); EIMS *m*/*z* (rel intensity) 384 (M^+ − 60, 100), 369 (17), 353 (5), 342 (3), 276 (8), 263 (22), 255 (17), 213 (10), 147 (19); HRMS calcd for $C_{26}H_{40}O_2$ (M − AcOH) 384.3028, found 384.3029. Anal. Calcd for $C_{28}H_{44}O_4$: C, 75.6; H, 10.0. Found: C, 75.6; H, 10.2%.

Silica gel (300–400 mesh, 6 g) flash chromatography of an aliquot of mother liquors (300 mg) using ethyl acetate (0.4–3.6%) in hexane gave unchanged homoester **2** (61 mg) and ene ester **3** (less polar by-product, 32 mg) as a white solid: UV λ_{max} 208 nm; IR 2940, 1725, 1629, 1367, 1244, 1041; ^1H NMR (200 MHz) δ 0.68 (3H, s, 18-CH_3), 0.98 (3H, d, *J*) 6 Hz, 21-CH_3), 1.02 (3H, s, 19-CH_3), 2.03 (3H, s, $COCH_3$), 3.7 (3H, s, $-COOCH_3$) 4.6 (1H, m, 3α-H), 5.38 (1H, br s, 4-H), 5.5, 6.1 (1H and 1H, br s and br s, 26-CH_2); EIMS *m*/*z* (rel intensity) 396 (M^+ – 60, 100), 381 (15), 354 (3), 275 (12), 255 (14), 213 (10), 145 (13); HRMS calcd for $C_{27}H_{40}O_2$ (M - AcOH) 396.3028, found 396.3013.

The Mechanism

The possible mechanism might involve the attack of dimethyl sulfoxide on the ester group that would generate intermediate **A**. The acidic proton at the malonate would be deprotonated to form ionic pair **B**, which would rearrange to ionic intermediate **C**. Finally, decarboxylation and elimination of methanethiol lead to the side product **3**.

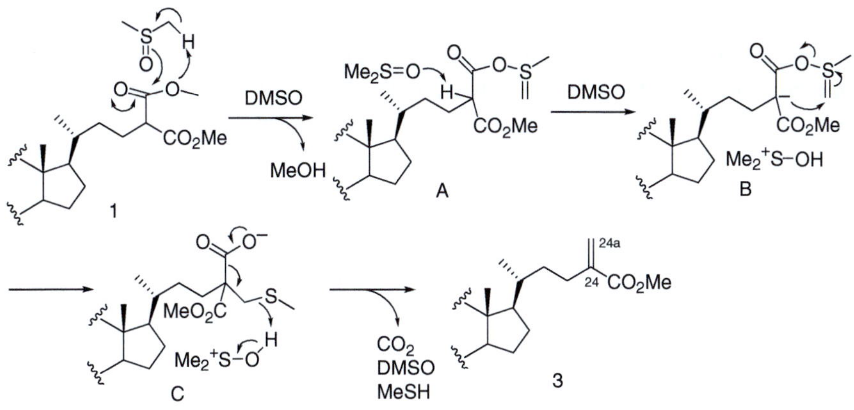

5.7.2 Terramycin

Problem 152: An Unusual Reaction of Terramycin with Methyl Iodide

Terramycin (**1**), a broad-spectrum antibiotic treatment for eye infections, was patented in 1949 and came into commercial use in 1950. During 1953–1954, Woodward and Zimmerman performed a methylation of terramycin **1** with methyl iodide, and a decomposition product **2** was identified. The results were published in 1981 [65].

Could you please propose a mechanism for the formation of **2**?

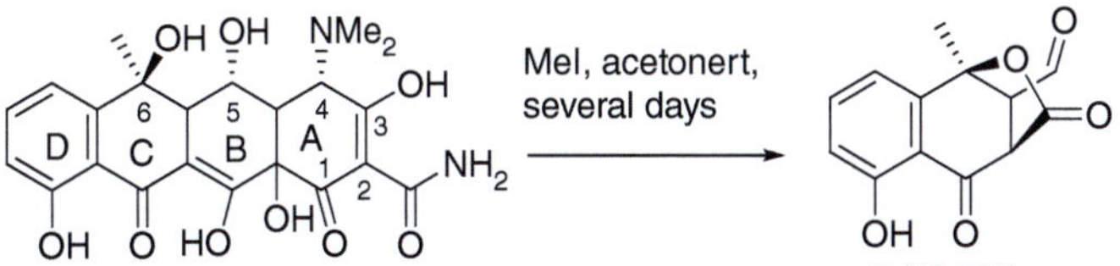

Experimental Procedure (from *Tetrahedron*, **1981**, *37*, 311)

Formation of Lactone 2. Terramycin (**1**, 50 g) was dissolved in 3500 mL analytical grade acetone and to this solution was added 200 mL MeI. The mixture was kept in a stoppered 4-L dark brown vessel. At the end of several days, crystals of tetramethylammonium iodide began to separate. The mixture was allowed to stand at room temp for a least 10 days. The best yield of **2** was obtained when the reaction was run for 14 days. One run, which lasted 50 days, gave a lower yield of product. At the end of the reaction period, the solution was filtered free of 7.1 g tetramethylammonium iodide. The filtrate was concentrated in ~~eucuo~~ vacuum to leave a tan glass. This was dissolved in 300 mL acetone and then treated with 1500 mL dry ether to give a yellow ppt, which was removed by centrifugation. The ether ppt fraction weighed 41.6 g. The ether filtrate was concentrated under vacuum to leave 14.08 g of brownish-yellow solid, which was dissolved in 150 mL acetone and treated with 1600 mL benzene. The ppt that formed was filtered off and the filtrate was concentrated *in vacuo*. The residue was extracted with three 70-mL portions of hot

EtOAc and the remaining solid was taken up in 20 mL hot acetone. Crystalline **2** was separated from all four extracts on cooling. The mother liquors yielded more **2** on concentration. The lactone was recrystallized from EtOAc containing a little acetone. Slow crystallization by evaporation of solvent from a saturated solution at room temperature proved most effective. The yield of product averaged 1.5 g; the highest yield was 2.9 g (12.7%). The ether-precipitated fraction isolated from short runs could be redissolved in acetone and treated with MeI to yield more **2**.

Lactone **2** decomposed at 148° turning green. The IR spectrum of **2** in dioxane showed maxima at 5.60, 5.85 m, 6.04, 6.20, and 6.32 μ. The spectrum gave much more detail in the 7–14 μ region but was very similar at shorter wavelengths except that the CO absorption band at 5.84 μ was considerably more intense. The UV spectrum in 0.001 N HCl in EtOH exhibited maxima at 267.5 ($\log\varepsilon = 3.86$) and 339.0 (3.67) nm. The lactone was sublimed in high vacuum before analysis. (Found: C, 63.62; H, 4.30. Calc. for $C_{12}H_{10}O_5$: C, 63.41: H, 4.09%.)

Lactone **2** readily formed a yellow 2,4-dinitrophenylhydrazone: m.p. 257-8° after extraction with boiling EtOH and recrystallization from EtOAc (Found: N. 12.90. Calc. for $C_{19}H_{14}O_8N_4$: N, 13.14%).

The Mechanism

Alkylation of the tertiary amine provides the quaternary ammonium salt **A**. Elimination of trimethylamine with β-eliminations of the C-5 carbinol group yields an aldehyde **B**, in which the enol is isomerized to the ketone **C**. This intermediate can readily aromatize by β-diketone fission occasioned by intramolecular attack of the tertiary OH group of ring C on the CO group of ring B to release lactone **2** and 2,3,6-trihydroxybenzamide **3**.

Compound **3** was not isolated.

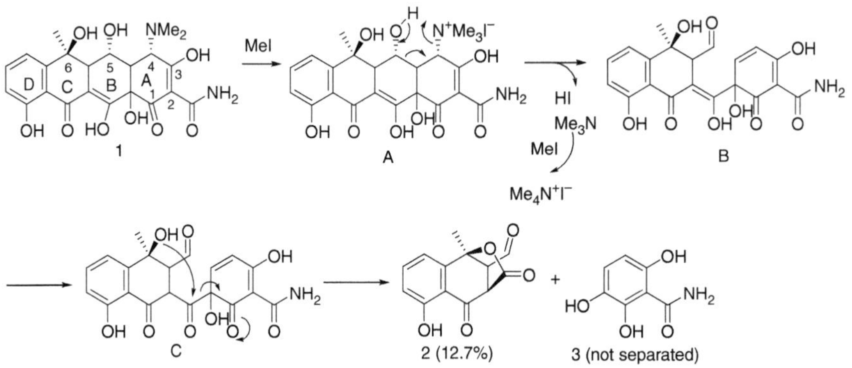

5.7.3 Cephem

Problem 153: Degradation of Cephem Intermediate

BMS-247243 (**1**) is an antimethicillin-resistant *Staphylococcus aureus* cephem. In a process chemistry of BMS-247243, the deprotection of the diester **2** to the diacid **3** is a key step. Treatment of **2** with aqueous TFA provided **3** in 70–80% yield along with a degradation product **4** in 3–10% yield [66]. The degradation of diester **2** to diacid

4 was caused by the presence of water in the TFA. Diester **2** actually was converted with 20% aqueous TFA to diacid **4** in over 70% yield.

Please provide a mechanism for the formation of **4**.

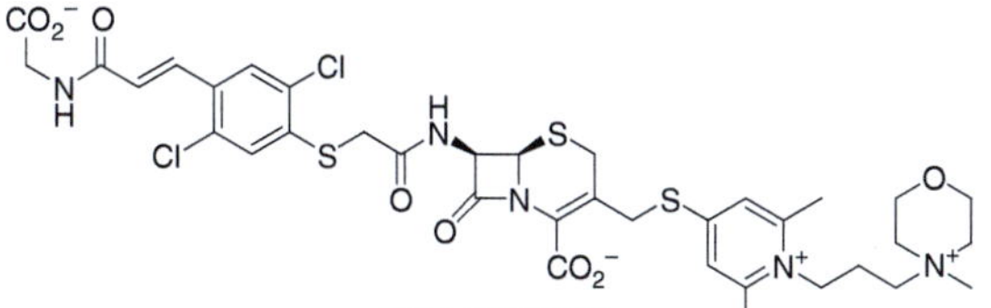

BMS-247243 (1)

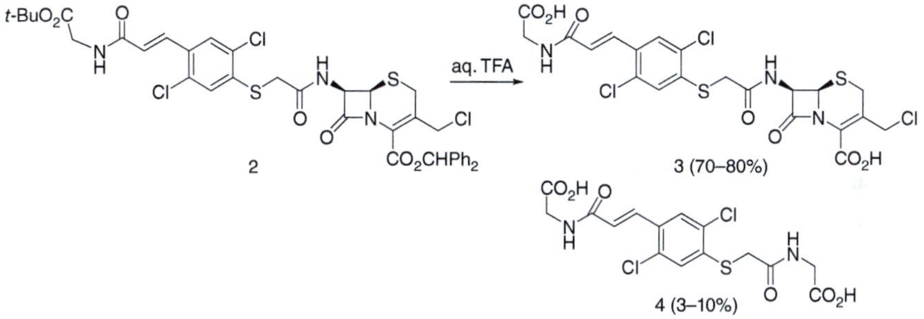

aq. TFA

2

3 (70–80%)

4 (3–10%)

Experimental Procedure (from *Org. Process Res. Dev.*, **2000**, *4*, 488, Supplementary material)

(*E*)-(3-(4-((2-((carboxymethyl)amino)-2-oxoethyl)thio)-2,5-dichlorophenyl) acryloyl)glycine (4). ^{1}H NMR (DMSO-d$_6$, 270 MHz) δ, 3.80 (d, J = 6.0 Hz, 2H), 3.89 (d, J = 6.0 Hz, 2H), 3.90 (d, J = 6.0 Hz, 2H), 6.80 (d, J = 15.8 Hz, 1H), 7.52 (s, 1H), 7.60 (d, J = 15.8 Hz, 1H), 7.79 (s, 1H), 8.40 (t, J = 6.0 Hz, 1H), 8.60 (t, J = 6.0 Hz, 1H). ^{13}C NMR (DMSO-d$_6$, 68 MHz) δ, 35.8, 42.1, 42.2, 126.2, 128.3, 128.5, 130.4, 131.5, 133.7, 134.0, 140.4, 165.9, 168.6, 172.0, 172.2. MS (high resolution) exact mass = 419.99.

The Mechanism

The degradation of diester **2** to diacid **4** was caused by the presence of water in the TFA. Diester **2** was actually converted with 20% aqueous TFA to diacid **4** in over 70%

R=

3

HCl

A

H_3^+O

B

C

H_2O

R

4

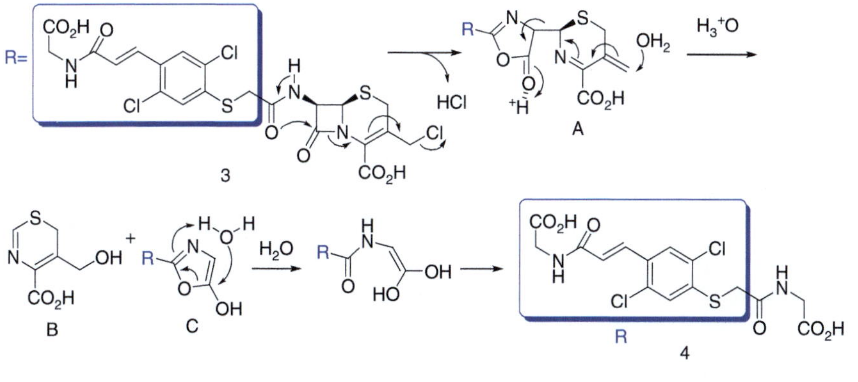

yield. Due to the leaving property of the Cl and the instability of the γ-lactam in **3**, under the acidic condition, elimination of HCl from **3**, as shown in the following scheme, will give an intermediate **A**. Addition of H_2O to **A** will cause the cleavage of the molecule and thus two compounds **B** and **C** are produced. Hydrolysis of **C** leads to the diglycine amide product **4**.

References

1 Fukui, Y.; Maegawa, Y.; Matsuura, T.; Kurita, T. WO2015/093586A1, **2015**.
2 Takahashi, T.; Hoshi, E.; Takagi, M.; Katsumata, N.; Kawahara, M.; Eguchi, K. *Cancer Sci.*, **2010,** *101*, 2455.
3 Sharma, G. G.; Mota, I.; Mologni, L.; Patrucco, E.; Gambacorti-Passerini, C.; Chiarle, R. *Cancers*, **2018,** *10*, 62.
4 Hosoya, M.; Nishijima, S.; Kurose, N.; *Org. Process Res. Dev.*, **2020,** *24*, 405.
5 Fu, Y.; Zhao, X. L.; Hügel, H.; Huang, D.; Du, Z.; Wang, K.; Hu, Y. *Org. Biomol. Chem.*, **2016,** *14*, 9720.
6 Hanselmann, R.; Johnson, G.; Reeve, M. M.; Huang, S.-T. *Org. Process Res. Dev.*, **2009,** *13*, 54.
7 Alimardanov, A.; Gontcharov, A.; Nikitenko, A.; Chan, A. W.; Ding, Z.; Ghosh, M.; Levent, M.; Raveendranath, P.; Ren, J.; Zhou, M.; Mahaney, P. E.; McComas, C. C.; Ashcroft, J.; Potoski, J. R. *Org. Process Res. Dev.*, **2009,** *13*, 880.
8 Li, W.; Towne, T. B.; Liang, G.; Haight, A. R.; Barnes, D. M. *Org. Process Res. Dev.*, **2014,** *18*, 1696.
9 Dannenberg, H.; Meyer, E. *Chem. Ber.*, **1969,** *102*, 2384.
10 Nonoyama, A.; Nakai, Y.; Lee, S.; Suzuki, S.; Ando, T.; Fukuda, N.; Tanaka, H.; Takahashi, K. *Org. Process Res. Dev.*, **2019,** *23*, 499.
11 Boyer, T.; Choudary, B. M.; Edwards, A. J.; Etridge, E.; Etridge, S.; Giddings, A.; Harvey, J.; Hodnett, N. S.; Druot-Houllemare, S.; Leahy, J. H.; Payne, K.; Reid, R.; Roberts, A. D.; Sasse, M.; Simpson, A.; Smith, S.; Stevenson, N. G.; Stonestreet, P.; Urquhart, M. W.; White, A. *Org. Process Res. Dev.*, **2020,** *24*, 371.
12 Bellingham, R.; Buswell, A. M.; Choudary, B. M.; Gordon, A. H.; Moore, S. O.; Peterson, M.; Sasse, M.; Shamji, A.; Urquhart, M. W. J. *Org. Process Res. Dev.*, **2010,** *14*, 1254.
13 Pasto, D. J.; Garves, K.; Sevenair, J. P. *J. Org. Chem.*, **1968,** *33*, 2975.
14 Saba, A. *Gazz. Chim. Ital.*, **1991,** *121*, 55.
15 Yasukata, T.; Masui, M.; Ikarashi, F.; Okamoto, K.; Kurita, T.; Nagai, M.; Sugata, Y.; Miyake, N.; Hara, S.; Adachi, Y.; Sumino, Y. *Org. Process Res. Dev.*, **2019,** *23*, 565.
16 Michida, M.; Ishikawa, H.; Kaneda, T.; Tatekabe, S.; Nakamura, Y. *Org. Process Res. Dev.*, **2019,** *23*, 524.
17 Li, X.; Russell, R. K.; Spink, J.; Ballentine, S.; Teleha, C.; Branum, S.; Wells, K.; Beauchamp, D.; Patch, R.; Huang, H.; Player, M.; Murray, W. *Org. Process Res. Dev.*, **2014,** *18*, 321.

18 Grunewald, G. L.; Caldwell, T. M.; Li, Q.; Criscione, K. R. *J. Med. Chem.*, **2001,** *44*, 2849.

19 Rowley, M.; Hallett, D. J.; Goodacre, S.; Moyes, C.; Crawforth, J.; Sparey, T. J.; Patel, S.; Marwood, R.; Patel, S.; Thomas, S.; Hitzel, L.; O'Connor, D.; Szeto, N.; Castro, J. L.; Hutson, P. H.; MacLeod. A. M. *J. Med. Chem.*, **2001,** *44*, 1603.

20 Hallett, D. J.; Gerhard, U.; Goodacre, S. C.; Hitzel, L.; Sparey, T. J.; Thomas, S.; Rowley, M. *J. Org. Chem.*, **2000,** *65*, 4984.

21 Kempson, J.; Zhang, H.; Wong, M. K. Y.; Li, J.; Li, P.; Wu, D.-R.; Rampulla, R.; Galella, M. A.; Dabros, M.; Traeger, S. C.; Muthalagu, V.; Gupta, A.; Arunachalam, P. N.; Mathur, A. *Org. Process Res. Dev.*, **2018,** *22*, 846.

22 *a)* Gentili, F.; Bousquet, P.; Brasili, L.; Caretto, M.; Carrieri, A.; Dontenwill, M.; Giannella, M.; Marucci, G.; Perfumi, M.; Piergentili, A.; Quaglia, W.; Rascente, C.; Pigini, M. *J. Med. Chem.*, **2002,** *45*, 32. *b)* Gentili, F.; Ghelfi, F.; Giannella, M.; Piergentili, A.; Pigini, M.; Quaglia, W.; Vesprini, C.; Crassous, P. A.; Paris, H.; Carrieri, A. *J. Med. Chem.*, **2004,** *47*, 6160. *c*) Galley, G.; Groebke, Z. K., Norcross, R., Stalder, H. WO2009003868 A2, **2009.**

23 Donnola, M.; Airoldi, A.; Barozza, A.; Roletto, J.; Paissoni, P. *Org. Process Res. Dev.*, **2021,** *25*, 1816.

24 Donnola, M.; Airoldi, A.; Barozza, A.; Roletto, J.; Paissoni, P. WO2020/254580, A1, **2020.**

25 Fernández, I.; Cossío, F. P.; Sierra, M. A. *Chem. Rev.*, **2009,** *109*, 6687.

26 Hansen, M. M.; Jarmer, D. J.; Arslantas, E.; DeBaillie, A. C.; Frederick, A. L.; Harding, M.; Hoard, D. W.; Hollister, A.; Huber, D.; Kolis, S. P.; Kuehne-Willmore, J. E.; Kull, T.; Laurila, M. E.; Linder, R. J.; Martin, T.; Martinelli, J. R.; McCulley, M. J.; Richey, R. N.; Ward, J. A.; Zaborenko, N.; Zweifel, T. *Org. Process Res. Dev.*, **2015,** *19*, 1214.

27 DeBaillie, A. C.; Jasper, P. J.; Li, S.; McCulley, M. J.; Vaidyaraman, S.; Zhang, Z. *Org. Process Res. Dev.*, **2015,** *19*, 1244.

28 Knapp, S.; Morriello, G. J.; Doss, G. A. *Tetrahedron Lett.*, **2002,** *43*, 5797.

29 Knapp, S.; Morriello, G. J.; Doss, G. A. *Org. Lett.*, **2002,** *4*, 603.

30 Prashad, M.; Liu, Y.; Shieh, W.-C.; Schaefer, F.; Hu, B.; Girgis, M. *Org. Process Res. Dev.*, **2015,** *19*, 290.

31 Li, B.; Kasthurikrishnan, N.; Ragan, J. A.; Young, G. R. *Org. Process Res. Dev.*, **2002,** *6*, 64.

32 Wang, H.; Goodman, S. N.; Dai, Q.; Stockdale, G, W.; Clark, Jr.W. M. *Org. Process Res. Dev.*, **2008,** *12*, 226.

33 Jiang, N.; Li, C. J. *Chem. Commun.*, **2004,** 394.

34 Vuluga, D.; Legros, J.; Crousse, B.; Bonnet-Delpon, D. *Green Chem.*, **2009,** *11*, 156.

35 Li, Y. Q.; Li, C. J.; Zheng, Y. L.; Wei, X. C.; Ma, Q. Q.; Wei, P.; Liu, Y. X.; Qin, Y. G.; Yang, N.; Sun, R. F.; Ling, Y.; Yang, X. L.; Wang, Q. M. *J. Agric. Food Chem.*, **2014,** *62*, 3064.

36 Yu, X. L.; Liu, Y. X.; Li, Y. Q.; Wang, Q. M. *J. Agric. Food Chem.*, **2015,** *63*, 9690.

37 Zhang, Y.; Chen, S.; Liu, Y.; Wang, Q. *Org. Process Res. Dev.*, **2020,** *24*, 216.

38 Hansen, M. M.; Borders, S. S. K.; Clayton, M. T.; Heath, P. C.; Kolis, S. P.; Larsen, S. D.; Linder, R. J.; Reutzel-Edens, S. M.; Smith, J. C.; Tameze, S. L.; Ward, J. A.; Weigel, L. O. *Org. Process Res. Dev.*, **2009,** *13*, 198.

39 Yang, Y.; Hansen, L.; Sjogren, J. K.; León, I. R.; Receveur, J.-M.; Badalassi, F. *Org. Process Res. Dev.*, **2021,** *25*, 1923.

40 Lu, J.; Shi, Y.; Li, X.; Liang, X.; Wang, Y.; Yuan, S.; Wu, T. *Org. Process Res. Dev.*, **2021,** *25*, 1112.

41 Conant, J. B.; Tuttle, N. *Org. Synth.*, **1921,** *1*, 53.

42 Sternbach, L. H.; Reeder, E. *J. Org. Chem.*, **1961,** *26*, 1111.

43 Haran, R.; Tuchagues, J. P. *J. Heterocyclic Chem.*, **1980,** *17*, 1483.

44 Abbaszadeh, H.; Motthagi, S. *Phytother. Res.*, **2021,** *35*, 6843.

45 Gurzadyan, G.; Mende, S.; Schmid, A.; Lindel, T. *Org. Process Res. Dev.*, **2023,** *27*, 890.

46 Baldwin, J. E. *J Chem. Soc., Chem. Commun.*, **1976,** 734.

47 Brodney, M. A.; Johnson, D. E.; Sawant-Basak, A.; Coffman, K. J.; Drummond, E. M.; Hudson, E. L.; Fisher, K. E.; Noguchi, H.; Waizumi, N.; McDowell, L. L.; Papanikolaou, A.; Pettersen, B. A.; Schmidt, A. W.; Tseng, E.; Stutzman-Engwall, K.; Rubitski, D. M.; Vanase-Frawley, M. A.; Grimwood, S. *J. Med. Chem.*, **2012,** *55*, 9240.

48 Noguchi, H.; Waizumi, N. PCT Int. Appl. WO2011/101774 A1, **2011.**

49 Widlicka, D. W.; Murray, J. C.; Coffman, K. J.; Xiao, C.; Brodney, M. A.; Rainville, J. P.; Samas, B. *Org. Process Res. Dev.*, **2016,** *20*, 233.

50 Pokhrel, L.; Kim, Y.; Nguyen, T. D. T.; Prior, A. M.; Lu, J.; Chang, K.-O.; Hua, D. H. *Bioorg. Med. Chem. Lett.*, **2012,** *22*, 3480.

51 Vaitilingam, B.; Nayyar, A.; Palde, P. B.; Monga, A.; Jain, R.; Kaur, S.; Singh, P. P. *Bioorg. Med. Chem.*, **2004,** *12*, 4179.

52 Kaupang, A.; Kase, E. T.; Vo, C. X. T.; Amundsen, M.; Vik, A.; Hansen, T. A. *Bioorg. Med. Chem.*, **2016,** *24*, 247.

53 Ivachtchenko, A. V.; Khvat, A. V.; Kobak, V. V.; Kysil, V. M.; Williams, C. T. *Tetrahedron Lett.*, **2004,** *45*, 5473.

54 El Ashry, E. S. H.; Ramadan, E. S.; Hamid, H. A.; Hagar, M. *Synth. Comm.*, **2005,** *35*, 2243.

55 Feldman, K. S.; Coca, A. *Tetrahedron Lett.*, **2008,** *49*, 2136.

56 Atkins, R. J.; Breen, G. F.; Crawford, L. P.; Grinter, T. J.; Harris, M. A.; Hayes, J. F.; Moores, C. J.; Saunders, R. N.; Share, A. C.; Walsgrove, T. C.; Wicks, C. *Org. Process Res. Dev.*, **1997,** *1*, 185.

57 Gopalsamy, A.; Shi, M.; Nilakantan, R. *Org. Process Res. Dev.*, **2007,** *11*, 450.

58 *a)* Boschelli, D. H.; Wu, B.; Sosa, A. B.; Durutlic, H.; Chen, J. J.; Wang, Y.; Golas, J. M.; Lucas, J.; Boschelli, F. *J. Med. Chem.*, **2005,** *48*, 3891. *b)* Wang, Y. D.; Boschelli, D. H.; Johnson, S.; Honores, E. *Tetrahedron*, **2004,** *60*, 2937. *c)* Boschelli, D. H.; Ye, F. *J. Heterocycl. Chem.*, **2002,** *39*, 783.

59 Guvvala, V.; Subramanian, V. C.; Anireddy, J.; Konda, M. *Org. Process Res. Dev.*, **2017,** *21*, 11.

60 https://www.synzeal.com/clobetasol-propionate-ep-impurity-f

61 Geng, L.; Li, Y.; Han, K.; Qi H. CN109206467, **2019.**

62 Tian, Y.; Zhang, H.; Zhu, Z.; Chen, L.; Wang, E.; Jin, J.; Li, D.; Zhu, W.; Li, M.; Fu, L. https://doi.org/10.26434/chemrxiv.11809077.v1

63 Guijarro, D.; Yus, M. *Current Org. Chem.*, **2005,** *9*, 1713.

64 Kutner, A.; Chodyn´ski, M.; Masnyk, M.; Wicha, J. *Org. Process Res. Dev.*, **1998,** *2*, 290.

65 Woodawrd, R. B.; Zimmerman, H. E. *Tetrahedron*, **1981,** *37*, 311.

66 Singh, J.; Kim, O. K.; Kissick, T. P.; Natalie, K. J.; Zhang, B.; Crispino, G. A.; Springer, D. M.; Wichtowski, J. A.; Zhang, Y.; Goodrich, J.; Ueda, Y.; Luh, B. Y.; Burke, B. D.; Brown, M.; Dutka, A. P.; Zheng, B.; Hsieh, D.-M.; Humora, M. J.; North, J. T.; Pullockaran, A. J.; Livshits, J.; Swaminathan, S.; Gao, Z.; Schierling, P.; Ermann, P.; Perrone, R. K.; Lai, M. C.; Gougoutas, J. Z.; DiMarco, J. D.; Bronson, J. J.; Heikes, J. E.; Grosso, J. A.; Kronenthal, D. R.; Denzel, T. W.; Mueller, R. H. *Org. Process Res. Dev.*, **2000,** *4*, 488.

Index

Overcoming Synthetic Challenges in Medicinal Chemistry: Mechanistic Insights and Solutions, First Edition. Tongshuang Li.

d

o

p

q

r

s

u

v

x

y